MOLECULAR IMPROVEMENT OF CEREAL CROPS

*Dedication*

With love, to

Vimla, Kavita, Charu and Ryan

Advances in Cellular and Molecular Biology of Plants

VOLUME 5

# Molecular improvement of cereal crops

*Edited by*

Indra K. Vasil

*Laboratory of Plant Cell and Molecular Biology, University of Florida, Gainesville, Florida, USA*

KLUWER ACADEMIC PUBLISHERS
DORDRECHT / BOSTON / LONDON

Library of Congress Cataloging-in-Publication data is available.

ISBN 0-7923-5471-0

---

Published by Kluwer Academic Publishers,
P.O. Box 17, 3300 AA Dordrecht, The Netherlands.

Sold and distributed in North, Central and South America
by Kluwer Academic Publishers,
101 Philip Drive, Norwell, MA 02061, U.S.A.

In all other countries, sold and distributed
by Kluwer Academic Publishers,
P.O. Box 322, 3300 AH Dordrecht, The Netherlands

*Printed on acid-free paper*



Printed and bound in Great Britain by Antony Rowe Limited.

# Contents

# Preface

Cereal grains have been the most important source of human nutrition since prehistoric times, dating back to 8,000 BC. Even today, eight cereals – wheat, rice, maize, barley, oats, rye, sorghum and millets – collectively account for 66% of the world food supply, forming the centerpiece of international agriculture. Innovative methods of breeding and selection, such as those used in the production of hybrid maize and the Green Revolution varieties of wheat and rice, have greatly increased the productivity and quality of cereals, with an average increase in grain yield of 2.1% per year over the past three decades. More recently, however, the average annual yield increase has declined to 0.5% owing to a variety of unfavorable biological and environmental factors. At the same time, there is a greater demand for cereal grains owing to increases in population, economic development and changing dietary habits in countries such as China and India. It has thus been estimated that world demand for food will double by the year 2025, and triple by 2050. Increases in food productivity of such a magnitude cannot be attained in such a short period of time by conventional methods of breeding and selection, on less per capita arable land, with deteriorating and decreasing supplies of fresh water, and under increasingly adverse environmental conditions. These already difficult problems are further complicated by the fact that in spite of the high input of agrochemicals, 42% of crop productivity is still lost owing to weeds, pathogens and pests.

Attainment of food security and protection of the environment are two of the greatest challenges facing humankind at the beginning of the third millennium. In view of the predominant role of cereal crops in human diet and international agriculture, food security cannot be attained without major increases in cereal productivity. Advances in the molecular genetic manipulation of plants during the last 15 years have provided convincing evidence of the usefulness of these powerful technologies in complementing and supplementing plant breeding programs for the genetic improvement of important crop species. Large-scale cultivation of transgenic crops will have the added benefits of reducing the use of agrochemicals, and of conservation of land and water supplies through increased productivity.

*I.K. Vasil (ed.), Molecular Improvement of Cereal Crops, vii–viii*

For more than 20 years, my own research interests have focused almost entirely on the molecular biology and genetic improvement of cereals. My commitment to this field of research began at a time when there was much discussion of the potential of plant biotechnology in crop improvement, but little substantive work on any of the important crop species. Monocotyledonous species in general, but cereals in particular, were considered to be either too difficult or not amenable to in vitro genetic manipulation. The rapid and remarkable advances in cereal biotechnology during the past two decades are a powerful demonstration of how sustained and collective efforts, and the exploration of innovative ideas, can lead to the resolution of long-standing and difficult scientific problems.

In this volume leading experts who have made significant and pioneering contributions to the molecular improvement of cereal crops describe the development of transgenic cereals for better protection against weeds, pathogens and pests, and for improved quality traits. They describe not only the latest developments, but also discuss future directions and prospects. Transgenic maize is already grown globally on 8.3 million hectares, including on 22% of the national maize area in the United States. Transgenic rice and wheat, which are undergoing field trials and evaluation, are expected to follow soon. Considering that the first transgenic cereal, rice, was produced only ten years ago, the progress has been truly impressive. Within the next 10–15 years, transgenic cereals are expected to be fully integrated into international agriculture.

I am grateful to each of the authors for providing first-hand, detailed and insightful accounts of the many difficulties and frustations, but also of the triumphs, of cereal biotechnology.

*Indra K. Vasil*

# 1. Molecular Improvement of Cereal Crops – An Introduction

INDRA K. VASIL
*Laboratory of Plant Cell and Molecular Biology, 1143 Fifield Hall, University of Florida, Gainesville, FL 32611-0690, USA. E-mail: ikv@gnv.ifas.ufl.edu*

ABSTRACT. From the pre-historic era to the modern times, cereal grains have been the most important source of human nutrition, and have helped sustain the increasing population and the development of human civilization. In order to meet the food needs of the 21st century, food production must be doubled by the year 2025, and nearly tripled by 2050. It is doubtful that such enormous increases in food productivity can be brought about by relying entirely on conventional breeding methods, especially on less per capita land, poorer quality and quantity of water, and under rapidly deteriorating environmental conditions. Complementing and supplementing the breeding of major food crops, such as the cereals which together account for 66% of the world food supply, with molecular breeding and genetic manipulation may well provide a grace period of about 50 years in which to control population growth and achieve sustainable development. Commercialization of the first generation of transgenic cereals that are designed to substantially reduce or prevent the enormous losses to cereal productivity caused by competition with weeds, and by various pests and pathogens, is an important first step in that direction.

## Introduction

Fossil records of grasses (Gramineae/Poaceae) indicate that they appeared early during the evolution of angiosperms, and had become prominent about 70 million years ago at the end of the Cretaceous. They have since evolved into a highly diversified group of species that are found in a wide variety of habitats in virtually all climatic zones of the world. One of the largest families of angiosperms, the Gramineae comprises nearly 785 genera, and more than 10,000 species, ranging from the smallest angiosperms to the giant bamboos.

Primitive forms of cereal grains were among the first plants domesticated by early humans more than 10,000 years ago during the Neolithic age when agriculture gradually replaced hunting and gathering. It is rather fortunate that the choices made by our Neolithic ancestors have proven not only to be wise and remarkably durable, but have also sustained the development of the human civilization by providing one of the most important sources of human nutrition over several millennia. It is remarkable also that in 1996

*I.K. Vasil (ed.), Molecular Improvement of Cereal Crops, 1–8*

the world production of just eight species of cereals (wheat, rice, maize, barley, oats, rye, sorghum, and pearl millet) totaled 2,050 million metric tons (FAO, 1996), together accounting for an impressive 66% of the world food supply (Borlaug, 1998). Cereals, therefore, have been justifiably described as being the centerpiece of international agriculture (Vasil and Vasil, 1994).

Enormous improvements in the productivity of cereal crops have been achieved during the past several decades by plant breeding, leading to the introduction of hybrid maize and the so-called Green Revolution varieties of wheat and rice (Borlaug and Dowswell, 1988). These advances helped to maintain an average increase in grain yield of 2.1% per year until 1990 (Vasil, 1998). As a result of a variety of unfavorable biological and environmental factors, the average annual yield increase has since declined to only 0.5%. At the same time, improved economic conditions and changing dietary habits in China and India, the two most populous nations in the world, have created a greater demand for cereal grains. Based on the projected increases in world population and food production, it has been estimated that the demand for food will double by the year 2025, and nearly triple by 2050. It has been also argued that such increases in food production can not be brought about in such a short period of time by conventional breeding alone, on less per capita land, with less water, and under increasingly challenging environmental conditions (Vasil, 1998).

Furthermore, in spite of the use of advanced agricultural technologies and the heavy application of agrochemicals, 42% of crop productivity is still lost to competition with weeds, and to pests and pathogens, with an additional 10–30% post-harvest losses caused by a variety of biological and physical factors (Oerke et al., 1994). Recent advances in plant tissue culture and molecular biology (biotechnology) have made it possible to produce engineered crops that are resistant to a variety of biotic and abiotic stresses, and that can help to substantially reduce or even eliminate the huge crop losses (Vasil, 1998). These technologies also overcome the most important biological barrier to crop improvement (that of sexual incompatibility between species) by making the entire global gene pool – whether it be plant, animal, bacterial, or viral – accessible for plant improvement. An added benefit of the cultivation of transgenic plants is the resultant reduction in the use of agrochemicals, which are a major source of soil and groundwater pollution.

Owing to the predominant role of cereal crops in the human diet, food security in the future can not be achieved without major increases in cereal productivity. The various chapters in this volume describe the advances made in the production of engineered cereal crops by combining the tools of plant tissue culture and molecular biology. This new generation of transgenic cereals is resistant to potent but environment friendly herbicides, and to pests and pathogens. Transgenic cereals with improved quality traits are also being produced. Maize is one of the first major food crops to be commercially produced (planted on 22% of the national acreage in the US in 1998) and is being used for a variety of products for humans and livestock. Other transgenic cereals, especially wheat and rice, are undergoing advanced field trials before commercial production.

The molecular improvement of cereals, like other plant species, involves the use of tissue culture and molecular technologies, which permit the introduction, integration and expression of genes of interest in cells that can then be regenerated into normal and fertile plants. This chapter provides a brief overview of the strategies used for the production of transgenic cereals. Leading experts provide more detailed and specific accounts for each species in their respective chapters.

## Plant Regeneration from Cultured Cells

Until about 1980, cereal species – along with legumes and woody tree species – were known for their recalcitrance to manipulation and regeneration in culture. Indeed, it was believed by some that monocotyledonous species in general, but the Gramineae in particular, were either not amenable to, or perhaps incapable of, sustained growth and regeneration in vitro. There were only a few scattered reports of sporadic and transient plant regeneration from tissue cultures of gramineous species. No protocols existed for reproducible and sustained recovery of large numbers of normal and fertile plants for any species. Four trend-setting discoveries were made in 1980–1981 that have formed the basis of much of the modern work on regeneration of almost all of the major species of grasses, including sugarcane, bamboos, and all of the economically important cereals: (a) That the culture of immature embryos, and segments of young inflorescences and bases of immature leaves in peak physiological condition at defined stages of development, (b) on simple nutrient media containing high concentrations of strong herbicidal auxins like 2,4-D, (c) gives rise to cultures in which plant regeneration takes place predominantly by the formation of somatic embryos, and that (d) embryogenic cell suspension cultures derived from embryogenic calli yield totipotent protoplasts (for detailed accounts see Vasil, 1994, Vasil and Vasil, 1992, 1994).

The physiological condition and developmental state of the explant, along with the genotype, were found to be the most significant factors in determining in vitro response. However, it was often possible to overcome the negative influence of the genotype by a judicious choice of the physiological condition, developmental stage and type of the explant, and the nutrient medium. Several studies have emphasized the critical role of endogenous plant growth substances in tissue culture response, especially in somatic embryogenesis (Rajasekaran et al., 1987a,b, Wenck et al., 1988, Centeno et al., 1997). Plants derived from somatic embryos are non-chimeric in nature owing to their single-cell origin, are easy to establish in soil because of the presence of a root-shoot axis, and are generally devoid of any major genetic abnormalities because of stringent selection during regeneration (Swedlund and Vasil, 1985, Vasil, 1994, Vasil and Vasil, 1994).

Until 1990, the only method available for the production of transgenic cereals was through the direct delivery of DNA into protoplasts. Although mesophyll protoplasts have been used successfully for the regeneration of plants in a wide variety of species, it has never been possible, inspite of extensive and exhaustive efforts, to induce sustained cell divisions in the mesophyll protoplasts of any of the cereal species. The problems faced in the culture of cereal protoplasts were finally overcome by the establishment of regenerable embryogenic cell suspension cultures, which to-date remain the only source of totipotent protoplasts for cereal species (Vasil and Vasil, 1992).

Although a few somatic hybrids have been produced in cereals by protoplast fusion (Terada et al., 1987, Takamizo et al., 1991, Kisaka et al., 1998), somatic hybridization has not yet proven to be of any particular benefit in cereal improvement. Even the practical use of the transfer of cytoplasmic male steriltiy by somatic hybridization, demonstrated in rice a decade ago (Kyozuka et al., 1989, Yang et al., 1989), has not led to any improved products. One of the better and more innovative uses of protoplast fusion has been the in vitro fusion of sperm and egg protoplasts, and sperm and central cell protoplasts of maize, leading to the development of fertile plants and endosperm tissue, respectively (Kranz and Lörz, 1993, Kranz et al., 1998). Such studies would be useful in understanding the biology of fertilization, and the early stages of the development of embryo and endosperm.

## Production of Transgenic Cereals

Early attempts toward the transformation of cereals began by the direct delivery of DNA by osmotic (polyethylene glycol, PEG) or electric (electroporation) shock into protoplasts derived from suspension cultures (see chapter by Shillito, this volume), and the demonstration of transient expression of reporter genes (Paszkowski et al., 1984, Fromm et al., 1985, Hauptmann et al., 1987). These studies helped to define and optimize the parameters for the efficient delivery of DNA into protoplasts, the maintenance of protoplast viability, and the choice of promoters to enhance the expression of transgenes and of selectable markers to obtain stably transformed cell lines (Lörz et al., 1985, Potrykus et al., 1985, Fromm et al., 1986, Hauptmann et al., 1988, see also chapter by Båga et al., this volume). The first transgenic grasses, including rice and maize, were obtained from the transformation of protoplasts (Horn et al., 1988, Rhodes et al., 1988, Shimamoto et al., 1989). Protoplasts have proven to be particularly useful for the transformation of forage grass species.

The difficulties inherent in the establishment and maintenance of embryogenic suspension cultures, and the regeneration of plants from protoplasts, encouraged the search for other more friendly methods for the transformation of cereals. Among the numerous methods of DNA delivery that have been developed during the past decade, the novel biolistics process invented by Sanford et al. (1987) has proven to be the most versatile and useful for cereal species (see chapter by Klein and Jones, this volume). The method is based on the bombardment of DNA-coated microprojectiles at high velocities into regenerable plant cells and tissues, and is currently the most widely used procedure for the production of transgenic cereals.

*Agrobacterium tumefaciens* has been used widely for the transformation of dicotyledonous species since the early 1980's, but monocotyledonous species have generally been considered to be outside the host range of the organism (see chapter by Komari and Kubo, this volume). Elaborate scenaios were presented by some as to why monocotyledons in general, and cereal species in particular, could not be transformed by *Agrobacterium* (Potrykus, 1991). However, as described by Komari and Kubo (this volume), normal and fertile transgenic rice, maize, barley and wheat have been produced in the last few years by *Agrobacterium*-mediated transformation.

The biolistics as well as the *Agrobacterium* procedures are likely to remain the most popular methods for the production of transgenic cereals. It is not entirely clear whether *Agrobacterium*-mediated transformation, as has been claimed, yields mostly stable transgenic plants with only one or a few copies of the transgene, while the plants derived from direct delivery of DNA are said to be comparatively less stable with multiple copies of the transgenes. These questions can not be fully resolved without direct comparative studies, but results obtained with wheat (see chapter by Vasil and Vasil, this volume), rice (see chapter by Datta, this volume) and maize (see chapter by Gordon-Kamm et al., this volume) clearly show that stably transformed plants with low copy number can be obtained with the biolistics procedure. At the same time, a high degree of variation and instability has been found in transgenic barley plants (see chapter by Lemaux et al., this volume). These differences in the stability of transgenic plants between different species may be related to the peculiarities of the different genomes, rather than the methods of transformation used. Another advantage of *Agrobacterium*-mediated transformation that has often been cited is that only defined DNA sequences, limited by the T-DNA borders, are integrated into the plant genome. This discrete nature of T-DNA integration has been brought into question by recent studies which have confirmed the presence of DNA outside the T-DNA borders in the genomes of plants obtained following *Agrobacterium*-mediated transformation (Ramanathan and Veluthambi, 1995, Kononov et al., 1997, Wenck et al., 1997, Smith, 1998).

The utility of transgenic plants is complicated by instances of gene silencing (Flavell, 1994, Matzke and Matzke, 1995) and the genetic variability encountered in some species (see chapter by Lemaux et al., this volume). The causes of both phenomena are not fully understood at this time, but further insights may allow the development of strategies, such as site-directed integration of single gene copies, to minimize or eliminate these undesirable events. Nevertheless, many independent, stably transformed, and normal fertile transgenic cereals with low copy number have been obtained (Srivastava et al., 1996, Arencibia et al., 1998, also see chapters by Vasil and Vasil, Datta, and Gordon-Kamm et al., this volume). It is such transformants that are selected for field trials, and are finally used for commercial production.

## Future Considerations

Attempts to improve the regeneration of normal and fertile plants from cultured tissues of cereal species should continue, along with efforts to improve the efficiency of transformation and stable gene expression. Almost all of the genes that are available and considered useful for the improvement of cereals have already been successfully introduced into one or more cereal species. Transgenic cereals with such genes will be commercially introduced within the next few years, as has already been done with maize. Results of on-going research with genes for abiotic stress are encouraging, and it is expected that plants expressing such genes will form the second generation of transgenic cereals. Search for other important genes, especially those that control more complex multi-genic characters, must continue vigorously. The discovery of syntney (conservation of gene content and order) in cereal genomes (Ahn and Tanksley, 1993, Bennetzen and Freeling, 1993, Moore et al., 1995, Devos and Dale, 1997, see also chapter by Bennetzen, this volume), and the various plant genome initiatives (for example, see Bennetzen et al., 1998) are expected to have considerable useful impact on these attempts.

## References


Ahn, S., and Tanksley, S.D. (1993) Comparative linkage maps of the rice and maize genomes. Proc. Nat. Acad. Sci. USA 90: 7980–7984.

Arencibia, A., Gentinetta, E., Cuzzoni, E., Castiglione, S., Kohli, A., Vain, P., Leech, M., Christou, P., and Sala, F. (1998) Molecular analysis of the genome of transgenic rice (*Oryza sativa* L.) plants produced via particle bombardment or intact cell electroporation. Mol. Breed 4: 99–109.

Bennetzen, J.L., and Freeling, M. (1993) Grasses as a single genetic system: genome composition, colinearity and compatibility. Trends Genet. 9: 259-261.

Bennetzen, J.L., Kellogg, E.A., Lee, M., and Messing, J. (1998) A plant genome initiative. Plant Cell 10: 488–493.

Borlaug, N.E. (1998) Feeding a world of 10 billion people: the miracle ahead. Plant Tiss. Cult. Biotech. 3: 119–127.

Borlaug, N.E., and Dowsell, C.R. (1988) World revolution in agriculture. In: Book of the Year 1988, pp. 5–14. Encyclopedia Brittanica, Chicago.
Centeno, M.L., Rodriguez, R., Berros, B., and Rodriquez, A. (1997) Endogenous hormonal content and somatic embryogenic capacity of *Corylus avellana* L. cotyledons. Plant Cell Rep. 17: 139–144.
Devos, K.M., and Dale, M.D. (1997) Comparative genetics in the grasses. Plant Mol. Biol. 35: 3–15.
FAO (1996) FAO Production Yearbook. FAO, Rome.
Flavell, R.B. (1994) Inactivation of gene expression in plants as a consequence of specific gene duplication. Proc. Nat. Acad. Sci. USA 91: 3490–3496.
Fromm, M.E., Taylor, L.P., and Walbot, V. (1985) Expression of genes transferred into monocot and dicot plant cells by electroporation. Proc. Nat. Acad. Sci. USA 82: 5824–5828.
Fromm, M.E., Taylor, L.P., and Walbot, V. (1986) Stable transformation of maize after gene transfer by electroporation. Nature 319: 791–793.
Hauptmann, R.M., Ozias-Akins, P., Vasil, V., Tabaeizadeh, Z., Rogers, S.G., Horsch, R.B., Vasil, I.K., and Fraley, R.T (1987) Transient expression of electroporated DNA in monocotyledonous and dicotyledonous species. Plant Cell Rep. 6: 265–270.
Hauptmann, R.M., Vasil, V., Ozias-Akins, P., Tabaeizadeh, Z., Rogers, S.G., Fraley, R.T., Horsch, R.B., and Vasil, I.K. (1988) Evaluation of selectable markers for obtaining stable transformants in the Gramineae. Plant Physiol. 86: 602–606.
Horn, M.E., Shillito, R.D., Conger, B.V., and Harms, C.T. (1988) Transgenic plants of orchardgrass (*Dactylis glomerata* L.) from protoplasts. Plant Cell Rep. 7: 469–472.
Kisaka, H., Kisaka, M., Kanno, A., and Kameya, T. (1998) Intergeneric somatic hybridization of rice (*Oryza sativa* L.) and barley (*Hordeum vulgare* L.) by protoplast fusion. Plant Cell Rep. 17: 362–367.
Kononov, M.E., Bassuner, B., and Gelvin, S.B. (1997) Integration of T-DNA binary vector 'backbone' sequences into the tobacco genome: evidence for multiple complex patterns of integration. Plant J. 11: 945–957.
Kranz, E., and Lörz, H. (1993) In vitro fertilization with isolated, single gametes results in zygotic embryogenesis and fertile maize plants. Plant Cell 5: 739–746.
Kranz, E., von Wiegen, P., Quader, H., and Lörz, H. (1998) Endosperm development after fusion of isolated, single maize sperm and central cells in vitro. Plant Cell 10: 511–524.
Kyozuka, J., Taneda, K., and Shimamoto, K. (1989) Production of cytoplasmic male sterile rice (*Oryza sativa* L.) by cell fusion. Bio/Technology 7: 1171–1174.
Lörz, H., Baker, B., and Schell, J. (1985) Gene transfer to cereal cells mediated by protoplast transformation. Mol. Gen. Genet. 199: 178–182.
Matzke, M.A., and Matzke, A.J.M. (1995) Why and how do plants inactivate homologous (trans)genes? Plant Physiol. 107: 679–685.
Moore, G., Devos, K.M., Wang, Z., and Gale, M.D. (1995) Grasses, line up and form a circle. Curr. Biol. 5: 737–739.
Oerke, E-C., Dehne, H-W., Schöbeck, F., and Weber, A. (1994) Crop Production and Crop Protection. Elsevier, Amsterdam.
Pazkowski, J., Shillito, R.D., Saul, M., Mandak, V., Hohn, T., Hohn, B., and Potrykus, I. (1984) Direct gene transfer to plants. EMBO J. 3: 2717–2722.
Potrykus, I. (1991) Gene transfer to plants: assessment of published approaches and results. Annu. Rev. Plant Physiol. Plant Mol. Biol. 42: 205–225.
Potrykus, I., Saul, M.W., Petruska, J., Paszkowski, J., and Shillito, R.D. (1985) Direct gene transfer to cells of a graminaceous monocot. Mol. Gen. Genet. 199: 183–188.
Rajasekaran, K., Hein, M.B., and Vasil, I.K. (1987a) Endogenous abscisic acid and indole-3-acetic acid and somatic embryogenesis in cultured leaf explants of *Pennisetum purpureum* Schum. Plant Physiol. 84: 47–51.
Rajasekaran, K., Hein, M.B., Davis, G.C., Carnes, M.G., and Vasil, I.K. (1987b) Endogenous

growth regulators in leaves and tissue cultures of *Pennisetum purpureum* Schum. J. Plant Physiol. 130: 13–25.

Ramanathan, V., and Veluthambi, K. (1995) Transfer of non-T-DNA portions of the *Agrobacterium tumefaciens* Ti plasmid pTiA6 from left termintus of $T_L$-DNA. Plant Mol. Biol. 28: 1149–1154.

Rhodes, C.A., Pierce, D.A., Mettler, I.J., Mascarenhas, D., and Detmer, J.J. (1988) Genetically transformed maize plants from protoplasts. Science 240: 204–207.

Sanford, J.C., Klein, T.M., Wolf, E.D., and Allen, N. (1987) Delivery of substances into cells and tissues using a particle bombardment process. J. Part. Sci. Technol. 5: 27–37.

Shimamoto, K., Terada, R., Izawa, T., and Fujimoto, H. (1989) Fertile transgenic rice plants regenerated from transformed protoplasts. Nature 338: 274–276.

Smith, N. (1998) More T-DNA than meets the eye. Trends Plant Sci. 3: 85.

Srivastava, V., Vasil, V., and Vasil, I.K. (1996) Molecular characterization of the fate of transgenes in transformed wheat (*Triticum aestivum* L.). Theor. Appl. Genet. 92: 1031–1037.

Swedlund, B., and Vasil, I.K. (1985) Cytogenetic characterization of embryogenic callus and regenerated plants of *Pennisetum americanum* (L.) K. Schum. Theor. Appl. Genet. 69: 575–581.

Takamizo, T., Spangenberg, G., Suginobu, K., and Potrykus, I. (1991) Intergeneric somatic hybridization in Gramineae: somatic hybrid plants between tall fescue (*Festuca arundinacea* Schreb.) and Italian ryegrass (*Lolium multiflorum* Lam.). Mol. Gen. Genet. 231: 1–6.

Terada, R., Kyozuka, J., Nishibayashi, S., and Shimamoto, K. (1987) Plantlet regeneration from somatic hybrids of rice (*Oryza sativa* L.) and barnyard grass (*Echinochloa oryzicola* Vasing.). Mol. Gen. Genet. 210: 39–43.

Vasil, I.K. (1994) Cellular and molecular genetic improvement of cereals. In: Terzi, M., Cella, A. and Falavgina, A., (eds), Current Issues in Plant Molecular and Cellular Biology, pp. 5–18. Kluwer Academic Publishers, Dordrecht.

Vasil, I.K. (1998) Biotechnology and food security for the 21st century: a real-world perspective. Nature Biotechnology 16: 399–400.

Vasil, I.K., and Vasil, V. (1992) Advances in cereal protoplast research. Physiol. Plant. 85: 279–283.

Vasil, I.K., and Vasil, V. (1994) In vitro culture of cereals and grasses. In: Vasil, I.K. and Thorpe, T.A. (eds), Plant Cell and Tissue Culture, pp. 293–312. Kluwer Academic Publishers, Dordrecht.

Wenck, A., Conger, B.V., Trigiano, R., and Sams, C.L. (1988) Inhibition of somatic embryogenesis in orchardgrass by endogenous cytokinins. Plant Physiol. 88: 990–992.

Wenck, A., Czakó, M., Kanevski, I., and Márton, L. (1997) Frequent collinear long transfer of DNA inclusive of the whole binary vector during *Agrobacterium*-mediated transformation. Plant Mol. Biol. 34: 913–922.

Yang, Z., Shikanai, T., Mori, K., and Yamada, Y. (1989) Plant regeneration from cytoplasmic hybrids of rice (*Oryza sativa* L.). Theor. Appl. Genet. 77: 305–310.

# 2. Methods of Genetic Transformation: Electroporation and Polyethylene Glycol Treatment

RAY SHILLITO
*Agrevo USA Co., 703 NOR-AM Road, P. O. Box 538, Pikeville, NC 27863, USA.*
*E-mail: shillir@agrgold.hcc.com*

ABSTRACT. Methods for direct gene transfer into protoplasts via polyethylene glycol (PEG) treatment and electroporation were developed in early 1980's. Genetic transformation of protoplasts led to the production of the first transgenic cereals. This paper describes events leading up to the stable transformation of protoplasts via these techniques, the possible mechanisms involved, the improvement of the methods, and their application to the molecular improvement of cereals. While no longer the predominant method for transforming cereal crops, protoplast transformation still plays an important role in basic studies of gene regulation and function, understanding of the transformation process, and in the production of transgenic crops, particularly of many grass species.

## Introduction

The challenge facing scientists in the early 1980's was of introducing cloned genes into plants to study their function, and to confer new and useful traits. By 1982, this had been achieved for bacteria (Avery et al., 1944, Cosloy and Oishi, 1973, Klebe et al., 1983), and animal cells (Colbere-Garapin et al., 1981). In plants, *Agrobacterium tumefaciens* had been shown to transfer DNA into cells of a limited number of species. By 1983 it was successfully used to transfer T-DNA genes into plant tissue cultures (Fraley et al., 1983) and into regenerating protoplast cultures of tobacco and petunia (Marton et al., 1979). However, the method had limitations because of its dependence on a unique biological association. Many species, particularly the important gramineous crops, were not known host plants for *Agrobacterium* (de Cleene and de Ley, 1979). The method by which *Agrobacterium* transferred the DNA, and its integration into the genome, were a black box. In addition, it was not known whether *Agrobacterium*-derived elements were required for integration into the genome. Even today there are steps in the process which are poorly understood. At the time, there was also discussion over whether gramineous plants were inherently incapable of transformation (Potrykus, 1990).

The advent of methods to introduce DNA directly into plant protoplasts

*I.K. Vasil (ed.), Molecular Improvement of Cereal Crops, 9–20*

via polyethylene glycol (or other polycation) treatment and electroporation – methods not dependent on *Agrobacterium* – gave new optimism to those trying to transform monocots. They were quickly adapted and used to transform cell cultures of *Lolium*, *Triticum* and *Zea mays* (Potrykus et al., 1985, Lörz et al., 1985, Fromm et al., 1986), and eventually to introduce genes into protoplasts capable of forming fertile plants. With the advent of biolistic transformation techniques, the use of polyethylene glycol (PEG) and electroporation for stable transformation has declined, but it still remains an important tool in studying transient gene expression, and a method of choice for rice and many grass species.

## Mechanism

### *Polyethylene Glycol (and polycation)-Mediated Direct Gene Transfer*

Transfer of DNA across membranes by treatment with polycations such as PEG is not a fully understood process. The method is similar to fusion of protoplasts and involves the incubation of DNA (circular or linear) with protoplasts, addition of a polycation together with a divalent ion such as calcium or magnesium, and the subsequent dilution of the solution with a buffer which also contains divalent ions. Protoplasts are unstable in the absence of divalent cations, and the choice of calcium or magnesium is governed by the goal of the experiment. In general, calcium promotes better transient expression, whereas magnesium gives better transformation rates (Shillito, unpublished, Negrutiu et al., 1987). The conformation of DNA is strongly affected by PEG (Salianov et al., 1978, Matsuzawa and Yoshikawa, 1993), and particularly by PEG of a molecular weight above 600 which is normally used for direct gene transfer. Thus it is generally believed that the polycation compacts the DNA and also allows it to associate with the membrane, due to neutralisation of charges on the DNA and the membrane. The polycation solution has a high osmotic pressure, and withdraws water from the protoplast, as well as promoting adhesion of protoplasts to each other. It is supposed that DNA uptake takes place during the dilution phase, but the mechanism is unknown. Also unclear is how the DNA makes its way to the nucleus (or chloroplast) through the cell. Loyter et al., (1982) studied the behaviour of DNA/calcium co-precipitates in animal cells. However, co-precipitates are not generally used in plant cells; there are few comparable studies (Gisel et al., 1996) and the mechanism may be different.

### *Electroporation-Mediated Direct Gene Transfer*

Electroporation is more clearly understood than polycation-mediated DNA uptake. A polycation such as PEG may be used to potentiate uptake and/or integration (Shillito et al., 1985), but is not required. The uptake of DNA into the protoplast (or cell) is induced by pore formation in the cell membrane. A voltage applied across the membrane promotes and/or increases the stability of micropores which continually form in a resting membrane. The likelyhood of a pore opening is related to the voltage and the time for which it is applied. It is supposed that the open pore allows DNA to cross the membrane, and the DNA then finds its way to the nucleus. There have been some suggestions that the DNA may be induced to cross the membrane by the electric field inherent in electroporation. Pores are inherently unstable, and are kept open by the presence of the electric field. They must close in order for the protoplasts to remain viable. During electroporation animal cells must be incubated on ice for a period of time in order to allow the pores to close (Neumann et al., 1982), and in spite of extensive research (Chang et al., 1992) the mechanisms remain unknown (Maccarrone et al., 1995). The mechanism has not been investigated closely in plant cells but pore closing appears to be a rapid process (Shillito, unpublished). This may be due to the less structural nature of plant cell membranes.

### *DNA Integration*

The way in which DNA enters the nucleus and is expressed and/or integrated is also not fully understood. It is clear that DNA may be acted on by recombination mechanisms before integration (Bates et al., 1990), leading to complex inserts (Takano et al., 1997). Recombination before integration was first suggested by the fact that cotranformation led to integration of a non-selected gene together with the gene of interest (Schocher et al., 1986). Furthermore, DNA introduced by either of these methods is integrated into the genome at one locus about 75% of the time, although many copies may be integrated at each locus. Experiments showing reconstitution of genes from fragments confirmed that recombination occurs at a high rate, both when double stranded (Baur et al., 1990) and single stranded DNA (Bilang et al., 1992) are introduced.

## Development of PEG-mediated Transformation and Electroporation

The further development of transformation methods depended on a number of factors. These were effective selectable markers, totipotent tissue cultures, and a method that did not require *Agrobacterium*. Before the first successful demonstration of non-*Agrobacterium*-mediated protoplast transformation, there were a large number of publications that attempted to show that DNA was taken up and/or incorporated into protoplasts. However, it was, and still is, very difficult to prove physical uptake of DNA into protoplasts via PEG treatment. It was known as early as the late 1970's that DNA could adsorb to cell components, and particularly to the outer membrane of protoplasts and to nuclear membranes (Ohyama et al., 1977, Lurquin et al., 1979). Plasmid DNA can remain associated with carrot protoplasts after PEG treatment and subsequent DNAase treament. When the protoplasts are lysed, this DNA can become bound to the nucleus, giving rise to misleading results. The large number of equivocal reports raised the barrier for proof of stable transformation. The lack of suitable markers for transformation was also a block to progress, as early workers had to rely on showing DNA uptake and the gene constructs they used were not optimised for expression in plant cells.

Thus hormone independence conferred by the Ti plasmid of *Agrobacterium tumefaciens* was the first marker that was convincingly used as a selectable marker in plant cells. This allowed DNA uptake into protoplasts to be demonstrated by Davey et al., (1980) and Krens et al., (1982) using Ti plasmids that were taken up and gave rise to tumorous cells. The advent of 'disarmed' Ti plasmids and binary vectors (Hoekema et al., 1983, Bevan, 1984) simplified the use of the method and the advent of good markers such as neomycin phosphotransferase (NPT; De Block et al., 1984) allowed the tumor genes to be discarded as markers. However, the only real test of whether *Agrobacterium*-derived elements were required for transformation was to use DNA that did not contain the critical *vir* genes and border elements of T-DNA.

The first demonstration of non-*Agrobacterium*-mediated transformation of tobacco protoplasts, with a simple *E.coli* plasmid, was published by Paszkowski. et al., (1984). The proofs for transformation were not only molecular (Southerns, border fragments), but also segregation in microspores and transmission of a phenotype to progeny, linked to the segregation of the same integrated DNA fragment. Once published, this method, which came to be known as 'direct gene transfer', was used by a number of laboratories and spurred a renewed interest in cereal protoplast culture.

Shortly afterwards, the stable transformation of plant protoplasts via electroporation was also demonstrated (Shillito et al., 1985). Around the same time, electroporation was used to show transient expression of genes in maize protoplasts (Howard et al., 1985, Fromm et al., 1985). This was followed by stable transformation of non-regenerable protoplasts of *Lolium multiflorum* and *Triticum monococcum* using PEG (Potrykus et al., 1985, Lörz et al., 1985), and non-regenerable protoplasts of maize by electroporation (Fromm et al., 1986).

Studies of transient expression tend to be dominated by electroporation with a low voltage pulse, rather than by PEG transformation. Low voltage electroporation has been used in a very large number of species to optimize conditions in order to carry out stable transformation, but most extensively to explore promoter function (e.g. Yang, 1985, Hauptmann et al., 1987, Junker et al., 1987). Great care must be used in using transient results to optimize a system for stable transformation. The best conditions for transient expression are not necessarily those that give the best yield of stable transformants.

**Improvement of the Techniques**

Most stable protoplast transformation has been carried out using PEG, probably because of the simplicity of the technique. Advances, such as the use of magnesium in the medium (Negrutiu et al., 1987), have increased efficiencies in some systems to the 1% range, close to that which can be achieved using a combination of electroporation and PEG (Shillito et al., 1985). There have been few reports of optimisation of protocols for PEG-mediated transformation (Maas and Werr, 1989, Armstrong et al., 1990). An efficiency of 1–3% of regenerable colonies can often be achieved, a level at which colonies can be screened for the gene of interest, and the selectable marker can be eliminated from the process. Typically $80 \times 10^6$ protoplasts can be used per experiment leading to $10^5$ to $10^6$ tranformants. These numbers are not approachable via other methods.

Over the years, it has been suggested that the transfer of DNA into protoplasts was due to impurities in the PEG (better results are obtained with particular sources or batches of PEG), or is potentiated by use of other polycationic agents. In addition, other treatments such as radiation (Köhler et al., 1989), and synchronisation of cultures have been suggested. However, the method remains little changed from that first used – application of between 8 and 20% PEG of a molecular weight of 4000–8000 to protoplasts, followed by slow dilution of the PEG with a buffer containing divalent cations (Johnson et al., 1989). The main criterion for choosing the concentration of PEG is simply to use the amount that will not kill too many (more than 50%) of the protoplasts (Vasil et al., 1988).

An important development is the use of PEG transformation to introduce DNA into the chloroplast genome. This technique was pioneered by Maliga and co-workers (Carrer et al., 1993, Bock et al., 1994) and shows great promise for the engineering of crops which outcross into wild populations. However, it is inherently difficult and inefficient, and requires transformation efficiencies of green tissue that are not yet available with gramineous species.

In the late 1980's there were two main schools of electroporation – those that used a short, high voltage pulse (e.g. Shillito et al., 1985, Schocher et al., 1986, Riggs and Bates, 1986) and those using a long, lower voltage pulse (Fromm et al., 1986, Guerche et al., 1987). This second approach also led to the use of square-wave pulses (Lindsey and Jones, 1987). In general, the longer pulse seems to lead to better transient expression, but the data is not clear on which approach gives the best stable transformation. The highest transformation efficiencies (% of surviving colonies which give rise to a selectable transformed colony) have been achieved using high voltage electroporation in the presence of PEG (Shillito et al., 1985) although simple electroporation with a long pulse can lead to comparable efficiencies in some cases (Guerche et al., 1987).

There have been a number of studies that compared the penetration of DNA and or dyes due to pulsing with different machines and voltages (Rathus and Birch, 1992, Penmetsa and Ha, 1994, Bates, 1994). However, there is little or no clear direction to be gained from the literature. Most protocols tend to be developed by trial and error. The best approach is to tailor the voltage of the applied pulse to the protoplast diameter. For high voltage electroporation, the voltage required is inversely proportional to the diameter of the protoplast. For example, a tobacco protoplast with an average diameter of 50 $\mu$m will transform optimally at about 1500 V/cm, with a t½ (exponential decay constant: time for the voltage to decay to one half of its original level) of 10 $\mu$s, whereas a maize or similar protoplast derived from an embryogenic cell culture typically has a diameter of 25 $\mu$m and therefore will require 3000 V/cm for the same pulse length. Pulse length is controlled by adjusting the resistance of the medium, and/or changing the size of capacitor discharged across the cell containing the protoplasts. For low voltage (~ 400 V) systems, the parameters do not appear to be as clearly defined, and most protocol development requires a review of the literature on similar protoplast systems and some optimisation experiments to tune the system.

Electroporation has only rarely been applied to regenerable cereals, but grasses such as *Dactylis* (Horn et al., 1988) and *Agrostis* (Lee et al., 1996) have been transformed, as have rice (Chamberlain et al., 1994, Xu and Li, 1994, Rao et al., 1995) and barley (Salmenkallio et al., 1995). A novel approach, that of tissue electroporation, described by D'Halluin et al., (1992) has seen limited use, as has electroporation of partially digested tissues (Yang et al., 1993).

## Application to Cereals

At the time when protoplast transformation was being developed, all the monocot protoplast systems available were non-regenerable, and this remained a major hurdle on the road to transformed maize and other gramineous plants (Vasil and Vasil, 1984). Rice was the first to be regenerated successfully from protoplasts (Fujimura et al., 1985), as was expected due to its excellent tissue culture characteristics. However, others remained recalcitrant. The first non-rice plants regenerated from protoplasts were those of *Dactylis glomerata* (Horn et al.1988) and other grasses (Dalton, 1988), wheat (Vasil et al., 1990), and maize (Rhodes et al., 1988). In 1989 the first fertile maize plants (Prioli and Sondahl, 1989, Shillito et al., 1989) were reported. Regeneration of fertile plants from these recalcitrant species opened them up to transformation by PEG and electroporation. Eventually, protoplast cultures were used to produce fertile transgenic plants from Japonica, Javanica (Toriyama et al., 1988, Shimamoto et al., 1989, Li et al., 1992) and Indica (Datta et al., 1990) rice and maize (Golovkin et al., 1993), as well as grasses such as fescue (Wang et al., 1992).

## Conclusions

Over the years, many different approaches have been tried to introduce DNA into plants. Zhou et al., (1988) claimed to have transformed cotton via injection into ovules, and De La Pena et al., (1987) described transformation of rye plants by injection into immature inflorescences. However, transmission to progeny was not demonstrated, and these methods have yet to be confirmed by independent laboratories. Sonication is another technique that has been used to introduce DNA into protoplasts (Joersbo and Brunstedt, 1990), but this has found little application.

In 1995, Jahne et al., stated that 'to date only three methods have been found to be suitable for obtaining transgenic cereals: transformation of totipotent protoplasts, particle bombardment of regenerable tissues and, more recently, tissue electroporation.' With the exception of *Agrobacterium*-mediated transformation of embryos and embryogenic callus (see chapter by Komari and Kubo, this volume), this statement remains essentially true today. The advent of direct gene transfer and electroporation gave new hope to those who wished to engineer cereals. This was a method which was independent of *Agrobacterium*, and it was quickly applied to cereal cultures. If an efficient totipotent protoplast culture could be produced, it promised an easy route to transformed crops (e.g., rice; Shimamoto et al., 1989, and maize, Morocz et al., 1990). While totipotent cell culture systems were developed fairly early for a number of dicotyledons, the monocots, and particularly the cereals, proved difficult to bring into culture and then retrieve as fertile plants (Vasil and Vasil, 1992). A concerted effort was therefore made by a number of groups to develop embryogenic suspension cultures which were, and to-date remain, the only source of totipotent grass protoplasts. Such cultures provided the first transgenic cereals, and have been most useful in rice and several forage grass species (Vasil, 1994).

Stable transformation via electroporation and PEG-mediated direct gene transfer has now been eclipsed in the most part by other means of transforming cereals, particularly the biolistics procedure (see Chapter 3). However, direct gene transfer to protoplasts has remained popular for transforming some grasses, and rice. The method remains particularly useful where large numbers of transformants are needed, where a particularly regenerable protoplast culture is available (e.g., rice) or where intellectual property issues dictate the approaches available. The method can also be used for screening gene banks where a selectable phenotype expressed in culture is available. An example would be the direct transfer of a herbicide or antibiotic resistance (Gallois et al., 1992). In addition, protoplast transformation, and particularly electroporation, can be used to deliver proteins and protein complexes to plant cells (Ashraf et al., 1993). The use of these techniques will remain vital in dissecting processes such as *Agrobacterium*-mediated transformation (Hansen et al., 1997), and the use of recombinases (Albert et al., 1995, Lyznik et al., 1996) and other approaches to control integration of genes into the genome.

While not the predominant method for transforming cereal crops, transformation of protoplasts via PEG and electroporation still remains an important weapon in our scientific armory.

## Acknowledgements

Special thanks to Martha Wright (Novartis Seeds) for her help and encouragement witht his article. Thanks are also due my many colleagues in tissue culture and transformation research with whom I have worked or discussed these topics over the years.

## References

Armstrong, C.L., Petersen, W.L., Buchholz, W.G., Bowen, B.A., and Sulc, S.L. (1990) Factors affecting PEG-mediated stable transformation of maize protoplasts. Plant Cell Rep. 9: 335–339.

Ashraf, M., Altschuler, M., Galasinski, S., and Griffiths, T.D. (1993) Alteration of gene expression by restriction enzymes electroporated into plant cells. Mut. Res. 302: 75–82.

Avery, O.T., MacLeod, C.M., and McCarty, M. (1944) Studies on the chemical nature of the substance inducing transformation of pneumococcal types. J. Exp. Med. 79: 137 (reprinted in J. Exp. Med. 149: 297–326, 1979).

Bates, G.W. (1994) Genetic transformation of plants by protoplast electroporation. Mol. Biotechnol. 2: 135–45.

Bates, G.W., Carle, S.A., and Piastuch, W.C. (1990) Linear DNA introduced into carrot protoplasts by electroporation undergoes ligation and recircularization. Plant Mol. Biol. 14: 899–908.

Baur, M., Potrykus, I., and Paszkowski, J. (1990) Intermolecular homologous recombination in plants. Mol. Cell. Biol. 10: 492–500.

Bergman, P., and Glimelius, K. (1993) Electroporation of rapeseed protoplasts: transient and stable transformation. Physiol. Plant 88: 604–611.

Bevan, M. (1984) Binary *Agrobacterium* vectors for plant transformation. Nuc. Acids Res. 12: 8711–8721.

Bilang, R., Peterhans, A., Bogucki, A., and Paszkowski, J. (1992) Single-stranded DNA as a recombination substrate in plants as assessed by stable and transient recombination assays. Mol. Cell Biol. 12: 329–36.

Bock, R., Kosse,l H., and Maliga, P. (1994) Introduction of a heterologous editing site into the tobacco plastid genome: the lack of RNA editing leads to a mutant phenotype. EMBO. J. 13: 4623–4628.

Carrer, H., Hockenberry, T.N., Svab, Z., and Maliga, P. (1993) Kanamycin resistance as a selectable marker for plastid transformation in tobacco. Mol. Gen. Genet. 241: 49–56.

Chamberlain, D.A., Brettell, R.I.S., Last, D.I., Witrzens, B., McElroy, D., Dolferus, R., and Dennis, E.S. (1994) The use of the Emu promoter with antibiotic and herbicide resistance genes for the selection of transgenic wheat callus and rice plants. Aust. J. Plant Physiol. 21: 95–112.

Chang, D.C., Chassy, B.M., Saunders, J.A., and Sowers, A.E. (eds) (1992) Guide to Electroporation and Electrofusion. Academic Press, San Diego.

Colbere-Garapin, F., Horodniceanu, F., Kourilsky, P., and Garapin, A.C. (1981) A new dominant hybrid selective marker for higher eukaryotic cells. J. Mol. Biol. 150: 1–14.

Cosloy, S.D., and Oishi, M. (1973) The nature of the transformation process in *E.coli* K12. Mol. Gen. Genet. 124: 1–10.

Dalton, S.J. (1988) Plant regeneration from cell suspension protoplasts of *Festuca arundinacea* Schreb. (tall fescue) and *Lolium perenne* L (perennial ryegrass). J. Plant Physiol. 132: 170–175.

Datta, S.K., Datta, K., and Potrykus, I. (1990) Fertile Indica rice plants regenerated from protoplasts isolated from microspore derived cell suspensions. Plant Cell Rep. 9: 253–256.

Davey, M.R., Cocking, E.C., Freeman, J., Pearce, N., and Tudor, I. (1980) Transformation of petunia protoplasts by isolated *Agrobacterium* plasmids. Plant Sci. Lett. 18: 307–313.

de Block, M., Herrera-Estrella, L., van Montagu, M., Schell, J., and Zambryski, P.M. (1984) Expression of foreign genes in regenerated plants and their progeny. EMBO. J. 3: 1681–1689.

de Cleene, M., and de Ley, J. (1976) The host range of crown gall. Bot. Rev. 42: 389–466.

De La Pena, A., Lörz, H., and Schell, J. (1987) Transgenic rye plants obtained by injection of DNA into young floral tillers. Nature 325: 274–276.

D'Halluin, K., Bonne, E., Bossut, M., DeBeukelaar, M., and Leemans, J. (1992) Transgenic maize plants by tissue electroporation. Plant Cell 4: 1495–1505.

Fraley, R.T., Rogers, S.G., Horsch, R.B., Sanders, P.R., Flick, J.S., Adams, S.P., Bittner, M.L., Brand, L.A., Fink, C.L., Fry, J.S., Galluppi, G.R., Goldberg, S.B., Hoffmann, N.L., and Woo, S.C. (1983) Expression of bacterial genes in plant cells. Proc. Nat. Acad. Sci. USA 80: 4803–4807.

Fromm, M.E., Taylor, L.P., and Walbot, V. (1985) Expression of genes transferred into monocot and dicot cells by electroporation. Proc. Nat. Acad. Sci. USA 82: 5824–5828.

Fromm, M.E., Taylor, L.P., and Walbot, V. (1986) Stable transformation of maize after gene transfer by electroporation. Nature 319: 791–793.

Fujimura, T., Sakurai, M., Negishi, T., and Hirose, A. (1985) Regeneration of rice plants from protoplasts. Plant Tiss. Cult. Lett. 2: 74–75.

Gallois, P., Lindsey, K., Malone, R., Kreis, M., and Jones, M.G. (1992) Gene rescue in plants by direct gene transfer of total genomic DNA into protoplasts. Nuc. Acids Res. 20: 3977–3982.

Gisel, A., Rothen, B., Iglesias, V.A., Potrykus, I., and Sautter, C. (1996) In vivo observation of large foreign DNA molecules in host plant cells. Eur. J. Cell Biol. 69: 368–72.

Golovkin, M.V., Abraham, M., Morocz, S., Bottka, S., Feher, A., and Dudits, D. (1993) Production of transgenic maize plants by direct DNA uptake into embryogenic protoplasts. Plant Sci. 90: 41–52.

Guerche, P., Bellini, C., Le Moullec, J.M., and Caboche, M. (1987) Use of a transient expression assay for the optimization of direct gene transfer into tobacco mesophyll protoplasts by electroporation. Biochimie. 69: 621–628.

Hansen, G., Shillito, R.D., and Chilton, M.D. (1997) T-strand integration in maize protoplasts after codelivery of a T-DNA substrate and virulence genes. Proc. Nat. Acad. Sci. USA 94: 11726–11730.

Hauptmann, R.M., Ozias-Akins, P., Vasil, V., Tabaeizadeh, Z., Rogers, S.G., Horsch, R.B.,

Vasil, I.K., and Fraley, R.T. (1987) Transient expression of electroporated DNA in monocotyledonous and dicotyledonous species. Plant Cell Rep. 6: 265–270.

Hiei, Y., Ohta, S., Komari, T., and Kumashiro, T. (1994) Efficient transforamtion of rice (*Oryza sativa*) mediated by *Agrobacterium* and sequence analysis of the boundaries of the T-DNA. Plant J. 6: 271–282.

Hoekema, A., Hirsch, P.R., Hooykaas, P.J.J., and Schilperoort, R.A. (1983) A binary plant vector strategy based on separation of vir and T-region of the *Agrobacterium tumefaciens* Ti-plasmid. Nature 303: 179–180.

Horn, M.E., Shillito, R.D., Conger, B.V., and Harms, C.T. (1988) Transgenic plants of orchardgrass (*Dactylis glomerata* L.) from protoplasts. Plant Cell Rep. 7: 469–472.

Howard, E.A., Danna, K.J., Dennis, E.S., and Peacock, W.J. (1985) Transient expression in maize protoplasts. UCLA Symp. Mol. Cell Biol. 35: 225–234.

Jahne, A., Becker, D., and Lörz, H. (1995) Genetic engineering of cereal crop plants: a review. Euphytica. 85: 35–44.

Joersbo, M., and Brunstedt, J. (1990) Direct gene transfer to protoplasts via mild sonication. Plant Cell Rep. 9: 207–210.

Johnson, C.M., Carswell, G.K., and Shillito, R.D. (1989) Direct gene transfer via polyethylene glycol. J. Tiss. Cult. Meth. 12: 127–133.

Junker, B., Zimny, J., Luehrs, R., and Lörz, H. (1987) Transient expression of chimeric genes in dividing and non-dividing cereal protoplasts after PEG-induced DNA uptake. Plant Cell Rep. 6: 329–332.

Klebe, R.J., Harriss, J.V., Sharp, Z.D., and Douglas, M.G. (1983) A general method for polyethylene-glycol-induced genetic transformation of bacteria (*Escherichia coli*) and yeast (*Saccharomyces cerevisiae*). Gene 25: 333–342.

Köhler, F., Cardon, G., Pöhlman, M., Gill, R., and Schieder, O. (1989) Enhancement of transformation rates in higher plants by low dose irradiation: Are DNA repair systems involved in the incorporation of exogenous DNA into the plant genome? Plant Mol. Biol. 12: 189–199.

Krens, F.A., Molendijk, L., Wullems, G.J., and Schilperoort, R.A. (1982) *In vitro* transformation of plant protoplasts with Ti plasmid DNA. Nature 296: 72–75.

Lee, L., Laramore, C.L., Day, P.R., and Tumer, N.E. (1996) Transformation and regeneration of creeping bentgrass (*Agrostis palustris* Huds.) protoplasts. Crop Sci. 36: 401–406.

Lindsey, K., and Jones, M.G.K. (1987) Transient gene expression in electroporated protoplasts and intact cells of sugar beet. Plant Mol. Biol. 10: 43–52.

Li, Z., Xie, Q., Rush, M.C., and Murai, N. (1992) Fertile transgenic rice plants generated via protoplasts from the U.S. cultivar labelle. Crop Sci. 32: 810–814.

Lörz, H., Baker, B., and Schell, J. (1985) Gene transfer to cereal cells mediated by protoplast transformation. Mol. Gen. Genet. 199: 179–182.

Loyter, A., Scango, G., Juricek, D., Keene, D., and Ruddle, F.H. (1982) Mechanisms of DNA entry into mammalian cells. II. Phagocytosis of calcium phosphate DNA co-precipitate visualized by electron microscopy. Exp. Cell Res. 139: 223–234.

Lurquin, P.F. (1979) Entrapment of plasmid DNA by liposomes and their interactions with plant protoplasts. Nuc. Acids Res. 3: 3773–3784.

Lyznik, A., Rao, K.V., and Hodges, T.K. (1996) FLP-mediated recombination of FRT sites in the maize genome. Nuc. Acids Res. 24: 3784–3789.

Maas, C., and Werr, W. (1989) Mechanism and optimized conditions for PEG mediated DNA transfection into plant protoplasts. Plant Cell Rep. 8: 148–151.

Maccarrone, M., Rosato, N., and Agro, A.F. (1995) Electroporation enhances cell membrane peroxidation and luminescence. Biochem. Biophys. Res. Comm. 206: 238–245.

Marton, L., Wullems, G.J., Molendijk, L., and Schilperoort, R.A. (1979) In vitro transformation of cultured cells from *Nicotiana tabacum* by *Agrobacterium tumefaciens*. Nature 277: 129–131.

Matsuzawa, Y., and Yoshikawa, K. (1993) Controlling the conformation of large DNA molecule in aqueous solution. Nuc. Acids Symp. Ser. 108: 147–148.

Morocz, S., Donn, G., Nemeth, J., and Dudits, D. (1990) An improved system to obtain fertile regenerants via maize protoplasts isolated from highly embryogenic suspension culture. Theor. Appl. Genet. 80: 721–726.

Negrutiu, I., Shillito, R., Potrykus, I., Biasini, G., and Sala, F. (1987). Hybrid genes in the analysis of transformation conditions: I. Setting up a simple method for direct gene transfer in plant protoplasts. Plant Mol. Biol. 8: 363–374.

Neumann, E., Schaefer-Ridder, M., Wang, Y., and Hofschneider, P.H. (1982) Gene transfer into mouse lyoma cells by electroporation in high electric fields. EMBO. J. 1: 841–845.

Ohyama, K., Pelcher, L.E., and Horn, D. (1979) DNA binding and uptake by nuclei isolated from plant protoplasts: factors affecting DNA binding and uptake. Plant Physiol. 60: 98–101.

Paszkowski, J., Shillito, R.D., Saul, M., Mandak, V., Hohn, T., Hohn, B., and Potrykus, I. (1984) Direct gene transfer to plants. EMBO. J. 3: 2717–2722.

Penmetsa, R.V., and Ha, S.B. (1994) Factors influencing transient gene expression in electroporated tall fescue protoplasts. Plant Sc. 100: 171–178.

Prioli, L.M., and Sondahl, M.R. (1989) Plant regeneration and recovery of fertile plants from protoplasts of maize (*Zea mays* L.). Bio/Technology 7: 589–595.

Potrykus, I., Saul, M.W., Petruska, J., Paszkowski, J., and Shillito, R.D. (1985) Direct gene transfer to cells of a graminaceous monocot. Mol. Gen. Genet. 199: 183–188.

Potrykus, I .(1990) Gene transfer to cereals: an assessment. Bio/technology 8: 535–542.

Rathus, C., and Birch, R.G. (1992) Optimization of conditions for electroporation and transient expression of foreign genes in sugarcane protoplasts. Plant Sci. 81: 65–74.

Rao, K.V., Rathore, K.S., and Hodges, T.K. (1995) Physical, chemical and physiological parameters for electroporation-mediated gene delivery into rice protoplasts. Transgen. Res. 4: 361–368.

Rhodes, C.A., Lowe, K.S., and Ruby, K.L. (1988) Plant regeneration from protoplasts isolated from embryogenic maize cell cultures. Bio/Technology 6: 56–60.

Riggs, C.D., and Bates, G.W. (1986) Stable transformation of tobacco (*Nicotiana tabacum* cultivar Xanthi) by electroporation: Evidence for plasmid concatenation. Proc. Nat. Acad. Sci. USA 83: 5602–5606.

Salianov, V.I., Pogrebniak, V.G., Skuridin, S.G., Lortkipanidze, G.B., and Chidzhavadze, Z.G. (1978) Relationship between the molecular structure of aqueous solutions of polyethylene glycol and the compactness of double-stranded DNA molecules. Mol. Biol. (Moskva) 12: 485–495.

Salmenkallio, M.M., Aspegren, K., Akerman, S., Kurten, U., Mannonen, L., Ritala, R., Teeri, T.H., and Kauppinen, V. (1995) Transgenic barley (*Hordeum vulgare* L.) by electroporation of protoplasts. Plant Cell Rep. 15: 301–304.

Schocher, R.J., Shillito, R.D., Saul, M.W., Paszkowski, J., and Potrykus, I. (1986) Co-transformation of unlinked foreign genes into plants by direct gene transfer. Bio/Technology 4: 1093–1096.

Shimamoto, K., Terada, R., Izawa, T., and Fujimoto, H. (1989) Fertile transgenic rice plants regenerated from transformed protoplasts. Nature 338: 274–276.

Shillito, R.D., Saul, M.W., Paszkowski, J., Müller, M., and Potrykus, I. (1985) High efficiency direct gene transfer to plants. Bio/Technology 3: 1099–1103.

Shillito, R.D., Carswell, G.K., Johnson, C.M., DiMaio, J.J., and Harms, C.T. (1989) Regeneration of fertile plants from protoplasts of elite inbred maize. Bio/Technology 7: 581–587.

Takano, M., Egawa, H., Ikeda, J.E., and Wakasa, K. (1997) The structures of integration sites in transgenic rice. Plant J. 11: 353–361.

Toriyama, K., Arimoto, Y., Uchimiya, H., and Hinata, K. (1988) Transgenic rice plants after direct gene transfer into protoplasts. Bio/Technology 6: 1072–1074.

Vasil, I.K. (1994) Molecular improvement of cereals. Plant Mol. Biol. 25: 925–937.

Vasil, I.K., and Vasil, V. (1992) Advances in cereal protoplast research. Physiol. Plant 85: 279–283.
Vasil, V., Redway, F., and Vasil, I.K. (1990) Regeneration of plants from embryogenic suspension culture protoplasts of wheat (*Triticum aestivum* L.). Bio/Technology 8: 429–434.
Vasil, V., and Vasil, I.K. (1984) Isolation and maintenance of embryogenic cell suspension cultures of gramineae. In: Vasil, I.K. (ed) Cell Culture and Somatic Cell Genetics of Plants, Vol 1, pp. 152–158. Academic Press, New York.
Vasil, V., Hauptmann, R.M., Morrish, F.M., and Vasil, I.K. (1988) Comparative analysis of free DNA delivery and expression into protoplasts of *Panicum maximum* Jacq. (Guinea grass) by electroporation and polyethylene glycol. Plant Cell Rep. 7: 499–503.
Wang, Z.Y., Takamizo, T., Iglesias, V.A., Osusky, M., Nagel, J., Potrykus, I., and Spangenberg, G. (1992) Transgenic plants of tall fescue (*Festuca arundinacea* Schreb) obtained by direct gene transfer to protoplasts. Bio/Technology 10: 691–699.
Xu, X.P., and Li, B.J. (1994) Fertile transgenic Indica rice plants obtained by electroporation of the seed embryo cells. Plant Cell Rep. 13: 237–242.
Yang, N. (1985) Transient gene expression in electroporated plant cells. Trends Biotech. 3: 191–192.
Yang, J.S., Ge, K.L., Wang, Y.Z., Wang, B., and Tan, C.C. (1993) Highly efficient transfer and stable integration of foreign DNA into partially digested rice cells using a pulsed electrophoretic drive. Transgen. Res. 2: 245–251.
Zhou, G.Y., Weng, J., Gong, Z., Zhen, Y., Yang, W., Shen, W., Wang, Z., Tao, Q., and Huang, J. (1988) Molecular breeding of agriculture: A technique for introducing exogenous DNA into plants after self-pollination. Sci. Agr. Sinica. 21: 1–6.

# 3. Methods of Genetic Transformation: The Gene Gun

THEODORE M. KLEIN and TODD J. JONES
*DuPont Agricultural Products, Experimental Station, Wilmington, DR 19880, USA.*
*E-mail: kleintm@al.esvax.umc.dupont.com*

ABSTRACT. In the decade since its introduction, the gene gun has gone from being a scientific novelty to an established tool for introducing DNA and other molecules into plant cells. The unique ability of biolistic technology to deliver macromolecules into intact cells has been exploited in studies ranging from control of gene expression to organelle transformation. Further, biolistics is the critical technology allowing for the genetic transformation of the world's most important crop species. Still, many of the factors that influence the efficiency of transformation using the gene gun are poorly understood. The purpose of this paper, then, is to review the current research addressing the physical and biological parameters that effect biolistic transformation.

## Introduction

Plant molecular biology is advancing at a breathtaking pace. Ten years ago, isolating and sequencing a specific gene represented a graduate student's thesis. The range of tissue-specific or constitutive promoters that were available to express transgenes was very limited. The ability to genetically transform the large majority of crop species was severely restricted. Today, sequences for many genes are available from a number of on-line database sources, derived in part from various genomics programs throughout the world. The ability to control the spatial and temporal expression of heterologous genes is fairly well advanced and several chemically inducible gene expression systems have been developed for plants. Of particular significance to the emerging field of agricultural biotechnology is the development of gene transfer methods for the genetic transformation of a number of important crop species.

The gene gun was originally conceived of by John Sanford at Cornell University sometime in 1984. He was interested in methods to introduce DNA into cells by physical means to overcome the biological limitations of *Agrobacterium* and the difficulties associated with regenerating whole plants from protoplasts. He attempted pollen transformation using lasers and microinjection before developing the novel concept of accelerating DNA-coated particles into cells. One of us (TK) worked on the early

*I.K. Vasil (ed.), Molecular Improvement of Cereal Crops, 21–42*

development of the particle bombardment process with Dr. Sanford. Several acceleration concepts were attempted including air guns, bb guns, and centrifugal force in several configurations. After much trial and error we settled on developing a gun powder propelled device that employed a vacuum and published some early work in a relatively obscure journal (Sanford et al., 1987). One of the most difficult aspects of this work was trying to stop the macroprojectile (the 'bullet' carrying the microprojectiles on it's front surface) as it crashed into the stopping plate at the end of the barrel. The macroprojectile travels down the barrel at about 1100 ft/sec, so one would expect stopping it would not be easy without causing damage to the apparatus. We quickly learned that a metal macroprojectile was not suitable and turned to trying different plastics. The most useful material turned out to be ultra-high molecular weight polyethylene. This material did not shatter upon impact but 'melted' as it partially passed through the orifice of the stopping plate. We also found that a plastic stopping plate could help cushion the impact of the macroprojectile. After trying a number of materials we settled on using a polycarbonate plate. This material (used in the fabrication of bullet proof glass) is tough and does not easily shatter. Trial and error was also used to optimize binding of DNA to particles, as well as the number, size and velocity to be used for bombardment of our onion epidermal model system (Klein et al., 1987). With the knowledge that cells survive and express foreign DNA following bombardment, the biolistic concept was applied to the transformation of maize (Klein et al., 1988). A number of researchers were attracted to this new method and it was applied to a number of systems, especially for the transformation of recalcitrant monocot species. Several groups also developed new gene gun designs.

Over the past decade the concept of accelerating DNA-coated particles into cells and tissues has evolved from a novelty to an established tool in plant molecular biology. The gene gun has played a pivotal role in the production of transgenic plants, allowing the transformation of species that were previously difficult or impossible to transform. This list includes maize, soybean, wheat, barley, oat and rice. The first transgenic maize and soybean varieties to be commercialized were produced by particle bombardment. In addition to applied applications, our understanding of the spatial, temporal and environmental regulation of plant genes has advanced as a result of transient expression studies using the gene gun (Russell and Fromm, 1995). Yet another application is the delivery of genes to organelles. The finding that the chloroplast of *Chlamydomonas* can be transformed by microprojectile bombardment (Boynton and Gillham, 1993) has been extended to higher plants (Zoubenko et al., 1994, Daniell et al., 1998). The ability to produce plants containing enormous numbers of transformed plastids has implications for high-level gene expression in crops (McBride et al.,1994, Daniell et al., 1998). Mitochondrial transformation of yeast (Fox et al., 1991) and *Chlamydomonas* (Boynton and Gillham, 1996) has also been achieved using the gene gun. In addition to delivering DNA, the gene gun should be a useful tool for delivering other molecules, such as RNA (Tanaka et al., 1995) into cells.

The success of biolistic technology for gene transfer to plants has been extended to a range of organisms. In fact, the gene gun is widely used for gene transfer to animal cells (most recent publications involve bombardment of animal cells) and may be an important tool for the emerging field of genetic immunization (Johnston and Tang, 1993, 1994). Following some very promising experiments in mice (Hui and Chia, 1997), clinical trials are now underway to determine if biolistic technology can be used in humans for treating several diseases including cancer. Powderject, a company based in England, is in the process of commercializing biolistic technology for medical applications. Although biolistics has matured into a technology with a wide range of applications, our understanding of many of the important factors that influence the process is incomplete. The purpose of this review is to summarize our understanding of the biolistic approach with an emphasis on the physical and biological variables that impact the process. This does not assume to be a complete examination of the topic. For this purpose, the reader is directed to several review articles (Casas et al., 1995, Klein and Fitzpatrick-McElligott, 1993) and books (Christou, 1996, Yang and Christou, 1994) on the subject.

## Targets for Transformation

The power of the biolistic concept is that genes can readily be delivered to intact tissues. Therefore, essentially any tissue can be a target. Particle bombardment has been utilized to examine transient gene expression in many plant organs, including leaves (Bansal et al., 1992, Denchev et al., 1997), roots (Warkentin et al., 1992), petals (Greisbach and Klein, 1993), pollen (Twell et al., 1989), and endosperm (Knudsen and Müller, 1991). Callus (Wan et al., 1995), suspension cultures (Gordon-Kamm et al., 1990, Fromm et al., 1991), meristems (McCabe et al., 1988, Cao et al., 1990), embryos, and immature inflorescences (Barcelo and Lazzeri, 1995, Casas et al., 1997) have all been used as recipient tissue for stable plant transformation. For transient expression studies of gene expression, it should be remembered that the particles generally do not penetrate beyond the first several layers of the tissue (Taylor and Vasil, 1991). Therefore, the appropriate controls that regulate the expression of the introduced gene must be present in these layers for the approach to be successful. There are some cases where the introduced gene does not exhibit proper regulation when compared to the native gene or when the transgene is stably introduced into the genome. For example, Frisch et al. (1995) created $\beta$-glucuronidase fusions under the control of the seed-specific $\beta$-phaseolin promoter. Stably transformed tobacco plants displayed the expected seed-specific expression of the chimeric gene. However, transient expression of the $\beta$-phaseolin:$\beta$-glucuronidase gene was detected in tobacco and soybean leaf tissue following introduction by particle bombardment. The authors speculated that chromatin structure plays an important role in $\beta$-phaseolin expression. This example notwithstanding, numerous other studies show tight tissue-specific regulation of transiently introduced genes. For example, proper regulation of *A1*and *Bz1* in maize was found following their delivery to aleurone and embryo tissue of maize (Klein et al., 1989). Bansal et al., (1992) found proper cell-specific and light-regulated expression from *cab* and *rbcS* promoter sequences when bombarded into maize leaves. Some researchers continue to utilize the gunpowder version of the gene gun instead of the newer helium version because it delivers particles to deeper layers in plant tissue. This has proved important in the study of light regulated genes in maize leaf tissue (Purcell et al., 1995).

Developing a transformation system for transgenic plant production requires careful consideration of the target tissue and selection scheme. Embryogenic cultures have been used for the stable transformation of a number of monocot species. However, regenerable embryogenic cultures can be difficult to maintain for long periods. For some species, prolific embryogenic cultures can be produced from only particular genotypes. Embryogenic cultures often lose their ability to produce fertile plants over time. This is true for soybean, where the useful life of an embryogenic suspension is limited to about six to eight months due to the accumulation of mutations (Singh et al., 1998). Researchers working on cereal transformation have overcome the necessity for establishing embryogenic cultures prior to bombardment by targeting immature embryos and allowing subsequent callus proliferation from the scutellum. This approach has been successful for maize, wheat, barley, rice and other species (reviewed by Vasil, 1994, Barcelo and Lazzeri, 1995). It should be remembered that the success of this method is still genotype dependent. For example, generation of transgenic maize plants from genotypes that grow well in tissue culture is highly efficient and routine (Songstad et al., 1996, Brettschneider et al., 1997). However, less than 0.1% of bombarded embryos give rise to transgenic callus when other, less well adapted genotypes are used (Koziel et al., 1993). Moore et al. (1994) present evidence that transient expression of genes introduced into immature soybean cotyledons is gentoype dependent. These differences may be due to variation in some physical characteristics of the cotyledons. Alternatively, there may be differences in biochemical characteristics that affect the expression of the $\beta$-glucuronidase gene or the activity or stability of the enzyme.

A critical parameter for embryo bombardment is the period that the embryos are incubated on culture initiation medium prior to bombardment (Altpeter et al., 1996, Songstad et al., 1996, Takumi and Shimada, 1996). For example, wheat embryos bombarded after 1 to 4 days in culture produced no transformants while about 1 to 10% of bombarded embryos produced transgenic plants if they were cultured for 5 to 9 days (Takumi and Shimada, 1996). One explanation for this finding is that embryos in culture for a short duration do not yet possess cells that can give rise to sustained callus proliferation. Alternatively, embryos bombarded during the first few days of culture are more prone to damage than those incubated for longer periods (Vasil et al., 1993). This was also observed by Brettschneider et al. (1997) for wheat and by Zimny et al. (1995) for triticale scutellum. Koprek et al. (1996) found that barley cultivars differed in their regeneration ability following bombardment of scutella. These researchers observed that bombardment with a particle inflow gun (Vain et al., 1993) was less injurious to some genotypes than the PDS-1000/He (Bio-Rad Laboratories, Hercules, CA), allowing regeneration and transformation of the more recalcitrant varieties. The particle inflow gun accelerates microprojectiles with a gentle burst of gas. Therefore, this gene gun may produce less of a shock wave than is produced when the macrocarrier impacts the stopping screen in the PDS-1000/He. However, it is not clear if the particle inflow gun has comparable velocities to the PDS-1000/He and therefore may not be appropriate for stable transformation in systems where dividing and regenerable cells reside below the surface of the tissue.

Meristems offer an attractive target in that problems with regeneration in tissue culture can be circumvented. Meristem transformation was first demonstrated in soybean (McCabe et al., 1988). Since then *Phaseolus* (Russell, 1993), peanut (Brar, 1994), and cotton (McCabe and Martinell, 1993) have been transformed by this approach. However, the rates of transformation reported by these researchers are extremely low requiring the bombardment of many meristems to produce one transformant. Another drawback is that a proprietary gene gun based on electrostatic discharge is used in these procedures. However, recent research by Aragão et al. (1996) demonstrates that transgenic bean can be produced by meristem bombardment using a helium-powered device. The frequency of transformation achieved by these researchers was comparable or better than that observed with an electrostatic discharge gun (Russel, 1993). Sautter et al. (1991) attempted to improve meristem transformation by designing a micro-targeting gene gun that can precisely bombard as small as 150 $\mu$m in diameter. This gene gun uses a burst of gas to accelerate particles through a capillary tube. Up to 3% of the cells in a target area (or about $35 \times 10^3$ cells per $cm^2$) express $\beta$-glucuronidase. This high rate of gene transfer is probably due to the relative gentleness of the technique and the precise control of the number of particles delivered to the target. In addition, this procedure does not use a precipitation step to adhere DNA to the particles, thus avoiding agglomeration. Although this approach has not yet proven useful for transgenic plant production, it has been used as a tool to study plant development (Kost et al., 1996).

Other researchers have used a combination of tissue culture techniques and meristem bombardment to produce transgenic plants. Zhong et al. (1996) placed maize shoot tips on a meristem multiplication medium to produce a meristem culture. This material was bombarded and bialophos selection applied. Transgenic shoots were recovered at a fairly high frequency. Lowe et al. (1995) bombarded the exposed apical meristem of immature maize embryos and, after germinating the embryos, excised and cultured the apical meristems on shoot multiplication medium. They were able to identify clonal sectors by utilizing $\beta$-glucuronidase staining or an antibiotic selection method (kanamycin or spectinomycin were used) that bleaches non-transgenic tissue.

Pollen offers a potential tissue-culture-free means of transgenic plant production and the allure has attracted many to attempt it, with generally less-than-convincing results. Despite the evidence that particle bombardment can deliver genes to pollen at high frequency (Twell et al., 1989, Hay et al., 1994, Martinussen et al., 1994), generating stably transformed plants via pollen bombardment is not straightforward. van der Leede-Plegt et al. (1995) introduced genes into *Nicotiana glutinosa* pollen and they recovered two transgenic plants from a series of experiments. For reasons that remain unclear, those two plants failed to transmit the transgenes to progeny. They postulated that this might be caused by a recombination event during the integration of the introduced DNA resulting in a gametic lethal deletion of part of the target chromosome. Recent results indicate that maize can be transformed by bombardment of pollen with magnetite particles (Horikawa et al., 1997). Pollinations were carried out with a magnetically enriched population of pollen containing microprojectiles. These pollinations resulted in the production of 570 seedlings, of which 3 appeared to carry the introduced basta resistance gene.

## Optimization of the Biolistic Process

### *Biological Parameters*

When using the gene gun for stable transformation, it is apparent that a balance must be reached between optimizing DNA delivery and minimizing cell damage. One means of enhancing cell survival involves incubating the target tissue in medium containing relatively high osmotic pressure. This approach was first utilized with yeast (Armaleo, 1990) and was demonstrated to increase transformation rates in plants by Russell (1992b). Vain et al. (1993), Leduc (1994), Altpeter et al. (1996) and many others have used this approach. Maize transformation was increased almost 7-fold by pretreatment with a medium that contained sorbitol and mannitol (Vain et al., 1993). Ye et al. (1990) showed that transient expression of genes introduced into the chloroplast is also enhanced by osmotic treatment. Histological examination of bombarded maize embryos indicated that they were protected from particle bombardment damage when precultured on high osmoticum medium (Kemper et al., 1996). Interestingly, these researchers also found that an osmotic treatment with mannitol led to an increase in the number of particles able to penetrate into maize embryos.

The duration of pre- and post-bombardment treatments and concentration of osmoticum must be optimized for each application. For example, a procedure for the transformation of fescue utilized a short (0.5 hr.) preincubation period and a long (4 day) post-bombardment incubation on media containing osmoticum (Spangenberg et al., 1995). It may be advantageous to gradually step down the osmotic strength of the medium following bombardment (Russell et al., 1992b). Again, most of the publications report conditions for optimal transient expression and do not consider effects on stable transformation. Osmotic treatments that yield maximal levels of transient gene expression may detrimentally impact subsequent regeneration as was found by Altpeter et al. (1996) with wheat embryos. Morrish et al. (1993) cite unpublished work with maize cells that show optimal transient gene expression following incubation with 6 per cent mannitol but optimal levels of stable transformation using 2 per cent mannitol. It is assumed that elevated levels of osmoticum plasmolyze the target cells and limit leakage of the cell's cytoplasm. The osmotic treatment may also induce certain cellular stress responses that protect the tissue from mounting an apoptotic reaction to particle penetration. The role of stress response is also suggested by the work of Perl et al. (1992). They found that in addition to an osmotic treatment, incubating wheat embryos in the presence of silver thiosulphate, an ethylene inhibitor, increased $\beta$-glucuronidase activity. Morrish et al. (1993) reported that transient gene expression in embryogenic maize callus is inversely related to ethylene levels. Adding activated charcoal to embryogenic grape suspensions increased long term expression of $\beta$-glucuronidase, possibly by absorbing oxidized substances excreted by the bombarded cells (Hébert et al., 1993). Other means for modulating the production of stress related compounds may have an impact on transformation efficiency.

Finer and McMullen (1991) used a simple drying treatment prior to bombardment of embryogenic suspensions of soybean. They suggested that this led to a decrease in the cell's osmotic pressure and thus higher rates of cell survival. Chen and Beversdorf (1994) desiccated microspore-derived embryos prior to bombardment. Following bombardment, these researchers imbibed the tissue in a DNA solution. They found that the combined treatment of particle bombardment and DNA imbibition led to about a 4-fold increase in stable transformation when compared to particle bombardment alone.

The relationship between cell division and biolistic transformation is a poorly understood parameter. Iida et al. (1991) reported optimal levels of transient expression when tobacco cells were bombarded during M or $G_2$ phase of growth. Bombarded wheat cells entering log phase exhibited higher levels of transient expression than cells in stationary phase (Vasil et al., 1991). *Larix* cells exhibited highest levels of transient expression when they were bombarded five days after subculture (Duchesne et al., 1993). Transient expression in soybean cells was found to be highest two days after subculture and was correlated with mitotic index (Hazel et al., 1997). These researchers also studied the relation of cell division to stable transformation in embryogenic suspensions. They found that cultures yielding large numbers of transformants had more dividing cells in the outermost cell layers than cultures yielding few transformants. Goldfarb et al. (1991) found that transient $\beta$-glucuronidase expression was stimulated by treatment of Douglas-fir cotyledons with cytokinins, presumably by stimulating cell division. Kausch et al. (1995) observed that bombardment depressed division and increased the number of S-phase cells in embryogenic maize tissue. The cause of this is not known but they suggest that DNA breakage and repair are stimulated in cells receiving exogenous DNA resulting in an inhibition of cell division. It is of interest to note that the number and arrangement of transgenes can be influenced by stage of the cell cycle at which transformation is carried out (Kartzke et al., 1990).

*Physical parameters*

Gene transfer to a particular cell or tissue type requires the optimization of various physical parameters (Sanford et al., 1993). Basically, these parameters impact the effective momentum of the particle, the number of particles that impact the tissue, and the amount of useful DNA that the particles carry into the cells. These parameters are summarized in Table 1. Of prime importance is that virtually all of these parameters interact with each other and can interact with some of the biological variables associated with the target tissue. However, it should be noted that employing the standard conditions recommended for use with the commercially available PDS-1000/He (Bio-Rad Labs) gives adequate results for many applications. Figure 1 shows a schematic diagram of the commercially available device.

There are a number of studies that detail the optimization of gene delivery to plant tissue using transient expression of a reporter gene such as $\beta$-glucuronidase (Hébert et al., 1993). However, optimal levels of transient gene expression do not necessarily correlate with optimal cell survival and maximal recovery of transgenic cell lines. For example, Russell et al. (1992a) monitored transient $\beta$-glucuronidase expression several days after bombardment of a field of tobacco cells spread onto filter paper. The highest numbers of $\beta$-glucuronidase expressing cells were found in a zone near the

TABLE 1
Physical Parameters that Influence Gene Transfer by Particle Bombardment.

| Parameter | Variable |
|---|---|
| Microprojectile | |
| Penetrating Ability (effective momentum) | size, density, shape. |
| DNA Carrying Capacity | Precipitation protocol, DNA quantity, DNA quality, particle size, shape and surface characteristics. |
| Gene Gun | |
| Microprojectile Velocity | Acceleration force, distance to target, vacuum in chamber. |
| Microprojectile No. / Unit Area of Tissue | Distance to target, Initial number loaded. |

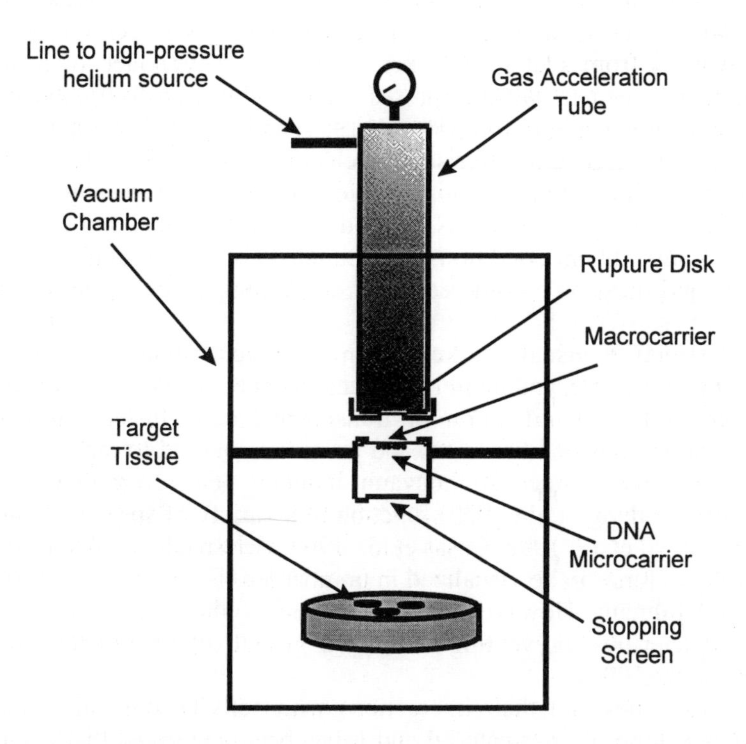


*Fig. 1.* Schematic representation of one gene gun design (PDS-1000/He, Bio-Rad Laboratories Inc., Hercules, California). In this design, the microcarriers are placed on one side of a plastic sheet (macrocarrier). The macrocarrier is propelled into a stopping screen by a shock wave created by a burst of compressed helium upon failure of the rupture disc. The microcarriers continue through a partial vacuum until impact with the target tissue.

center of the filter paper. However, when attempts were made to recover stably transformed colonies from these filters, none were recovered from the central zone. Highest numbers of stable transformants were found in a 'ring' outside of the central zone. Apparently, the cells in the center survived long enough to express $\beta$-glucuronidase but were too damaged to continue to divide. Similarly, Christou and Ford (1995) found that optimizing for maximum levels of transient expression in rice embryos was detrimental to the recovery of stable transformants. For example, they found that optimal levels of $\beta$-glucuronidase activity were found when a 16 kV accelerating voltage was used while the highest number of stable transformants were recovered when 12 kV was employed. Brettschneider et al. (1997) observed similar levels of transient gene expression when a helium pressure of 900 or 1200 psi was used to transfer genes into maize embryos. However, rates of stable transformation were higher when bombardments were carried out with 1200 psi. Bombardments using 30 $\mu$g of gold particles produced more transformants from wheat embryos than when 100 $\mu$g was used, although transient expression was higher with the latter gold quantity (Altpeter et al., 1996). The larger gold quantities have a deleterious effect on somatic embryogenesis from scutella (Becker et al., 1994). The lack of a direct correlation between transient expression and stable transformation holds true for bombarded animal cells (Heiser, 1994) and is probably true for other physical means of gene transfer such as electorporation. The establishment of optimal conditions for generating stable transformants typically involves determining that point where DNA introduction is maximal and cell death from the bombardment conditions is minimal. This is an extremely important issue and must be considered when developing a new transformation system.

Non-destructive visual markers such as green fluourescent protein (Haseloff et al., 1997), and luciferase (Raemakers et al., 1996, Bower et al., 1996) permit the identification of transformed cell clusters or sectors relatively soon after bombardment and provide a method to rapidly assess stable transformation rates. Anthocyanin-inducing genes from maize (Goff et al., 1990, Ludwig et al., 1990) function in a variety of species (Wong et al., 1991, Lloyd et al., 1992, Casas et al., 1993, Griesbach and Klein, 1993) and purple sectors can be visualized in bombarded tissue one or two weeks after bombardment. However, overexpression of these regulatory genes may be deleterious, preventing prolonged growth of transformed tissue (Bowers et al., 1996).

Microprojectiles and their interactions with cells.Theoretically, microprojectiles can be of any material and it has been suggested that even ice crystals or water droplets can be accelerated to sufficient velocity for penetration of cells. Bacteria have been used as microprojectiles (Rasmussen et al., 1994). Magnetite from magnetotactic bacteria may also make interesting microprojectiles (Matsunaga, 1991). They have been used to deliver genes into lily pollen and timothy grass cells and magnetic separation was used to enrich for transformed cells (Yoshizumi et al., 1996). However, the most commonly used materials are powders of metals that have high density such as gold or tungsten. The first materials to be used as microprojectiles were tungsten powders obtained from General Electric Corp. and Sylvania, Inc. These powders are used in light bulb production and are available in lots of different average diameters. For example, the M10 tungsten powder (available from Bio-Rad Labs, Hercules, CA) has an average diameter of about 0.94 $\mu$m. About 15 per cent of the particles have diameters of over 1 $\mu$m and less than 2.5 $\mu$m (Klein and Maliga, 1991). The first gold microprojectiles had very wide ranges in average diameter. For example, the gold powder used for the transformation of soybean had a range of diameter of 1.5 to 3 $\mu$m (McCabe et al., 1988). The Electronics Department at DuPont manufactures gold powders for the electronics industry and these powders are available through Bio-Rad. Gold powders can also be produced using a protocol described by Sautter et al. (1991). There is data that shows that gold microprojectiles are better for stable transformation. It has been demonstrated that long-term exposure to tungsten may be toxic to plant cells (Russell et al., 1992a) thus reducing the yield of stable transformants relative to gold particles. However, in another study with further optimized bombardment conditions (Russell et al., 1992b), comparable numbers of stable tobacco transformants were obtained with M10 tungsten and 1.0 $\mu$m gold particles. Jain et al. (1996) found that gold generally produced more rice transformants than tungsten particles. It should be remembered that in these comparisons particle size and number are also being varied. Thus differences in transformation efficiencies between particles of different materials may be due to other factors than the metal's toxicity. One important reason to use gold is that, under some conditions, tungsten can catalytically degrade DNA (Sanford et al., 1993).

The size of the microprojectile particle is a critical variable. However, comparing particles of different sizes is not simple. Particle size interacts with other variables such as the DNA carrying capacity of the particle, the number of particles that impact the target tissue, and momentum of the particle. Particle size will also influence the impact velocity since particles of different sizes will have different aerodynamic characteristics. Thus, if particle size is changed, the influence of other variables such as the number of particles, the amount of DNA, and distance to the target may also need adjustment. It should also be remembered that transient expression is probably not the optimal means for assessing the effect of particle size if the eventual goal is to obtain stable transformants. Recent results indicate that relatively small particles may be optimal for stable transformation. For example, Randolph-Anderson et al. (1997) bombarded maize callus with either 0.6 or 1.0 $\mu$m gold particles coated with pBC17 (Klein and Zhang, 1994), a plasmid which encodes genes for the C1 and B-Peru regulatory genes in the maize anthocyanin biosynthetic pathway (Goff et al., 1990). After six weeks, about 4- to 7-fold more multicellular sectors were found in the tissue bombarded with 0.6 $\mu$m particles than with the 1.0 $\mu$m particles. Apparently the smaller gold particles are less damaging to the cells and this is probably the main explanation for the higher efficiencies. A hint of the damage caused by relatively large particles came from work with $\beta$-glucuronidase expression in black Mexican sweet maize cells (Klein et al., 1988). Expressing cells often occurred in clusters of 5 to 50 cells following bombardment with 1.2 $\mu$m or 2.4 $\mu$m particles. Often a highly expressing cell was observed in the center of the cluster. In contrast, single discreet cells expressing $\beta$-glucuronidase were observed following bombardment with 0.6 $\mu$m particles. It now appears that the larger particles caused damage to the cell resulting in loss of $\beta$-glucuronidase or reaction products to the surrounding medium. 0.6 $\mu$m gold particles were also found to be better than 1.0 $\mu$m particles for the transformation of *Chlamydomonas* chloroplasts and yeast mitochondria. Bacterial transformation is enhanced nearly 4-fold with the use of the smaller M5 tungsten particles (mean of 0.77 $\mu$m) compared to the larger M10 tungsten (Smith et al., 1992).

Particle size also influences the depth to which particles can penetrate a tissue. Particles 1 $\mu$m in diameter and smaller rarely penetrated below the first layer of cells in maize callus while 2.0–3.5 $\mu$m particles were often two to three layers below the surface (Kausch et al., 1995). Interestingly, Leduc et al. (1994) found that particles with average diameters of 1.4, 1.6 or 2.2 $\mu$m penetrated into the L2 layer of immature wheat inflorescences with equal efficiency. Using a 'micro-targeting' gene gun, Iglesias et al. (1994) reported that 1.2 $\mu$m gold routinely penetrated wheat meristems to the L4 layer. Kemper et al. (1996) found that about 55% of the $\beta$-glucuronidase expressing cells were located below the second cell layer when maize embryos were bombarded with 1.6 $\mu$m particles. However, bombardment with 1.0 $\mu$m particles resulted in far fewer expressing cells in the third layer and below. Control of the depth of penetration is important since cells capable of division and reiteration of a culture may reside below the first cell layer. For example, Dunder et al. (1995) recommend accelerating microprojectiles into maize embryos that give rise to Type I callus at higher velocity than embryos that produce Type II callus. This is probably because Type I callus originates from internal meristems (Vasil et al., 1985). Plant transformation by bombardment of meristems relies on introducing particles into lower cell layers (McCabe et al., 1988, McCabe and Christou, 1993).

Kausch et al. (1995) studied particle aggregation as influenced by DNA precipitation. They found that the M10 tungsten particles had a mean diameter of 1.9 $\mu$m before addition of DNA. DNA was then precipitated onto the particles and they were accelerated into oil with a helium-powered gene gun. The recovered material had an average diameter of 2.3 $\mu$m. Relatively large aggregates can cause damage to bombarded tissue (Taylor and Vasil, 1991, Kausch et al., 1995). Modulating the amount of DNA as well as the use of screens between the tissue and the incoming particles helps to limit particle aggregation and the resulting damage (Gordon-Kamm et al., 1990, Franks and Birch, 1991, Taylor and Vasil, 1991, Dunder et al., 1995).

Staining particles with a DNA-specific dye such as bisbenzimide can be used to assess the precipitation of DNA onto microprojectiles (Ratnayaka and Oard, 1995). These researchers found that fluorescence of the particles is correlated with transient gene expression and that a precipitation protocol using PEG performed better than one using glycerol. The amount of DNA probably varies greatly between particles and the proportion of particles that carry no DNA into cells is not known. Sautter (1993) reported that about 10 percent of the particles that penetrate wheat meristem result in transient expression. Early work by Armaleo et al. (1990) with yeast indicated that large amounts of exogenous DNA can be observed within cells and that this DNA maintains a clumpy, insoluble appearance even after its initial rapid detachment from the particle.

Kausch et al. (1995) found many non-expressing maize cells containing one or more particles. Using ethidium bromide-stained particles, these researchers found that most, if not all, of the particles within cells still had DNA associated with them. There are several possible explanations for these observations. One is that not all of the DNA associated with the particle becomes free to travel to the nucleus. Alternatively, not all of the cells are physiologically capable of expressing the introduced DNA or are damaged to the extent that expression is not possible. Another possibility is that the subcellular location of the implanted particle is important. Results from both Hunold et al. (1994) and Yamashita et al. (1991) indicate that, in general, the particle must penetrate the nucleus (or at least be in close association with the nucleus) for the cell to express an introduced gene. However, Kausch et al. (1995) found no such correlation with maize cells expressing the anthocyanin regulatory genes and observed expression even when the particle was in the vacuole. Using a *Populus* culture, Devantier et al. (1993) found that about 60% of $\beta$-glucuronidase expressing cells had microprojectiles in or close to the nucleus. The available information indicates that there is a preference for expressing cells to have a particle associated with the nucleus. This suggests that DNA transport from the particle to nucleus limits gene transfer.

Most cells expressing an introduced marker gene contain a single particle although cells can tolerate the insertion of multiple microprojectiles (Hunold et al., 1994, Yamashita et al., 1991, Kausch et al., 1995). However, it is unlikely that many of the cells that receive multiple particles remain competent for cell division. Yeast cells that contain more than one particle generally fail to divide (Armaleo et al., 1990). Observations by Hunold et al. (1994) indicate that about 99% of tobacco and maize cells that receive even a single particle die after several days. This is long enough to express a marker gene transiently but, unfortunately, the cell is then doomed. The number of stable transformants that they recovered corresponded well to the predicted numbers of surviving cells. These researchers suggest that virtually every cell that exhibits prolonged survival after penetration can become a stably transformed clone. If true, this has important implications and suggests several directions for improving the biolistic process. It is interesting that the ratio of stable transformants to transiently expressing cells is relatively high using the microtargeting gene gun described by Sautter (1993). He estimated conversion rates of one stable transformant for every 10 transiently expressing tobacco cells and attributed this to the relative mildness of the microtargeting system. However, conversion frequencies can be relatively high using more conventional particle delivery systems. For example, Franks and Birch (1991) reported a 4% conversion rate for sugarcane cells. Russel (1993) reported 10% conversion rates for tobacco cells using the PDS-1000/He.

Koprek et al. (1996) observed higher transformation frequencies, greater regeneration capacity and less physical injury when bombarding barley scutella with the particle inflow gun as compared to the PDS 1000/He. There is evidence that the shock wave in a conventional helium-driven device produced by the impact of the macroprojectile upon impact with the stopping screen can cause damage to cells (Russell et al., 1992a). Apparently, further means of protecting cells from damage during bombardment and to enhance the survival of cells penetrated by microprojectiles should increase stable transformation rates.

The method of precipitating DNA onto the particles is another important factor. Several researchers have found improvements that appear to enhance DNA delivery. Perl et al. (1992) developed a method that uses calcium nitrate and lacks spermidine, which they claim performs better than standard methods. Loeb and Reynolds (1994) also used a method that employed calcium nitrate. In addition, they found that raising the pH of the calcium nitrate to 10 increased rates of gene delivery into wheat tissue.

There is little information concerning the amount of DNA that a particle can carry into a cell or the limits to the length of DNA that can be introduced. Co-transformation frequencies are generally high. Using yeast and a plasmid dilution scheme, Armaleo et al. (1990) estimated that a single particle can carry about 20 biologically active plasmids into a cell. In plants, the number and sites of integration of a transgene can vary greatly and copy number typically ranges from 1 to more than 20 (Pawlowski and Somers, 1996). Qu et al. (1996) analyzed 50 independent gene gun-derived rice transformants and determined that 22% integrated one copy, 52% had between 2 and 5 copies, 18% had between 6 and 10 copies and 8% had more than 10 copies. However, regardless of the number of insertions, all copies cosegregated indicating insertion at a single locus. Maize transformants also carry the introduced genes at a single genetic locus. This is in contrast to soybean where about one third of the transformants produced from bombardment of embryogenic suspensions carry the transgene at 2 genetic loci (personal observation). Bombardment can also produce transgenic cell lines with extremely high copy number. Allen et al. (1996) recovered some tobacco cell lines with approximately 300 copies. Hadi et al. (1996) reported that microprojectiles can deliver as many as 10 to 15 copies each of 12 different plasmids into maize cells.

Studies directed at optimization of the quantity of DNA have generally relied on transient expression. In general, transient expression increases with increasing amounts of DNA until agglomeration of the particles becomes severe (Klein et al., 1988). However, the influence of DNA concentration on the number of copies of the transgene that integrate into the genome is not clear. This is of concern because integration of multiple copies of a transgene is often associated with suppression of their expression (Flavell, 1994). Gene silencing has been observed in transgenic maize produced by particle bombardment (Register et al., 1994).

Integration of multiple copies of a transgene is observed also in transformants produced by other methods of direct DNA delivery and in fact can be true for *Agrobacterium*-mediated delivery as well. However, extremely high copy numbers are only observed following direct DNA delivery. In an effort to produce simple transgene integration, some researchers reduce the amount of DNA added to the particles. This can reduce transformation efficiency but can also increase the chances of recovering transformants with simple insertion patterns. Hansen and Chilton (1996) are developing an 'agrolistic' system that may lead to the routine recovery of simple transgene insertions. They used particle bombardment to deliver a gene flanked by T-DNA border sequences. The *Agrobacterium* virulence genes *vir*D1 and *vir*D2 were placed under the control of the 35S promoter and co-introduced with the target plasmid containing the border sequences. About 20 per cent of the transformants contained inserts with junctions expected for *Agrobacterium*-mediated delivery. Therefore the transgene was properly spliced from the plasmid and precisely inserted into the plant genome. These agrolistic transformation events inserted fewer copies of the transgene (1 or 2) than did comparable biolistic events (1 to 10), nor did they carry extraneous vector DNA. It will be of interest to see if this method can be routinely utilized to avoid the integration of unwanted sequences and to produce transgenic plants that are less prone to gene silencing.

A beneficial feature of particle bombardment is that there is no obvious biological constraint to the size of the DNA molecule that can be coated onto the surface of the microprojectile. Consequently, particle bombardment can be used to introduce long DNA fragments that can, hopefully, be integrated into the plant genome. As an example, yeast artificial chromosomes have been delivered into plant cells by particle bombardment (Van

Eck, 1995, Adam et al., 1997). These findings indicate that the biolistic process can be used to deliver very large DNA fragments although extent of damage to the DNA during delivery and integration is not known. The discovery that bacteria and phage can be used as microprojectiles (Sanford et al., 1993) suggests alternative approaches for the introduction of large fragments of DNA. The ability to transform plants with large pieces of DNA will greatly facilitate complementation studies and positional gene cloning and can provide researchers the means for introducing multi-gene traits into crop plants.

## Conclusion

The gene gun has clearly evolved into an important tool for plant research. It has enabled researchers to investigate gene expression in plant tissues and cellular organelles that were heretofore inaccessible. Most importantly, biolistic technology has facilitated the routine production of transgenic plants from species not amenable to *Agrobacterium* or protoplast transformation techniques. It is clear, however, that many interacting factors, both biological and physical, must be taken into consideration in order to achieve routine transformation. The various parameters need to be optimized empirically for each species or system attempted and, in many cases, the conditions for maximal transient expression are not the same conditions that are optimal for stable transformation. It is apparent that achieving efficient stable transformation requires the maintenance of a delicate balance between effective DNA delivery and minimal cellular damage. While significant strides have been made since the first reports of biolistic transformation a little over a decade ago, there are still numerous areas for research.

## References

Adam, G., Mullen, J.A., and Kindle, K.L. (1997) Retrofitting YACs for direct DNA transfer into plants. Plant J. 11: 1349–1358.

Allen, G.C., Hal, Jr. G., Michalowski, S., Newman, W., Spiker, S., Weissinger, A.K., and Thompson, W.F. (1996) High-level transgene expression in plant cells: effects of a strong scaffold attachment region from tobacco. Plant Cell 8: 899–913.

Altpeter, F., Vasil, V., Srivastava, V., Stöger, E., and Vasil, I.K. (1996) Accelerated production of transgenic wheat (*Triticum aestivum* L.) plants. Plant Cell Rep. 16: 12–17.

Aragão, F.J.L., Barros, L.M.G., Brasileiro, A.C.M., Ribeiro, S.G., Smith, F.D., Sanford, J.C., Faria, J.C., and Rech, E.L. (1996) Inheritance of foreign genes in transgenic bean (*Phaseolus vulgaris* L.) co-transformed via particle bombardment. Theor. Appl. Genet. 93: 142–150.

Armaleo, D., Ye, G.N., Klein, T.M., Shark, K.B., Sanford, J.C., and Johnston, S.A. (1990) Biolistic nuclear transformation of Saccharomyces cerevisiae and other fungi. Curr. Genet. 17: 97–103.

Bansal, K.C., Viret, J.-F., Khan, B.M., Schantz, R., and Bogorad, L. (1992) Transient expressin from *cab-m1* and *rbcS-m3* promoter sequences is different in mesophyll and bundle sheath cells in maize leaves. Proc. Nat. Acad. Sci. USA 89: 3654–3658.

Barcelo, P., and Lazzeri, P.A. (1995) Transformation of cereals by microprojectile bombardment of immature inflorescence and scutellum tissues. Meth. Mol. Biol. 49: 113–123.

Becker, D., Brettschneider, R., and Lörz, H. (1994) Fertile transgenic wheat from microprojectile bombardment of scutellar tissue. Plant J. 5: 299–307.

Bowers, R., Elliott, A.R., Potier, B.A.M., and Birch, R.G. (1996) High-efficiency, microprojectile-mediated cotransformation of sugarcane, using visible or selectable markers. Mol. Breed. 2: 239–249.

Boynton, J.E., and Gillham, N.W. (1993) Chloroplast transformation in *Chlamydomonas*. Meth. Enzym. 217: 510–536.

Boynton, J.E., and Gillham, N.W. (1996) Genetics and transformation of mitochondria in the green alga *Chlamydomonas*. Meth. Enzym. 264: 279–296.

Brar, G.S., Cohen, B.A., Vick, C.L., and Johnson, G.W. (1994) Recovery of transgenic peanut (*Arachis hypogaea* L.) plants from elite cultivars utilizing ACCELL technology. Plant J. 5: 745–753.

Brettschneider, R., Becker, D., and Lörz, H. (1997) Efficient transformation of scutellar tissue of immature maize embryos. Theor. Appl. Genet. 94: 737–748.

Cao, J., Wang, Y.-C., Klein, T.M., Sanford, J., and Wu, R. (1990) Transformation of rice and maize using the biolistic process. UCLA Symp. Mol. Cell. Biol., New Ser. 129: 21–33.

Casas, A.M., Kononowicz, A.K., Zehr, U.B., Tomes, D.T., Axtell, J.D., Butler, L.G., Bressan, R.A., and Hasegawa, P.M. (1993) Transgenic sorghum plants via microprojectile bombardment. Proc. Nat. Acad. Sci. USA 90: 11212–11216.

Casas, A.M., Kononowicz, A.K., Bressan, R.A., and Hasegawa, P.M. (1995) Cereal transformation through particle bombardment. In: Janick, J. (ed.). Plant Breed. Rev. 13: 235–264.

Casas, A.M., Kononowicz, A.K., Zhang, L., Tomes, D.T., Bressan, R.A., and Hasegawa, P.M. (1997) Transgenic sorghum plants obtained after microprojectile bombardment of immature inflorescences. In Vitro Cell Dev. Biol. 33: 92–100.

Chen, J.L., and Beversdorf, W.D. (1994) A combined use of microprojectile bombardment and DNA imbibition enhances transformation frequency of canola. (*Brassica napus* L.). Theor. Appl. Genet. 88: 187–192.

Christou, P. (1996) Particle bombardment for genetic engineering of plants. Academic Press, San Diego, CA.

Christou, P., and Ford, T. (1995) Parameters influencing stable transformation of rice immature embryos and recovery of transgenic plants using electric discharge particle acceleration. Ann. Bot. 75: 407–413.

Daniell, H., Datta, R., Varma, S., Gray, S., and L, S.-B. (1998) Containment of herbicide resistance through genetic engineering of the chloroplast genome. Nature Biotechnology 16: 345–348.

Devantier, Y.A., Moffatt, B., Jones, C., and Charest, P.J. (1993) Microprojectile-mediated DNA delivery to the Salicaceae family. Can. J. Bot. 71: 1458–1466.

Denchev, P.D., Songstad, D.D., McDaniel, J.K., and Conger, B.V. (1997) Transgenic orchardgrass (*Dactylis glomerata*) plants by direct embryogenesis from microprojecticle bombarded leaf. Plant Cell Rep. 16: 813–819.

Duchesne, L.C., Lelu, M.-A., von Aderkas, P., and Charest, P.J. (1993) Microprojectile-mediated DNA delivery in haploid and diploid embryogenic cells of Larix spp. Canad. J. For. Res. 23: 312–316.

Dunder, E., Dawson, J., Suttie, J., and Page, G. (1995) Maize transformation by microprojectile bombardment of immature embryos. In: Potrykus, I., and Spangenberg, G. (eds), Gene Transfer to Plants, pp. 127–138. Springer, Berlin.

Finer, J.J., and McMullen, M.D. (1991) Transformation of soybean via particle bombardment of embryogenic suspension culture tissue. In Vitro Cell Dev. Biol. Plant. 27P: 175–82.

Flavell, R.B. (1994) Inactivation of gene expression in plants as a consequence of specific sequence duplication. Proc. Nat. Acad. Sci. USA 91: 3490–3496.

Fox, T.D., Folley, L.S., Mulero, J.J., McMullin, T.W., Thorsness, P.E., Hedin, L.O., and

Costanzo, M.C. (1991) Analysis and manipulation of yeast mitochondrial genes. Meth. Enzym. 194: 149–165.

Franks, T., and Birch, R.G. (1991) Gene transfer to intact sugarcane cells using microprojectile bombardment. Aust. J. Plant. Physiol. 18: 833–839.

Frisch, D.A., van der Geest, A.H.M., Dias, K., and Hall, T.C. (1995) Chromosomal integration is required for spatial regulation of expression from the $\beta$-phaseolin promoter. Plant J. 7: 503–512.

Fromm, M.E., Morrish, F., Armstrong, C., Williams, R., Thomas, J., and Klein, T.M. (1990) Inheritance and expression of chimeric genes in the progeny of transgenic maize plants. Bio/Technology 8: 833–839.

Goff, S.A., Klein, T.M., Roth, B.A., Fromm, M.E., Cone, K.C., Radicella, J.P., and Chandler, V.L. (1990) Transactivation of anthocyanin biosynthetic genes following transfer of B regulatory genes into maize tissues. EMBO J. 9: 2517–2522.

Goldfarb, B., Strauss, S.H., Howe, G.T., and Zaerr, J.B. (1991) Transient gene expression of microprojectile-introduced DNA in Douglas-fir cotyledons. Plant Cell Rep. 10: 517–521.

Gordon-Kamm, W.J., Spencer, T.M., Mangano, M.L., Adams, T.R., Daines, R.J., Start, W.G., O'Brien, J.V., Chambers, S.A., Adams, W.R., Willetts, N.G., Rice, T.B., Mackery, C.J., Krueger, R.W., Kausch, A.P., and Lemaux, P.G. (1990) Transformation of maize cells and regeneration of fertile transgenic plants. Plant Cell 2: 603–618.

Griesbach, R.J., and Klein, T.M. (1993) In situ complementation of a flower color mutant in *Doritis pulcherrima* (orchidaceae). Lindleyana 8: 223–226.

Hadi, M.Z., McMullen, M.D., and Finer, J.J. (1996) Transformation of 12 different plasmids into soybean via particle bombardment. Plant Cell Rep. 15: 500–505.

Hansen, G., and Chilton, M.-D. (1996) 'Agrolistic' transformation of plant cells: integration of T-strands generated in planta. Proc. Nat. Acad. Sci. USA 93: 14978–14983.

Haseloff, J., Siemering, K.R., Prashe, D.C., and Hodge, S. (1997) Removal of a cryptic intron and subcellular localization of green fluorescent protein are required to mark transgenic Arabidopsis plants brightly. Proc. Nat. Acad. Sci. USA 94: 2122–2127.

Hay, I., Lachance, D., Von Aderkas, P., and Charest, P.J. (1994) Transient chimeric gene expression in pollen of five conifer species following microparticle bombardment. Canad. J. For. Res. 24: 2417–2423.

Hazel, C.B., Klein, T.M., Anis, M., Wilde, H.D., and Parrott, W.A. (1998) Growth characteristics and transformability of soybean embryogenic cultures. Plant Cell Rep. In press.

Hébert, D., Kikkert, J.R., Smith, F.D., and Reisch, B.I. (1993) Optimization of biolistic transformation of embryogenic grape cell suspensions. Plant Cell Rep. 12: 585–589.

Heiser, W.C. (1994) Gene transfer into mammalian cells by particle bombardment. Anal. Biochem. 217: 185–196.

Horikawa, Y., Yoshizumi, T., and Kakuta, H. (1997) Transformants through pollination of mature maize (*Zea mays* L.) pollen delivered bar gene by particle gun. Grassl. Sci. 47: 117–123.

Hui, K.M., and Chia, T.F. (1997) Eradication of tumor growth via biolistic transformation with allogeneic MHC genes. Gene Ther. 4: 762–767.

Hunold, R., Bronner, R., and Hahne, G. (1994) Early events in microprojectile bombardment: Cell viability and particle location. Plant J. 5: 593–604.

Iida, A., Yamashita, T., Yamada, Y., and Morikawa, H. (1991) Efficiency of particle-bombardment-mediated transformation is inflenced by cell cycle stage and in sychronized cultured cells of tobacco. Plant Physiol. 97: 1585–1587.

Iglesias, A., Gisel, A., Bilang, R., Leduc, N., Potrykus, I., and Sautter, C. (1994) Transient expression of visible marker genes in meristem cells of wheat embryos after ballistic micro-targeting. Planta 192: 84–91.

Jain, R.K., Jain, S., Wang, B., and Wu, R. (1993) Optimization of biolistic method for transient gene expression and production of agronomically useful transgenic Basmati rice plants. Plant Cell Rep. 15: 963–968.

Johnston, S.A., and Tang, D.C. (1993) The use of microparticle injection to introduce genes into animal cells in vitro and in vivo. Genet. Eng. (N.Y.) 15: 225–236.

Johnston, S.A., and Tang, D.C. (1994) Gene gun transfrection of animal cells and genetic immunizaiton. Meth. Cell Biol. 43: 353–365

Kartzke, S., Saedler, H., and Meyer, P. (1990) Molecular analysis of transgenic plants derived from transformations of protoplasts at various stages of the cell cycle. Plant. Sci. 67: 63–72.

Kausch, A.P., Adams, T.R., Mangano, M., Zachwieja, S.J., Gordon-Kamm, W., Daines, R., Willetts, N.G., Chambers, S.A., Adams, Jr W., Anderson, A., Williams, G., and Haines, G. (1995) Effects of microprojectile bombardment on embryogenic suspension cell cultures of maize (*Zea mays* L.) used for genetic transformation. Planta 196: 501–509.

Kemper, E.L., da Silva, M.J., and Arruda, P. (1996) Effect of microprojectiles bombardment parameters and somotic treatment on particl penetration and tissue damage in transiently transformed culture immature maize (*Zea mays* L.) embryos. Plant Sci. 121: 85–93.

Klein, T.M., Wolf, E.D., Wu, R., and Sanford, J.C. (1987) High-velocity microprojectiles for delivering nucleic acids into living cells. Nature 327: 70–73.

Klein, T.M., Fromm, M., Weissinger, A., Tomes, D., Schaaf, S., Sletten, M., and Sanford, J.C. (1988) Transfer of foreign genes into intact maize cells with high-velocity microprojectiles. Proc. Nat. Acad. Sci. USA 85: 4305–4309.

Klein, T.M., Gradziel, T., Fromm, M.E., and Sanford, J.C. (1988) Factors influencing gene delivery into Zea mays cells by high-velocity microprojectiles. Bio/Technology 6: 559–563.

Klein, T.M., Roth, B.A., and Fromm, M.E. (1989) Regulation of anthocyanin biosynthetic genes intorduced into intact maize tissues by microprojectiles. Proc. Nat. Acad. Sci. USA 86: 6681–6685.

Klein, T.M., and Maliga, P. (1991) Plant transformation by particle bombardment. In: Maramorosch K (ed), Biotechnology for Biological Control of Pests and Vectors, pp. 105–117. CRC Press, Boca Raton, FL.

Klein, T.M., and Fitzpatrick-McElligott, S. (1993) Particle bombardment: a universal approach for gene transfer to cells and tissues. Curr. Opin. Biotech. 4: 583–590.

Klein, T.M., and Zhang, W. (1994) Progress in the genetic transformation of recalcitrant crop species. Aspect Appl. Biol. 39: 35–44.

Knudsen, S., and Müller, M. (1991) Transformation of the developing barley endosperm by particle bombardment. Planta 185: 330–336.

Koprek, T., Hänsch, R., Nerlich, A., Mendel, R.R., and Schulze, J. (1996) Fertile transgenic barley of different cultivars obtained by adjustment of bombardment conditions to tissue response. Plant Sci. 119: 79–91.

Kost, B., Leduc, N., Sautter, C., Potrykus, I., and Neuhaus, G. (1996) Transient marker-gene expression during zygotic in-vitro embryogenesis of *Brassica juncea* (Indian mustard) following particle bombardment. Planta 198: 211–220.

Koziel, M.G., Beland, G.L., Bowman, C., Carozzi, N.B., Crenshaw, R., Crossland, L., Dawson, J., Desai, N., and Hill, M. (1993) Field performance of elite transgenic maize plants expressing an insecticidal protein derived from *Bacillus thuringiensis*. Bio/Technology 11: 194–200.

Leduc, N., Iglesias, V.A., Bilang, R., Gisel, A., Potrykus, I., and Sautter, C. (1994) Gene transfer to inflorescence and flower meristems using ballistic micro-targeting. Sex Plant Reprod. 7: 135–143.

Lloyd, A.M., Walbot, V., and Davis, R.W. (1992) Arabidopsis and Nicotiana anthocyanin production activated by; maize regulators R and C1. Science 258: 1773–1775.

Loeb, T.A., and Reynolds, T.L. (1994) Transient expression of the uidA gene in pollen embryoids of wheat following microprojectile bombardment. Plant Sci. 104: 81–91.

Lowe, K., Bowen, B., Hoerster, G., Ross, M., Bond, D., Pierce, D., and Gordon-Kamm, W. (1995) Germline transformation of maize following manipulation of chimeric shoot meri-

stems. Bio/Technology 13: 677–682.
Ludwig, S.R., Bowen, B., Beach, L., and Wessler, S.R. (1990) A regulatory gene as a novel visible marker for maize transformation. Science 247: 449–450.
Martinussen, I., Junttila, O., and Twell, D. (1994) Optimization of transient gene expression in pollen of Norway spruce (*Picea abies*) by particle acceleration. Physiol. Plant 92: 412–416.
Matsunaga, T. (1991) Applications of bacterial magnets. Trends Biotech. 9: 91–95.
McBride, K.E., Schaaf, D.J., Daley, M., and Stalker, D.M. (1994) Controlled expression of plastid transgenes in plants based on a nuclear DNA-encoded and plastid-targeted T7 RNA polymerase. Proc. Nat. Acad. Sci. USA 91: 7301–7305.
McCabe, D., and Christou, P. (1993) Direct DNA transfer using electric discharge particle acceleration accell technology. Plant Cell Tiss. Org. Cult. 33: 227–236.
McCabe, D., and Martinell, B.J. (1993) Transformation of elite cotton cultivars via particle bombardment of meristems. Bio/Technology 11: 596–598.
McCabe, D.E., Swain, W.F., Martinell, B.J., and Christou, P. (1988) Stable transformation of soybean (*Glycine max*) by particle bombardment. Bio/Technology 6: 923–926.
Moore, P.J., Moore, A.J., and Collins, G.B. (1994) Genotypic and developmental regulation of transient expression of a reporter gene in soybean zygotic cotyledons. Plant Cell Rep. 13: 556–560.
Morrish, F., Songstad, D.D., Armstrong, C.L., and Fromm, M. (1993) Microprojectile bombardment: a method for the production of transgenic cereal crop plants and the functional analysis of genes. In: Hiatt, A. (ed.), Trangenic Plants: Fundamentals and Applications, pp. 133–171. Marcel Dekker, Inc., New York.
Pawlowski, W.P., and Somers, D.A. (1996) Transgene inheritance in plants genetically engineered by microprojectile bombardment. Mol. Biotech. 6: 1730.
Perl, A., Kless, H., Blumenthal, A., Galili, G., and Galun, E. (1992) Improvement of plant regeneration and GUS expression in scutellar wheat calli by optimization of culture conditions and DNA-microprojectile delivery procedures. Mol. Gen. Genet. 235: 279–284.
Purcell, M., Mabrouk, Y.M., and Bogorad, L. (1995) Red/far-red and blue light-responsive regions of maize rbcS-m3 are active in bundle sheath and mesophyll cells, respectively. Proc. Nat. Acad. Sci. USA 92: 11504–11508.
Qu, R., De Kochko, R., Zhang, L., Marmey, P., Li, L., Tian, W., Zhang, S., Fauquet, C.M., and Beachy, R.N. (1996) Analysis of a large number of independent transgenic rice plants produced by the biolistic method. In Vitro Cell Dev. Biol. 32P: 233–240.
Raemakers, C.J.J.M., Sofiari, E., Taylor, N., Henshaw, G., Jacobsen, E., and Visser, R.G.F. (1996) Production of transgenic cassava (*Manihot esculenta Crantz*) plants by particle bombardment using luciferase activity as selection marker. Molec. Breed. 2: 339–349.
Randolph-Anderson, B., Boynton, J.E., Dawson, J., Dunder, E., Eskes, R., Gillham, N.W., Johnson, A., Perlman, P.S., Suttie, J., and Heiser, W.C. (1997) Sub-micron gold particles are superior to larger particles for efficient Biolistic® transformation of organelles and some cell types. Bio-Rad. Tech. Bull. 2015 (95–0699).
Rasmussen, J.L., Kikkert, J.R., Roy, M.K., and Sanford, J.C. (1994) Biolistic transformation of tobacco and maize suspension cells using bacterial cells as microprojectiles. Plant Cell Rep. 13: 212–217.
Ratnayaka, I.J.S., and Oard, J.H. (1995) A rapid method to monitor DNA precipitation onto microcarries before particle bombardment. Plant Cell Rep. 14: 794–798.
Register, III, J.C., Peterson, D.J., Bell, P.J., Bullock, W.P., Evans, I.J., Frame, B., Greenland, A.J., Higgs, N.S., Jepson, I., Jiao, S., Lewnau, C.J., Sillick, J.M., and Wilson, H.M. (1994) Structure and function of selectable and non-selectable transgenes in maize after introduction by particle bombardment. Plant Mol. Biol. 25: 951–961.
Russell, J.A., Roy, M.K., and Sanford, J.C. (1992a) Physical trauma and tungsten toxicity reduce the efficiency of biolistic transformation. Plant Physiol. 98: 1050–1056.

Russell, J.A., Roy, M.K., and Sanford, J.C. (1992b) Major improvements in biolistic transformation of suspension-cultured tobacco cells. In Vitro Cell Dev. Biol. 28P: 97–105.

Russel, J.R. (1993) The Biolistic® PDS-1000/He device. Plant Cell Tiss. Org. Cult. 33: 221–226.

Russell, D.A., and Fromm, M.E. (1995) Transient gene expression studies with the biolistic system. In: Potrykus, I., and Spangenberg, G. (eds), Gene Transfer to Plants, pp. 118–126. Springer-Verlag, Berlin.

Sanford, J.C., Klein, T.M., Wolf, E.D., and Allen, N. (1987) Delivery of substances into cells and tissues using a particle bombardment process. Part. Sci. Technol. 5: 27–37.

Sanford, J.C., Smith, F.D. and Russell, J.A. (1993) Optimizing the biolistic process for different biological applications. Meth. Enzymol. 217: 483–509.

Sautter, C. (1993) Development of a microtargeting device for particle bombardment of meristems. Plant Cell Tiss. Org. Cult. 33: 251–257.

Sautter, C., Waldner, H., Heuhaus-Url, G., Galli, A., Neuhaus, G., and Potrykus, I. (1991) Micro-targeting: high efficiency gene transfer using a novel approach for the acceleration of microprojectiles. Bio/Technology 9: 1080–1085.

Singh, R.J., Klein, T.M., Mauvais, C.J., Knowlton, S., Hymowitz, T., and Kostow, C.M. (1998) Cytological characterization of transgenic soybean. Theor. Appl. Genet. 96: 319–324.

Songstad, D.D., Armstrong, C.L., Peterson, W.L., Hairston, B., and Hinchee, M.A.W. (1996) Production of transgenic maize plants and progeny by bombardment of Hi-II immature embryos. In Vitro Cell Dev. Biol. 32P: 179–183.

Spangenberg, G., Wang, Z.Y., Wu, X.L., Nagel, J., Iglesias, V.A., and Potrykus, I. (1995) Transgenic tall fescue (*Festuca arundinacea*) and red fescue (*F. Rubra*) plants from microprojectile bombardment of embryogenic suspension cells. J. Plant. Physiol. 145: 693–701.

Smith, F.D., Harpending, P.R., and Sanford, J.C. (1992) Biolistic transformation of prokaryotes: factors that affect biolistic transformation of very small cells. J. Gen. Microbiol. 138: 239–248.

Takumi, S., and Shimada, T. (1996) Production of transgenic wheat through particle bombardment of scutellar tissues: frequency is influenced by culture duration. J. Plant. Physiol. 149: 418–423.

Tanaka, T., Nishihara, M., Seki, M., Sakamoto, A., Tanaka, K., Irifune, K., and Morikawa, H. (1995) Successful expression in pollen of various plant species of *in vitro* synthesized mRNA introduced by particle bombardment. Plant. Mol. Biol. 28: 337–341.

Taylor, M.G., and Vasil, I.K. (1991) Histology of, and physical factors affecting, transient GUS expression in pearl millet (*Pennisetum glaucum* (L.) R. Br.) embryos following microprojectile bombardment. Plant. Cell. Rep. 10: 120–125.

Twell, D., Klein, T.M., Fromm, M.E., and McCormick, S. (1989) Transient expression of chimeric genes delivered into pollen by microprojectile bombardment. Plant. Physiol. 91: 1270–1274.

Vain, P., Keen, N., Murillo, J., Rathus, C., Nemes, C., and Finer, J.J. (1993) Development of the particle inflow gun. Plant Cell Tiss. Org. Cult. 33: 237–246.

Vain, P., McMullen, M.D., and Finer, J.J. (1993) Osmotic treatment enhances particle bombardment-mediated transient and stable transformation of maize. Plant Cell Rep. 12: 84–88.

van der Leede-Plegt, E.M., van de Ven, B.C.E., Schilder, M., Fraken, J., and van Tunen, A.J. (1995) Development of a pollen-mediated transformation method for *Nicotiana glutinosa*. Transgen. Res. 4: 77–86.

Van Eck, M., Blowersm, A.D., and Earle, E.D. (1995) Stable transformation of tomato cell cultures after bombardment with plasmid and YAC DNA. Plant Cell Rep. 14: 299–304.

Vasil, V., Lu, C., and Vasil, I.K. (1985) Histology of somatic embryogenesis in cultured immature embryos of maize (*Zea mays* L.). Protoplasma 127: 1–8.

Vasil, V., Brown, S.M., Re, D., Fromm, M.E., and Vasil, I.K. (1991) Stably transformed callus lines from microprojectile bombardment of cell suspension cultures of wheat Bio/

Technology 9: 743–747.

Vasil, V., Castillo, A.M., Fromm, M.E., and Vasil, I.K. (1992) Herbicide-resistant fertile transgenic wheat plants obtained by microprojectile bombardment of regenerable embryogenic callus. Bio/Technology 10: 667–674.

Vasil, V., Srivastava, V., Castillo, A.M., Fromm, M.E., and Vasil, I.K. (1993) Rapid production of transgenic wheat plants obtained by microprojectile bombardment of cultured immature embryos. Bio/Technology 11: 1553–1558.

Vasil, I.K. (1994) Molecular improvement of cereals. Plant. Mol. Biol. 25: 925–937.

Wan, Y., Widholm, J.M., and Lemaux, P.G. (1995) Type I callus as a bombardment target for generating fertile transgenic maize (*Zea mays* L.). Planta 196: 7–14.

Warkentin, T.D., Jordan, M.C. and Hobbs, S.L.A. (1992) Effect of promoter-leader sequences on transient reporter gene expression in particle bombarded pea (*Pisum sativum* L.) tissues. Plant Sci. 87: 171–177.

Wong, J.R., Walker, L.S., Krikeilis, H., and Klein, T.M. (1991) Anthocyanin regulatory genes from maize (*B-peru* and *C1*) activate the anthocyanin pathway in wheat, barley, and oat cells. J. Cell. Biochem. 15A: 159.

Yamashita, T., Iida, A., and Morikawa, H. (1991) Evidence that more than 90% of $\beta$-glucuronidase-expressing cells after particle bombardment directly receive the foreign gene in the nucleus. Plant. Physiol. 97: 829–831.

Yang, N.-S., and Christou, P. (1994) Particle Bombardment Technology for Gene Transfer. Oxford University Press, New York.

Ye, G., Daniell, H., and Sanford, J.C. (1990) Optimization of delivery of foreign DNA into higher-plant chloroplasts. Plant. Mol. Biol. 15: 809–819.

Yoshizumi ,T., Horikawa, Y., and Kakuta, H. (1996) Gene delivery into plant cells and magnetic separation using magnetite particles. Res. Bull. Obihiro Univ. Nat. Sci. 20: 5–9.

Zhong, H., Sun, B., Warkentin, D., Zhang, S., Wu, R., Wu, T., and Sticklen, B. (1996) The competence of maize shoot meristems for integrative transformation and inherited expression of transgenes. Plant Physiol. 110: 1097–1107.

Zimny, J., Becker, D., Brettschneider, R., and Lörz, H. (1995) Fertile transgenic Triticale (x Triticosecale Wittmack). Mol. Breed. 1: 155–164.

Zoubenko, O.V., Allison, L.A., Svab, Z., and Maliga, P. (1994) Efficient targeting of foreign genes into the tobacco plastid genome. Nuc. Acids Res. 22: 3819–3824.

# 4. Methods of Genetic Transformation: *Agrobacterium tumefaciens*

TOSHIHIKO KOMARI* and TOMOAKI KUBO
(*corresponding author)
*Plant Breeding and Genetics Research Laboratory, Japan Tobacco Inc. 700 Higashibara, Toyoda, Iwata, Shizuoka 438–0802, JAPAN. E-mail: Toshihiko.Komari@pbgrl.jti.co.jp*

ABSTRACT. The soil phytopathogen *Agrobacterium tumefaciens* induces tumors, known as crown galls, mainly on dicotyledonous plants. Such tumors are generated by a complex, multi-step transformation process. *Agrobacterium* has been routinely utilized for the transfer of genes to dicotyledonous plants, and monocotyledonous plants, including important cereals, were thought until recently to be outside the range of this technology since they are generally not considered within the host range of crown gall. Various attempts to infect monocotyledons with *Agrobacterium* were made in the 1970's and 1980's, but conclusive evidence of integrative transformation was not obtained until recently. This delay occurred in part because many methods for the transformation of dicotyledons depend heavily on the cell divisions that are induced by wounding. Similar approaches in monocotyledons failed, because they do not exhibit active responses to wounding. Wounding is necessary for formation of a crown gall and the main roles of wounding are to produce compounds that activate the virulence genes of *Agrobacterium* and to induce DNA synthesis in host cells. Efficient protocols for *Agrobacterium*-mediated transformation have recently been developed for rice, maize, barley and wheat. A key point in these protocols is the use of tissues that consist of actively dividing, embryonic cells, such as immature embryos and calli induced from scutella, which are co-cultivated with *Agrobacterium* in the presence of acetosyringone, which is a potent inducer of the virulence genes. The advantages of *Agrobacterium*-mediated transformation include the transfer of pieces of DNA with defined ends with minimal rearrangement, the transfer of relatively large segments of DNA, the integration of small numbers of copies of genes into plant chromosomes, and the high quality and fertility of both monocotyledonous and dicotyledonous transgenic plants. Stable inheritance and expression of transgenes in the progeny has also been demonstrated. It is now clear that *Agrobacterium* is capable of transferring DNA to monocotyledons if tissues that contain 'competent' cells are infected. The success with cereals highlights the critical importance of numerous experimental factors, which include the genotype of the plants, the type and age of the tissue that is inoculated, the vector, the strain of *Agrobacterium*, the selectable marker genes and selection agents, and the various conditions of tissue culture.

## Introduction

*Agrobacterium* is a remarkable microorganism: it is the only prokaryotic organism known that is capable of transferring DNA to eukaryotic cells (Bundock et al., 1995, Ream, 1989). This naturally occurring mechanism of

*I.K. Vasil (ed.), Molecular Improvement of Cereal Crops, 43–82*

the inter-kingdom transfer of genes has been extensively exploited by researchers in plant biotechnology and molecular biology (Fraley et al., 1986). The *Agrobacterium*-mediated transfer of genes is a biological process that exhibits a number of advantages over other methods of transformation. *Agrobacterium* efficiently integrates small numbers of copies of large segments of DNA with defined ends into plant chromosomes with little rearrangement. Until recently, most monocotyledonous species were believed to be outside the host range of *Agrobacterium* (Smith and Hood, 1995). One of the most exciting recent developments in plant biotechnology has been the development of techniques by which important cereals, such as rice, maize, barley and wheat, joined the long list of the plants that can be routinely modified by *Agrobacterium*-mediated transformation (Hiei et al., 1994, Ishida et al., 1996, Cheng et al., 1997, Tingay et al., 1997).

## *Agrobacterium* and Crown Gall

*A. tumefaciens*, a soil bacterium, can genetically transform plant cells with a segment of DNA (transfer DNA, abbreviated as T-DNA) from a tumor-inducing plasmid (Ti plasmid) with the resultant production of a crown gall, which is a plant tumor (Nester and Kosuge, 1981). Another species of *Agrobacterium*, *A. rhizogenes*, causes hairy root disease in higher plants via an identical process (Nester et al., 1984). The plasmid in *A. rhizogenes*, which is a variant of the Ti plasmid, is called the Ri plasmid (for root-inducing plasmid; Huffman et al., 1984). *A. tumefaciens* is the most extensively studied species of *Agrobacterium* and the term '*Agrobacterium*-meditated transformation' usually refers to transformation mediated by *A. tumefaciens*.

Crown gall is a disease that causes considerable damage to perennial crops, such as young rose bushes and chrysanthemums, peach and apple trees, and other flower and fruit trees. Studies of crown gall have a long history (Braun, 1982). Crown gall was the first tumor in any organism, for which the causative agent was identified. The finding preceded the discovery of Rous' sarcoma virus by several years. The formation of a crown gall involves a complex molecular process that consists of a number of discrete, essential steps. The mechanism of crown gall formation has been extensively reviewed by a number of authors (Binns and Thomashow, 1988, Hooykaas, 1989, Hooykaas and Beijersbergen, 1994, Hooykaas and Shilperoort, 1992, Klee et al., 1987, Ream, 1989, Sheng and Citovsky, 1996, Zambryski, 1988, Zambryski, 1992, Zupan and Zambryski, 1995), and we shall summarize each step only briefly here.

## The Mechanism of Crown Gall Tumorigenesis

Recent evidence suggests that the machinery for transfer of T-DNA by

*Agrobacterium* evolved from bacterial conjugation systems, which mobilize plasmids for transport between bacterial cells (Lessl and Lanka, 1994). Proteins involved in the transfer of T-DNA are closely related, both structrally and functionally, to the proteins in bacterial conjugation systems.

### *Contacts between Plant Cells and Bacterial Cells*

Wounding of a plant, which often occurs in the region of the stem-root interface, allows the entry of bacteria. Bacteria are attracted to a wounded plant in response to signal molecules that are released by the plant cells; they multiply in the wound sap and attach to the walls of plant cells (Hohn et al., 1989, Zambryski, 1992). Specific attachment of *Agrobacterium* to plant cells is a prerequisite for the subsequent transfer of DNA. Products of several chromosomal genes (*chvA, chvB, pacA*, and *att*) in *Agrobacterium* and surface receptors of plant cells, which might include both proteins and carbohydrates, are believed to be involved in this process (Sheng and Citovsky, 1996).

### *Activation of Virulence (*vir*) Genes*

Genes responsible for the transfer of T-DNA are clustered in a region called the virulence region outside the T-DNA in the Ti plasmid (Klee et al., 1983). At least six operons, namely, *virA* (1 cistron), *virB* (11 cistrons), *virC* (2 cistrons), *virD* (4 cistrons), *virE* (2 cistrons) and *virG* (1 cistron), encode proteins with important functions in T-DNA transfer (Stachel and Nester, 1986). The expression of these genes is tightly regulated and they are expressed when the bacteria infect a plant.

Plant wound sap is an acidic solution that contains various substances, which include phenolic compounds, such as derivatives of compounds in the biosynthetic pathways to lignin and flavonoid compounds. The expression of virulence genes is induced by such wound sap. Bolton et al., (1986) identified seven phenolic compounds that induce the expression of virulence genes. The most effective inducers are monocyclic phenolic compounds, such as acetosyringone (Stachel et al., 1985).

VirA and VirG form a two-component regulatory system. *Vir*-inducing signals are first recognized by VirA protein and then VirG is activated, most likely by phosphorylation (Citovsky et al., 1992). Activated VirG is a transcriptional factor that up-regulates the expression of all the virulence genes.

### *Generation of the T-complex*

Both ends of T-DNA are flanked by 25-bp imperfect direct-repeat sequences (Yadav et al., 1982, Zambryski et al., 1982). VirD2 protein recognizes these border repeats and makes nicks between the third and forth bases in the 'bottom strand' of the T-DNA, with the help of VirD1, VirC1 and, possibly, VirD3 (Zambryski, 1992). VirD2 attaches covalently to the 5′ end of the bottom strand of the T-DNA, which is then released from the Ti plasmid by replacement DNA synthesis to form the T-strand, which is an intermediate in the transfer of the T-DNA. VirE2 proteins then coat the T-strand to make the T-complex. VirE2 is a protein that binds to single-stranded DNA.

Only the 25-bp direct repeats at the ends of the T-DNA are required, in *cis*, for the mobilization of the T-DNA. Virtually any DNA sequence flanked by such border repeats can be transferred to plant cells by the virulence system. These findings have allowed the modification of the *Agrobacterium* system to yield various plant-transformation vectors.

### *Transport of the T-complex*

VirB proteins and VirD4 are thought to form a channel between the bacterium and the plant cell that it is infecting (Ream, 1989). The T-complex travels through this channel into the cytoplasm of the plant cell. Many of these proteins have characteristics of membrane proteins and some of them are ATPases that probably provide the energy required for the passage of the T-complex (Beaupré et al., 1997, Dang and Christie, 1997, Rashkova et al., 1997).

Transfer of the T-complex from the cytoplasm to the plant nucleus is mediated by nuclear localization signals (NLS) that are located in VirD2 and VirE2 (Sheng and Citovsky, 1996). Presumably, these signals are recognized by proteins related to importin (Görlich et al., 1996, Görlich and Mattaj, 1996), which guide the T-complex to the nucleus.

### *Integration of T-DNA into the Plant Chromosome and Expression of T-DNA Genes*

Soon after entry of the T-complex into the nucleus, many of the T-strands are converted to double-stranded DNA, and genes encoded by the T-DNA are expressed without integration into plant chromosomes. This phenomenon is referred to as transient expression (Janssen and Gardner, 1989, Kapila et al., 1997). It is unclear whether the integration process occurs before or after the conversion of T-strands to double-stranded DNA. In either case, it has been proposed that integration is initiated at the left border region (Koncz et al., 1994). Integration reactions are catalyzed primarily by host enzymes and most probably by components of the DNA replication/ repair machinery. The integration of T-DNA into a plant chromosome is a type of illegitimate recombination and the sites of integration appear to be random (Hooykaas and Schilperoort, 1992, Koncz et al., 1994). It has been suggested that VirD2 and VirE2 might be involved in this process (Rossi et al., 1996, Tinland et al., 1995). Mutants of *Arabidopsis* that are deficient in T-DNA integration have been identified (Sonti et al., 1995) and it is expected that this process will be dissected in further detail in the near future.

The T-DNA encodes enzymes that are involved in the biosynthesis of phytohormones, such as auxin and cytokinin, as well as amino acid derivatives known as opines (Nester et al., 1984). Overproduction of phytohormones eventually results in the neoplastic growth that is visible as a crown gall. Opines are catabolized specifically by the strain of *Agrobacterium* that has induced a particular tumor. The enzymes for the catabolism of opines are encoded in the Ti plasmid.

## Host Range of Crown Gall

*Agrobacterium* is a phytopathogen with an exceptionally wide host range. In 1976, De Cleene and De Lay reported a list of natural host plants for *Agrobacterium* and their list contained 42 gymnosperms, 596 dicotyledons plants, and 5 monocotyledons. The number of monocotyledons in the list was very small and no agronomically important cereals were included. Thus, the general statement that *Agrobacterium* does not infect monocotyledons is not literally true, but it reflects general beliefs over a long period of time (Van Wordragen and Dons, 1992).

The statement is misleading in two ways. First, the diversity of monocotyledonous species is very great and monocotyledons should not be treated as a single group. The nature of non-host responses in different monocotyledons can vary considerably. Second, the host range in phytopathological studies should not be confused with the host range of engineered transformation vectors. As discussed in the previous section, formation of a crown gall by *Agrobacterium* is a complex, multistep process that involves entry of bacteria into a wound, attachment of bacteria to plant cells, transduction of virulence genes, processing of the T-DNA, intercellular transport of the T-complex, nuclear import of the T-complex, expression of T-DNA genes, overproduction of phytohormones and, finally, the neoplastic growth of plant cells. A plant is a non-host if one of the steps cannot be completed. Inefficient attachment of bacteria to plant cells and variations in hormone metabolisms were suggested as possible roadblocks in the formation of crown galls in monocotyledons (Lippincott and Lippincott, 1978, Rao et al., 1982). As discussed below, the nature of the wound response of a plant is a very important factor in crown gall formation (Potrykus, 1990a). By contrast, some of the steps might not be so critical or might be controllable by the tools exploited by contemporary genetic engineers. For example, if a T-DNA promoter is inefficient in a particular species, it can, possibly, be replaced with a more powerful promoter. Thus, there is always a good chance that a non-host plant can be converted to a host for the *Agrobacterium* vector system. In fact, successful transformation of yeast has been achieved with *Agrobacterium* (Bundock et al., 1995, Bundock and Hooykaas, 1996, Piers et al., 1996).

## Vector Systems for *Agrobacterium*-mediated Transformation

As soon as it had been demonstrated that the formation of a crown gall represented the genetic transformation of plant cells, attempts to develop vector systems were initiated. Various sophisticated plant-transformation vectors, designed on the basis of this naturally occurring gene-transfer mechanism, have been developed, and such vectors are widely employed in plant molecular biology and in the genetic engineering of plants (Fraley et al., 1986). Two problems had to be overcome initially: 1) oncogenes cause tumors and inhibit plant regeneration; and 2) Ti plasmids are more than 190 kb long and they are too big for standard molecular-cloning techniques. The first remarkable development in this field was the removal of wild type T-DNA from Ti plasmids to create 'disarmed strains' such as LBA4404 (Hoekema et al., 1983) and C58Cl(pGV3850) (Zambryski et al., 1983). Then two strategies were invented for the introduction of engineered T-DNA into *A. tumefaciens*. In one case, *E. coli* vectors are used that can be integrated into a disarmed Ti plasmid. Various genes, including selection marker genes and genes of interest, can be placed in the vector and eventually introduced into a disarmed strain. This strategy is referred to as the intermediate vector system (Fraley et al., 1983, 1985, Zambryski et al., 1983). The second strategy exploits the fact that the process for transfer of T-DNA is active even if the virulence genes and the T-DNA are located on separate replicons in an *Agrobacterium* cell (Hoekema et al., 1983). This strategy employs small plasmids that can be replicated in both *Agrobacterium* and *E. coli*. Artificial T-DNA is constructed within the plasmids and then introduced into disarmed strains, in a so-called binary vector system (Figure 1). Vectors such as pBIN19 (Bevan, 1984), pBI121 (Jefferson, 1987) and pGA482 (An et al., 1988) have been extensively utilized in various attempts to exploit this strategy.

## Super-binary Vectors

Strain A281 (Watson et al., 1975) is a 'super-virulent' strain, and its host range is wider and its transformation efficiency is higher than those of other strains (Hood et al., 1987, Komari, 1989). These characteristics are due to the Ti plasmid, pTiBo542, that is harbored by this strain (Hood et al., 1984, Jin et al., 1987, Komari et al., 1986). Two new types of systems based on pTiBo542 have been developed. The first involves strain EHA101 (Hood et

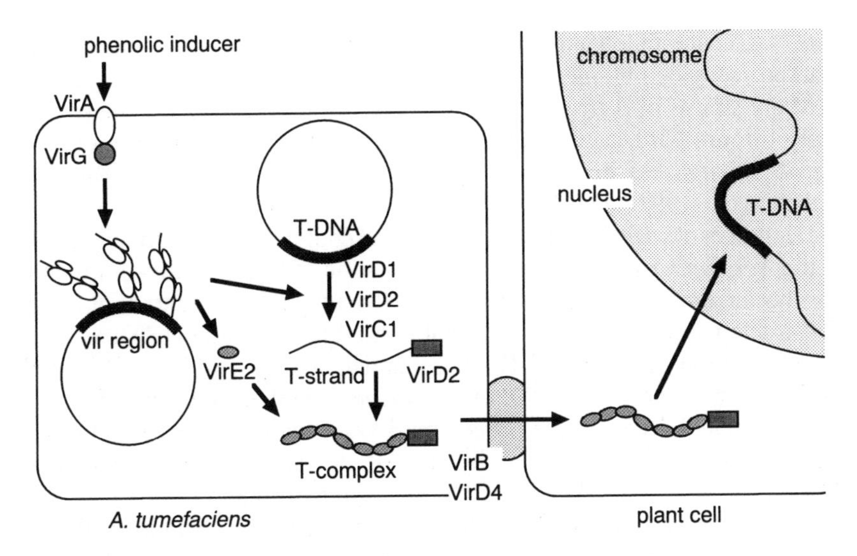


*Fig. 1* A schematic representation of the transformation of a higher plant by a binary vector

al., 1986) or EHA105 (Hood et al., 1993), which carries a 'disarmed' version of pTiBo542. The second involves a 'super-binary' vector. In this system, a DNA fragment that includes *virB*, *virC* and *virG* from the virulence region of pTiBo542 is introduced into a small T-DNA-carrying plasmid (Komari, 1990) in a binary vector system, in which the disarmed Ti plasmid has its own full set of virulence genes. *virB* and *virG* are the genes responsible for the hypervirulence of A281, and extra copies of these genes might also enhance the virulence. Super-binary vectors are very efficient in the transformation of various plants (Komari, 1990, Li et al., 1996, Saito et al., 1992), and such systems have played important roles in studies of the transformation of rice and other monocotyledons.

## A Co-transformation System for the Generation of Marker-free Transformants

In direct transformation methods, such as microprojectile bombardment and electroporation, co-transformation, with marker genes and other genes being placed on different DNA molecules, is a common practice (Christou et al., 1991, Lyznik et al., 1989, Schocher et al., 1986, Shimamoto et al., 1989). Co-transformation provides a number of advantages. For example, construction of molecules may be simplified since, at least, reactions to link marker genes and other genes are unnecessary. In addition, in the subsequent generations of co-transformants, marker genes and other genes may segregate independently and transgenic plants, free from selection markers, may be obtained (Yoder and Goldsbrough, 1994).

Many strains of *Agrobacterium* carry multiple T-DNAs and many crown gall tumors are actually co-transformants. Co-transformation by *Agrobacterium* has been examined in several laboratories (De Block and Debrouwer, 1991, De Framond et al., 1986, Depicker et al., 1985, McKnight et al., 1987, Simpson et al., 1986) but it is not a routine procedure in the *Agrobacterium*-mediated transformation of higher plants, for the most part because suitable vectors have not been available. Komari et al., (1996) constructed super-binary vectors that carried two separate T-DNAs for co-transformation. The first T-DNA contained a drug-resistance, selectable marker gene, and various DNA fragments could be introduced into the second T-DNA by a simple procedure. Large numbers of rice and tobacco transformants were produced by use of *A. tumefaciens* LBA4404 that carried the vectors. The frequency of co-transformation with the two T-DNAs was greater than 47%. Moreover, progeny that contained the non-selectable T-DNA and were also free of the selectable marker were obtained from more than half of the co-transformants. Delivery of two T-DNAs to plants from mixtures of *Agrobacterium* was also tested but the frequency of co-transformation was relatively low. The efficiency of co-transformation and the frequency of unlinked integration of the two T-DNAs were both reasonably high. Since multiple fragments of DNA were integrated at one locus in many of the cases of co-transformation when direct transformation methods were applied (Christou et al., 1992, Rathore et al., 1993, Spencer et al., 1992), super-binary vectors appear to provide a suitable co-transformation system for the production of marker-free transformants that is more efficient than direct transformation systems. Since various segments of DNA can easily be inserted into the vectors, co-transformation is an option in *Agrobacterium*-mediated transformation of cereals.

## Attempts at the *Agrobacterium*-mediated Transformation of Monocotyledons

### *Induction of Tumors on Monocotyledons*

Ti plasmids were recognized as useful natural vectors for transformation of higher plants in the early 1980's. Subsequently, application to monocotyledonous plants of methods based on Ti plasmids has been extensively explored. The capacity to serve as a host plant for crown gall disease was thought to be a prerequisite for any host of vector systems based on *Agrobacterium*. Various monocotyledonous plants were infected with wild-type *Agrobacterium* strains and the formation of tumors and/or the biosynthesis of opines were examined. For example, Hernalsteens et al., (1984) infected stem segments of *Asparagus officinalis* with *A. tumefaciens* and observed induction of tumors and detected nopaline in the tumor cells. Similar observations were made in *Chlorophytum capense* and *Narcissus* cv. 'Paperwhite' (Hooykaas-Van Slogteren, et al., 1984), *Gladiolus* spp. (Graves and Goldman, 1987), yam (*Dioscorea bulbifera*; Schäfer et al., 1987), onion (*Allium cepa*; Dommisse et al., 1990), and lily (Langeveld et al., 1995).

Cereal species were also examined. Graves and Goldman (1986) wounded seedlings of *Zea mays* and infected them with *A. tumefaciens.* They detected the activity of lysopine dehydrogenase, which is encoded by transcripts of T-DNA. Baba et al., (1986) introduced spheroplasts of *A. tumefaciens* into rice protoplasts by treatment with polyethylene glycol. They observed the hormone-free growth of rice cells and detected opines in the protoplasts. These studies showed that the host range of crown gall was wider than previously anticipated. However, data from Southern hybridization to prove the integration of T-DNA into the genomes of monocotyledons were presented only for tumors in *Asparagus officinalis* (Bytebier et al., 1987) and yam (Schäfer et al., 1987).

### *Attachment of* Agrobacterium *to Monocotyledonous Cells*

Absence of the ability to attach to plant cells was suggested as one of the possible barriers to infection by *Agrobacterium* (Lippincott and Lippincott, 1978). Therefore, considerable attention has been paid to the attachment of *Agrobacterium* to monocotyledonous cells. Douglas et al., (1985) reported that *A. tumefaciens* attaches specifically to bamboo cells in suspension culture. Subsequently, studies using the scanning electron microscope showed that strains of *A. tumefaciens* could attach to cells on the surfaces of maize and wheat seedlings, gladiolus corm disks (Graves et al., 1988) and immature wheat embryos (Mooney and Goodwin, 1991) in the same manner as they attach to dicotyledonous plants. These studies suggested that *Agrobacterium* might be capable of attaching to the cells of at least some of the plants on which crown gall tumors cannot naturally be formed.

### *Production of* vir-*inducing Substances by Monocotyledons*

Another possible barrier to infection by *Agrobacterium* is the absence of *vir*-inducing compounds. Various monocotyledons have been examined for their *vir*-inducing abilities. Ashby et al., (1988) found that *vir* inducers are also chemoattractants for *Agrobacterium* and observed the chemotaxis of *Agrobacterium* toward homogenates of wheat shoots and roots. Usami et al., (1988) found that wheat and oat plants exuded *vir*-inducing substances, which were not phenolics, from wounds. Messens et al., (1990) identified a potent *vir*-inducing compound, ethyl ferulate, in cell cultures of *Triticum monococcum*. Primich-Zachwieja and Minocha (1991) found that tissues from eight monocotyledonous species (*Asparagus densiflorus*, *Cyperus papyrus*, *Alopecurus* sp., *Quesnelia testudo*, *Triticum aestivum*, *Avena sativa*, *Hordeum vulgare*, *Zea mays*) induced expression of the *virE* gene of *A. tumefaciens*. Vijayachandra et al., (1995) found that the scutellum of four-day-old rice seedlings induced the expression of *vir* genes while other rice tissues that they tested did not.

Negative data have also been reported. Usami et al., (1987) reported the absence of diffusible plant factors that could induce *vir* genes from seedlings of seven monocotyledonous species (*Zea mays, Avena sativa, Oryza sativa, Lilium longiflorum, Asparagus officinalis, Tradescantia reflexa, Allium fistulosum*). Sahi et al., (1990) found that homogenates of maize seedlings inhibited the growth of *Agrobacterium* and the induction of *vir* genes. They identified one inhibitory compound as 7-methoxy-2H-1,4-binzoxazin-3(4H)-one.

Thus, at least some monocotyledonous plants do seem to produce *vir*-inducing compounds, but the extent of induction by these compounds might be quite limited.

### *Agroinfection*

Grimsley et al., (1986) developed a new technique, designated 'agro-infection', by which the sequence of a viral genome can be introduced into a higher plant via *Agrobacterium* with the resultant systemic infection of the plant by the virus. They demonstrated that maize could be infected with maize streak virus that was delivered by agroinfection (Grimsley et al., 1987) and they characterized various parameters that affected the agro-infection of maize (Escudero et al., 1996, Grimsley et al., 1988, Grimsley et al., 1989, Jarchow et al., 1991, Schläppi and Horn, 1992). During the course of these studies, the transfer of T-DNA to maize cells from intracellularly injected *Agrobacterium* was observed (Escudero et al., 1996). Raineri et al., (1993) studied the role of the *virA* gene in agroinfection, and the technique was extended to studies of the infection of wheat with wheat dwarf virus (Dale et al., 1989) and of rice with rice tungro bacilliform virus (Dasgupta et al., 1991).

This method allows an extremely sensitive assay of the delivery of DNA to plant cells since the viral replication machinery generates multiple copies of sequences derived from the introduced DNA with high efficiency. These studies provided clear evidence that DNA is transferred from *Agrobacterium* to maize, wheat and rice cells. Since infectious viral genomes can be produced from unintegrated T-DNA, the integration of T-DNA in chromosomes was not demonstrated in these studies.

*Expression of Marker Genes in Monocotyledons*

In the early 1990's, introduction of a gene for $\beta$-glucuronidase (GUS; Jefferson, 1987) and of drug-resistance genes to cereal species was attempted in many laboratories. For example, segments of young seedlings of maize (Ritchie et al., 1993, Shen et al., 1993) and rice (Li et al., 1992, Liu et al., 1992) were infected with *Agrobacterium* and transient expression of GUS was observed. The *intron-gus* marker, in which there is an intron in the coding sequence of GUS, has been an especially powerful tool since the presence of this gene does not result in the expression of GUS activity in *Agrobacterium* (Ohta et al., 1990, Vancanneyt et al., 1990).

Kanamycin- or G418-resistant cells that expressed GUS were obtained from embryos of rice (Raineri et al., 1990), from enzymatically digested embryos of wheat (Mooney et al., 1991), and from root segments of rice (Chan et al., 1992) that had been co-cultivated with *Agrobacterium.* Regenerated plants expressing GUS were obtained from shoot apices of maize (Gould et al., 1991), immature embryos of rice (Chan et al., 1993), and embryogenic callus of *Asparagus officinalis* L. (Delbreil et al., 1993) that had been co-cultivated with *Agrobacterium.* The cited studies included results of Southern hybridization. Gould et al., (1991) and Chan et al., (1993) also observed the inheritance and expression of foreign genes in the progeny of a very limited number of plants. However, the efficiency of transformation was very low.

Some results have generated considerable controversy. Potrykus (1990a, b, 1991) and Langridge et al., (1992) published critical reviews of methods of *Agrobacterium*-mediated transformation. The criteria for firm proof of integrative transformation that they proposed can be summarized as follows. 1) Expression of marker genes in plant cells must be distinguished from expression of the genes in *Agrobacterium* or other contaminating microorganisms, as well as from the endogenous background expression of the activity in the plant. 2) Credible results of Southern hybridization analysis should be presented to confirm the integration of T-DNA in the plant chromosomal DNA. 3) Mendelian inheritance of transgenes should be demonstrated and confirmed by Southern hybridization. 4) The data should be obtained from a number of independent transgenic plants.

Although the cited studies showed that *Agrobacterium*-mediated transformation is potentially useful for modification of cereal plants, none of them met all of the requirements summarized above. Furthermore, the low frequency of transformation in these studies is also a problem. In addition to the criteria that prove that transformation has occurred, if the technique is to be widely applicable, the efficiency of the process is very important. Even if complete evidence of integration of foreign genes were obtained, the technique would not be very useful unless the frequency of transformation was reasonably high. The efficiency or frequency of transformation is often expressed as the number of independent transformants obtained as a percentage of the number of pieces of tissue infected with *Agrobacterium*. For routine use of these techniques, frequencies greater than 10% are obviously desirable.

### *Development of Efficient Methods for Transformation*

The efficient transformation of a cereal species, with sufficient data to satisfy all of the above listed criteria, was first reported in rice by Hiei et al., (1994) They inoculated calli, induced from scutella of mature seeds of rice, with *Agrobacterium* that carried a hygromycin-resistance gene and the *intron-gus* gene. They obtained regenerated rice transformants from 10% to 30% of infected calli using three Japonica cultivars.

The evidence for transformation that they provided was as follows. 1) The *intron-gus* gene, which is not expressed in *Agrobacterium* cells, was used to monitor the expression of foreign genes. This gene is expressed strongly in rice, and background expression in rice is negligible. 2) In Southern blotting analysis, they primarily targeted DNA fragments that contained junctions between the T-DNA and plant DNA. More than 20 independent transgenic plants ($R_0$ generation) were analyzed, and the sizes of the detected DNA fragments varied from plant to plant. Thus, random integration of the T-DNA into rice chromosomes was clearly demonstrated. 3) Inheritance of the T-DNA was followed to the $R_2$ generation, with demonstration of the Mendelian segregation of transgenes. Both assays of gene expression and Southern hybridization were performed, and the results corresponded perfectly. 4) Junctions between the T-DNA and rice DNA were isolated by polymerase chain reactions and sequenced. The borders of the T-DNA in the rice transformants were essentially identical to those found in dicotyledonous transformants. Furthermore, the high efficiency of transformation was remarkable (Table 1). Calli induced from scutella can be prepared in abundance, and the reported frequency of transformation ranged between 10% and 30%. Thus, even small-scale experiments should be able to produce a large number of transformants. Mature seeds are good sources of tissue for *Agrobacterium* infection and only three or four weeks are required for induction of calli from rice seeds. Thus, suitable materials are available at virtually any time of the year.

Similar methods also proved effective in Javanica rice (Dong et al., 1996) and in Indica rice (Rashid et al., 1996). Aldemita and Hodges, (1996) reported that transgenic plants were efficiently produced from immature embryos of Japonica and Indica rice that had been co-cultivated with *Agrobacterium*. We have also succeeded in the efficient transformation of ten elite cultivars of Indica rice, including IR64 and IR72, by the immature-embryo method (unpublished results). By contrast, Park et al., (1996)

TABLE 1
Typical frequencies of transformation of cereal plants mediated by *A. tumefaciens*

| Plant | Cultivar | Tissue | Strain | Number of tissue pieces inoculated | Number of independent transformants | Frequency (%) | Reference |
|---|---|---|---|---|---|---|---|
| Rice (Japonica) | Tsukinohikari | Callus | LBA4404(pTOK233) | 540 | 141 | 26.1 | Hiei et al. 1994 |
| Rice (Japonica) | Koshihikari | Callus | LBA4404(pTOK233) | 283 | 65 | 23.0 | Hiei et al. 1994 |
| Rice (Javanica) | Gulfmont | Callus | LBA4404(pTOK233) | 70 | 14 | 20.0 | Dong et al. 1996 |
| Rice (Indica) | Basmati 370 | Callus | EHA101(pIG121Hm) | 118 | 26 | 22.0 | Rashid et al. 1996 |
| Rice (Indica) | IR72 | Immature embryo | LBA4404(pTOK233) | 60 | 5 | 8.3 | Aldemita and Hodges 1996 |
| Maize | A188 | Immature embryo | LBA4404(pSB131) | 121 | 20 | 16.5 | Ishida et al. 1996 |
| Maize | A188 | Immature embryo | LBA4404(pTOK233) | 72 | 7 | 9.7 | Ishida et al. 1996 |
| Barley | Golden Promise | Immature embryo | AGL1(pDM805) | 387 | 14 | 3.6 | Tingay et al. 1997 |
| Wheat | Bobwhite | Immature embryo | C58(ABI)(pMON18365) | 140 | 2 | 1.4 | Cheng et al. 1997 |
| Wheat | Bobwhite | Callus | C58(ABI)(pMON18365) | 110 | 3 | 2.7 | Cheng et al. 1997 |

produced transgenic rice from isolated shoot apices that had been inoculated with *Agrobacterium* at a somewhat lower frequency.

Ishida et al., (1996) developed an efficient *Agrobacterium*-mediated transformation method for another important cereal, maize. They inoculated immature embryos with *Agrobacterium* and the frequency of transformation was very high, between 5% and 30%. They also presented evidence to prove the integration of T-DNA, which included results of an assay of expression of the *intron-gus* gene; Southern hybridization data that showed the random integration of T-DNA into the maize genome (an example the Southern blotting analysis of transformants is shown in Figure 2); inheritance of T-DNA, based on both the expression of transgenes and on Southern blotting analysis; and sequencing of the junctions between the T-DNA and maize DNA. They analyzed more than 30 independent transgenic plants in their study.

Ishida et al., (1996) also showed that many factors, including the genotype of maize, the type of tissue inoculated, the developmental stage of the immature embryos, the concentration of the inoculum, the composition of the tissue culture media, the selection marker genes, the strains of *Agrobacterium* and the vectors, are all of critical importance.

Tingay et al., (1997) obtained transgenic barley plants from immature embryos that had been infected with *Agrobacterium* and presented molecular and genetic evidence to prove stable integration of foreign genes. The frequency of transformation, 4.2%, was somewhat lower than the studies with rice and maize but is likely to be enhanced in the future. Tingay et al., (1997) did not include acetosyringone in their co-cultivation method,

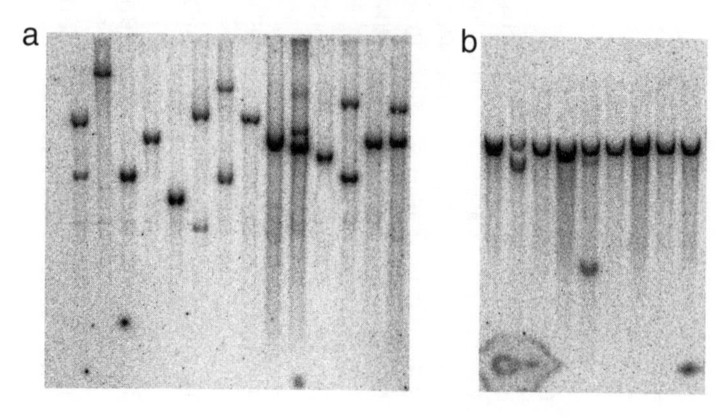


*Fig. 2* An example of the results of Southern blotting analysis of transgenic maize produced by *Agrobacterium*-mediated transformation. **a**: Fragments containing junctions between T-DNA and plant DNA were targeted, and the random integration of T-DNA in the chromosomes of maize was demonstrated. The majority of transformants carried one or two copies of the T-DNA. **b**: An internal fragment of the T-DNA was targeted. All of the transformants contained at least one intact copy of this fragment

and they used particle bombardment to wound immature barley embryos. This method could serve as another option in the procedure.

Cheng et al., (1997) reported an *A. tumefaciens*-mediated transformation system for wheat. They inoculated freshly isolated immature embryos, precultured immature embryos and embryogenic calli with *A. tumefaciens* and produced transgenic wheat plants from the three types of tissues, with a transformation frequency of 0.3%–4.3%. They also conducted extensive molecular and genetic analyses of transformants. They identified various factors including surfactants in the inoculation medium that affected the efficiency of T-DNA delivery.

With respect to other monocotyledons, May et al., (1995) obtained transgenic banana plants from shoot tips and corm slices that were first wounded by particle bombardment and then infected with *Agrobacterium*. Thus, the controversy over the possible transformation of monocotyledonous plants by *Agrobacterium* has finally been resolved and it is now clear this method may prove both efficient and reliable for the modification of various important monocotyledonous plants.

### *The Recalcitrance of some Monocotyledonous Plants to Infection by* Agrobacterium

It is useful to consider why monocotyledonous plants remained resistant to infection by *Agrobacterium* for a long time and why it suddenly became possible to transform these plants using *Agrobacterium*. In considering the several roadblocks in incompatible *Agrobacterium*-plant interactions, special attention should be paid to wounding.

Early experiments with *Agrobacterium* indicated that wounding of the host plant is essential for subsequent transformation (Lippincott and Lippincott, 1975). Obviously, one of the functions of wounding is to permit the penetration of the bacteria into the host plant. Another role of wounding is the production of phenolic compounds that induce *vir* genes. Furthermore, a very important putative role of wounding is to render host cells transiently 'competent' or 'conditioned' for transformation (Kahl, 1982). Binns and Thomashow (1988) pointed out the distinct correlation in the reports in the literature between the wound-induced division of cells and the competence of such cells to be transformed by *Agrobacterium*. They proposed that processes related to the synthesis of DNA and cell division might be required for the incorporation of foreign DNA into a host genome. Evidence supporting this hypothesis has been reported in the case of both *Agrobacterium*-mediated transformation and direct transformation (An 1985, Iida et al., 1991, Kudirka et al., 1986, Okada et al., 1986, Valvekens et al., 1988, Wullems et al., 1981). This hypothesis is consistent with our understanding that integration into the host's genome of pieces of foreign

DNA is catalyzed by host enzymes, most probably enzymes that are part of the DNA replication/repair machinery.

The wound responses of many monocotyledons differ from those of many dicotyledons (Baron and Zambryski, 1995). Three main types of wound reaction can be observed in higher plants (Kahl, 1982). Some plants, including many monocotyledons, use the damaged cells or their debris as a protective layer, which physically prevents access by plant pathogens, as well as extensive loss of water (type 1). The cells underneath this protective layer usually respond to the wound stimulus with the lignification or impregnation with phenol of their cell walls but never by cell division. By contrast, wound responses of many dicotyledons are characterized by cell divisions and proliferations that result in formation of 'wound callus' (type 2) or 'wound periderm' (type 3) for reestablishment of the normal shape of the wounded organ.

In many routine protocols for transformation of dicotyledons, explants, such as leaf disks, stem segments or tuber sections, are co-cultivated with *Agrobacterium*. Then the explants are cultivated on media suitable for the proliferation of cells and the regeneration of plants. Such explants provide ideal materials for transformation, since they include cells that are actively dividing or are about to divide. These processes depend strongly on an active wound response, which is the biological basis of proliferation and regeneration from somatic cells (Potrykus, 1991).

Apparently, routine methods of transformation are not useful for dicotyledons or monocotyledons that do not exhibit an active wound response. Potrykus, (1991) stated that 'transformation of 'monocots' is of no importance in this context: it is not because they are monocots that cereals are difficult to transform but because they do not have the proper wound response'. However, these methods in dicotyledons had proved so successful and so efficient that much attention was paid and much effort was made to the possible exploitation of similar approaches in monocotyledonous plants. This is one of the reasons why development of efficient methods for *Agrobacterium*-mediated transformation of cereals was delayed for so long.

The main roles of wounding in the transformation process can be summarized as the production of *vir*-inducing molecules, the induction of DNA synthesis, and the induction of the rapid division of cells. Therefore, it would seem likely that actively growing, regenerable tissues from monocotyledons might be transformable in the presence of *vir*-inducing compounds. Reports of the successful transformation of cereals clearly support this possibility (Hiei et al., 1994, Ishida et al., 1996). Tissues consisting of actively dividing, embryonic cells were co-cultivated with *Agrobacterium* in the presence of 100 $\mu M$ acetosyringone in successful studies. However, numerous factors are of critical importance in the *Agrobacterium*-mediated transformation of rice and maize and this multiplicity of factors probably also explains why it was initially so difficult to apply this technology to cereals.

## Key Factors involved in the *Agrobacterium*-mediated Transfer of Genes to Cereals

A typical procedure for the transformation of cereals by *Agrobacterium* is shown schematically in Figure 3. We shall discuss here the key factors at each of the critical steps. Many of the parameters have been adjusted on the basis of the so-called transient expression of a transgene, which is most often *intron-gus*. Stable transformants have never been efficiently obtained under conditions that yield only limited transient expression. It should be noted that such adjustments are less straightforward in maize. Conditions that allow high-level expression of markers after co-cultivation are generally associated with high-frequency production of stable transformants in rice (Hiei et al., 1994). In maize, by contrast, it proved relatively easy to find conditions for high-level transient expression but stable transformants were obtained in only a few instances (Ishida et al., 1996). It is possible that the main hurdle in transformation is not the delivery of DNA fragments to the plant cells but, rather, the recovery of cells that have acquired the T-DNA in their chromosomes via tissue culture systems. The phenomenon might also be related to the suggestion by Narasimhulu et al., (1996) that the failure to transform maize might involve integration of T-DNA and not entry of T-DNA into cells or targeting to the nucleus.

### *Choice of Tissues for Infection*

The type and quality of the starting material is a very important factor, as is always the case in tissue culture. Callus cultures are excellent sources of cells for the production of transgenic rice (Dong et al., 1996, Hiei et al., 1994, Rashid et al., 1996). The use of actively growing, embryogenic calli is very important for efficient transformation of rice. Such calli can be obtained from mature or immature embryos. Other tissues, including shoot apices, immature inflorescences and young roots, might also produce embryogenic calli but they have not yet been tested extensively. The production of transgenic rice plants from isolated shoot apices has been reported (Park et al., 1996) but the frequency of transformation was about 10% of that when callus was used. Since shoot apices from various cereal cultivars can rapidly regenerate into plants, this method should be potentially useful if the efficiency can be improved. Long-term culture does not significantly affect the efficiency of transformation, but the risk of so-called somaclonal variation might be enhanced during such long-term culture.

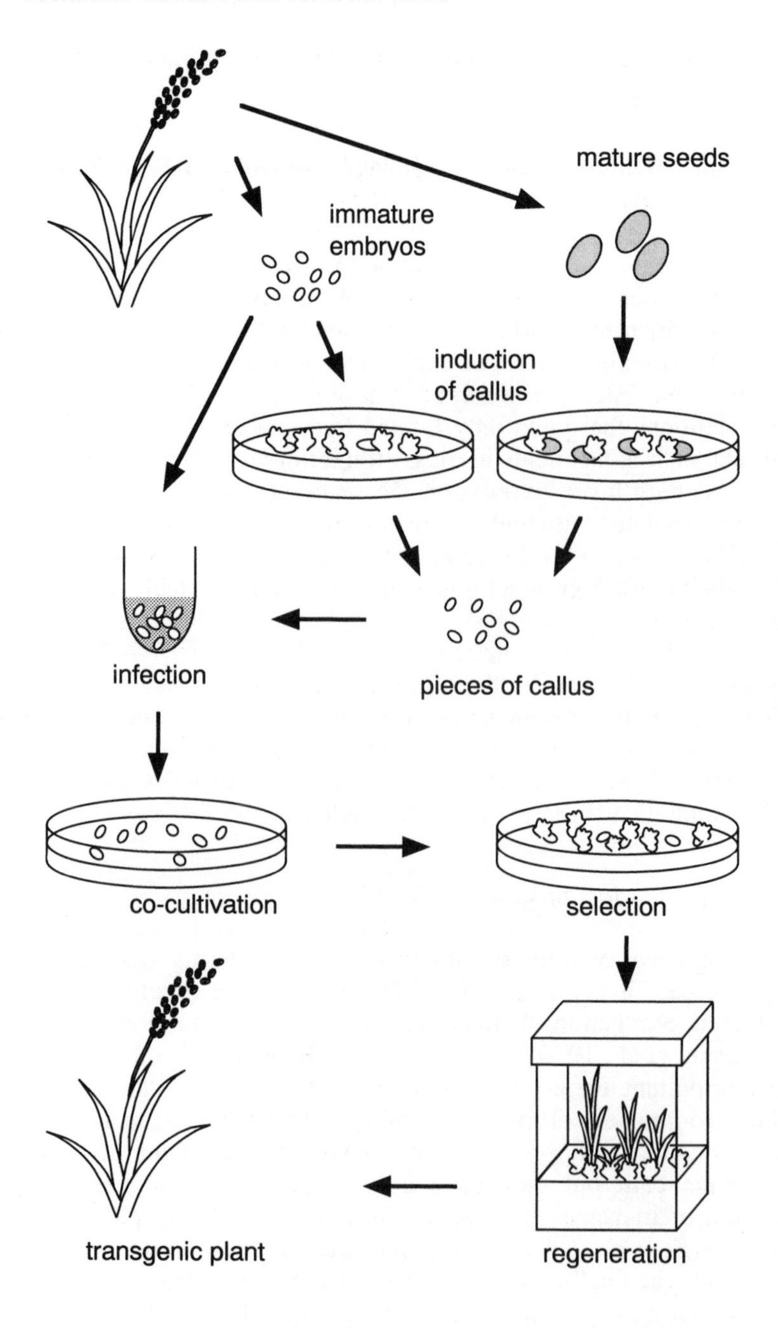


*Fig. 3* A schematic representation of the procedure for the *Agrobacterium*-mediated transformation of a cereal plant

Unlike the embryogenic calli of Japonica rice, those of many Indica varieties and of maize have been poor starting materials for *Agrobacterium*-mediated transformation. By contrast, freshly isolated immature embryos consist of actively dividing cells. Although immature embryos proved to be poor materials in early studies in rice (Chan et al., 1993, Hiei et al., 1994), transformants can now be produced efficiently from immature embryos in rice, as a result of optimization of the procedure (Aldemita and Hodges, 1996), and also in maize (Ishida et al., 1996). Further improvements in the procedure might allow the use of mature embryos for efficient transformation.

Use of rice cells in suspension culture for co-cultivation yielded low frequencies of transient expression of GUS activity and low frequencies of stable transformation, even though such cells were actively proliferating (Hiei et al., 1994). Preculture of cells on solid medium for several days resulted in the high-frequency transfer of genes. Thus, liquid cultures of rice cells seem unsuitable for *Agrobacterium*-mediated transformation.

The most promising starting materials for the transformation of cereal plants are currently immature embryos and callus cultures that have been induced from mature or immature embryos. Immature embryos, in particular, should be prepared from healthy plants and the stages of the embryos should be examined carefully. If calli are used, 'good' pieces of callus should be selected under a dissecting microscope. The choice of materials at this step is really critical, and it directly affects the efficiency of transformation. It is at this step that knowledge and experience of experts in tissue culture of the species of interest are indispensable.

The choice of material involves various factors. If 'easy' cultivars of Japonica rice such as Nipponbare are targeted, mature seeds are the material of choice. If access is available to a greenhouse, in which developing embryos can be generated in quantity all year round, immature embryos are the best choice. If maize or Indica rice are the plants of interest, immature embryos are the only choice at the present time.

### *Vectors*

Various binary vectors can be used for transformation of cereals. For example, pIG121Hm (Hiei et al., 1994), which is a derivative of the most commonly used vectors pBI121 (Jefferson, 1987) and pBIN19 (Bevan, 1984), is efficient in rice transformation (Hiei et al., 1984, Rashid et al., 1996). Other choices are super-binary vectors, which are particularly efficient.

Hiei et al., (1994) and Dong et al., (1996) compared a normal vector, pIG121Hm, with a super-binary vector, pTOK233, in rice transformation. In their transformation experiments, pTOK233 was as effective or slightly more effective than pIG121Hm with 'easy' genotypes, and it was definitely much more effective with 'difficult' genotypes, which included the Koshihikari cultivar. These data suggest that choices of vectors and strains are important for the transformation of difficult cultivars. In maize, efficient transformation has proved possible only with super-binary vectors (Ishida et al., 1996). Therefore, super-binary vectors seem to be particularly useful when recalcitrant genotypes or species are to be transformed.

pTOK233 is a convenient vector for introduction of *intron-gus*, a kanamycin-resistance gene and a hygromycin-resistance gene into various plants. This plasmid has, however, numerous restriction sites and it is not suitable for further manipulation. Another super-binary vector, which can transfer various genes to higher plants, and a procedure for generation of an appropriate construct using this system are shown schematically in Figure 4. pSB11 (Komari et al., 1996) is a derivative of pBR322 (Bolivar et al., 1977) and carries multiple cloning sites between the two borders of the T-DNA. Selection marker genes and other genes can be inserted to this 'intermediate vector' to create T-DNAs for plant transformation. This vector is not replicated in *Agrobacterium* but has a fragment homologous to the 'acceptor vector' pSB1. The intermediate vector can be introduced into a strain of *Agrobacterium* that carries pSB1, such as LBA4404(pSB1) and, finally, the vector for plant transformation is generated by homologous recombination in *Agrobacterium*.

### *Strains of* Agrobacterium

The most commonly used strains of *Agrobacterium* for transformation of higher plants are LBA4404 (Hoekema et al., 1983) and EHA101 (Hood et al., 1986). Both strains efficiently mediate transformation of rice (Hiei et al., 1994, Rashid et al., 1996, Aldemita and Hodges, 1996, Dong, 1996), and LBA4404 is also useful for transformation of maize (Ishida et al., 1986). Therefore, it would be a good idea to test these strains in attempts to transform other cereals.

To date, numerous strains have been isolated and characterized (Anderson and Moore, 1979) but only a few of them have been modified for use in the transformation of higher plants. It is potentially possible to identify strains suitable for a particular cereal species by screening a large number of strains. It is advisable to consider genetic variations in Ti plasmids and in main chromosomes separately. Several chromosomal genes are involved in the recognition of host plant cells and in the attachment of bacteria to plant cells. It is very likely that genetic variations that are useful for transformation of cereals might be identified in such genes and in other genes that are involved in the development of the extracellular structures of *Agrobacterium*.

The genetic information on Ti-plasmids has been characterized in greater detail than that on main chromosomes. Manipulation of the structure or

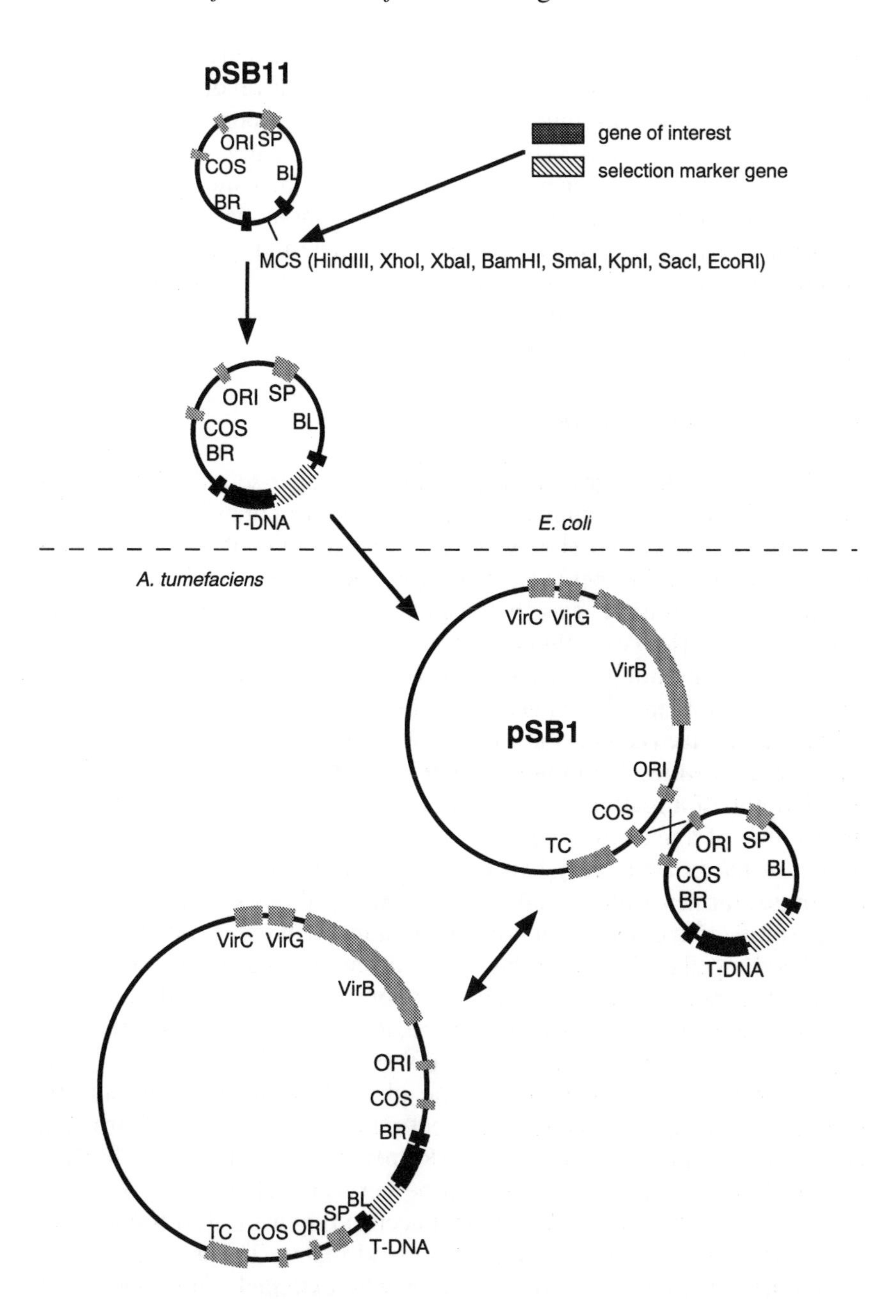


*Fig. 4* A schematic representation of the procedure for construction of a super-binary vector for transformation of cereal plants. Abbreviations: SP, spectinomycin-resistance gene; TC, tetracycline-resistance gene; BR, right border; BL, left border; cos, cos site of bacteriophage lambda; ORI, origin of replication of ColE1; MCS, multiple cloning sites

regulation of the expression of known genes might be more productive than screening for genetic variations. For example, if the level of induction of *vir* genes in a strain is a limiting factor, the level can be controlled by adding acetosyringone to the co-cultivation medium, or by modification of promoters of *vir* genes. Genes of particular interest here are the genes of the *virB* operon and *virD*4, whose products are thought to make the channel for the passage of the T-complex between the bacterium and the plant cell. Different *virB* and *virD*4 genes might provide more efficient channels to a particular cereal species, and a search for less conserved *virB* and *virD*4 genes might produce useful tools for transformation.

### *Infection and Co-cultivation*

*Agrobacterium* can be grown either in liquid or agar-solidified media. The concentration of the inoculum affects the efficiency of transformation. Transformation of rice was possible when the concentration of *A. tumefaciens* in the inoculum ranged between $1.0 \times 10^6$ and $1.0 \times 10^{10}$ cfu/ml, and the optimal concentration was between $1.0 \times 10^8$ and $1.0 \times 10^{10}$ cfu/ml (Aldemita and Hodges, 1996, Dong et al., 1996, Hiei et al., 1994). The optimal concentration for maize transformation was between $5.0 \times 10^8$ and $2.0 \times 10^9$ cfu/ml, and the range of suitable concentrations was narrower in maize than in rice (Ishida et al., 1996). We normally culture *Agrobacterium* on AB agar medium (Chilton et al., 1974) at 28°C for 3 days and suspend one loopful of bacteria in 1 ml of infection medium. This method for preparation of a uniform inoculum is reasonably reproducible; the concentration of bacteria can vary two-fold but never more in such preparations. Of course, further precise control can be exercised based on optical density.

After infection of tissues in liquid inoculum, plant tissues and bacteria are cultivated together for 3 to 5 days. One of the techniques most commonly used in the transformation of dicotyledonous plants is the addition of phenolic compounds, such as acetosyringone, to co-cultures or bacterial cultures (Godwin et al., 1991, Sheikholeslam and Weeks, 1987, Van Wordragen, 1992). Their addition is indispensable if the plant does not produce sufficient amounts of signal molecules. Rice cells might be capable of producing certain amounts of signal molecules (Raineri et al., 1990, Vijayachandra et al., 1995) but these levels appear to be very low. Hiei et al., (1994) and Ishida et al., (1996) demonstrated that acetosyringone at 100 $\mu M$ is critical to the successful transformation of rice and maize. The level of transient expression of GUS after co-cultivation was extremely low when acetosyringone was omitted. Chan et al., (1993) included the medium from a suspension culture of potato cells, which is a rich source of phenolic compounds, in co-cultures of rice and *Agrobacterium*. The transfer of T-DNA is initiated at an early stage of co-cultivation in the presence of acetosyringone, and pre-treatment of bacteria with acetosyringone is not essential. However, pre-treatment might slightly increase the efficiency of gene transfer (Aldemita and Hodges, 1996).

The temperature of co-cultivation is another important factor. The optimal temperature for the growth of *Agrobacterium* is between 28°C and 30°C, and tissue cultures of cereals are usually incubated at between 25°C and 30°C. Induction of *vir* genes is, however, more efficient at temperatures between 15°C and 25°C (Alt-Moerbe et al., 1988, Alt-Mörbe et al., 1989). Fullner and Nester (1996) found that the mobilization of the plasmid by the T-DNA transfer machinery is most active at temperatures as low as 19 °C. Therefore, temperatures lower than 25°C might also be optimal for the transfer of T-DNA from bacteria to plant cells. The temperature of co-cultivation should be carefully adjusted to balance the effects on the activity of the T-DNA transfer machinery and the viability of both the plant cells and *Agrobacterium*. In rice, the efficiency of transformation was high between 22°C and 28°C (Hiei et al., 1994)

Other factors during co-cultivation, such as an acidic pH (Alt-Moerbe et al., 1988, Turk et al., 1991) and high osmotic pressure (Usami et al., 1988), have also been reported to be important for the expression of *vir* genes. Veluthambi et al., (1989) found that opines stimulated the induction of *vir* genes. In addition, Shimoda et al., (1990) and Cangelosi et al., (1990) reported that a group of aldoses, including D-glucose and some non-catabolizable sugars, such as 2-deoxy-D-glucose and 6-deoxy-D-glucose, markedly enhanced the expression of *vir* genes. The effects of sugars were clear when levels of phenolic inducers were limited. Hiei et al., (1994) confirmed the importance of these factors. Transient expression of GUS in calli derived from scutella after co-cultivation with *Agrobacterium* was strongest when the pH of the co-cultivation medium was between 4.8 and 6.2. Exogenously added D-glucose, D-galactose or L-arabinose and nopaline had no significant synergistic effects in the presence of 100 $\mu M$ acetosyringone.

The composition of the co-cultivation medium affects the efficiency of transformation, and media that can support the active division of cells are preferable. Media solidified with a gelling agent are better for co-cultivation than liquid media (Hiei et al., 1994, Ishida et al., 1996). When co-cultivation with rice calli is carried out in liquid medium for several days, the level of transient expression of GUS after co-cultivation is extremely low. Optimization of the medium seems to be necessary, depending on the genotype or the type of material, and it can be achieved by monitoring the transient expression of marker genes.

### *Selection of Transformants*

In general, the choice of selection marker genes and of selective agents strongly influence the efficiency of transformation (Wilmink and Dons,

1993). The gene for neomycin phosphotransferase [*nptII*, also called *aph(3′)II or neo*] was used in many early attempts at the direct transformation of cereals. This gene confers resistance to the amino-glycoside antibiotic kanamycin. Kanamycin can be used as a selective agent during the regeneration of protoplasts but it is not effective for the selection of transformed calli in rice. In addition, many of the calli that are recovered after kanamycin selection are unable to regenerate green plants (Ayers and Park, 1994, Toriyama et al., 1988). G418 is a related amino-glycoside antibiotic that is also inactivated by neomycin phosphotransferase. Not only is selection of transformed calli more efficient with G418 than with kanamycin, but calli selected on G418 regenerate into green plants at a higher frequency than those selected on kanamycin (Ayers and Park, 1994). In *Agrobacterium*-mediated transformation, stably transformed calli can be produced efficiently with G418, but the absence of regenerants suggests that long-term exposure of cells to G418 might inhibit regeneration (Aldemita and Hodges, 1996).

Another widely used and more effective selectable marker is hygromycin phosphotransferase (*hpt, aphIV* or *hph*), which confers resistance to the amino-glycoside antibiotic hygromycin (Bilang et al., 1991, Walters et al., 1992). Hygromycin allows clear discrimination between transformed and non-transformed rice tissues and no problems with albinos or the fertility of regenerants have been reported (Ayers and Park, 1994). The *hpt* gene has been used as an efficient marker gene for selection after *Agrobacterium*-mediated transformation in rice (Aldemita and Hodges, 1996, Dong et al., 1996, Hiei et al., 1994, Rashid et al., 1996) and in maize (Ishida et al., 1996).

Other selective agents include herbicides. Genes have been isolated that endow plants with resistance to various commercially important herbicides. Among them, the *bar* gene for phosphinothricin acetyltransferase might be the most valuable both in rice (Ayers and Park, 1994, Dekeyser et al., 1989) and in maize (Gordon-Kamm et al., 1990, Spencer et al., 1990). The *bar* gene confers resistance to L-phosphinothricin (PPT), gluphosinate (an ammonium salt of PPT), and bialaphos (a derivative of PPT). In the *Agrobacterium*-mediated transformation of maize, very efficient transformation was achieved with the *bar* gene (Ishida et al., 1996). The *bar* gene has been very useful for the efficient transformation of various cereals (Vasil, 1996, see chapter by Baga et al., this volume).

In many successful attempts to transform cereals, genes for selectable markers have been driven by the constitutively expressed 35S promoter of cauliflower mosaic virus (CaMV). This promoter appears to direct sufficiently strong expression of genes for the selection of transformants at high frequency. The level of expression of this promoter in cereals is, however, as much as 100-fold lower than in dicotyledons (Fromm et al., 1985, Hauptmann et al., 1987). Different promoters, such as those of the gene for ubiquitin 1 (*Ubi1*) of maize (Toki et al., 1992) and the gene for actin (*Act1*) of rice (Zhang et al., 1991), and elements that enhance gene expression, such as introns, have been successfully tested in cereals (Vasil, 1994, see chapter by Baga et al., this volume). Use of more powerful promoters and elements for the expression of marker genes should further improve the efficiency of transformation of cereals by *Agrobacterium*.

Transgenic plants can be regenerated via somatic embryogenesis after *Agrobacterium*-mediated transformation by the methods described by Hiei et al., (1994) and Ishida et al., (1996). Maintenance of the embryogenic status of cells on the selection medium is important for the efficient recovery of regenerants. The conditions of tissue culture and the composition of media that support the active growth of plant cells facilitate efficient selection and shorten the time required for the selection and regeneration steps (Toki, 1997). Inclusion of a selective agent in the regeneration and rooting media greatly favors the production of transgenic plants that stably express transgenes (Hiei et al., 1994, Ishida et al., 1996, Komari et al., 1996).

### *Genotypes of Plants*

Hiei et al., (1994) reported that various Japonica rice cultivars, including Koshihikari, can be transformed efficiently by *Agrobacterium*. Koshihikari is well known for the high quality of its grain and many of the new, elite Japanese cultivars are related to Koshihikari. Koshihikari and related cultivars are also known for their poor responses in tissue culture and for the difficulties associated with their transformation. The establishment of a routine and efficient method for the transformation of Koshihikari would contribute considerably to efforts aimed at the improvement of Japanese rice cultivars by genetic engineering. Dong et al., (1996) reported the *Agrobacterium*-mediated transformation of the Javanica cultivars Gulfmont and Jefferson, which are, respectively, widely cultivated or about to be commercially cultivated in the southern U.S.A.

Rashid et al., (1996) obtained transgenic plants from Basmati cultivars of Indica rice. However, most Indica varieties belong to a different group, group I, when rice cultivars are classified on the basis of results of isozyme analysis (Glaszmann, 1987). These latter Indica varieties are generally recalcitrant both in tissue culture and in attempts at transformation. Aldemita and Hodges, (1996) reported the successful transformation by *Agrobacterium* of two cultivars in group I, TSC10 and IR72. They infected immature embryos with a super-binary vector and the frequency of transformation ranged between 1% and 5%. We have also succeeded in the efficient production of transformants of group I varieties of Indica rice, including Nan Jin 11, Xin Qing Ai 1, Suewon 258, IR8, IR24, IR26, IR36, IR54, IR64 and IR72, by a similar method (unpublished results). Thus, various genotypes of rice can be now transformed by *Agrobacterium*.

In maize, only a few genotypes, inbred A188 and $F_1$ hybrids between A188 and other inbreds, have so far been transformed with *Agrobacterium*, and the frequencies of transformation of the hybrids were lower than that of A188 (Ishida et al., 1996). These varieties are very useful in tissue culture and basic studies of the molecular biology of maize but they are not used commercially. Further attempts to develop efficient procedures for more genotypes are definitely needed for the genetic modification of elite maize cultivars.

## Advantages of *Agrobacterium*-mediated Transformation

For the transformation of various dicotyledons, the advantages of *Agrobacterium*-mediated transformation include the high efficiency of transformation, the transfer of pieces of DNA with defined ends with minimal rearrangement, the transfer of relatively large segments of DNA, the integration of small numbers of copies of genes into plant chromosomes, and the absence of a requirement for techniques for protoplast culture (Klee et al., 1987, Zambryski, 1988). In addition, foreign genes delivered by this method are usually transmitted to progeny plants in Mendelian manner (Budar, 1986, Feldmann and Marks, 1987). All of these advantages also apply to the *Agrobacterium*-mediated transformation of rice and maize.

Reported frequencies of *Agrobacterium*-mediated transformation range between 10% and 30% in rice (Hiei et al., 1994) and between 5% and 30% in maize (Ishida et al., 1996). These values indicate the really high efficiency of transformation. It is now rather easy to produce hundreds of independent transgenic plants.

Hiei et al., (1994) found that almost all the transgenic Japonica rice plants that they produced by the *Agrobacterium*-mediated transformation had normal morphology and 70% were fully fertile. Similar observations were made with transgenic plants derived from Indica varieties and with maize (Ishida et al., 1996). Since then, such observations have been extended to over 5,000 transgenic rice plants and over 1,000 transgenic maize plants produced in our laboratory. Transgenic plants showing certain abnormalities, such as reduced height and fertility, were only obtained in experiments in which the culture period during the selection of transformants was prolonged for months, prior to full optimization of all parameters and conditions.

The genomes of 60 independent, GUS-positive, hygromycin-resistant, transgenic Japonica rice plants that had been transformed with two strains of *Agrobacterium* were analyzed by Southern hybridization (Hiei and Komari, 1996). These strains contained relatively large T-DNAs of more than 10 kb with three functional genes. The number of copies of the integrated genes ranged from one to six, and forty of the transformants contained one or two copies of the integrated genes. Transgenes were inherited by the progeny in Mendelian fashion in all cases and 41 of the transformants contained one locus of the T-DNA.

The genomes of 33 independent, GUS-positive, phosphinothricin-resistant transformants of maize were analyzed by Southern hybridization (Ishida et al., 1996), and 31 of them were found to contain one or two copies of the transgenes. Intact pieces of T-DNA were detected in 31 transformants with another restriction enzyme. Mendelian segregation of the transgenes was observed in the progeny of all of the plants and 31 of the transformants contained one locus only of the T-DNA.

Rice plants transformed with LBA4404(pTOK233) that gave negative results in a histochemical assay of GUS activity were occasionally found among hygromycin-resistant regenerants. Some other transformants showed mosaicism in the expression of GUS. In typical experiments, between 10% and 20% of rice transformants had these undesirable characteristics. We analyzed these transformants by Southern hybridization. All of the plants contained intact copies of the *hpt* gene. Intact copies of the gene for GUS were not found in 13 of the 20 plants that gave negative results in the assay for GUS activity. Thus, part of the gene appeared to have been lost during transformation. The other seven negative plants and 30 plants with mosaic expression of GUS had one or more copies of the intact gene for GUS. These phenomena were probably related to 'gene silencing', which has been observed in various transformation systems (Finnegan and McElroy, 1994, Flavell, 1994, Kilby et al., 1992, Matzke and Matzke, 1995, Park et al., 1996). Mosaic expression of GUS was also observed in leaves of rice plants transformed by the particle-bombardment method (Kohli et al., 1996).

The stable inheritance and expression of foreign genes are of critical importance in the application of genetically engineered rice to agriculture, but neither has been studied extensively. Only $R_1$ progeny were examined in many cases. Hiei and Komari, (1996) analyzed the inheritance of transgenes as far as the $R_4$ generation for eighteen primary transgenic rice plants that had been produced by *Agrobacterium*-mediated methods. Mendelian transmission of the two marker genes, *intron-gus* and *hpt*, was confirmed and offspring homozygous for the two genes were identified in the $R_2$ lines. $R_3$ plants were obtained from the homozygous $R_2$ offspring, and subsequently $R_4$ plants were produced. The two genes were expressed in the $R_3$ and $R_4$ generations and none of the plants in these generations had lost either gene. The DNA from selected offspring was analyzed by Southern hybridization and a close correlation between phenotype and genotype was demonstrated.

## Future Prospects

### *Transformation of other Cereals*

It is now clear that the key problem in the successful *Agrobacterium*-mediated transformation of cereals was not related to the host range of crown gall disease. It is very important to understand that every plant tissue in any plant consists of a heterogeneous population of cells. Only a small fraction of cells is equipped for integrative transformation even in the tissues that are most suitable for infection by *Agrobacterium*. We need not now ask whether a plant is a host for *Agrobacterium*. *Agrobacterium* is probably capable of transferring T-DNA to most plant species, if not to all. It is important to identify sites of competent cells in a plant and to determine how a large number of competent cells can be prepared. Dedifferentiated, regenerable cells derived from embryos or regenerable cell cultures originating from embryos are likely sources of competent cells in cereals. Obviously, the key technology for obtaining such materials is tissue culture, and the expertise of tissue culture specialists who are familiar with the target crops is definitely needed. Key factors in the transformation of rice and maize will very likely prove also to be key factors in the transformation of other important cereal species.

### *Tissue Culture-free Transformation*

Transformation that does not depend on tissue culture techniques is a fantasy. If it were possible, the time for production of transgenic plants would be shortened dramatically, and problems associated with tissue culture, such as somaclonal variation, would be eliminated. Furthermore, dependence on particular genotypes might be significantly reduced. Variations among species of plants or among genotypes of species in terms of responses to manipulations *in vitro* are an inherent problem in tissue culture technology.

Active cell divisions are important for successful transformation. Thus, one might think that meristems of intact plants would be ideal tissues for transformation. Tissue culture-free transformation might be possible by targeting cells in shoot apical meristems, for example. Strong transient expression of GUS was observed in apical tissues of rice shoots that had been co-cultivated with *Agrobacterium* (Hiei et al., 1994, Li et al., 1992, Liu et al., 1992). Thus, transfer of T-DNA to these tissues might be quite efficient. However, Potrykus (1990a, 1991) pointed out that meristematic cells might not be equipped for integrative transformation. Another problem associated with these tissues is that the number of cells that divide further and eventually give rise to reproductive organs is very limited. *Agrobacterium*-mediated transformation is a very efficient process, but only a small fraction of cells is actually transformed even in the most efficient transformation. Therefore, unless methods are developed that allow *Agrobacterium* cells to penetrate a very large number of shoot apical meristems simultaneously, meristem transformation is not likely to be a realistic approach.

At this time, efficient transformation without tissue culture is possible only in the so-called *in planta* transformation of *Arabidopsis thaliana* (Bouchez et al., 1993, Chang et al., 1994, Feldmann and Marks, 1987, Katavic et al., 1994). In a typical experiment, inflorescence shoots are cut at their bases and wound sites are inoculated with *Agrobacterium* with concurrent vacuum infiltration. Transformants are recovered among the progeny at a reasonably high frequency. Selection of transformants is usually achieved by germinating the seeds on media that contain antibiotics. All transformants recovered in such experiments have been hemizygous (Katavic et al., 1994), suggesting that transformation occurs after pollination and, probably, at a very early stage of embryogenesis. We suspect that undifferentiated cells at the proembryonic stage are the target of *in planta* transformation. This possibility suggests that dedifferentiated or undifferentiated cells are the most suitable cells for *Agrobacterium*-mediated transformation.

Development of similar systems in other species, including cereals, is of course desirable. In addition to a requirement for methods for delivery of *Agrobacterium* to intact tissues of plants, *Agrobacterium* cells that can persist in cereal tissues for several weeks are needed. *Agrobacterium* strain C58Cl can persist in transgenic tobacco for as long as 12 months (Matzk et al., 1996). Thus, strains including EHA101, which has the same chromosomal background as C58Cl, may be useful in this context. Since an external supply of inducers of *vir* genes might be less effective, strains whose virulence genes can be induced by natural metabolites in cereal tissues are required. In addition, vector plasmids need to be stable in *Agrobacterium* cells, without selective pressure, for a long period of time. Therefore, it is important to screen for the best combinations of strains and vectors. Cereal transformants obtained in this way will probably be selected by spraying of progeny plants with herbicides.

### *Integration of Large Segments of DNA*

One of the advantages in *Agrobacterium*-mediated transformation is that large segments of DNA can be transferred to plants. So the question is, how big can such fragments be? Typical lengths of T-DNAs in wild-type *Agrobacterium* strains range between 10 and 23 kb. *Agrobacterium* often carries two distinctive T-DNAs, and the presence of two small T-DNAs might be more advantageous than just one large T-DNA. Working with tobacco, Hamilton et al., (1996) demonstrated that DNA segments of as much as 150 kb in length can be transferred to the plant genome by *Agrobacterium*. They reported that the 150-kb fragment in transformed plants was intact and inherited by the progeny. This size is close to the upper limit of current molecular-cloning techniques. They used specially designed vectors known as binary bacterial artificial chromosome (BIBAC) vectors that allow stable maintenance of large fragments of DNA both in *E. coli* and in *Agrobacterium*. If such vectors can be further improved, larger fragments, such as mega base pair fragments might be transferable. Such technology would certainly open up opportunities for the genetic engineering of cereals and the efficient identification of useful genes in cereals.

### *'Agrolistic' Transformation*

The combination of particle-bombardment methods and *Agrobacterium*-mediated technology is of considerable interest. Classic examples include the wounding of plants by particle bombardment before infection by *Agrobacterium* (Bidney et al., 1992), which resulted in an elevated frequency of transformation. *Agrobacterium* can also be delivered to plant tissues by particle bombardment. Recently, Hansen and Chilton, (1996) reported a new technique that they named 'agrolistic' transformation. They observed that T-strands were generated and integrated into plant chromosomes from plasmids delivered by particle bombardment, with the help of the enzymes that process T-DNA and are expressed in plant cells. They claimed that this technique combines the advantages of *Agrobacterium*-mediated transformation, which include low numbers of integrated copies of transgenes and the high efficiency of the ballistic delivery of DNA. *Agrobacterium*-mediated transformation is a complex process and some of the steps may not always be efficient. Thus, ballistic methods might complement less efficient steps in *Agrobacterium*-mediated transformation in various ways. Such approaches might provide novel options in attempts to develop a technology for transformation of cereals.

## Conclusion

Delivery of foreign DNA to rice and maize plants using *Agrobacterium* has become a routine technique in a growing number of laboratories. This technique should allow the genetic improvement of rice and maize, as well as studies of many aspects of the molecular biology of these crops. Studies to date have provided strong support for the hypothesis that T-DNA is transferred from *Agrobacterium* to dicotyledons and monocotyledons by an identical molecular mechanism. Therefore, the monocotyledonous nature of the plants no longer prevents the application of *Agrobacterium*-mediated techniques to the transfer of genes to certain important cereal crops. The host range of *Agrobacterium*-based vector systems is probably much wider than the host range of crown gall disease, and it might include most higher plants. The first priority in any approach to the transformation of a recalcitrant plant should be the identification of methods for preparation of 'compatible' cells. Within the next few years, protocols for efficient *Agrobacterium*-mediated transformation of many major cereals should become available.

The studies of *Agrobacterium*-mediated transformation of cereals have contributed to our understanding of the nature of non-host responses of plants to infection by phytopathogens. With respect to responses to a plant pathogen, plants can be divided into three categories, namely, susceptible plants, resistant plants, and non-host plants. Molecular mechanisms underlying resistance are now being dissected and quite a few genes responsible for resistance to bacterial, viral and fungal diseases have been cloned from a number of plant species (Bent, 1996). Such studies have focused primarily on the difference between susceptible and resistant plants. By contrast, the biology of the non-host response appears to be far more complex and very little is known about it. *Agrobacterium* is definitely the best-studied plant pathogen, and it provides a rare example in which we already have a quite good understanding of the basis for the difference between host plants and non-host plants. A plant species might be a host when all of the steps in tumorigenesis that we discussed in this chapter are operative and *vice versa*. If the molecular basis of the host/non-host differences for other phytopathogens can be understood to this extent, it might be possible to convert susceptible plants to non-host plants by genetic engineering, which would provide a superior method for the control of diseases in important crops.

## Acknowledgements

The authors thank Yukoh Hiei, Yuji Ishida, Hideaki Saito, Shozo Ohta and Takashi Kumashiro for their valuable advice and for helpful discussions.

## References

Aldemita, R.R., and Hodges, T.K. (1996) *Agrobacterium tumefaciens*-mediated transformation of japonica and indica rice varieties. Planta. 199: 612–617.

Alt-Moerbe, J., Neddermann, P., Von Lintig, J., Weiler, E.W., and Schröder, J. (1988) Temperature-sensitive step in Ti plasmid *vir* region induction and correlation with cytokinin secretion by *Agrobacteria*. Mol. Gen. Genet. 213: 1–8.

Alt-Mörbe, J., Kühmann, H., and Schröder, J. (1989) Differences in induction of Ti-plasmid virulence genes *virG* and *virD*, and continued control of *virD* expression by four external factors. Mol. Plant-Microbe. Interact. 2: 301–308.

An, G. (1985) High efficiency of transformation of cultured tobacco cells. Plant Physiol. 79: 568–570.

An, G., Evert, P.R., Mitra, A., and Ha, S.B. (1988) Binary vectors. In: Gelvin, S.B., and Schilperoort, R.A. (eds), Plant Molecular Biology Manual A3 (pp. 1–19). Kluwer Academic Publishers, Dordrecht.

Anderson, A.R., and Moore, I.W. (1979) Host specificity in the genus *Agrobacterium*. Phyopathol. 69: 320–323.

Ashby, A.M., Watson, M.D., Loake, G.J., and Shaw, C.H. (1988) Ti plasmid-specified chemotaxis of *Agrobacterium tumefaciens* C58Cl toward *vir*-inducing phenolic compounds and soluble factors from monocotyledonous and dicotyledonous plants. J. Bacteriol. 170: 4181–4187.

Ayres, N.M., and Park, W.D. (1994) Genetic transformation of rice. Critical Rev. Plant Sci. 13: 219–239.

Baba, A., Hasezawa, S., and Syono, K. (1986) Cultivation of rice protoplasts and their transformation mediated by *Agrobacterium* spheroplasts. Plant Cell Physiol. 27: 463–471.

Baron, C., and Zambryski, P.C. (1995) The plant response in pathogenesis, symbiosis, and wounding: Variations on a common theme? Annu. Rev. Genetics 29: 107–129.

Beaupré , C.E., Bohne, J., Dale, E.M., and Binns, A.N. (1997) Interaction between VirB9 and VirB10 membrane proteins involved in movement of DNA from *Agrobacterium tumefaciens* into plant cells. J. Bacteriol. 179: 78–89.

Bent, A.F. (1996) Plant disease resistance genes: Function meets structure. Plant Cell 8: 1757–1771.

Bevan, M. (1984) Binary *Agrobacterium* vectors for plant transformation. Nucleic Acids Res. 12: 8711–8721.

Bidney, D., Scelonge, C., Martich, J., Burrus, M., Sims, L., and Huffman, G. (1992) Microprojectile bombardment of plant tissues increases transformation frequency by *Agrobacterium tumefaciens*. Plant Mol. Biol. 18: 301–313.

Bilang, R., Iida, S., Peterhans, A., Potrykus, I., and Paszkowski, J. (1991) The 3'-terminal region of the hygromycin-B-resistance gene is important for its activity in *Escherichia coli* and *Nicotiana tabacum*. Gene 100: 247–250.

Binns, A.N., and Thomashow, M.F. (1988) Cell biology of *Agrobacterium* infection and transformation of plants. Ann. Rev. Microbiol. 42: 575–606.

Bolivar, F., Rodriguez, R.L., Greene, P.J., Betlach, M.C., Heyneker, H.L., Boyer, J.H., Crosa, J.H., and Falkow, S. (1977) Construction and characterization of new cloning vehicles, II. A multi-purpose cloning system. Gene 2: 95–113.

Bolton, G.W., Nester, E.W., and Gordon, M.P. (1986) Plant phenolic compounds induce expression of the *Agrobacterium tumefaciens* loci needed for virulence. Science 232: 983–984.

Bouchez, D., Camilleri, C., and Caboche, M. (1993) A binary vector based on Basta resistance for *in planta* transformation of *Arabidopsis thaliana*. C. R. Acad. Sci. Paris, Life Sciences 316: 1188–1193.

Braun, A.C. (1982) A history of the crown gall problem. In: Kahl, G., and Schell, J. (eds), Molecular Biology of Plant Tumors (pp. 155–210). Academic Press, New York.

Budar, F., Thia-Toong, S., van Montagu, M., and Hernalsteens, J.P. (1986) *Agrobacterium*-mediated gene transfer results mainly in transgenic plants transmitting T-DNA as a single Mendelian factor. Genetics 114: 303–313.

Bundock, P., Den Dulk-Ras, A., Beijersbergen, A., and Hooykaas, P.J.J. (1995) Trans-kingdom T-DNA transfer from *Agrobacterium tumefaciens* to *Saccharomyces cerevisiae*. EMBO J. 13: 3206–3214.

Bundock, P., and Hooykaas, P.J.J. (1996) Integration of *Agrobacterium tumefaciens* T-DNA in the *Saccharomyces cerevisiae* genome by illegitimate recombination. Proc. Natl. Acad. Sci. USA 93: 15272–15275.

Bytebier, B., Deboeck, F., De Greve, H., Van Montagu, M., and Hernalsteens, J.P. (1987) T-DNA organization in tumor cultures and transgenic plants of monocotyledon *Asparagus officinalis*. Proc. Natl. Acad. Sci. USA 84: 5345–5349.

Cangelosi, G.A., Ankenbauer, R.G., and Nester, E.W. (1990) Sugars induce the *Agrobacterium* virulence genes through a periplasmic binding protein and transmembrane signal protein. Proc. Natl. Acad. Sci. USA 87: 6708–6712.

Chan, M.T., Chang, H.H., Ho, S.L., Tong, W.F., and Yu, S.M. (1993) *Agrobacterium*-mediated production of transgenic rice plants expressing a chimeric $\alpha$-amylase promoter/$\beta$-glucuronidase gene. Plant Mol. Biol. 22: 491–506.

Chan, M.T., Lee, T.M., and Chang, H.H. (1992) Transformation of indica rice (*Oryza sativa* L.) mediated by *Agrobacterium tumefaciens*. Plant Cell Physiol. 33: 577–583.

Chang, S.S., Park, S.K., Kim, B.C., Kang, B.J., Kim, D.U., and Nam, H.G. (1994) Stable genetic transformation of *Arabidopsis thaliana* by *Agrobacterium* inoculation *in planta*. Plant J. 5: 551–558.

Cheng, M., Fry, J.E., Pang, S., Zhou, H., Hironaka, C.M., Duncan, D.R., Conner, T.W., and Wan, Y. (1997) Genetic transformation of wheat mediated by *Agrobacterium tumefaciens*. Plant Physiol. 115: 971–980.

Chilton, M.D., Currier, T.C., Farrand, S.K., Bendich, A.J., Gordon, M.P., and Nester, E.W. (1974) *Agrobacterium tumefaciens* DNA and PS8 bacteriophage DNA not detected in crown gall tumors. Proc. Natl. Acad. Sci. USA 71: 3672–3676.

Christou, P., Ford, T.L., and Kofron, M. (1992) The development of a variety-independent gene-transfer method for rice. TIBTECH 10: 239–246.

Christou, P., Ford, T.L., and Kofron, M. (1991) Production of transgenic rice (*Oryza sativa* L.) from agronomically important indica and japonica varieties via electric discharge particle acceleration of exogenous DNA into immature zygotic embryos. Bio/technology 9: 957–962.

Citovsky, V., Zupan, J., Warnick, D., and Zambryski, P. (1992) Nuclear localization of *Agrobacterium* VirE2 protein in plant cells. Science 256: 1802–1805.

Dale, P.J., Marks, M.S., Brown, M.M., Woolston, C.J., Gunn, H.V., Mullineaux, P.M., Lewis, D.M., Kemp, J.M., Chen, D.F., Gilmour, D.M., and Flavell, R.B. (1989) Agroinfection of wheat: Inoculation of *in vitro* grown seedlings and embryo. Plant Sci. 69: 237–245.

Dang, T.A.T., and Christie, P.J. (1997) The VirB4 ATPase of *Agrobacterium tumefaciens* is a cytoplasmic membrane protein exposed at the periplasmic surface. J. Bacteriol. 179: 453–462.

Dasgupta, I., Hull, R., Eastop, S., Poggi-Pollini, C., Blakebrough, M., Boulton, M.I., and Davies, J.W. (1991) Rice tungro bacilliform virus DNA independently infects rice after *Agrobacterium*-mediated transfer. J. Gen. Virology 72: 1215–1221.

De Block, M., and Debrouwer, D. (1991) Two T-DNA's co-transformed into *Brassica napus* by a double *Agrobacterium tumefaciens* infection are mainly integrated at the same locus. Theor. Appl. Genet. 82: 257–263.

De Cleene, M., and De Ley, J. (1976) The host range of crown gall. Bot. Rev. 42: 389–466.

De Framond, A.J., Back, E.W., Chilton, W.S., Kayes, L., and Chilton, M.-D. (1986) Two unlinked T-DNAs can transform the same tobacco plant cell and segregate in the F1 generation. Mol. Gen. Genet. 202: 125–131.

Dekeyser, R., Claes, B., Marichal, M., Van Montagu, M., and Caplan, A. (1989) Evaluation of selectable markers for rice transformation. Plant Physiol. 90: 217–223.

Delbreil, B., Guerche, P., and Jullien, M. (1993) *Agrobacterium*-mediated transformation of *Asparagus officinalis* L. long-term embryogenic callus and regeneration of transgenic plants. Plant Cell Rep. 12: 129–132.

Depicker, A., Herman, L., Jacobs, A., Schell, J., and Van Montagu, M. (1985) Frequencies of simultaneous transformation with different T-DNAs and their relevance to the *Agrobacterium*/plant cell interaction. Mol. Gen. Genet. 201: 477–484.

Dommisse, E.M., Leung, D.W.M., Shaw, M.L., and Conner, A.J. (1990) Onion is a monocotyledonous host for *Agrobacterium*. Plant Sci. 69: 249–257.

Dong, J., Teng, W., Buchholz, W.G., and Hall, T.C. (1996) *Agrobacterium*-mediated transformation of Javanica rice. Molecular Breeding 2: 267–276.

Douglas, C., Halperin, W., Gordon, M., and Nester, E. (1985) Specific attachment of *Agrobacterium tumefaciens* to bamboo cells in suspension cultures. J. Bacteriol. 161: 764–766.

Escudero, J., Neuhaus, G., Schläppi, M., and Hohn, B. (1996) T-DNA transfer in meristematic cells of maize provided with intracellular *Agrobacterium*. Plant J. 10: 355–360.

Feldmann, K.A., and Marks, M.D. (1987) *Agrobacterium*-mediated transformation of germinating seeds of *Arabidopsis thaliana*: A non-tissue culture approach. Mol. Gen. Genet. 208: 1–9.

Finnegan, J., and McElroy, D. (1994) Transgenic inactivation: plants fight back! Bio/technology 12: 883–888.

Flavell, R.B. (1994) Inactivation of gene expression in plants as a consequence of specific sequence duplication. Proc. Natl. Acad. Sci. USA 91: 3490–3496.

Fraley, R.T., Rogers, S.G., and Horsch, R.B. (1986) Genetic transformation in higher plants. Crit. Rev. Plant Sci. 4: 1–46.

Fraley, R.T., Rogers, S.G., Horsch, R.B., Eicholtz, D.A., and Flick, J.S. (1985) The SEV system: a new disarmed Ti plasmid vector for plant transformation. Bio/technology 3: 629–635.

Fraley, R.T., Rogers, S.G., Horsch, R.B., Sanders, P.R., Flick, J.S., Adams, S.P., Bittner, M.L., Brand, L.A., Fink, C.L., Fry, J.S., Galluppi, G.R., Goldberg, S.B., Hoffmann, N.L., and Woo, S.C. (1983) Expression of bacterial genes in plant cells. Proc. Natl. Acad. Sci. USA 80: 4803–4807.

Fromm, M.E., Taylor, L.P., and Walbot, V. (1985) Expression of genes transferred into monocot and dicot plant cells by electroporation. Proc. Natl. Acad. Sci. USA 882: 5824–5828.

Fullner, K.J., and Nester, E.W. (1996) Temperature affects the T-DNA transfer machinery of *Agrobacterium tumefaciens*. J. Bacteriol. 178: 1498–1504.

Glaszmann, J.C. (1987) Isozymes and classification of Asian rice varieties. Theor. Appl. Genet. 74: 21–30.

Godwin, I., Todd, G., Ford-Lloyd, B., and Newbury, H.J. (1991) The effects of acetosyringone and pH on *Agrobacterium*-mediated transformation vary according to plant species. Plant Cell Rep. 9: 671–675.

Gordon-Kamm, W.J., Spencer, T.M., Mangano, M.L., Adams, T.R., Daines, R.J., Start, W.G., O'Brien, J.V., Chambers, S.A., Adams, J., W.R., Willetts, N.G., Rice, T.B., Mackey, C.J., Krueger, R.W., Kausch, A.P., and Lemaux, P.G. (1990) Transformation of maize cells and regeneration of fertile transgenic plants. Plant Cell 2: 603–618.

Gould, J., Devey, M., Hasegawa, O., Ulian, E.C., Peterson, G., and Smith, R.H. (1991) Transformation of *Zea mays* L. using *Agrobacterium tumefaciens* and the shoot apex. Plant Physiol. 95: 426–434.

Graves, A.C.F., and Goldman, S.L. (1987) *Agrobacterium tumefaciens*-mediated transformation of the monocot genus *Gladiolus*: Detection of expression of T-DNA-encoded genes. J. Bacteriol. 169: 1745–1746.

Graves, A.C.F., and Goldman, S.L. (1986) The transformation of *Zea mays* seedlings with *Agrobacterium tumefaciens*. Plant Mol. Biol. 7: 43–50.

Graves, A.E., Goldman, S.L., Banks, S.W., and Graves, A.C.F. (1988) Scanning electron microscope studies of *Agrobacterium tumefaciens* attachment to *Zea mays*, *Gladiolus* sp., and *Triticum aestivum*. J. Bacteriol. 170: 2395–2400.

Grimsley, N., Hohn, B., Hohn, T., and Walden, R. (1986) 'Agroinfection,' an alternative route for viral infection of plants by using the Ti plasmid. Proc. Natl. Acad. Sci. USA 83: 3282–3286.

Grimsley, N., Hohn, B., Ramos, C., Kado, C., and Rogowsky, P. (1989) DNA transfer from *Agrobacterium* to *Zea mays* or *Brassica* by agroinfection is dependent on bacterial virulence functions. Mol. Gen. Genet. 217: 309–316.

Grimsley, N., Hohn, T., Davies, J.W., and Hohn, B. (1987) *Agrobacterium*-mediated delivery of infectious maize streak virus into maize plants. Nature 325: 177–179.

Grimsley, N.H., Ramos, C., Hein, T., and Hohn, B. (1988) Meristematic tissues of maize plants are most susceptible to agroinfection with maize streak virus. Bio/technology 6: 185–189.

Görlich, D., Kraft, R., Kostka, S., Vogel, F., Hartmann, E., Laskey, R.A., Mattaj, I.W., and Izaurralde, E. (1996) Importin provides a link between nuclear protein import and U snRNA export. Cell 87: 21–32.

Görlich, D., and Mattaj, I.W. (1996) Nucleocytoplasmic transport. Science 271: 1513–1518.

Hamilton, C.M., Frary, A., Lewis, C., and Tanksley, S.D. (1996) Stable transfer of intact high molecular weight DNA into plant chromosomes. Proc. Natl. Acad. Sci. USA 93: 9975–9979.

Hansen, G., and Chilton, M.D. (1996) 'Agrolistic' transformation of plant cells: Integration of T-strands generated *in planta*. Proc. Natl. Acad. Sci. USA 93: 14978–14983.

Hauptmann, R.M., Ozias-Akins, P., Vasil, V., Tabaeizadeh, Z., Rogers, S. G., Horsch, R.B., Vasil, I.K., and Fraley, R.T. (1987) Transient expression of electroporated DNA in monocotyledonous and dicotyledonous species. Plant Cell Rep. 6: 265–270.

Hernalsteens, J.P., Thia-Toong, L., Schell, J., and Van Montagu, M. (1984) An *Agrobacterium*-transformed cell culture from the monocot *Asparagus officinalis.* EMBO J. 13: 3039–3041.

Hiei, Y., and Komari, T. (1996) Stable inheritance of transgenes in rice plants transformed by *Agrobacterium tumefaciens*. In: Khush, G.S. (ed), Rice Genetics III. Proceedings of the Third International Rice Genetics Symposium. Manila (Philippines) (pp. 131–142). International Rice Research Institute, Los Baños, Philippines.

Hiei, Y., Ohta, S., Komari, T., and Kumashiro, T. (1994) Efficient transformation of rice (*Oryza sativa* L.) mediated by *Agrobacterium* and sequence analysis of the boundaries of the T-DNA. Plant J. 6: 271–282.

Hoekema, A., Hirsch, P.R., Hooykaas, P.J.J., and Schilperoort, R.A. (1983) A binary plant vector strategy based on separation of *vir*- and T-region of the *Agrobacterium tumefaciens* Ti-plasmid. Nature 303: 179–180.

Hohn, B., Koukolíková-Nicla, Z., Bakkeren, G., and Grimsley, N. (1989) *Agrobacterium*-mediated gene transfer to monocots and dicots. Genome 31: 987–991.

Hood, E.E., Fraley, R.T., and Chilton, M.D. (1987) Virulence of *Agrobacterium tumefaciens* strain A281 on legumes. Plant Physiol. 83: 529–534.

Hood, E.E., Gelvin, S.B., Melchers, L.S., and Hoekema, A. (1993) New *Agrobacterium* helper plasmids for gene transfer to plants. Transgenic Res. 2: 208–218.

Hood, E.E., Helmer, G.L., Fraley, R.T., and Chilton, M.D. (1986) The hypervirulence of *Agrobacterium tumefaciens* A281 is encoded in a region of pTiBo542 outside of T-DNA. J. Bacteriol. 168: 1291–1301.

Hood, E.E., Jen, G., Kayes, L., Kramer, J., Fraley, R.T., and Chilton, MD. (1984) Restriction endonuclease map of pTiBo542, a potential Ti-plasmid vector for genetic engineering of plants. Bio/technology 2: 702–709.

Hooykaas, P.J.J. (1989) Transformation of plant cells via *Agrobacterium.* Plant Mol. Biol. 13: 327–336.

Hooykaas, P.J.J., and Beijersbergen, A.G.M. (1994) The virulence system of *Agrobacterium tumefaciens*. Ann. Rev. Phytopathol. 32: 157–179.

Hooykaas, P.J.J., and Schilperoort, R.A. (1992) *Agrobacterium* and plant genetic engineering. Plant Mol. Biol. 19: 15–38.

Hooykaas-Van Slogteren, G.M.S., Hooykaas, P.J.J., and Schilperoort, R.A. (1984) Expression of Ti plasmid genes in monocotyledonous plants infected with *Agrobacterium tumefaciens*. Nature 311: 763–764.

Huffman, G.A., White, F.F., Gordon, M.P., and Nester, E.W. (1984) Hairy-root-inducing plasmid: Physical map and homology to tumor-inducing plasmids. J. Bacteriol. 157: 269–276.

Iida, A., Yamashita, Y., Yamada, Y., and Morikawa, H. (1991) Efficiency of particle-bombardment-mediated transformation is influenced by cell stage in synchronized cultured cells of tobacco. Plant Physiol. 97: 1585–1587.

Ishida, Y., Saito, H., Ohta, S., Hiei, Y., Komari, T., and Kumashiro, T. (1996) High efficiency transformation of maize (*Zea mays* L.) mediated by *Agrobacterium tumefaciens*. Nature Biotechnol. 14: 745–750.

Janssen, B.J., and Gardner, R.C. (1989) Localized transient expression of GUS in leaf discs following cocultivation with *Agrobacterium*. Plant Mol. Biol. 14: 61–72.

Jarchow, E., Grimsley, N.H., and Hohn, B. (1991) *virF*, the host-range-determining virulence gene of *Agrobacterium tumefaciens*, affects T-DNA transfer to *Zea mays*. Proc. Natl. Acad. Sci. USA 88: 10426–10430.

Jefferson, R.A. (1987) Assaying chimeric genes in plants: the GUS gene fusion system. Plant Mol. Biol. Rep. 5: 387–405.

Jin, S., Komari, T., Gordon, M.P., and Nester, E.W. (1987) Genes responsible for the supervirulence phenotype of *Agrobacterium tumefaciens* A281. J. Bacteriol. 169: 4417–4425.

Kahl, G. (1982) Molecular biology of wound healing: The conditioning phenomenon. In: Kahl, G., and Schell, J. (eds), Molecular Biology of Plant Tumors (pp. 211–267). Academic Press, New York.

Kapila, J., Rycke, R.D., Van Montagu, M., and Angenon, G. (1997) An *Agrobacterium*-mediated transient gene expression system for intact leaves. Plant Sci. 122: 101–108.

Katavic, V., Haughn, G.W., Reed, D., Martin, M., and Kunst, L. (1994) *In planta* transformation of *Arabidopsis thaliana*. Mol. Gen. Genet. 245: 363–370.

Kilby, N.J., Leyser, H.M.O., and Furner, I.J. (1992) Promoter methylation and progressive transgenic inactivation in *Arabidopsis*. Plant Mol. Biol. 20: 103–112.

Klee, H.J., Horsch, R., and Rogers, S. (1987) *Agrobacterium*-mediated plant transformation and its further applications to plant biology. Ann. Rev. Plant Physiol. 38: 467–486.

Klee, H.J., White, F.F., Iyer, V.N., Gordon, M.P., and Nester, E.W. (1983) Mutational analysis of the virulence region of an *Agrobacterium tumefaciens* Ti plasmid. J. Bacteriol. 153: 878–883.

Kohli, A., Ghareyazie, B., Kim, H.S., Kush, G.S., and Bennett, J. (1996) Cytosine methylation implicated in silencing of $\beta$-glucuronidase genes in transgenic rice. In: Khush, G.S. (ed), Rice Genetics III. Proceedings of the Third International Rice Genetics Symposium. Manila (Philippines) (pp. 825–828). International Rice Research Institute, Los Baños, Philippines.

Komari, T. (1989) Transformation of callus cultures of nine plant species mediated by *Agrobacterium*. Plant Sci. 60: 223–229.

Komari, T. (1990) Transformation of cultured cells of *Chenopodium quinoa* by binary vectors that carry a fragment of DNA from the virulence region of pTiBo542. Plant Cell Rep. 9: 303–306.

Komari, T., Halperin, W., and Nester, E.W. (1986) Physical and functional map of supervirulent *Agrobacterium tumefaciens* tumor-inducing plasmid pTiBo542. J. Bacteriol. 166: 88–94.

Komari, T., Hiei, Y., Saito, Y., Murai, N., and Kumashiro, T. (1996) Vectors carrying two separate T-DNAs for co-transformation of higher plants mediated by *Agrobacterium tumefaciens* and segregation of transformants free from selection markers. Plant J. 10: 165–174.

Koncz, C., Németh, K., Pédei, G.P., and Schell, J. (1994) Homology recognition during T-DNA integration into the plant genome. In: Paszkowski, J. (ed), Homologous Recombination and Gene Silencing in Plants (pp. 167–189). Kluwer Academic, Dordrecht.

Kudirka, D.T., Colburn, S.M., Hinchee, M.A., and Wright, M.S. (1986) Interactions of *Agrobacterium tumefaciens* with soybean (*Glycine max* Merr.) leaf explants in tissue cultures. Can. J. Genet. Cytol. 28: 808–817.

Langeveld, S.A., Gerrits, M.M., Derks, A.F.L.M., Boonekamp, P.M., and Bol, J.F. (1995) Transformation of lily by *Agrobacterium*. Euphytica 85: 97–100.
Langridge, P., Brettschneider, R., Lazzeri, P., and Lörz, H. (1992) Transformation of cereals via *Agrobacterium* and the pollen pathway: a critical assessment. Plant J. 2: 631–638.
Lessl, M., and Lanka, E. (1994) Common mechanisms in bacterial conjugation and Ti-mediated T-DNA transfer to plant cells. Cell 77: 321–324.
Li, H.Q., Sautter, C., Potrykus, I., and Puonti-Kaerlas, J. (1996) Genetic transformation of cassava (*Manihot esculenta* Crantz). Nature Biotechnol. 14: 736–740.
Li, X.Q., Liu, C.N., Ritchie, S.W., Peng, J.Y., Gelvin, S.B., and Hodges, T. K. (1992) Factors influencing *Agrobacterium*-mediated transient expression of *gusA* in rice. Plant Mol. Biol. 20: 1037–1048.
Lippincott, J.A., and Lippincott, B.B. (1978) Cell walls of crown-gall tumors and embryonic plant tissues lack *Agrobacterium* adherence sites. Science 199: 1075–1078.
Lippincott, J. A., and Lippincott, B.B. (1975) The genus *Agrobacterium* and plant tumorigenesis. Ann. Rev. Microbiol. 29: 377–405.
Liu, C.N., Li, X.Q., and Gelvin, S.B. (1992) Multiple copies of *virG* enhance the transient transformation of celery, carrot and rice tissues by *Agrobacterium tumefaciens*. Plant Mol. Biol. 20: 1071–1087.
Lyznik, L.A., Ryan, R.D., Ritchie, S.W., and Hodges, T.K. (1989) Stable co-transformation of maize protoplasts with *gusA* and *neo* genes. Plant Mol. Biol. 13: 151–161.
Matzk, A., Mantell, S., and Schiemann, J. (1996) Localization of persisting Agrobacteria in transgenic tobacco plants. Mol. Plant-Microbe Interact. 9: 373–381.
Matzke, M.A., and Matzke, A.J.M. (1995) How and why do plants inactivate homologous (trans) genes? Plant Physiol. 107: 679–689.
May, G.D., Afza, R., Mason, H.S., Wiecko, A., Novak, F.J., and Arntzen, C.J. (1995) Generation of transgenic banana (*Musa acuminata*) plants via *Agrobacterium*-mediated transformation. Bio/technol. 13: 486–492.
McKnight, T.D., Lillis, M.T., and Simpson, R.B. (1987) Segregation of genes transferred to one plant cell from two separate *Agrobacterium* strains. Plant Mol. Biol. 8: 439–445.
Messens, E., Dekeyser, R., and Stachel, S.E. (1990) A nontransformable *Triticum monococcum* monocotyledonous culture produces the potent *Agrobacterium vir*-inducing compound ethyl ferulate. Proc. Natl. Acad. Sci. USA 87: 4368–4372.
Mooney, P.A., and Goodwin, P.B. (1991) Adherence of *Agrobacterium tumefaciens* to the cells of immature wheat embryos. Plant Cell Tissue Organ Cult. 25: 199–208.
Mooney, P.A., Goodwin, P.B., Dennis, E.S., and Llewellyn, D.J. (1991) *Agrobacterium tumefaciens*-gene transfer into wheat tissues. Plant Cell Tissue Organ Cult. 25: 209–218.
Narasimhulu, S.B., Deng, X.B., Sarria, R., and Gelvin, S.B. (1996) Early transcription of *Agrobacterium* T-DNA genes in tobacco and maize. Plant Cell 8: 873–886.
Nester, E.W., Gordon, M.P., and Yanofsky, M.F. (1984) Crown gall: A molecular and physiological analysis. Annu. Rev. Plant Physiol. 35: 387–413.
Nester, E.W., and Kosuge, T. (1981) Plasmids specifying plant hyperplasias. Annu. Rev. Microbiol. 35: 531–565.
Ohta, S., Mita, S., Hattori, T., and Nakamura, K. (1990) Construction and expression in tobacco of a $\beta$-glucuronidase (GUS) reporter gene containing an intron within the coding sequence. Plant Cell Physiol. 31: 805–813.
Okada, K., Takebe, I., and Nagata, T. (1986) Expression and integration of genes introduced into highly synchronized plant protoplasts. Mol. Gen. Genet. 205: 398–403.
Park, S.H., Pinson, S.R.M., and Smith, R.H. (1996) T-DNA integration into genomic DNA of rice following *Agrobacterium* inoculation of isolated shoot apices. Plant Mol. Biol. 32: 1135–1148.
Park, Y.D., Papp, I., Moscone, E.A., Iglesias, V.A., Vaucheret, H., Matzke, A.J.M., and Matzke, M.A. (1996) Gene silencing mediated by promoter homology occurs at the level

of transcription and results in meiotically heritable alterations in methylation and gene activity. Plant J. 9: 183–194.

Piers, K.L., Heath, J.D., Liang, X., Stephens, K.M., and Nester, E.W. (1996) *Agrobacterium tumefaciens*-mediated transformation of yeast. Proc. Natl. Acad. Sci. USA 93: 1613–1618.

Potrykus, I. (1990b) Gene transfer to cereals: An assessment. Bio/technology 8: 535–542.

Potrykus, I. (1990a) Gene transfer to plants: Assessment and perspectives. Physiol. Plant. 79: 125–134.

Potrykus, I. (1991) Gene transfer to plants: Assessment of published approaches and results. Annu. Rev. Plant Physiol. Plant Mol. Biol. 42: 205–225.

Primich-Zachwieja, S., and Minocha, S.C. (1991) Induction of virulence response in *Agrobacterium tumefaciens* by tissue explants of various plant species. Plant Cell Rep. 10: 545–549.

Raineri, D.M., Bottino, P., Gordon, M.P., and Nester, E.W. (1990) *Agrobacterium*-mediated transformation of rice (*Oryza sativa* L.). Bio/technology 8: 33–38.

Raineri, D.M., Boulton, M.I., Davies, J.W., and Nester, E.W. (1993) VirA, the plant-signal receptor, is responsible for the Ti plasmid-specific transfer of DNA to maize by *Agrobacterium*. Proc. Natl. Acad. Sci. USA 90: 3549–3553.

Rao, S.S., Lippincott, B.B., and Lippincott, J.A. (1982) *Agrobacterium* adherence involves the pectic portion of the host cell wall and is sensitive to the degree of pectin methylation. Physiol. Plant. 56: 374–380.

Rashid, H., Yokoi, S., Toriyama, K., and Hinata, K. (1996) Transgenic plant production mediated by *Agrobacterium* in *Indica* rice. Plant Cell Rep. 15: 727–730.

Rashkova, S., Spudich, G.M., and Christie, P.J. (1997) Characterization of membrane and protein interaction determinants of the *Agrobacterium tumefaciens* VirB11 ATPase. J. Bacteriol. 179: 583–591.

Rathore, K.S., Chowdhury, V.K., and Hodges, T.K. (1993) Use of *bar* as a selectable marker gene and for the production of herbicide-resistant rice plants from protoplasts. Plant Mol. Biol. 21: 871–884.

Ream, W. (1989) *Agrobacterium tumefaciens* and interkingdom genetic exchange. Annu. Rev. Phytopathol. 27: 583–618.

Ritchie, S.W., Lui, C.N., Sellmer, J.C., Kononowicz, H., Hodges, T.K., and Gelvin, S.B. (1993) *Agrobacterium tumefaciens*-mediated expression of *gusA* in maize tissues. Transgenic Res. 2: 252–265.

Rossi, L., Hohn, B., and Tinland, B. (1996) Integration of complete transferred DNA units is dependent on the activity of virulence E2 protein of *Agrobacterium tumefaciens*. Proc. Natl. Acad. Sci. USA 93: 126–130.

Sahi, S.V., Chilton, M.D., and Chilton, W. S. (1990) Corn metabolites affect growth and virulence of *Agrobacterium tumefaciens*. Proc. Natl. Acad. Sci. USA 87: 3879–3883.

Saito, Y., Komari, T., Masuta, C., Hayashi, Y., Kumashiro, T., and Takanami, Y. (1992) Cucumber mosaic virus-tolerant transgenic tomato plants expressing a satellite RNA. Theor. Appl. Genet. 83: 679–683.

Schläppi, M., and Hohn, B. (1992) Competence of immature maize embryos for *Agrobacterium*-mediated gene transfer. Plant Cell 4: 7–16.

Schocher, R.J., Shillito, R.D., Saul, M.W., Paszkowski, J., and Potrykus, I. (1986) Co-transformation of unlinked foreign genes into plants by direct gene transfer. Bio/technology 4: 1093–1096.

Schäfer, W., Görz, A., and Kahl, G. (1987) T-DNA integration and expression in a monocot crop plant after induction of *Agrobacterium*. Nature 327: 529–532.

Sheikholeslam, S.N., and Weeks, D.P. (1987) Acetosyringone promotes high efficiency transformation of *Arabidopsis thaliana* explants by *Agrobacterium tumefaciens*. Plant Mol. Biol. 8: 291–298.

Shen, W.H., Escudero, J., Schläppi, M., Ramos, C., Hohn, B., and Koukolíková-Nicola, Z. (1993) T-DNA transfer to maize cells: Histochemical investigation of $\beta$-glucuronidase activity in maize tissues. Proc. Natl. Acad. Sci. USA 90: 1488–1492.

Sheng, J., and Citovsky, V. (1996) *Agrobacterium*-plant cell DNA transport: Have virulence proteins, will travel. Plant Cell 8: 1699–1710.

Shimamoto, K., Terada, R., Izawa, T., and Fujimoto, H. (1989) Fertile transgenic rice plants regenerated from transformed protoplasts. Nature 338: 274–276.

Shimoda, N., Toyoda-Yamamoto, A., Nagamine, J., Usami, S., Katayama, M., Sakagami, Y., and Machida, Y. (1990) Control of expression of *Agrobacterium vir* genes by synergistic actions of phenolic signal molecules and monosaccharides. Proc. Natl. Acad. Sci. USA 87: 6684–6688.

Simpson, R.B., Spielmann, A., Margossian, L., and McKnight, T.D. (1986) A disarmed binary vector from *Agrobacterium tumefaciens* functions in *Agrobacterium rhizogenes*. Plant Mol. Biol. 6: 403–415.

Smith, R.H., and Hood, E.E. (1995) *Agrobacterium tumefaciens* transformation of monocotyledons. Crop Sci. 35: 301–309.

Sonti, R.V., Chiurazzi, M., Wong, D., Davies, C.S., Harlow, G.R., and Mount, D.W. (1995) *Arabidopsis* mutants deficient in T-DNA integration. Proc. Natl. Acad. Sci. USA 92: 11786–11790.

Spencer, T.M., Gordon-Kamm, W.J., Daines, R.J., Start, W.G., and Lemaux, P.G. (1990) Bialaphos selection of stable transformants from maize cell culture. Theor. Appl. Genet. 79: 625–631.

Spencer, T.M., O'Brien, J.V., Start, W.G., Adams, T.R., Gordon-Kamm, W. J., and Lemaux, P.G. (1992) Segregation of transgenes in maize. Plant Mol. Biol. 18: 201–210.

Stachel, S.E., Messens, E., Van Montagu, M., and Zambryski, P. (1985) Identification of the signal molecules produced by wounded plant cells that activate T-DNA transfer in *Agrobacterium tumefaciens*. Nature 318: 624–628.

Stachel, S.E., and Nester, E.W. (1986) The genetic and transcriptional organization of the *vir* region of the A6 Ti plasmid of *Agrobacterium tumefaciens*. EMBO J. 5: 1445–1454.

Tingay, S., McElroy, D., Kalla, R., Fieg, S., Wang, M., Thornton, S., and Brettell, R. (1997) *Agrobacterium tumefaciens*-mediated barley transformation. Plant J. 11: 1369–1376.

Tinland, B., Schoumacher, F., Gloeckler, V., Bravo-Angel, A.M., and Hohn, B. (1995) The *Agrobacterium tumefaciens* virulence D2 protein is responsible for precise integration of T-DNA into the plant genome. EMBO J. 14: 3585–3595.

Toki, S. (1997) Rapid and efficient *Agrobacterium*-mediated transformation in rice. Plant Mol. Biol. Rep. 15: 16–21.

Toki, S., Takamatsu, S., Nojiri, C., Ooba, S., Anzai, H., Iwata, M., Christensen, A. H., Quail, P.H., and Uchimiya, H. (1992) Expression of a maize ubiquitin gene promoter-*bar* chimeric gene in transgenic rice plants. Plant Physiol. 100: 1503–1507.

Toriyama, K., Arimoto, Y., Uchimiya, H., and Hinata, K. (1988) Transgenic rice plants after direct gene transfer into protoplasts. Bio/technology 6: 1072–1074.

Turk, S.C.H.J., Melchers, L.S., Den Dulk-Ras, H., Regensburg-Tuïnk, A.J.G., and Hooykaas, P.J.J. (1991) Environmental conditions differentially affect *vir* gene induction in different *Agrobacterium* strains. Role of the VirA sensor protein. Plant Mol. Biol. 16: 1051–1059.

Usami, S., Morikawa, S., Takebe, I., and Machida, Y. (1987) Absence in monocotyledonous plants of the *diffusible* plant factors inducing T-DNA circularization and *vir* gene expression in *Agrobacterium*. Mol. Gen. Genet. 209: 221–226.

Usami, S., Okamoto, S., Takebe, I., and Machida, Y. (1988) Factor inducing *Agrobacterium tumefaciens vir* gene expression is present in monocotyledonous plants. Proc. Natl. Acad. Sci. USA 85: 3748–3752.

Valvekens, D., van Montagu, M., and van Lijsebettens, M. (1988) *Agrobacterium tumefaciens*-mediated transformation of *Arabidopsis thaliana* root explants by using kanamycin selection. Proc. Natl. Acad. Sci. USA 85: 5536–5540.

Van Wordragen, M.F., and Dons, H.J.M. (1992) *Agrobacterium tumefaciens*-mediated transformation of recalcitrant crops. Plant Mol. Biol. Rep. 10: 12–36.

Vancanneyt, G., Schmidt, R., O'Connor-Sanchez, A., Willmtzer, L., and Rocha-Sosa, M. (1990) Construction of an intron-containing marker gene: Splicing of the intron in transgenic plants and its use in monitoring early events in *Agrobacterium*-mediated plant transformation. Mol. Gen. Genet. 220: 245–250.

Vasil, I.K. (1994) Molecular improvement of cereals. Plant Mol. Biol. 25: 925–937.

Vasil, I.K. (1996) Phosphinothricin-resistant crops. In: Duke, S.O. (ed), Herbicide-resistanct Crops (pp. 85–91). CRC Press, Inc.

Veluthambi, K., Krishnan, M., Gould, J.H., Smith, R.H., and Gelvin, S.B. (1989) Opines stimulate induction of the *vir* genes of the *Agrobacterium tumefaciens* Ti plasmid. J. Bacteriol. 171: 3696–3703.

Vijayachandra, K., Palanichelvam, K., and Veluthambi, K. (1995) Rice scutellum induces *Agrobacterium tumefaciens vir* genes and T-strand generation. Plant Mol. Biol. 29: 125–133.

Walters, D.A., Vetsch, C.S., Potts, D.E., and Lundquist, R.C. (1992) Transformation and inheritance of a hygromycin phosphotransferase gene in maize plants. Plant Mol. Biol. 18: 189–200.

Watson, B., Currier, T.C., Gordon, M.P., Chilton, M.D., and Nester, E.W. (1975) Plasmid required for virulence of *Agrobacterium tumefaciens*. J. Bacteriol. 123: 255–264.

Wilmink, A., and Dons, J.J.M. (1993) Selective agents and marker genes for use in transformation of monocotyledonous plants. Plant Mol. Biol. Rep. 11: 165–185.

Wullems, G.J., Molendijk, L., Ooms, G., and Schilperoort, R.A. (1981) Differential expression of crown gall tumor markers in transformants obtained after *in vitro Agrobacterium tumefaciens*-induced transformation of cell wall-regenerating protoplasts derived from *Nicotiana tabacum*. Proc. Natl. Acad. Sci. USA 78: 4344–4348.

Yadav, N.S., Vanderleyden, J., Bennett, D.R., Barnes, W.M., and Chilton, M.D. (1982) Short direct repeats flank the T-DNA on a nopaline Ti plasmid. Proc. Natl. Acad. Sci. USA 79: 6322–6326.

Yoder, J.I., and Goldsbrough, A.P. (1994) Transformation systems for generating marker-free transgenic plants. Bio/technology 12: 263–267.

Zambryski, P. (1988) Basic processes underlying *Agrobacterium*-mediated DNA transfer to plant cells. Ann. Rev. Genet. 22: 1–30.

Zambryski, P. (1992) Chronicles from the *Agrobacterium*-plant cell DNA transfer story. Ann. Rev. Plant Physiol. 43: 465–490.

Zambryski, P., Depicker, A., Kruger, K., and Goodman, H.M. (1982) Tumor induction by *Agrobacterium tumefaciens*: analysis of the boundaries of T-DNA. J. Mol. Appl. Genet. 1: 361–370.

Zambryski, P., Joos, H., Genetello, C., Leemans, J., van Montagu, M., and Schell, J. (1983) Ti plasmid vector for the introduction of DNA into plant cells without alteration of their normal regeneration capacity. EMBO J. 2: 2143–2150.

Zhang, W., McElroy, D., and Wu, R. (1991) Analysis of rice *Act1* 5' region activity in transgenic rice plants. Plant Cell 3: 1155–1165.

Zupan, J.R., and Zambryski, P. (1995) Transfer of T-DNA from *Agrobacterium* to Plant Cell. Plant Physiol. 107: 1041–1047.

# 5. Expression and Regulation of Transgenes for Selection of Transformants and Modification of Traits in Cereals*

MONICA BÅGA[1], RAVINDRA N. CHIBBAR and KUTTY K. KARTHA

*Plant Biotechnology Institute, National Research Council of Canada, 110 Gymnasium Place, Saskatoon, Saskatchewan, Canada S7N 0W9. E-mail: rchibbar@pbi.nrc.ca*

[1]*Department of Crop Science and Plant Ecology, University of Saskatchewan, 51 Campus Drive, Saskatoon, Saskatchewan, Canada S7N 5A8.*

ABSTRACT. To effectively use genetic transformation strategies to improve agricultural crops, it is vital to understand the underlying factors that control transgene expression and stability. This information will be required to generate transgenic crops with a stable, predictable and consistent performance. We will in this review discuss factors, carried or encoded by the transgene expression cassette, that affect gene expression levels and patterns in transgenic cereals.

## 1. Introduction

The development of genetic transformation techniques for most of the important cereals has opened up new possibilities for genetic improvement of these crops. Among the various conditions that affect the outcome of a plant transformation process are factors involved in control and stability of transgenes. Efficient regulation of gene expression is important during the entire course of a transformation procedure in order to: (i) efficiently select transformed cells or tissues, (ii) regenerate healthy transgenic plants with desired phenotype, and (iii) maintain stability and expression pattern of the inserted gene. Temporal and spatial programs of transgene expression are mainly directed by interactions between host regulatory factors and different control elements present on or encoded by the transgene. However, the behavior of foreign genes when inserted into the plant genome can be very complex and unpredictable during development and in response to environmental stimuli. The future application of molecular biology for improvement of agronomic performance in crops will depend largely on revealing the molecular mechanisms involved in different aspects of gene regulation.

*NRCC number: 40725

*I.K. Vasil (ed.), Molecular Improvement of Cereal Crops, 83–131*

We will, in the following sections, discuss factors that are currently known to be of importance for regulation of gene expression in cereals. These factors include choice of genes, promoters, regulatory sequences, leader sequences, codon usage, introns, and polyadenylation sequences. In addition, we will also consider conditions that may affect the stability and maintenance of the transgene when it is integrated into the host genome.

## 2. Marker Genes

The first two steps of a genetic transformation protocol involve delivery of a gene cassette into recipient cells, followed by expression of delivered genes. The result of these events can be assessed by assays of transient expression of a reporter gene introduced into plant cell cultures or intact plant tissues. These analyses do not require insertion of the transgene into the host genome, and are frequently used to test promoter and gene functions. The ideal reporter gene encodes a protein, preferably with an enzymatic activity, that is easy to quantify within a reasonable time (1–48 hours) after DNA delivery. It is also preferable that the marker protein shows low background activity in plants, does not affect plant development, is relatively resistant to proteases, and is stable over a wide range of pH values and temperatures. Analysis of gene expression in stably transformed plants requires a marker gene system with a relatively low detection limit. When markers are monitored for changes in gene expression, it is required that the marker gene transcript and encoded protein have short half-lives. Only a few of the currently available reporter genes encode proteins that meet all of the above mentioned criteria. In the following paragraphs we will describe some of the commonly used reporter gene systems for development of cereal transformation protocols, and for expression analysis of promoter function and tissue-specificity in transgenic cereals.

*The chloramphenicol acetyltransferase gene*

Early studies on cereal transformation used the chloramphenicol acetyltransferase (CAT; EC 2.3.1.28) enzyme as a scorable marker to monitor promoter function, and to optimize gene delivery parameters (for examples see Bruce et al., 1989, Callis et al., 1987, Fromm et al., 1985, Hauptmann et al., 1987, Kartha et al., 1989, Lee et al., 1989). The CAT gene is isolated from the *Escherichia coli* transposon Tn*9* and encodes conversion of the antibiotic chloramphenicol into acetylated derivatives, which all lack antibiotic activity. Both enzymatic and immuno-chemical assays are available to measure CAT activity and amount of CAT protein, respectively, as an index of CAT gene expression (Chibbar et al., 1991, Seed and Sheen 1988). These assays are sensitive, but suffer from the drawbacks of being tedious and destructive. Plants contain low amounts of endogenous CAT-like activities, or CAT-inhibiting substances, which interfere with the CAT assay (Chibbar et al., 1991).

*The β-glucuronidase gene*

The *Eschericia coli* gene, *uidA* (Jefferson et al., 1986), encoding the enzyme β-glucuronidase (GUS; EC 3.2.1.31), has been the most widely used reporter gene for the analysis of function and tissue-specificity of promoters in plant systems including cereals. Protein targeting studies with cereal genes have also employed the GUS marker (Chan et al., 1994). The usefulness of the GUS protein is partly due to the ability of the enzyme to tolerate extensions at both the amino- and the carboxy-termini (Datla et al., 1991, Jefferson 1987), which facilitates construction of hybrid genes. The GUS enzyme hydrolyses β-glucuronide compounds, many of which are commercially available, and gives reaction products that can be quantified by spectrophotometric or spectrofluorometric methods (Jefferson et al., 1986, 1987). Histochemical detection of *uidA* expression in plant material is performed by staining with the indigogenic substrate 5-bromo-4-chloro-3-indolyl-1-glucuronide, which is converted into a clearly visible blue precipitate as a result of GUS activity (Jefferson et al., 1986). However, GUS assays with intact tissues or organs have been reported to be leaky, because production of the blue color is not always restricted to cells where *uidA* is expressed (Bowen 1992). This limits the use of the GUS marker for cell-type expression studies. Another complication with the GUS system is related to the high stability of the enzyme, which causes a non-linear correlation between protein accumulation and rates of transcription and translation (Hensgens et al., 1992, Jefferson et al., 1987). The initial studies performed using the GUS marker gene suggested that plants contain very little or no endogenous glucuronidase activity (Jefferson et al., 1986). However, over the years, it has become evident that most plant species have intrinsic glucuronidase-like activities, especially in reproductive tissues, that obscure measurements of low GUS activity levels (Hänsch et al., 1995, Hu et al., 1990, Terada and Shimamoto, 1990). Various procedures to eliminate these interfering activities from plant extracts have been presented, and include removal of low molecular weight fluorescent compounds by gel filtration (Chibbar et al., 1991, 1993), or specific inactivation of endogenous GUS-like activities by a 55°C heat-treatment (Hänsch et al., 1995) or incubation of the enzymatic reaction in the presence of 20% methanol (Kosugi et al., 1990). Despite the above-mentioned drawbacks with the GUS system, it has been very useful for optimizing genetic transformation parameters in order to develop stable transformation protocols for most cereals.

*The anthocyanin genes*

A set of marker systems used in cereals include genes that stimulate the endogenous anthocyanin accumulation that results from deposition of red and purple pigments in the vacuoles of plant tissues (Ludwig and Wessler, 1990, Marrs et al., 1995). The colored phenotype can be visualized *in vivo* and followed throughout development. This marker system needs a functional anthocyanin pathway in the host plant, and cereals such as maize and barley have so far been shown to fulfill this requirement (Klein et al., 1989, Olsen et al., 1993). At least 20 loci are involved in the anthocyanin production in maize, and among these are genes encoding *trans*-acting factors of the *myc*-like and the *myb*-like classes (Coe et al., 1988, Dooner et al., 1991). The anthocyanin biosynthetic genes of transgenic plants are induced by introduction and expression of one or more members of the regulatory genes, which will confer a distinct, tissue-specific and developmentally regulated pigmentation pattern in maize (Lloyd et al., 1992, Ludwig and Wessler, 1990, Lusardi et al., 1994). For example, production of the *C1* and the *R* gene products stimulate transcription of a UDP glucose flavonol 3-*O*-glucosyl transferase gene, denoted *Bronze-1 (Bz1)* in maize, which encodes one of the last enzymes in the anthocyanin biosynthesis pathway. The *Bz1* gene product induces production of a pigment in tissues that are normally not colored in maize (Goff et al., 1990), giving a pigmentation pattern that is determined by the specific combination of expressed *trans*-acting factors. The advantage of the anthocyanin marker gene lies with its non-destructive detection system and lack of a substrate requirement, making it a suitable marker for tissue-specific and cell fate studies during development (Ludwig and Wessler, 1990, Lusardi et al., 1994). However, the detection limit for anthocyanin production is higher than for GUS activity.

*The firefly luciferase gene*

The gene encoding luciferase activity (LUC; EC 1.13.12.7) in the common North American firefly, *Photinus pyralis* (de Wet et al., 1987), has been the most frequently used luciferase marker for monitoring gene expression in cereals (Callis et al., 1987, Fromm et al., 1990). The firefly luciferase catalyses adenosine triphosphate (ATP)-dependent oxidation of the substrate luciferin to give a concomitant release of photons at 560 nm (McElroy and DeLuca, 1978). Light production is quantified with high sensitivity in a luminometer or a liquid scintillation counter, upon addition of $Mg^{2+}$-ATP and luciferin to cell extracts. Alternatively, luciferase activity can be determined *in vivo* by non-destructive methods using a photon-counting low light video camera (Millar et al., 1992). The luciferase is normally targeted to peroxisomes, but targeting can be blocked by mutations of the carboxy-terminal three amino acids without loss of enzymatic activity (Barnes

1990). The enzyme can also be directed to other sub-cellular compartments by linkage to alternative targeting signals (Schneider et al., 1990). Variability in uptake and distribution of the luciferin substrate in different plant tissues or organelles may affect the luciferase-mediated light production (Schneider et al., 1990). Other drawbacks with the luciferase marker system include the rather expensive equipment needed to quantify light production and the relatively large variability in luciferase assays caused by the short half-life of the enzyme. Maize plants transformed with the luciferase gene driven by the cauliflower mosaic virus (CaMV) 35S promoter-*Adh1* intron 1 cassette have been generated and shown to produce luciferase activity in leaves (Fromm et al., 1990). The luciferase marker could also be used to study temporal or tissue-specific gene expression patterns, as demonstrated in tobacco and *Arabidopsis* (Schneider et al., 1990, Millar et al., 1992).

*The green fluorescent protein gene*

A reporter gene system employing the green-fluorescent protein (GFP) from the jellyfish, *Aequorea victoria*, has recently been developed (Chalfie et al., 1994, Sheen et al., 1995) and found to function in plants (Pang et al., 1996, Sheen et al., 1995). The bioluminescent jellyfish produces green light when GFP is energized by blue light emitted by a $Ca^{2+}$-aequorin complex (Shimomura et al., 1962). The GFP is highly stable for 24 hours or more in maize protoplasts (Sheen et al., 1995) and shows tolerance to temperatures up to 65°C, pH values from 2 to 11, 1% sodium dodecyl sulfate, 6*M* guanidium chloride, and is relatively resistant to most proteases (Cubitt et al., 1995).

Production of the marker protein in transgenic plants can be visualized by fluorescence microscopy as bright green fluorescence after excitation of the plant tissues with near UV or blue light. GFP absorbs light maximally at 395 nm, with a minor peak at 470 nm, and emits maximum green light at 509 nm, with a shoulder peak at 540 nm. No exogenously added substrates or co-factors are needed for detection of fluorescence, which is very stable and gives virtually no photo-bleaching. Emission of fluorescence does not interfere with cell viability and the marker gene can, therefore, be used to study gene expression *in vivo*. The active site of the GFP is composed of a $Ser_{65}$-$Tyr_{66}$-$Gly_{67}$ tripeptide, which is cyclized and oxidized, to form a *p*-hydroxybenzylidene-imidazo-lidinone chromophore. The reaction occurs in many cellular environments and appears to be independent of enzymes or co-factors (Heim et al., 1994). A gene encoding a modified GFP with excitation maximum at 382 nm and emission maximum at 448 nm has been produced by mutating the $Tyr_{66}$ codon to a His codon (Heim et al., 1994). Thus, two forms of GFP exist giving green or blue fluorescence, respectively, which permits simultaneous use of two GFP reporter genes.

The first transgenic plants expressing the GFP gene produced a low level of fluorescence, which limited the full exploitation of the marker for plant research. Recently, modified GFP genes have been developed and reported to be highly active in plants (Haseloff et al., 1997, Pang et al., 1996, Reichel et al., 1996, Rouwendal et al., 1997). A 150-fold improvement in GFP production was achieved by replacement of the $Ser_{65}$ codon with a threonine or cysteine codon, and incorporation of the potato ST-LS1 second intron into the *gfp* coding region in the vicinity of a 5′ cryptic splicing site (Pang et al., 1996). Fluorescence produced by expression of the modified GFP genes in transgenic wheat, maize, tobacco, and *Arabidopsis* could readily be detected using a long-wave ultraviolet lamp (Pang et al., 1996).

To summarize, GFP has great potential as a marker for gene expression or protein transport studies in plants during plant growth and development, because the emitted fluorescence can easily be monitored. Therefore, this marker could also be useful for identification and selection of transformed cells or tissues, eliminating the need for selectable marker genes. Among all the reporter genes discussed above, the modified GFP marker gene seems to be the most useful, because it encodes most of the desired features of an ideal marker.

## 3. Selectable Marker Genes

Identification and selection of transformed cells and tissues can be based on two different characteristics conferred by the transformant (Bowen, 1993). These are: (i) traits that confer cell viability to transformed cells in the presence of a selective agent, and (ii) traits that confer a distinguishing phenotype to transformed cells or tissues. Both strategies require a high level of selectable marker gene expression during the early stages of the regeneration program to expeditiously select transformed cells or tissues, and to minimize the time spent in tissue culture. A high level of marker gene expression is provided by the use of constitutive promoters, which have played an important role in the development of genetic transformation protocols for plants in general, and cereals in particular. A number of selectable marker genes have been used to produce transgenic plants (Bowen, 1993), and of these, only a few have been effective in selection of transformed cereals. Some of the commonly used selectable marker genes will be discussed in the following sections.

### *3.1 Markers that Confer Cell Viability under Selective Conditions*

Selection programs based on viability of transformed cells are generally more efficient than selection based on phenotype, because the transformed cells have a growth advantage over non-transformed cells. Two different types of selectable marker genes can be used to confer cell viability to transformed cells: (i) genes encoding antibiotic resistance, and (ii) genes encoding herbicide resistance. Antibiotics and herbicides used in regeneration schemes cause the death of untransformed cells by interfering with basic mechanisms of cell division and growth. The selectable marker gene product provides protection of transformed cells by any of three strategies: (i) detoxification of the selective agent by enzymatic modification, (ii) production of a modified target with reduced affinity for the selective agent, or (iii) over-expression of the target. In practice, selectable markers that use the first two mechanisms to alleviate the effect of the selective agent have been more effective than markers that use the third mechanism. The usefulness of each agent for selection of transformed cells or tissues varies between different cereals. Some cereals show high levels of resistance or tolerance to certain antibiotics or herbicides, rendering these agents unsuitable for a selection-regeneration protocol. Most selective agents have, to varying extents, a negative effect on the regeneration of transformed cells into plants. Therefore, the selection system most useful for a certain cereal needs to be tested empirically.

*Antibiotic resistance markers*

The aminoglycoside group of antibiotics that includes kanamycin, geniticin (G418), hygromycin and paromomycin has been widely used in cereal transformation. These antibiotics bind to the 30S ribosomal subunit, thus inhibiting translation in both bacteria and plastids of eukaryotic cells. Prokaryotes carrying resistance genes for these antibiotics inactivate the drugs by acetylation of amino groups or modification of hydroxyl groups by phosphorylation or adenylation (Benveniste and Davies, 1973). Only resistance genes encoding phosphorylation of aminoglycosidic antibiotics (Beck et al., 1982, Tenover et al., 1989) have been used as selectable marker genes in the production of transgenic plants.

The neomycin phosphotransferase (APHII; EC 2.7.1.95) gene (*nptII*) of the Tn*5* transposon is frequently used as a selectable marker gene in dicot plant transformation, and has also been employed to develop transgenic cereals. Kanamycin, as a selection agent has been used in the production of transformed maize plants (D'Halluin et al., 1992, Rhodes et al., 1988). Several factors have made kanamycin less useful in selection of transformed wheat, barley and rice, such as inherent high levels of natural resistance (Dekeyser et al., 1989, Hauptmann et al., 1988), negative effects on regeneration of rice plants (Battraw and Hall, 1992, Zhang et al., 1988) and high levels of albinism in barley (Ritala et al., 1994). Because of these problems, other aminoglycosidic antibiotics inactivated by NPTII have been tested as selection agents. Geneticin was found to be effective in selection of *nptII*-transformed wheat (Nehra et al., 1994), barley (Funatsuki et al., 1995, Weir et al., 1996), rice (Chan et al., 1993) and tritordeum (Barcelo et al., 1994). Paromomycin has successfully been used to select transgenic oat (Torbert et al., 1995) and wheat (Båga & Chibbar, unpublished) plants.

Resistance to the antibiotic hygromycin B is conferred by the *Escherichia coli hph* gene product (EC 2.7.1.119), which detoxifies the drug by 4-O-phosphorylation (Rao et al., 1983). The hygromycin selection system has been very effective in rice transformation, where plants resistant to the antibiotic have been regenerated from electroporated protoplasts (Shimamoto et al., 1989), bombarded immature zygotic embryos (Christou et al., 1991) and embryogenic callus transformed by *Agrobacterium tumefaciens* (Hiei et al., 1994). Transgenic maize (Walters et al., 1992) and wheat (Ortiz et al., 1996) have also been produced through selection on media containing hygromycin.

The use of the methotrexate/dihydrofolate reductase (MTX/DHFR) selection system in plant transformation was first demonstrated in tobacco and turnip cells (Brisson et al., 1984, De Block et al., 1984), and the grass *Panicum maximum* (Hauptmann et al., 1988). The folate analogue MTX binds to the catalytic sites of DHFR (EC 1.5.1.3.), which catalyzes the NADPH-dependent reduction of folate to tetrahydrofolate, an essential cofactor in the synthesis of glycine, purines and thymidine (Blakley, 1969). Protection from MTX action in transgenic plants is provided by expression of bacterial or animal DHFR genes. Transgenic maize resistant to MTX was obtained by transformation with a modified mouse DHFR gene (Golovkin et al., 1993).

*Herbicide resistance markers*

Several selection strategies based on resistance to commonly used broad-spectrum herbicides have been developed for transformation of plants. The use of herbicide resistance genes for cereal transformation may be preferred, with regard to the concerns about presence of antibiotic-resistance genes in human food products. This selection strategy also provides the regenerated plant with a useful agronomic trait for the control of weeds. However, the use of herbicide resistant genes as selectable markers is not advised for oat and sorghum (Vasil, 1994), which easily cross-fertilize with their respective wild relatives, wild oat species and Johnson grass (*Sorghum halipanse*).

Bialaphos (L-phosphinothricin; Glufosinate-ammonium) is a naturally occurring antibiotic synthesized by *Streptomyces hygroscopicus* (Ogawa et al., 1973). The drug consists of a tripeptide made up from two L-alanine molecules and a L-glutamic acid analogue called phosphinothricin (PPT). The antibiotic effect is mediated by PPT, which inhibits glutamine synthetase, a key enzyme involved in conversion of L-glutamate to L-glutamine in plant cells. Prevention of L-glutamine synthesis in plants leads to a concomitant accumulation of ammonia and cell death (Tachibana et al.,

1986). Resistance to L-PPT or bialaphos is mediated by the enzyme phosphinothricin acetyl transferase (PAT), which inactivates the PPT molecule by acetylation. The *bar* gene isolated from *Streptomyces hygroscopicus* encodes PAT, and has successfully been used as a selectable marker gene for most cereals (Vasil, 1996), including wheat (Becker et al., 1994, Nehra et al., 1994, Vasil et al., 1992, Weeks et al., 1993), rice (Christou et al., 1991), maize (Fromm et al., 1990, Gordon-Kamm et al., 1990), rye (Castillo et al., 1994), barley (Wan and Lemaux, 1994), sorghum (Casas et al., 1993), oats (Somers et al., 1992) and triticale (Zimny et al., 1995). Selection of plants resistant to herbicide application is most efficient at moderate L-PPT concentrations and under conditions that lower the metabolic rate of explants (De Block et al., 1995). This selection strategy also reduces the number of escapes or non-transformed cell lines, which may be a problem with L-PPT or Bialaphos selection (Christou et al., 1991, Dennehey et al., 1994). Selection and regeneration on L-PPT media was reported to result in a higher number of transformed maize plants as compared to using kanamycin resistance as a selection strategy (Omirulleh et al., 1993, Register et al., 1994).

Glyphosate (phosphomethylglycine) inhibits the activity of the plastid-localized enzyme 5-enolpyruvylshikimate-3-phosphate synthase (EPSPS); EC 2.5.1.19), a key enzyme in the biosynthesis of the aromatic amino acids phenylalanine, tyrosine and tryptophan. Resistance to glyphosate is obtained by modification of the target site or by inactivation of the herbicide. A gene coding for a glyphosate-tolerant form of EPSPS has been isolated from the *Agrobacterium* strain CP4 (Barry et al., 1992, Padgette et al., 1991). Inactivation of the glyphosate molecule is catalyzed by glyphosate oxido-reductase (GOX) encoded by a gene from *Achromobacter* sp. strain LBAA. GOX catalyses cleavage of the C-N bond of glyphosate to produce amino-methylphosphonic acid (AMPA) and glyoxylate (Barry et al., 1992). A synthetic GOX gene with an optimized sequence for plant expression is used as a herbicide resistance gene in plant transformation. Glyphosate tolerance in plants is achieved by targeting CP4 EPSPS and GOX to chloroplasts.

Transgenic wheat callus expressing a glyphosate-resistant EPSPS was first obtained by Vasil et al., (1991). Transformation of wheat cells with the CP4 EPSPS gene and/or the GOX gene, followed by selection on glyphosate-containing media, allows for rapid production of transgenic cell lines (Qureshi, Jordan and Chibbar, personal communication). Glyphosate has also been used as a selective agent to produce glyphosate-resistant transgenic wheat plants (Zhou et al., 1995). A limitation with the use of glyphosate as both a selectable and scorable marker is that a complex procedure is required for assay of EPSPS and GOX activities.

The sulfonylureas and imidazolinones class of herbicides inhibit acetolactate synthase (ALS; EC 4.1.3.18) activity in bacteria, yeast and plants (Chaleff and Mauvais 1984, Falco and Dumas, 1985, Ray, 1984, Shaner et al., 1984). The ALS enzyme is involved in biosynthesis of the branched chain amino acids leucine, isoleucine and valine. Resistance to sulfonylureas in plants can be induced by overexpression of the ALS gene, or a gene encoding a sulfonylurea-resistant form of ALS. Expression of sulfonylurea-resistant ALS (Haughn and Somerville, 1986) was used to select and regenerate sulfonylurea-resistant maize (Fromm et al., 1990) and rice (Li et al., 1992).

The number of genes that can be successfully introduced into cereals is currently restricted, due to the limitation in available selectable marker genes. Therefore, to allow pyramiding of multiple traits into transgenic cereals, it would be desirable to develop transformation methods that do not require selectable markers, or methods that will allow removal of the selectable marker gene from the genome of the transgenic plant. Such techniques will also satisfy the public concern about safety and the environmental impact of antibiotic- or herbicide-resistance genes in crop plants (Flavell et al., 1992, Yoder and Goldsbrough, 1994). Production of transgenic cereals without selection is possible (Ritala et al., 1994, Zhang and Wu, 1988), but requires an intensive screening of regenerated plants. This approach is, therefore, only feasible for small-scale production of transgenic cereals, or when the transgenic and the non-transgenic phenotypes can be distinguished. Removal of transgene sequences from dicot plant genomes has been demonstrated by introduction and expression of gene recombination systems, based on the action of transposable elements (Ebinuma et al., 1997; Goldsbrough et al., 1993) or site-specific recombinases (see review by Kilby et al., 1993). Recently, Lyznik et al., (1996) demonstrated removal of an inserted NPTII gene from transgenic maize cells by the use of the yeast FLP/FRT recombination system. The FLP-mediated recombination frequency was 2–3% when the recombinase was transiently expressed in protoplasts of kanamycin-resistant callus.

Another strategy to produce marker-free transgenic cereals involves placement of the marker gene and the gene of interest on two different vectors, which are then simultaneously introduced into plants by direct DNA delivery methods. Although co-transformation frequencies of 50% or higher have been reported in cereals (Barcelo et al., 1994, Vasil et al., 1991, Wan and Lemaux, 1994), it has been difficult to separate the two introduced genes in subsequent generations (Peng et al., 1995, Spencer et al., 1992). This is likely due to insertion of introduced genes at a single locus or at closely linked loci.

Recently, Komari et al., (1996) reported a very promising approach for producing marker-free rice plants. Their strategy is based on separation of the selectable and the non-selectable marker gene by placing them on two separate T-DNA cassettes. Both gene cassettes are carried by the same super binary vector, and introduced into rice by *Agrobacterium*-mediated transformation. Transgenic rice plants showed a 47% co-transformation frequency of the two T-DNA cassettes, and the selectable marker and non-selectable marker were separated in about half of the progeny of the co-transformants. The high frequency of marker-free transgenic rice plants obtained reinforces the notion that *Agrobacterium*-mediated transformation, in contrast to direct DNA delivery methods, does not favor insertion of transgenes as linked loci. This strategy has good potential to gain wide usage as the use of *Agrobacterium*-mediated transformation method is extended to cereals other than wheat, rice, maize, and barley (see chapter by Komari and Kubo, this volume).

### *3.2 Markers that Confer a Distinct Phenotype to Transformed Cells*

Identification of transformed cells by a non-destructive method would be advantageous, as it would not require exposure of transformed cells to a selectable agent, which may negatively affect regeneration into plants. In addition, regenerated plants are free from antibiotic- or herbicide-resistance genes. Cornejo et al., (1993) and Omirulleh et al., (1993) have successfully recovered transgenic rice and maize, respectively, through identification of transformed cells by a gentle GUS staining. Although this approach appears promising, it suffers from very low survival frequencies (1–5%) of identified GUS positive cells (Omirulleh et al., 1993). The main problem lies with accumulation of the indigogenic dye crystals, which reach lethal levels before GUS-positive cells have been identified. The use of this selection strategy also requires that the GUS marker gene is driven by a strong promoter, which may be a disadvantage in a fully regenerated plant. A promising approach would be to use the GFP marker system instead of the GUS marker for screening of putative transformants (Haseloff et al., 1997).

## 4. Promoters

### *4.1 Constitutively Active Promoters used in Development of Transgenic Cereals*

Efficient selection of transgenic cells or plants during a regeneration program requires a high level production of the marker protein to protect transformed cells from the action of the selective agent. This is ensured by using a strong and constitutively active promoter to drive transcription of the selectable marker gene. Constitutive promoters are also used to produce herbicide-, insect- or pathogen-resistant crops, which require ubiquitous expression of the transgene during all stages of plant development. The highly efficient promoters employed in animal or dicot plant systems have generally been isolated from genes involved in fundamental cell processes, or from viral genomes. This is also true for the strong promoters functional in cereal systems, exemplified by the CaMV 35S promoter (Odell et al.,

1985) and the 5′ regions of the rice actin *Act1* (McElroy et al., 1991) and the maize polyubiquitin *Ubi1* (Christensen et al., 1992) genes. DNA fragments encoding these promoter functions are frequently linked to antibiotic- or herbicide-resistance genes to construct selection vectors for cereals (Christensen et al., 1992, Christensen and Quail, 1996, Fromm et al., 1986, 1990, McElroy et al., 1990, 1991). Efficient gene expression in cereals generally requires the presence of homologous or heterologous introns positioned between the promoter and the selectable marker gene. Assays of transient gene expression in cell cultures or intact plant tissues can evaluate the usefulness of an expression vector for monocot species, before the construct is used in stable transformation experiments. These assays are often of value for predicting the promoter strength when inserted into the plant genome (Caplan et al., 1992, Li et al., 1997).

*The rice Act1 promoter*

The rice *Act1 5′* region commonly used to construct chimeric genes includes the 1.3-kb 5′ upstream region, the 5′ non-coding exon 1, intron 1 and the 5′ part of the first coding exon of the *Act1* gene (McElroy et al., 1990). Removal of the first intron from the *Act1* gene eliminates reporter gene expression driven by the *Act1* promoter (McElroy et al., 1990). Localization of *Act1*-GUS gene expression in transgenic rice, wheat and maize has revealed that the *Act1* 5′ region is active in both vegetative and reproductive tissues (Nehra et al., 1994, Zhang et al., 1991), thus reflecting the universal production of actin in all types of plant cells. The *Act1* 5′ region is one of the most constitutive and active promoters identified for cereal species, and so has been used to express a number of genes conferring resistance to antibiotics and herbicides. For example, transgenic wheat (Nehra et al., 1994), barley (Funatsuki et al., 1995), rice (Zhang et al., 1991) and maize (Zhong et al., 1996) have all been obtained using the *Act1* 5′ region to drive a selectable marker gene.

*The maize Ubi1 promoter*

The *Ubi1* and *Ubi2* promoters are part of two different maize polyubiquitin genes, that are regulated by cell cycle and induced by stress factors such as heat (Christensen et al., 1992, Cornejo et al., 1993, Takimoto et al., 1994). The *Ubi1* 5′ region has similar structural organization to that of the rice *Act1* 5′ region, where the first coding exon is preceded by a non-coding exon and an intron. Expression constructs carrying the *Ubi1* 5′ region include the first exon and intron, although no direct evidence for their role in gene expression has been provided. *Ubi* promoters have been shown to drive high levels of transient gene expression in several cereal and grass species (Taylor et al., 1993, Schledzewski and Mendel, 1994). The *Ubi1* 5′ region is very active in transiently transformed rice protoplasts, and drives gene expression levels that are much higher than those provided by expression cassettes carrying the rice *Act1-Act1* intron 1, the CaMV 35S-*Adh* intron 1 or the maize *Adh-Adh* intron 1 (Cornejo et al., 1993). The very strong activity from the *Ubi1* 5′ region in rapidly dividing cells and regenerable tissues of rice (Cornejo et al., 1993), made selection cassettes based on the *Ubi1* 5′ region very useful for stable transformation of rice (Cornejo et al., 1993, Li et al., 1997, Toki et al., 1992). In addition to rice, transgenic wheat (Vasil et al., 1993, Weeks et al., 1993), rye (Castillo et al., 1994), and barley (Wan and Lemaux, 1994) have also been produced using the *Ubi1* 5′ region linked to selectable marker genes. Rice plants with the *Ubi1* 5′ region linked to the GUS reporter gene show very high levels of GUS activity in young roots, but only moderate levels of marker activity in pollen, leaf vascular tissues and stomata (Cornejo et al., 1993). The promoter activity decreases dramatically when the plants mature, which may be of an advantage for development of crops where the *Ubi1* 5′ region drives expression of antibiotic-resistance genes.

*The CaMV 35S promoter*

Some of the early experiments performed with promoters that were known to be very effective in dicot systems demonstrated very low activity in cereals (Fromm et al., 1985, Hauptmann et al., 1987, Keith and Chua, 1986). For example, the CaMV 35S promoter was 10 times less active in electroporated protoplasts of *Triticum monococcum* as compared to electroporated dicot protoplasts (Hauptmann et al., 1987). The gene expression pattern directed by the CaMV 35S promoter was also reported to differ between monocot and dicot plants (Battraw and Hall, 1990). Expression of the GUS gene under the control of the CaMV 35S promoter was on average 10 times higher in rice leaves as compared to leaves of tobacco transformed with the same construct. In contrast, there was no difference in GUS activity levels in the roots of transgenic rice and tobacco. It was also noted that the CaMV 35S promoter, which is constitutively active in tobacco (Jefferson et al., 1987, Odell et al., 1985), was nonfunctional in certain parts of rice flowers such as the palea, stigma and anthers (Terada and Shimamoto, 1990). The different behavior of the CaMV 35S promoter in dicot and monocot systems is probably due to differences in quantity and/or quality of regulatory factors.

The modular architecture of many promoters (see review by Dynan, 1989) makes it possible to adjust promoter strength and activity pattern, by altering *cis*-acting regulatory DNA elements. The addition of strong enhancer elements, leader sequences or introns to an expression cassette often leads to an increase in promoter activity. For example, the activity from the CaMV 35S promoter in maize protoplasts was strengthened three-fold by a duplication of the 162-bp enhancer element in the 5′ upstream region of the promoter (Omirulleh et al., 1993). A six-fold improvement in CaMV 35S promoter strength in maize cells was obtained when 131 nucleotides of the 35S untranslated leader sequence was included in the expression cassette (Pierce et al., 1987). An even higher activation of the CaMV 35S promoter was observed when various introns were included in the expression cassette (Callis et al., 1987, Chibbar et al., 1993, Clancy et al., 1994, Maas et al., 1991, McElroy et al., 1991, Tanaka et al., 1990, Vasil et al., 1989). For example, positioning the *Act1* intron 1 between the CaMV 35S promoter and the reporter gene increased expression 40-fold in rice protoplasts and 65-fold in maize cells (McElroy et al., 1991). Improved versions of the CaMV 35S promoter containing a duplicated enhancer and/or an intron are often used for construction of cereal vectors (Hayakawa et al., 1992, Pang et al., 1996).

*The pEmu promoter*

A strong and constitutively active promoter, pEmu, was assembled by adding a set of enhancer elements to the 5′ end of a truncated *Adh1* promoter linked to its first intron (Last et al., 1991). The enhancer elements of the reconstituted promoter included six copies of the 41-bp anaerobic responsive element (ARE) from the maize *Adh1* promoter (Walker et al., 1987) and four copies of the 40-bp octopine synthase (OCS) enhancer (Ellis et al., 1987). The pEmu promoter was 10- to 50-fold stronger than the CaMV 35S promoter when tested for transient expression in protoplasts of several cereals (Last et al., 1991) and 400 times stronger than the CaMV 35S promoter in sugarcane protoplasts (Rathus et al., 1993). The pEmu promoter has been used to develop transgenic rice (Chamberlain et al., 1994, Li et al., 1997) and sugarcane (Bower and Birch, 1992).

*Other constitutive promoters*

Other constitutive, but less used, promoters in stable cereal transformation include the cassava vein mosaic virus (CVMV) promoter (Verdaguer et al., 1996), the rice GOS2 (Hensgens et al., 1993), the maize *ZmdJ1* (Baszczynski et al., 1997), and the mannopine synthase (MAS) and the nopaline synthase (NOS) promoters isolated from the Ti plasmid of *Agrobacterium tumefaciens* (Chan et al., 1993, Hensgens et al., 1993, Meijer et al., 1991).

Wilmink et al., (1995) have compiled information on promoter performance in two dicot and six monocot systems. As might be expected, these data suggest that monocot promoters perform better in cereals than in tobacco or carrot. However, an objective comparison of promoter strengths determined in different plant systems is difficult, due to differences in cell lines or tissues used for the analysis, differences in vector constructions, variations in transformation frequencies, etc. Schledzewski and Mendel, (1994) compared promoter strengths by normalizing transient expression assays by including a second reporter gene as an internal standard for transformation efficiency. It was found from their study that the monocot promoters, pEmu, the rice *Act1* 5′ region and the maize *Ubi1* 5′ region, were consistently stronger (6- to 15-fold) than the CaMV 35S promoter in barley and maize cells. The best performance was obtained from pEmu and the *Ubi1* 5′ region in barley cells, whereas the promoter strengths from the *Ubi1* 5′ region and the *Act1* 5′ regions were highest in maize cells. In contrast, the CaMV 35S promoter was more than 10 times stronger than the monocot promoters in tobacco cells.

### *4.2 Tissue-specific Promoters for Cereal Improvement*

Development of new traits in plants by genetic engineering often requires access to promoters that will direct transgene expression in a tissue-specific, developmental or inducible manner. The use of highly specific promoters, as compared to constitutive promoters, will conserve energy needed to express the transgene in tissues where the transgene product is desired. Tissue-specific gene expression patterns are usually established as a result of several interacting processes executed at various levels of gene regulation. Due to the complexity of gene regulatory mechanisms, it may be preferable to use promoters from homologous or closely related plant species for genetic engineering purposes in cereals. The late development of cereal transformation techniques has delayed analysis of cereal promoters in monocot systems. Most of the characterized cereal promoters have, therefore, been analyzed by transformation of dicot plants, or studied by transient gene expression assays of transformed cereal protoplasts or particle-bombarded cereal tissues. However, not all regulated promoters can be analyzed by transient expression (Matsuoka and Sanada, 1991, Russell and Fromm, 1997). Although the expression pattern of many monocot promoters appears to be faithfully reproduced in dicot species (see for example Colot et al., 1987, Lamppa et al., 1985, Marcotte et al., 1989), the study of promoter performance in cereals will, nevertheless, give more relevant information about regulatory sequences, *trans*-acting factors, and temporal and spatial gene expression patterns during plant development. Promoters shown to confer a tissue-specific expression pattern in transgenic cereals are listed in Table 1.

*Seed-specific promoters*

The obvious target for improvement of agricultural performance in cereals are various aspects of grain quality. Most of the main storage proteins and enzymes involved in production or degradation of starch are highly tissue-specific, and display a developmentally regulated expression program during grain filling or germination. The expression of promoters of the main

TABLE 1
Tissue-specific promoters functional in transgenic cereals.

| Promoter | Source | Reporter gene | Transgenic plant | Expression pattern | Reference |
|---|---|---|---|---|---|
| **Seed-specific:** | | | | | |
| Wx | Rice granule-bound starch synthase gene (*Wx*) | *wx cDNA* | Rice | Endosperm, pollen | Itoh et al., 1997 |
| ZmGBS | Maize granule-bound starch synthase gene (*Wx*) | *uidA* | Maize | Endosperm, pollen | Russell & Fromm 1997 |
| OsAGP | Rice small subunit of ADP-glucose pyrophosphorylase gene | *uidA* | Rice | Endosperm | Russell & Fromm 1997 |
| ZmZ27 | Maize zein gene | *uidA* | Maize | Endosperm | Russell & Fromm 1997 |
| Gt1 | Rice glutelin gene | *uidA* | Rice | Endosperm | Zheng et al., 1993 |
| 1Ax1 | Wheat high-molecular-weight glutenin gene | *1Ax1* | Wheat | Endosperm | Altpeter et al., 1996 |
| Dy10 | Wheat high-molecular-weight glutenin gene | *Dy10:Dx5* | Wheat | Endosperm | Blechl & Anderson 1996 |
| Ltp2 | Barley lipid transfer protein gene | *uidA* | Rice | Aleurone | Kalla et al., 1994. |
| **Stress-induced:** | | | | | |
| Adh1-Adh1 intron 1 | Maize alcohol dehydrogenase 1 gene | *uidA* | Rice | Root caps, anthers, scutellum, endosperm, shoot and root meristems of embryo; induced by anaerobisis in roots | Kyozuka et al., 1991 |
| RC24 | Rice basic chitinase gene | *uidA* | Rice | Root and stem; wound-induced | Xu et al., 1996 |
| Osgrp1 | Rice glycine-rich cell-wall protein gene | *uidA* | Rice | Cell differentiation and elongation regions of roots, young leaves and stem; wound-induced | Xu et al., 1995 |

*Continued on next page*

TABLE 1
Continued

| Promoter | Source | Reporter gene | Transgenic plant | Expression pattern | Reference |
|---|---|---|---|---|---|
| Pin2-Act1 intron 1 | Potato proteinase inhibitor II gene | *uidA* | Rice | Vascular tissue; systemically induced by wounding, methyl jasmonate and abscisic acid | Xu et al., 1993 |
| COMT | Maize caffeic acid *O*-methyltransferase gene | *uidA* | Maize | Lignin producing tissues; induced by wounding and by elicitors | Capellades et al., 1996 |
| Vst1 | Grapewine stilbene synthase gene | *vst1* | Rice | Induced by wounding, elicitor treatment and UV irradiation. | Stark-Lorenzen et al., 1997 |
| **Light-regulated:** | | | | | |
| PhyA | Oat phytochrome A gene (*phyA*) | *phyA* | Rice | Etiolated tissues | Clough et al., 1995 |
| LHCP | Rice light harvesting chlorophyll *a/b*-binding protein gene of photosystem II | *uidA* | Rice | Green tissues of leaves, stems and flowers; light-induced | Tada et al., 1991 |
| PEPC | Maize phosphoenolpyruvate carboxylate gene | *uidA* | Rice | Mesophyll cells; light-induced | Matsuoka et al., 1994 |
| PPDK | Maize pyruvate orthophosphate dikinase gene | *uidA* | Rice | Mesophyll cells; light-induced | Matsuoka et al., 1993 |
| RbcS | Maize Rubisco small subunit gene | *uidA* | Rice | Mesophyll cells; light-induced | Matsuoka et al., 1994 |
| RbcS | Rice Rubisco small subunit gene (*rbcS3C*) | uidA | Rice | Mesophyll cells; light-induced | Kyozuka et al., 1993 |
| RbcS | Tomato Rubisco small subunit gene | *uidA* | Rice | Mesophyll cells; light-induced | Kyozuka et al., 1993. |
| **Other:** | | | | | |
| H3 | Wheat histone H3 gene | *uidA* | Rice | Meristems of shoots, roots and young leaves, anther wall, pistil, coleoptile and embryo | Terada et al., 1993 |

*Continued on next page*

TABLE 1
Continued

| Promoter | Source | Reporter gene | Transgenic plant | Expression pattern | Reference |
|---|---|---|---|---|---|
| RolC | Ri plasmid of *Agrobacterium rhizogenes* | *uidA* | Rice | Vascular tissues | Matsuki et al., 1989 |
| RTBV | Rice tungro bacilliform virus | *uidA* | Rice | Phloem | Bhattacharyya-Pakrasi et al., 1993 |
| pca55 | Maize gene | *barnase* | Wheat | Tapetum | De Block et al., 1997 |
| pE1 | Rice gene | *barnase* | Wheat | Tapetum | De Block et al., 1997 |
| pT72 | Rice gene | *barnase* | Wheat | Tapetum | De Block et al., 1997 |
| Osg6B | Rice gene (*osg6B*) | *uidA* | Rice | Tapetum | Yokoi et al., 1997 |
| $\alpha$Amy8 | Rice $\alpha$-amylase gene | *uidA* | Rice | Mature leaves, stems, sheaths and roots | Chan et al., 1993 |
| Cab-6 | Pine chlorophyll *a/b* (Cab) binding protein | *uidA* | Rice | Photosynthetic tissues; light-independent | Yamamoto et al., 1994 |

storage protein genes, such as the wheat high-molecular-weight (HMW) glutenin, the wheat low-molecular-weight (LMW) glutenin, the barley hordein B, the maize zein, and the rice glutelin gene was first studied in heterologous transgenic plants, such as tobacco and petunia (Colot et al., 1987, Marris et al., 1988, Schernthaner et al., 1988, Takaiwa et al., 1991, Thomas and Flavell, 1990, Ueng et al., 1988). Most of the promoters tested directed an endosperm-specific expression pattern that was identical to that in their natural host. However, the rice glutelin *Gt3* promoter or the maize zein gene introduced into tobacco and petunia, respectively, showed some activity in vegetative tissues in addition to seeds (Leisy et al., 1989, Ueng et al., 1988), indicating that the spatial control of these promoters was not functional in the dicot plants. The seed-specific monocot promoters also appeared to be less active in dicot plants as compared to their natural host (Colot et al., 1987, Schernthaner et al., 1988, Zheng et al., 1993). Promoter deletion analysis performed in dicot plants have revealed several DNA elements with possible role in tissue-specific expression (see review by Vellanoweth and Okita, 1993). Most of the *cis*-acting DNA elements were located within 500 bp upstream of the transcriptional start site.

A more accurate determination of regulatory sequences in the 5′ upstream region of the rice *Gt1* promoter active in transgenic rice was reported by Zheng et al., (1993). Their data showed that high level activity, temporal and spatial regulation of the *Gt1* promoter required 5.1 kb of the 5′ upstream region. The *Gt1* promoter was used to direct expression of a phytoene synthase gene from *Narcissus pseudonarcissus* to the endosperm of transgenic rice (Burkhardt et al., 1997). Phytoene synthase is involved in biosynthesis of provitamin A, and development of transgenic rice expressing this enzyme constitutes one of the steps towards improvement of the nutritional quality of milled rice. Promoters from HMW glutenin genes in wheat have been used to express a hybrid glutenin gene (Blechl and Anderson, 1996) and the HMW glutenin subunit 1Ax1 gene (Altpeter et al., 1996) in wheat.

*Stress-induced promoters*

Promoters induced by various environmental factors, such as temperature stress, anaerobic growth conditions or pathogen attack, all have potential for use in the development of crops resistant to these various stress conditions. Temperature-induced promoters found functional in transgenic cereals include a cold-induced promoter from barley (Molina et al., 1996) and a heat-shock-induced promoter (*Gmhsp 17.5-E*) from soybean (Lyznik et al., 1995). Promoter induction by anaerobic stress has been identified for the maize alcohol dehydrogenase-1 (*Adh1*) promoter, which has been intensively studied (Howard et al., 1987, Kyozuka et al., 1991, Zhang and Wu, 1988) and used for construction of various cereal expression vectors (Fromm et al., 1990, Klein et al., 1988). Reporter gene expression driven by the maize *Adh1* promoter linked to its first intron in transgenic rice shows the same tissue-specificity as the endogenous maize *Adh1* gene (Zhang and Wu 1988), which reinforces the notion that promoters from one cereal species are likely to be expressed in the same way in other cereals. The *Adh1* promoter, when linked to a GUS gene, is very active in roots of transgenic rice exposed to anaerobic stress, whereas a low level of marker gene activity is present in maize pollen, root caps and seeds. Very little *Adh1* promoter activity has been observed in leaves under anoxic conditions. Induction of *Adh1* promoter activity in various plant tissues is mediated by an array of sequence elements present within the 5′-flanking region of the promoter (Kyozuka et al., 1991, 1994, Walker et al., 1987). An anaerobic responsive element (ARE) was identified in the –140 to –99 region of the *Adh1* promoter and found to confer induced activity from a truncated CaMV 35S promoter, in response to anoxia in maize protoplasts (Walker et al., 1987). The *Adh1* promoter linked to various marker genes has been introduced into different cereals (Fromm et al., 1990, Nehra et al., 1994; Vasil et al., 1992, Zhang and Wu, 1988).

Promoters induced by elicitors, wounding or pathogen infection are attractive for genetic engineering strategies to enhance plant disease resistance. Wound-induced promoters functional in transgenic cereals have been isolated from a rice basic chitinase gene (*RC24*; Xu et al., 1996), a rice glycine-rich cell-wall protein gene (*Osgrp1*; Xu et al., 1995), a lignin biosynthetic enzyme gene (*COMT*; Capellades et al., 1996) and a potato proteinase inhibitor II gene (*pin2*; Thornburg et al., 1987). The *pin2* promoter linked to the first intron of *Act1* is highly induced in transgenic rice by wounding, methyl jasmonate and abscisic acid (Xu et al., 1993), which are all signals of pathogen attack. Therefore, this promoter could be used to drive expression of insect and/or pathogen resistance genes in cereals (Xu et al., 1993). The *rolC* promoter, which is active in vascular plant tissues, could be part of a strategy to control virus propagation in plants (Matsuki et al., 1989).

*Light-regulated promoters*

Several light-responsive promoters of photosynthetic genes, such as those encoding the light harvesting chlorophyll *a/b* binding protein (LHCP), phosphoenolpyruvate carboxylate (PEPC), pyruvate orthophosphate dikinase (PPDK), and ribulose-1,5-bisphosphate carboxylase (Rubisco) small subunit (RbcS) have been isolated from several plants and studied in transgenic cereals (Kyozuka et al., 1993, Matsuoka et al., 1993, 1994, Tada et al., 1991). Rubisco is the most abundant protein found in plant leaves, which has attracted the use of the *rbcS* 5′ regions for high-level transgene expression in plant photosynthetic tissues, but also for targeting of proteins to the chloroplast. Both the tomato and the rice *rbcS* promoters are able to confer the same light-induced expression pattern on a GUS reporter gene in transgenic rice plants, where the GUS activity is confined to the photosynthetic mesophyll cells (Kyozuka et al., 1993). The level of light-induction is much higher from the monocot promoter, which suggests there are differences in regulatory mechanisms controlling *rbcS* expression in monocot and dicot plants. Due to the differences in distribution of photosynthetic cells in $C_3$ and $C_4$ plants (see review by Nelson and Langdale, 1992), some promoters of photosynthetic genes in $C_4$ plants confer a different cell-type expression pattern or expression level on reporter genes in $C_3$ plants, like rice (Matsuoka et al., 1994). One promoter sensitive to light has so far been isolated and studied in transgenic cereals (Bruce et al., 1989, Clough et al., 1995).

## 5. Modification of Heterologous Genes for Optimization of Gene Expression in Cereals

Gene expression levels in eukaryotes are affected by several factors that act after transcription of the pre-mRNA. These factors are involved in processing and modification of the pre-mRNA, nucleocytoplasmic mRNA transport, translation, and eventually protein modification, transport and folding. The transcriptional and translational machineries in animals and plants are essentially the same when function, regulatory sequences and factors involved are compared. However, different organisms have adapted specific mechanisms for gene regulation. These distinct features may cause problems when genes are transferred from one organism to another, and/or when chimeric constructs are assembled. Genes introduced into plants do not always behave as predicted, because sequences or structures on, or encoded by the transgene are inadvertently misinterpreted by the host. In the following section, we will discuss the influence from various components of a gene expression cassette on post-transcriptional events in cereals, and how these factors should be considered for genetic engineering strategies.

### *5.1 Intron Influence on Gene Expression in Cereals*

The majority of all protein-coding genes in higher eukaryotes contain one or more introns, which play an important role in gene regulation (Buchman and Berg 1988, Simpson et al., 1992). Several studies in cereals have shown that inclusion of an intron between the promoter and the reporter gene can have a positive effect on levels of gene expression (Callis et al., 1987, Clancy et al., 1994, Luehrsen and Walbot, 1991, Mascarenhas et al., 1990, Maas et al., 1991, McElroy et al., 1990, Tanaka et al., 1990, Vasil et al., 1989). The stimulatory effect mediated by certain introns appears to be more prominent in monocot than in dicot plants (Tanaka et al., 1990). Therefore, most of the cereal transformation vectors contain introns, and the most frequently used introns are listed in Table 2. One of the first reports on intron-enhanced gene expression in cereals came from studies on the *Adh1* gene (Callis et al., 1987). These experiments showed that expression of a CAT reporter gene linked to the *Adh1* 3′ polyadenylation region and driven by the *Adh1* promoter, was enhanced 100-fold in stably transformed maize cells, when the *Adh1* intron 1 was included in the expression cassette. Stimulated reporter gene expression was obtained only with a subset of the *Adh1* introns, and only when the stimulatory intron was part of the transcriptional unit, and preferably placed at the 5′ end of the marker gene. The fact that orientation and position of an intron within the expression cassette is crucial for enhancement of gene expression, suggests that introns differ from transcriptional enhancer elements. This hypothesis has been supported by the finding that extensive deletions of intron sequences do not impair intron-mediated enhancement of gene expression (Clancy et al., 1994, Luehrsen and Walbot, 1991, 1994a). The stimulatory effect of the *Adh1* intron 1 on gene expression varies with promoter and reporter gene used (Callis et al., 1987), and is dependent on maize cell type (Gallie and Young, 1994). Some studies have reported negative effects from the presence of an intron on an expression cassette (Last et al., 1991, Maas et al., 1991, McElroy et al., 1991, Rathus et al., 1993, Wilmink et al., 1995). Therefore, intron function cannot always be predicted, but needs to be tested empirically for each expression cassette.

Current evidence in animal and plant systems indicates that intron function is mediated by the splicing process *per se*, or combined with an interrelated mechanism like RNA capping or polyadenylation, that cause an increase in mRNA stability and transport from the nucleus (Callis et al., 1987, Huang and Gorman, 1990, Luehrsen and Walbot, 1991, Mascarenhas et al., 1990). Thus, intron-stimulated gene expression requires that introns are accurately recognized and excised, and these processes may be regulated by various factors. For example, splicing of the stress-induced transcripts of the soybean *Gmhsp26-A* and the maize *Bronze-2 (Bz2)* genes is inhibited when plants are treated with heavy metals (Czarnecka et al., 1988, Marrs and Walbot, 1997).

A few studies in cereals have suggested that certain exon and intron sequences act together to stimulate gene expression (Clancy et al., 1994, Luehrsen and Walbot, 1991, Maas et al., 1991, Mascarenhas et al., 1990). Mascarenhas et al., (1990) demonstrated that the second intron of the maize *Adh1* gene could stimulate heterologous gene expression, only when cognate exon sequences surrounding the intron were present on the expression cassette. The combination of the first exon and intron of the maize *Shrunken-1* gene stimulated reporter gene expression 1000-fold in rice and maize protoplasts (Maas et al., 1991). Since the intron alone increased pro-

TABLE 2
Introns used in cereal expression vectors

| Intron | Promoter | Reporter gene | Cereal tested in | Fold increase[1] | Reference |
|---|---|---|---|---|---|
| Maize *Adh1* intron 1 | CaMV 35S | *cat* | Maize protoplasts | 8- to 21-fold | Callis et al., 1987 |
| exon 1 + intron 2 | CaMV 35S | *cat* | Maize protoplasts | 12-fold | Mascarenhas et al., 1990 |
| intron 6 | CaMV 35S | *cat* | Wheat protoplasts | 31-fold | Oard et al., 1989 |
| Rice *Act1* intron 1 | CaMV 35S | *uidA* | Rice cells | 40-fold | McElroy et al., 1991 |
| | CaMV 35S | uidA | *Maize cells* | 65-fold | McElroy et al., 1991 |
| *Shrunken-1* intron 1 | CaMV 35S | *cat* | Maize protoplasts | 23- to 28-fold | Vasil et al., 1989 |
| *Bronze-1* intron | CaMV 35S | *cat* | Maize protoplasts | 6.2-fold | Callis et al., 1987 |
| Maize *hsp70* intron 1 | CaMV E35S | *pgfp* | Maize protoplasts | ND[2] | Pang et al., 1996 |
| Potato ST-LS1 intron 2 | CaMV E35S | *pgfp* | Maize protoplasts | 1.4-fold | Pang et al., 1996 |
| Castor bean catalase intron 1 | CaMV 35S | *uidA* | Rice calli | 10- to 40-fold | Tanaka et al., 1990 |
| | CaMV 35S | *uidA* | Rice plants | 80- to 90-fold | Tanaka et al., 1990 |
| Soybean phaseolin intron 3 | CaMV 35S | *nptII* | Rice plants | ND[2] | Peterhans et al., 1990. |

[1]Measured relative activity from control constuct lacking intron. [2]Not determined

moter strength 100-fold, and only a 10-fold increase in promoter strength was mediated by the exon, the combined effect from the exon and the intron appeared to be multiplicative. These data support the cumulative evidence from animal systems implying an essential role for exon sequences in splicing of introns (Berget, 1995, Watakabe et al., 1993). RNA processing studies in tobacco protoplasts have suggested that a high AU content and the presence of 5′ and 3′ splice sites are both necessary and sufficient for accurate recognition by the splicing machinery (Goodall and Filipowicz, 1989, 1991). In contrast to animals and yeast, these experiments did not reveal any role for polypyrimidine or conserved branchpoint sequences in splicing of plant introns. However, this suggestion has been challenged by later studies (Liu and Filipowicz, 1996, Simpson et al., 1996). The observation that plant introns have an AU content that is usually 11–19% higher than those of surrounding exons (Goodall and Filipowicz, 1989, White et al., 1992) has led to the hypothesis that the differential AU content of exons and introns, together with the presence of splice sites, serve as recognition signals for RNA processing in plants (Goodall and Filipowicz, 1989, Lou et al., 1993, Luehrsen and Walbot, 1994b). This model has been supported by studies where splicing efficiency is improved when the AU content of the intron, or the GC content of surrounding exons, are increased (Carle-Urioste et al., 1994, 1997, Luehrsen and Walbot, 1994a). Thus, recognition of introns in plants appears to be determined by different factors, such as the AT-content of the intron, the GC-content of exons and the sequence of the splice sites, all acting in concert with each other (Carle-Urioste et al., 1994). This model may also explain how some GC-rich cereal introns (Goodall and Filipowicz, 1991), or introns with unusual splice sites (Båga et al., 1995) are accurately recognized by the splicing machinery in cereals.

Studies on pre-mRNA processing have revealed some incompatibilities between splicing in monocot and dicot species. Some monocot genes are poorly expressed in dicot systems, due to failure or slow processing of the encoded pre-mRNA (Goodall and Filipowicz, 1989, 1991, Keith and Chua, 1986, Tanaka et al., 1990). On the other hand, many dicot introns are perfectly recognized and spliced in monocot systems, exemplified by efficient splicing of the first intron of the castor bean catalase gene in rice plants (Tanaka et al., 1990), and the third intron of the soybean phaseolin gene in rice cells (Peterhans et al., 1990). The observation that the dicot splicing machinery is more discriminating appears to be associated with a higher requirement for AT-rich introns to compensate for the overall lower GC content of these genes, as compared to monocot genes (Goodall and Filipowicz, 1991).

As discussed above, the intron effect is dependent on several interacting factors, such as cell type, type of promoter, reporter gene, exon, intron and splice site sequences. The variation in intron stimulation of gene expression could be the result of the foreign and chimeric nature of the expression cassette. For example, sequences that resemble splice sites or contain a high

AU content, both of which are recognized as RNA processing signals in the plant, may be carried by the expression vector. Luehrsen and Walbot (1994b) showed that the *Bz2* transcript containing an insertion of an AU-rich sequence was improperly processed. New introns, alternative splicing, and premature polyadenylation of the mRNA were obtained as a result of insertion of AU-rich, but not GC-rich, sequences into various regions of the *Bz2* mRNA. To optimize gene expression in cereals, construction of expression cassettes could be aided by employment of computer programs designed to detect possible splice sites and introns (Kleffe et al., 1996).

### *5.2 Enhanced Translation by Leader and Polyadenylation Sequences*

Studies in animals, yeast and plants have shown that mRNA stability, mRNA transport to cytoplasm and translation efficiency are influenced by the 5′- and 3′-untranslated regions (UTR) of the mRNA (see reviews by Abler and Green, 1996, Kozak, 1991, Rothnie, 1996, Sonenberg, 1994). Therefore, the sequences surrounding a coding region on an expression cassette can play an important role in determining the amount of transgene product produced in transgenic plants. The 5′ untranslated regions of several plant RNA viruses have been found to enhance translation of reporter RNA in both animal and plant systems, presumably by recruiting ribosomes to the mRNA and promoting initiation of translation (Gallie et al., 1987a,b). The presence of the 67-bp tobacco mosaic virus (TMV) leader sequence (Ω), at the 5′-end of the GUS mRNA, increases GUS activity three-fold to 11-fold in electroporated maize or rice protoplasts (Gallie et al., 1989). Similarly, enhanced translation is also mediated by the wheat *Em* 5′-UTR, which stimulates gene expression 10-fold in rice protoplasts, when placed between the CaMV 35S promoter and the GUS reporter gene (Marcotte et al., 1989).

Translation in eukaryotes is usually initiated from the first AUG codon on a transcript, except when the first AUG codon is in a poor sequence context (Kozak, 1991). Hensgens et al., (1992) observed that translation of a GUS coding region in tobacco could start at two different AUG codons, separated by 23 codons. The translational start at the first AUG codon was 10 times more efficient than initiation at the second start site in tobacco leaves, whereas in roots, translation was only initiated from the second AUG codon. When the second AUG codon was preceded by an out-of-frame AUG codon, a six times lower level of GUS activity was produced in both roots and leaves of the transgenic tobacco. The presence of short open reading frames in front of the coding region of a reporter gene has also caused negative affects on gene expression levels in other studies (Damiani and Wessler, 1993, Putterill & Gardner, 1989).

Recognition of initiation codon on the leader sequence is affected by sequence context of the AUG codon in animal systems (Kozak, 1989), whereas sequences surrounding the AUG codon in plants have been suggested to have a less important role (Luehrsen and Walbot, 1994c). Compilation of sequences around the start codons of 967 maize, rice, wheat and barley genes currently in the TransTerm database (Dalphin et al., 1997) produces a ${}^{G}/_{A}C^{G}/_{C}AUGGCG$ consensus sequence (Table 3), which corresponds closely to the ${}^{G}/_{A}{}^{C}/_{A}{}^{G}/_{C}AUGG^{C}/_{A}G$ consensus sequence derived from analysis of 85 maize genes (Luehrsen and Walbot, 1994c). As previously noted, U residues around the AUG start codon of cereal genes are avoided and are very rare at the −1 position (Table 3; Luehrsen and Walbot, 1994c). The consensus sequence for 1837 dicot genes present in the TransTerm database yields the $A^{A}/_{C}AAUGGC^{G}/_{U}$ consensus sequence (Table 3), which resembles the consensus AACAAUGGC proposed for plant genes by Lütcke et al., (1987). The consensus sequences of both dicot and monocot genes show that the AUG codon is likely to be followed by an alanine codon (GCN). This was also confirmed by inspection of peptides deduced from the 967 cereal sequences in the TransTerm database, where 52% of the barley and 46% of the rice, maize and wheat encoded protein sequences were initiated by a Met-Ala dipeptide. Inspection of stop codons used in cereal and dicot genes reveals a difference in the preferred termination codon (Table 4). The UGA translational stop codon is most frequent for cereal genes (45.5%), whereas dicot genes prefer the UAA stop codon (45.5%). A cytosine residue following the stop codon is avoided in both monocot and dicot genes.

The signals for mRNA processing and polyadenylation differ significantly between yeast, animals and plants, suggesting that 3′ end formation in these systems occurs by different mechanisms (see review by Rothnie, 1996). This is supported by studies that show that animal poly(A) sequences are non-functional in plants (Hunt et al., 1987). Wu et al., (1994) compared the polyadenylation profile of a CAT mRNA expressed from the CaMV 35S promoter in protoplasts of maize endosperm and tobacco leaves. No difference in polyadenylation in either system was observed when the reporter gene was flanked by the 3′ end of the endosperm-specific zein gene or the 3′ end from the CaMV transcription unit. The authors concluded that the 3′ processing mechanism is the same in different tissues of both monocot and dicot species. So far, there has been no data to suggest any difference between monocot and dicot poly(A) signals in cereals. In fact, most cereal expression vectors carry functional poly(A) sequences, that are derived from genes from the *Agrobacterium* Ti plasmid or dicot genes.

The effect of different polyadenylation sequences on expression levels of a NPTII gene driven by the CaMV 35S promoter in transgenic tobacco was compared by Ingelbrecht et al., (1989). Depending on which polyadenylation sequence was used, a 60-fold difference in NPTII activity was observed. It is interesting to note that different levels of gene stimulation by the polyadenylation sequences were found when the constructs were tested by transient versus stable expression. Very little is known about the effect

TABLE 3

DNA sequence context of dicot and cereal start codons

| | Dicot start codon context | | | | | | | | | Cereal start codon context | | | | | | | | |
|---|---|---|---|---|---|---|---|---|---|---|---|---|---|---|---|---|---|---|
| | **-3** | **-2** | **-1** | **+1** | **+2** | **+3** | **+4** | **+5** | **+6** | **-3** | **-2** | **-1** | **+1** | **+2** | **+3** | **+4** | **+5** | **+6** |
| **G** | 20 | 7 | 16 | 0 | 0 | 100 | 69 | 15 | 27 | 42 | 13 | 31 | 0 | 0 | 100 | 69 | 15 | 51. |
| **A** | 61 | 46 | 48 | 100 | 0 | 0 | 13 | 20 | 22 | 40 | 26 | 24 | 100 | 0 | 0 | 14 | 19 | 12. |
| **U** | 10 | 18 | 13 | 0 | 100 | 0 | 13 | 11 | 41 | 9 | 9 | 4 | 0 | 100 | 0 | 10 | 7 | 18. |
| **C** | 9 | 29 | 23 | 0 | 0 | 0 | 5 | 54 | 10 | 9 | 52 | 41 | 0 | 0 | 0 | 7 | 59 | 19 |
| | **A** | $^{A}/_{C}$ | **A** | **A** | **U** | **G** | **G** | **C** | $^{G}/_{U}$ | $^{G}/_{A}$ | **C** | $^{G}/_{C}$ | **A** | **U** | **G** | **G** | **C** | **G** |

Summary of start codon context of dicot and monocot genes from the TransTerm database (Dalphin et al., 1997). The analysis of the dicot start codons included 932 *Arabidopsis thaliana*, 228 *Glycine max*, 257 *Nicotiana tabacum*, 203 *Pisum sativum* and 217 *Solanum lycopersicum* genes. Cereal start codons were analysis from 332 *Zea mays*, 275 *Oryza sativa*, 149 *Triticum aestivum* and 211 *Hordeum vulgare* genes. DNA sequence of consensus start codon context is shown below

TABLE 4

Stop codon usage in cereal and dicot genes

| | UAAA | UAAC | UAAG | UAAU | UAGA | UAGC | UAGG | UAGU | UGAA | UGAC | UGAG | UGAU |
|---|---|---|---|---|---|---|---|---|---|---|---|---|
| **Cereal** | 8.0 | 2.7 | 10.0 | 7.0 | 11.6 | 3.9 | 5.5 | 5.8 | 11.5 | 2.3 | 14.6 | 17.1 |
| **genes** | | **27.7** | | | | **26.8** | | | | **45.5.** | | |
| **Dicot** | 16.8 | 3.4 | 13.0 | 12.3 | 10.6 | 1.2 | 3.2 | 4.4 | 13.4 | 2.0 | 9.4 | 10.4 |
| **genes** | | **45.5** | | | | **19.4** | | | | **35.1** | | |

Summary of four base stop codon context of dicot and cereal genes from the TransTerm database (Dalphin et al., 1997). The analysis of the dicot stop codons included 1019 *Arabidopsis thaliana*, 278 *Glycine max*, 297 *Nicotiana tabacum*, 235 *Pisum sativum* and 257 *Solanum lycopersicum* genes. Cereal stop codons were analysis from 375 *Zea mays*, 294 *Oryza sativa*, 178 *Triticum aestivum* and 237 *Hordeum vulgare* genes different polyadenylation sequences have on gene expression levels in cereals. An eight-fold increase in transient CAT expression from an *Adh1* promoter-*Adh1* intron 1-CAT fusion was obtained in electroporated maize protoplasts, when the reporter gene was linked to the NOS 3′ end instead of the *Adh1* 3′ end. For some cereal genes, the 3′ sequences have been implicated in tissue-specific expression. For example, it has been suggested that the 3′ region of maize *rbcS* has a role in repression of *rbcS* expression in mesophyll cells (Viret et al., 1994). The 5′ leader and the 3′ UTR of a barley $\alpha$-amylase have both been suggested to direct gene expression to aleurone and endosperm tissues of barley (Gallie and Young, 1994).

### *5.3 Modification of Gene Sequences for Optimal Gene Expression*

Plants genes have, like many genes from other organisms, a biased codon usage and base composition (Campbell and Gowri, 1990, Murray et al., 1989), which may be of importance for expression of foreign genes. Monocot nuclear genes use a set of 38 preferred codons, with the exception of the highly expressed genes, which favor only 32 codons (Campbell and Gowri, 1990). The bias in codon usage for highly active genes in yeast and *Escherichia coli* matches the level of isoaccepting tRNA species in the cells (Bennetzen and Hall, 1982, Sharp and Li, 1986). A correlation has also been found between the tRNA population of the maize endosperm and the codon usage of the zein gene, which encodes the main storage protein in this tissue (Viotti et al., 1978). The high level of glutamine, alanine and leucine codons on the zein transcript is well adapted to the most abundant isoaccepting tRNA species in the maize endosperm, but not to the tRNA population of the embryo. However, little is known about the distribution of tRNA species in various plant tissues or species. Experiments in yeast have shown that the occurrence of many rare codons on a transcript causes destabilization of the transcribed mRNA (Hoekema et al., 1987), but there is no direct evidence in plants that the presence of rare codons renders the transcript unstable (Abler and Green, 1996).

The G + C content of plant genes is high in comparison to genes from most other sources. A 44 to 70% G + C value has been reported for monocot genes, of which the majority are in the 60 to 70% range (Matassi et al., 1989). Genes from dicots have a more narrow distribution of G + C residues, with values ranging from 40 to 56% (Matassi et al., 1989). The main difference in base content of monocot and dicot coding regions relates to an overall higher G + C content at the third position of the codon, especially for highly expressed genes (Fennoy and Bailey-Serres, 1993, Murray et al., 1989). This difference in codon usage between monocot and dicot genes is also seen when homologous sequences are compared (Matassi et al., 1989). It has been explained that the high G + C content in monocot genes correlates with the isochore pattern of the genome (Matassi et al., 1989).

The importance of codon usage and G + C content of coding regions, for optimal gene expression in plants, has been demonstrated by the development of transgenic dicot and monocot plants expressing the *Bacillus thuringiensis* δ-endotoxin (*Bt* toxin) gene. The first transformation experiments with dicot plants using a truncated variant of the AT-rich *Bt* toxin gene under the control of strong promoters, resulted in production of an unstable toxin transcript, and consequently, a low production level of the insect control protein (Fischhoff et al., 1987, Perlak et al., 1990). Perlak et al., (1990) showed that toxin production encoded by a hybrid truncated *Bt* toxin gene, *CryIA(c)*, could be increased more than 50-fold by changing the codon usage and raising the G + C content of the coding region, without altering the encoded protein sequence. These modifications also removed potential RNA processing sequences, such as ATTTA sequences (Ohme-Takagi et al., 1993) and polyadenylation signals. Similar modifications to other *Bt* toxin genes were made to allow efficient transcription and translation in monocot systems (Fujimoto et al., 1993, Koziel et al., 1993, Nayak et al., 1997). A modified *CryIA(b)* gene driven by the CaMV 35S promoter fused to the first intron of the castor bean catalase 1 gene was used to develop insect-resistant japonica rice (Fujimoto et al., 1993). Nayak et al., (1997) obtained insect-resistant indica rice by introduction of a reconstructed *CryIA(c)* gene driven by the *Ubi1* 5′ region. Transgenic maize expressing an improved *CryIA(b)* gene under the control of the constitutive CaMV 35S promoter and tissue-specific promoters, respectively, have also been obtained (Koziel et al., 1993).

Development of transgenic barley expressing a heat-stable (1,3-1,4)-β-glucanase gene during germination was possible by modification of the coding sequence of the transgene (Jensen et al., 1996). The improvement of the hybrid bacterial (1,3-1,4)-β-glucanase gene involved adaptation of the codon usage to that of the barley (1,3-1,4)-β-glucanase isoenzyme EII coding region. Similarly, several different modifications of the AT-rich GFP gene have been made to improve expression levels in plants (Haseloff et al., 1997, Pang et al., 1996, Reichel et al., 1996, Rouwendal et al., 1997). For example, Rouwendal and coworkers (1997) increased GFP expression by removing a cryptic splice site, and by changes at the third position of the codons to increase the G + C content of the gene from 32 to 60 %. It is possible that some of the sequence changes made to the GFP gene had a stabilizing effect on the GFP transcript, similar to that which occurred with modifications made to the *Bt* toxin genes.

## 6. Silencing of Gene Expression in Cereals

The introduction of transgenes into animals or plants has led to some unanticipated findings, where total or variegated silencing of inserted or endogenous genes has been observed (see recent reviews by Baulcombe, 1996, Bingham, 1997, Stam et al., 1997). The unpredictable behavior of transgenes has become a serious concern for development of transgenic cereals, in which gene silencing appears to be a problem (Finnegan and McElroy, 1994, Itoh et al., 1997, Register et al., 1994). To date, no detailed studies on gene silencing mechanisms in monocots have been reported, but it is presumed to occur by similar mechanisms as in animals and dicot plants. Nevertheless, it will be of vital importance to learn more about the phenomenon in order to develop effective strategies for generation of transgenic crops, which require uniform and predictable agricultural performance. The regulatory mechanisms may also reveal how multicopy genes, that are part of polyploid genomes in cereals, have evolved to escape gene inactivation and how gene expression is regulated in tissues of different ploidy levels. It is interesting to note that in bread wheat, a hexaploid species, stability of transgene integration and expression has been shown for several sexual generations (Altpeter et al., 1996, Blechl and Anderson, 1996, Srivastava et al., 1996, Vasil and Anderson, 1997).

The gene silencing phenomenon was discovered in transgenic tobacco, in which expression of a transgene was lost after going through a second transformation round (Matzke et al., 1989). Inactivation of transgene expression was only seen in plants containing sequences homologous to introduced genes. Later experiments have shown that gene silencing occurs at both the transcriptional and post-transcriptional levels, and in general involves homology restricted to promoter or coding regions (Park et al., 1996). It is still unclear whether the two types of homology-dependent silencing are overlapping or whether they use different mechanisms to down-regulate gene expression (Stam et al., 1997).

Homologous gene silencing involving promoter sequences is frequently correlated with increased methylation of cytosine residues of the coding region or promoter sequences of suppressed genes. Transgenes that lack corresponding sequences on the host genome may become inactivated through methylation caused by insertion into a heavily methylated region on the chromosome and spreading of the methylation pattern into the transgene (Pröls and Meyer, 1992). In addition, altered methylation status of transgenes or host genes may result from tissue culture effects occurring during regeneration of transgenic plants (Brown, 1989, Phillips et al., 1994). It was recently reported that tissue culture media containing the antibiotics hygromycin, kanamycin or cefotaxime induce DNA hypermethylation in tobacco plants (Schmitt et al., 1997). The effect of these antibiotics on the DNA methylation pattern was found to be both time- and dose-dependent. It is likely that the antibiotic agents studied will cause similar effects during regeneration of transformed cereals.

The involvement of methylation in gene silencing in cereals has been verified by restriction analysis of genomic DNA using different methylation-sensitive restriction enzymes (Ronchi et al., 1995). Further evidence that methylation causes gene silencing in cereals was provided by experiments showing reactivation of suppressed genes by treatment of transgenic plants with the demethylating agent, 5-azacytidine (Meijer et al., 1991, Ronchi et al., 1995). Once the methylation pattern of the genomic DNA is established, it is generally maintained through cell divisions, thereby resulting in an epigenic gene regulation. The silenced phenotype can sometimes display an unstable inheritance and become gradually lost after several generations (Matzke et al., 1989, Neuhuber et al., 1994). The degree of methylation and silencing appears to be correlated with the copy number of the inserted gene, and is therefore particularly serious when direct transformation methods have been used to generate the transgenic plants. These gene delivery methods, such as particle bombardment, generally lead to insertion of several copies of the transgene at a single locus (D'Halluin et al., 1992, Register et al., 1994, Spencer et al., 1992) and could thereby, to a greater extent than *Agrobacterium*-based methods, provoke gene silencing. Recent developments of transformation methods for cereals using *Agrobacterium* (Chan et al., 1993, Cheng et al., 1997, Hiei et al., 1994, Ishida et al., 1996, Tingay et al., 1997) may lessen this problem.

Several models involving direct or indirect DNA:DNA, DNA:RNA and RNA:RNA interactions have been put forward in an attempt to explain gene inactivation occurring at the post-transcriptional level (see reviews by Baulcombe, 1996, Stam et al., 1997). A hypothesis is that gene shut-down is triggered by an initial high expression level from the transgene (Elmayan and Vaucheret, 1996). This may explain why the strong CaMV 35S promoter is often subject to post-transcriptional inactivation in dicot plants. It also raises the question of whether a promoter that is very active during the early stages of plant development actually increases the recovery rate of transgenic plants compared to less active promoters. It has so far been impossible to demonstrate any such correlation due to the limited number of transgenic plants generated and the wide variety of constructs and plant systems used.

Several experiments performed with dicot plants and animals suggest that the severity of gene silencing can be decreased by placing matrix attachment regions (MAR) or scaffold attachment regions (SAR) on each side of the expression cassette (Allen et al., 1993, 1996, Poljak et al., 1994, Schöffl et al., 1993). It is believed that the MAR or the SAR sequences will allow the enclosed DNA to form an independent loop domain that will be insulated from negative effects exerted by the surrounding chromatin (see review by Spiker and Thompson, 1996). A positive effect of these elements in plants was demonstrated when a 140-fold increase in GUS gene expression was noted in transgenic tobacco cell lines when tobacco SAR elements were added to a GUS expression vector (Allen et al., 1996). In these experiments, a more prominent effect of the SAR sequences was found in stably transformed cells as compared to transiently transformed cells, which suggested that SAR-mediated gene enhancement required integration of the transgene into the chromosome. Although most data from dicot plants show an average increase in marker gene expression when SAR or MAR sequences are used, this enhancement is more evident in transgenic plants with a low copy number of the transgene. Thus, the variability in transgene performance from different transformants does not appear to be diminished by the presence of the SAR/MAR elements. The effect of these sequences in monocot plants has not been assessed so far.

## 7. Effects of Transgene Methylation on Gene Expression and Stability

Methylation of plant genes has an active role in genome defense, differentiation and development (see reviews by Richards, 1997, Stam et al., 1997, Yoder and Bestor, 1996). In cereals, there is convincing evidence that the tissue-specific expression of the seed storage proteins genes, hordein in barley and zein in maize, is regulated by cytosine methylation (Bianchi and Viotti, 1988, Lund et al., 1995, Sørensen et al., 1996).

Transient expression assays in wheat tissues and barley cell lines have suggested that the methylation pattern of introduced genes affects reporter gene expression levels (Graham and Larkin, 1995, Rogers and Rogers, 1995). In these studies, methylation of adenine residues at *dam* sites, in contrast to deoxycytosine methylation, had a stimulatory effect on gene expression. The enhancement of expression levels varied between the promoters studied and was independent of the initial promoter strength. For example, transient GUS gene expression driven by the *Act1* 5′ region in wheat embryos was stimulated 50-fold by adenine methylation at *dam* sites on the introduced DNA. In contrast, no increase in GUS activity by *dam* methylation of a CaMV 35S promoter-*uidA* cassette was obtained, which the authors correlated to fewer *dam* sites in the promoter region.

Transient transformation studies in barley have indicated that the DNA methylation status of introduced genes affects transgene stability in the host plant (Rogers and Rogers, 1992). It was observed that dA-methylation of *dam* sites (GATC) on introduced DNA was associated with increased instability of transgenes, whereas a lack of dA methylation combined with the presence of dC methylation increased transgene stability over two generations. Instability or poor transmission of transgenes has been observed in transgenic cereals (Peng et al., 1995, Spencer et al., 1992); however, these studies did not correlate these problems to the methylation status of the introduced DNA. Selective loss of transgene in successive generations was reported in *Arabidopsis thaliana* to be due to change in ploidy level of the progeny plants (Mittelsten Scheid et al., 1996). Therefore, whether specific methylation patterns of introduced genes trigger transgene instability remains to be investigated.

## 8. Future Prospects for Cereal Improvement by Genetic Engineering

Research in the early part of this decade was focused on expanding and improving the transformation technology for elite cultivars of important cereals such as wheat, maize, rice and barley. This objective has been achieved for most of the important cereal crops. We have now reached a stage where incorporation of useful traits into cereals has started to produce plants with improved agronomic performance and/or end-use quality. The grain producers and food-processing industries demand a uniform and consistent performance from agricultural crops. This is a challenge for biotechnology to overcome for the benefit of agriculture. Therefore, it is important to focus cereal research towards the molecular aspects of transgene technology in order to gain understanding of what a plant phenotype means in genetic terms.

A number of issues need to be addressed so that improvements can be made in the transgene technology for cereals. Promoters play an important role in the production of transgenic plants, because these DNA sequences direct transgene expression in a spatial, temporal or inducible manner. Due to the differences in gene regulatory mechanisms in different plants, it is desirable that promoters are analyzed in the plant species in which they are intended for use. It is now possible to perform a detailed study of promoter function by using transgenic cereals. These studies will reveal relevant information about promoter function in cereals during development and upon exposure to various environmental stress conditions, which may have a great influence on crop performance. Studies on interactions between host regulatory factors and *cis*-acting DNA regulatory elements on expression cassettes can provide valuable information which can be used to construct synthetic promoters. Promoters directing a desired gene expression pattern in the plant could be obtained by linking of various modules of well-defined regulatory elements.

The experience with development of insect-resistant plants expressing a modified *Bt* toxin has demonstrated that insertion of a foreign sequence into plants may cause several problems at the level of gene regulation, that will affect the amount of transgene product produced. Introduction of heterologous genes into cereals will require optimization of DNA sequences of the genes for expression in cereals. These alterations may include removal of RNA instability determinants, changes in codon usage, insertion or removal of intron sequences and optimization of regulatory sequences. To avoid unforeseen complications with unrelated gene sequences, it may be preferable to use promoters and gene sequences isolated from cereals. However, the problem with homologous sequences causing transgene silencing and/or co-suppression in cereals needs to be solved. Studies on gene expression and regulation in cereals that are polyploid may provide some answers to the mechanisms that have naturally evolved to prevent co-suppression and/or gene silencing in these plants. Replacement of direct DNA delivery techniques with *Agrobacterium*-mediated transformation methods will lower the number of inserted transgene copies, thus reducing the level of gene silencing. The effect of MAR or SAR sequences in expression cassettes has so far not been studied in cereals, but data from dicot plants suggest that these elements may overcome some of the problems associated with multiple copy insertion as well as transgene silencing. Development and optimization of site-specific recombination systems for plants to target genes to specific locations on the genome could also reduce the number of inactive transgenes. For example, transgenes could be targeted to regions with active genes, which are known to be organized in clusters on the plant chromosomes (Gill et al., 1996). Promising results in transgenic cell lines using the yeast FLP/FRT site-specific recombination system (Lyznik et al., 1996) indicate that this may be a realistic possibility. Despite the various factors related to gene expression and regulation that need to be elucidated, transgenic cereals will play an important role in agriculture in the next century.

## Acknowledgments

We thank Drs Patrick Covello and Pierre Fobert for critical review of this manuscript and Karen Caswell for editorial assistance.

## References

Abler, M.L., and Green, P.J. (1996) Control of mRNA stability in higher plants. Plant Mol. Biol. 32: 63–78.

Allen, G.C., Hall, G.E., Childs, L.C., Weissinger, A.K., Spiker, S., and Thompson, W.F. (1993) Scaffold attachment regions increase reporter gene expression in stably transformed plant cells. Plant Cell 5: 603–613.

Allen, G.C., Hall, G., Michalowski, S., Newman, W., Spiker, S., Weissinger, A.K., and Thompson, W.F. (1996) High-level transgene expression in plant cells: effects of a strong scaffold attachment region from tobacco. Plant Cell 8: 899–913.

Altpeter, F., Vasil, V., Srivastava, V., and Vasil, I.K. (1996) Integration and expression of the high-molecular-weight glutenin subunit 1Ax1 gene into wheat. Nature Biotechnology 14: 1155–1159.

Båga, M., Chibbar, R.N., and Kartha, K.K. (1995) Molecular cloning and expression analysis of peroxidase genes from wheat. Plant Mol. Biol. 29: 647–662.

Barcelo, P., Hagel, C., Becker, D., Martin, A., and Lörz, H. (1994) Transgenic cereal (tritordeum) plants obtained at high efficiency by microprojectile bombardment of inflorescence tissue. Plant J. 5: 583–592.

Barnes, W.M. (1990) Variable patterns of expression of luciferase in transgenic tobacco leaves. Proc. Nat. Acad. Sci. USA 87: 9183–9187.

Barry, G., Kishore, G., Padgette, S., Taylor, M., Kolacz, K., Weldon, M., Re, D., Eichholtz, D., Fincher, K., and Hallas, L. (1992) Inhibitors of amino acid biosynthesis: Strategies for imparting glyphosate tolerance to crop plants. In: Singh, B.K., Flores, H.C., and Shannon, J.C. (eds), Biosynthesis and Molecular Regulation of Amino Acids in Plants, pp. 139–145.

American Society of Plant Physiologists, Bethesda, MD.
Baszczynski, C.L., Barbour, E., Zeka, B.L., Maddock, S.E., and Swenson, J.L. (1997) Characterization of a genomic clone for a maize *DnaJ*-related gene, *ZmdJ1*, and expression analysis of its promoter in transgenic plants. Maydica 42: 189–201.
Battraw, M., and Hall, T.C. (1990) Histochemical analysis of CaMV 35S promoter-$\beta$-glucuronidase gene expression in transgenic rice plants. Plant Mol. Biol. 15: 527–538.
Battraw, M.J., and Hall, T.C. (1992) Expression of a chimeric neomycin phosphotransferase II gene in first and second generation transgenic rice plants. Plant Sci. 86: 191–202.
Baulcombe, D.C. (1996) RNA as a target and an initiator of post-transcriptional gene silencing in transgenic plants. Plant Mol. Biol. 32: 79–88.
Beck, E., Ludwig, G., Auerswald, E.A., Reiss, B., and Schaller, H. (1982) Nucleotide sequence and exact localization of the neomycin phosphotransferase gene from transposon Tn5. Gene 19: 327–336.
Becker, D., Brettschneider, R., and Lörz, H. (1994) Fertile transgenic wheat from microprojectile bombardment of scutellar tissue. Plant J. 5: 299–307.
Bennetzen, J.L., and Hall, B.D. (1982) Codon selection in yeast. J. Biol. Chem. 257: 3026–3031.
Benveniste, R., and Davies, J. (1973) Mechanisms of antibiotic resistance in bacteria. Ann. Rev. Biochem. 42: 471–506.
Berget, S.M. (1995) Exon recognition in vertebrate splicing. J. Biol. Chem. 270: 2411–2414.
Bhattacharyya-Pakrasi, M., Peng, J., Elmer, J.S., Laco, G., Shen, P., Kaniewska, M.B., Kononowicz, H., Wen, F., Hodges, T.K., and Beachy, R.N. (1993) Specificity of a promoter from the rice tungro bacilliform virus for expression in phloem tissues. Plant J. 4: 71–79.
Bianchi, M.W., and Viotti, A. (1988) DNA methylation and tissue-specific transcription of the storage protein genes of maize. Plant Mol. Biol. 11: 203–214.
Bingham, P.M. (1997) Cosuppression comes to the animals. Cell 90: 385–387.
Blakley, R.L. (1969) The Biochemistry of Folic Acid and Related Pteridines. North-Holland, Amsterdam, The Netherlands.
Blechl, A.E., and Anderson, O.D. (1996) Expression of a novel high-molecular-weight glutenin subunit gene in transgenic wheat. Nature Biotechnology 14: 875–879.
Bowen, B. (1992) Anthocyanin genes as visual markers in transformed maize tissues. In: Gallager, S.R. (ed), GUS Protocols: Using the GUS Gene as a Reporter of Gene Expression, pp. 163–177. Academic Press, San Diego.
Bowen, B.A. (1993) Markers for plant gene transfer. In: Kung, S., and Wu, R. (eds) Transgenic Plants, 1: 89–123. Academic Press, San Diego.
Bower, R., and Birch, R.G. (1992) Transgenic sugarcane plants via microprojectile bombardment. Plant J. 2: 409–416.
Brisson, N., Paszkowski, J., Penswick, J.R., Gronenborn, B., Potrykus, I., and Hohn, T. (1984) Expression of a bacterial gene in plants by using a viral vector. Nature 310: 511–514.
Brown, P.T.H. (1989) DNA methylation in plants and its role in tissue culture. Genome 31: 717–729.
Bruce, W.B., Christensen, A.H., Klein, T., Fromm, M., and Quail, P.H. (1989) Photoregulation of a phytochrome gene promoter from oat transferred into rice by particle bombardment. Proc. Nat. Acad. Sci. USA 86: 9692–9696.
Buchman, A.R., and Berg, P. (1988) Comparison of intron-dependent and intron-independent gene expression. Mol. Cell Biol. 8: 4395–4405.
Burkhardt, P.K., Beyer, P., Wünn, J., Klöti, A., Armstrong, G.A., Schledz, M., von Lintig, J., and Potrykus, I. (1997) Transgenic rice (*Oryza sativa*) endosperm expressing daffodil (*Narcissus pseudonarcissus*) phytoene synthase accumulates phytoene, a key intermediate of provitamin A biosynthesis. Plant J. 11: 1071–1078.
Callis, J., Fromm, M., and Walbot, V. (1987) Introns increase gene expression in cultured

maize cells. Genes Dev. 1:1183–1200.
Campbell, W.H., and Gowri, G. (1990) Codon usage in higher plants, green algae, and cyanobacteria. Plant Physiol. 92: 1-11.
Capellades, M., Torres, M.A., Bastisch, I., Stiefel, V., Vignols, F., Bruce, W.B., Peterson, D., Puigdomenech, P., and Rigau, J. (1996) The maize caffeic acid *O*-methyltransferase gene promoter is active in transgenic tobacco and maize plant tissues. Plant Mol. Biol. 31: 307–322.
Caplan, A., Dekeyser, R., and van Montagu, M. (1992) Selectable markers for rice transformation. Meth. Enzymol. 216: 426–441.
Carle-Urioste, J.C., Brendel, V., and Walbot, V. (1997) A combinatorial role for exon, intron and splice site sequences in splicing in maize. Plant J. 11: 1253–1263.
Carle-Urioste, J.C., Ko, C.H., Benito, M., and Walbot, V. (1994) *In vivo* analysis of intron processing using splicing-dependent reporter gene assays. Plant Mol. Biol. 26: 1785–1795.
Casas, A.M., Kononowicz, A.K., Zehr, U.B., Tomes, D.T., Axtell, J.D., Butler, L.G., Bressan, R.A., and Hasegawa, P.M. (1993) Transgenic sorghum plants via microprojectile bombardment. Proc. Nat. Acad. Sci. USA 90: 11212–11216.
Castillo, A.M., Vasil, V., and Vasil, I.K. (1994) Rapid production of fertile transgenic plants of rye (*Secale cereale* L.) Bio/Technology 12: 1366–1371.
Chaleff, R.S., and Mauvais, C.J. (1984) Acetolactate synthase is the site of action of two sulfonylurea herbicides in higher plants. Science 224: 1443–1445.
Chalfie, M., Tu, Y., Euskirchen, G., Ward, W.W., and Prasher, D.C. (1994) Green flourescent protein as a marker for gene expression. Science 263: 802–805.
Chamberlain, D.A., Brettell, R.I.S., Last, D.I., Witrzens, B., McElroy, D., Dolferus, R., and Dennis, E.S. (1994) The use of Emu promoter with antibiotic and herbicide resistance genes for the selection of transgenic wheat callus and rice plants. Aust. J. Plant Physiol. 21: 95–112.
Chan, M., Chang, H., Ho, S., Tong, W., and Yu, S. (1993) *Agrobacterium*-mediated production of transgenic rice plants expressing a chimeric $\alpha$-amylase promoter/$\beta$-glucuronidase gene. Plant Mol. Biol. 22: 491–506.
Chan, M., Chao, Y., and Yu, S. (1994) Novel gene expression system for plant cells based on induction of $\alpha$-amylase promoter by carbohydrate starvation. J. Biol. Chem. 269: 17635–17641.
Cheng, M., Fry, J.E., Pang, S., Zhou, H., Hironaka, C.M., Duncan, D.R., Conner, T.W., and Wan, Y. (1997) Genetic transformation of wheat mediated by *Agrobacterium tumefaciens*. Plant Physiol. 115: 971–980.
Chibbar, R.N., Kartha, K.K., Leung, N., Qureshi, J., and Caswell, K. (1991) Transient expression of marker genes in immature zygotic embryos of spring wheat (*Triticum aestivum*) through microprojectile bombardment. Genome 34: 453–460.
Chibbar, R.N., Kartha, K.K., Datla, R.S.S., Leung, N., Caswell, K., Mallard, C.S., and Steinhauer, L. (1993) The effect of different promoter-sequences on transient expression of gus reporter gene in cultured barley (*Hordeum vulgare* L.) cells. Plant Cell Rep. 12: 506–509.
Christensen, A.H., and Quail, P.H. (1996) Ubiquitin promoter-based vectors for high-level expression of selectable and/or screenable marker genes in monocotyledonous plants. Transgenic Res. 5: 213–218.
Christensen, A.H., Sharrock, R.A., and Quail, P.H. (1992) Maize polyubiquitin genes: structure, thermal perturbation of expression and transcript splicing, and promoter activity following transfer to protoplasts by electroporation. Plant Mol. Biol. 18: 675–689.
Christou, P., Ford, T.L., and Kofron, M. (1991) Production of transgenic rice (*Oryza sativa* L.) plants from agronomically important indica and japonica varieties via electric discharge particle acceleration of exogenous DNA into immature zygotic embryos. Bio/Technology 9: 957–962.
Clancy, M., Vasil, V., Hannah, L.C., and Vasil, I.K. (1994) Maize *Shrunken-1* intron and exon

regions increase gene expression in maize protoplasts. Plant Sci. 98: 151–161.

Clough, R.C., Casal, J.J., Jordan, E.T., Christou, P., and Vierstra, R.D. (1995) Expression of functional oat phytochrome A in transgenic rice. Plant Physiol. 109: 1039–1045.

Coe, E.H. Jr, Neuffer, M.G., and Hoisington, D.A. (1988) The genetics of corn. In: Sprague, G.F. and Dudley, J.W. (eds), Corn and Corn Improvement. Agronomy Monograph No. 18, 3rd edition, pp. 81–258. Amer. Soc. Agron. Inc./Crop Sci. Soc. Amer. Inc./Soil Sci. Soc. Amer. Inc, Madison, WI.

Colot, V., Robert, L.S., Kavanagh, T.A., Bevan, M.W., and Thompson, R.D. (1987) Localization of sequences in wheat endosperm protein genes which confer tissue-specific expression in tobacco. EMBO J. 6: 3559–3564.

Cornejo, M., Luth, D., Blankenship, K.M., Anderson, O.D., and Blechl, A.E. (1993) Activity of a maize ubiquitin promoter in transgenic rice. Plant Mol. Biol. 23: 567–581.

Cubitt, A.B., Heim, R., Adams, S.R., Boyd, A.E., Gross, L.A., and Tsien. R.Y. (1995) Understanding, improving and using green fluorescent proteins. Trends Biochem. Sci. 20: 448–455.

Czarnecka, E., Nagao, R.T., Key, J.L., and Gurley, W.B. (1988) Characterization of *Gmhsp26-A*, a stress gene encoding a divergent heat shock protein of soybean: heavy-metal-induced inhibition of intron processing. Mol. Cell Biol. 8: 1113–1122.

Dalphin, M.E., Brown, C.M., Stockwell, P.A., and Tate, W.P. (1997) The translational signal database, TransTerm: more organisms, complete genomes. Nuc. Acids Res. 25: 246–247.

Damiani, Jr R.D., and Wessler, S.R. (1993) An upstream open reading frame represses expression of *Lc*, a member of the R/B family of maize transcriptional activators. Proc. Nat. Acad. Sci. USA 90: 8244–8248.

Datla, R.S.S., Hammerlindl, J.K., Pelcher, L.E., Crosby, W.L., and Selvaraj, G. (1991) A bifunctional fusion between $\beta$-glucuronidase and neomycin phosphotransferase: a broad-spectrum marker enzyme for plants. Gene 101: 239–246.

De Block, M., Debrouwer, D., and Moens, T. (1997) The development of a nuclear male sterility system in wheat. Expression of the *barnase* gene under the control of tapetum specific promoters. Theor. Appl. Genet. 95: 125–131.

De Block, M., De Sonville, A., and Debrouwer, D. (1995) The selection mechanism of phosphinothricin is influenced by the metabolic status of the tissue. Planta. 197: 619–626.

De Block, M., Herrera-Estrella, L., van Montagu, M., Schell, J., and Zambryski, P. (1984) Expression of foreign genes in regenerated plants and their progeny. EMBO J. 3: 1681–1689.

Dekeyser, R., Claes, B., Marichal, M., van Montagu, M., and Caplan, A. (1989) Evaluation of selectable markers for rice transformation. Plant Physiol. 90: 217–223.

Dennehey, B.K., Petersen, W.L., Ford-Santino, C., Pajeau, M., and Armstrong, C.L. (1994) Comparison of selective agents for use with selectable marker gene *bar* in maize transformation. Plant Cell Tiss. Org. Cult. 36: 1–7.

de Wet, J.R., Wood, K.V., DeLuca, M., Helinski, D.R., and Subramani, S. (1987) Firefly luciferase gene: structure and expression in mammalian cells. Mol. Cell Biol. 7: 725–737.

D'Halluin, K., Bonne, E., Bossut, M., De Beuckeleer, M., and Leemans, J. (1992) Transgenic maize plants by tissue electroporation. Plant Cell 4: 1495–1505.

Dooner, H.K., Robbins, T.P., and Jorgensen, R.A. (1991) Genetic and developmental control of anthocyanin biosynthesis. Ann. Rev. Genet. 25: 173–199.

Dynan, W.S. (1989) Modularity in promoters and enhancers. Cell 58: 1–4.

Ebinuma, H., Sugita, K., Matsunaga, E., and Yamakado, M. (1997) Selection of marker-free transgenic plants using the isopentenyl transferase gene. Proc. Nat. Acad. Sci. USA 94: 2117–2121.

Ellis, J.G., Llewellyn, D.J., Walker, J.C., Dennis, E.S., and Peacock, W.J. (1987) The *ocs* element: a 16 base pair palindrome essential for activity of the octopine synthase enhancer. EMBO J. 6: 3203–3208.

Elmayan, T., and Vaucheret, H. (1996) Expression of single copies of a strongly expressed

35S transgene can be silenced post-transcriptionally. Plant J. 9: 787–797.

Falco, S.C., and Dumas, K.S. (1985) Genetic analysis of mutants of *Saccharomyces cerevisiae* resistant to the herbicide sulfometuron methyl. Genetics 109: 21–35.

Fennoy, S.L., and Bailey-Serres, J. (1993) Synonymous codon usage in *Zea mays* L. nuclear genes is varied by levels of C and G-ending codons. Nuc. Acids Res. 21: 5294–5300.

Finnegan, J., and McElroy, D. (1994) Transgene inactivation: Plants fight back! Bio/Technology 12: 883–888.

Fischhoff, D.A., Bowdish, K.S., Perlak, F.J., Marrone, P.G., McCormick, S.M., Niedermeyer, J.G., Dean, D.A., Kusano-Kretzmer, K., Mayer, E.J., Rochester, D.E., Rogers, S.G., and Fraley, R.T. (1987) Insect tolerant transgenic tomato plants. Bio/Technology 5: 807–813.

Flavell, R.B., Dart, E., Fuchs, R.L., and Fraley, R.T. (1992) Selectable marker genes: safe for plants? Bio/Technology 10: 141–144.

Fromm, M.E., Morrish, F., Armstrong, C., Williams, R., Thomas, J., and Klein, T.M. (1990) Inheritance and expression of chimeric genes in the progeny of transgenic maize plants. Bio/Technology 8: 833–839.

Fromm, M., Taylor, L.P., and Walbot, V. (1985) Expression of genes transferred into monocot and dicot plant cells by electroporation. Proc. Nat. Acad. Sci. USA 82: 5824–5828.

Fromm, M.E., Taylor, L.P., and Walbot, V. (1986) Stable transformation of maize after gene transfer by electroporation. Nature 319: 791–793.

Fujimoto, H., Itoh, K., Yamamoto, M., Kyozuka, J., and Shimamoto, K. (1993) Insect resistant rice generated by introduction of a modified δ-endotoxin gene of *Bacillus thuringiensis*. Bio/Technology 11: 1151–1155.

Funatsuki, H., Kuroda, H., Kihara, M., Lazzeri, P.A., Müller, E., Lörz, H., and Kishinami, I. (1995) Fertile transgenic barley generated by direct DNA transfer to protoplasts. Theor. Appl. Genet. 91: 707–712.

Gallie, D.R., Lucas, W.J., and Walbot, V. (1989) Visualizing mRNA expression in plant protoplasts: factors influencing efficient mRNA uptake and translation. Plant Cell 1: 301–311.

Gallie, D.R., Sleat, D.E., Watts, J.W., Turner, P.C., and Wilson, T.M.A. (1987a) A comparison of eukaryotic viral 5′-leader sequences as enhancers of mRNA expression *in vivo*. Nuc. Acids Res. 15: 8693–8711.

Gallie, D.R., Sleat, D.E., Watts, J.W., Turner, P.C., and Wilson, T.M.A. (1987b) The 5′-leader sequence of tobacco mosaic virus RNA enhances the expression of foreign gene transcripts *in vitro* and *in vivo*. Nuc. Acids Res. 15: 3257–3273.

Gallie, D.R., and Young, T.E. (1994) The regulation of gene expression in transformed maize aleurone and endosperm protoplasts. Analysis of promoter activity, intron enhancement, and mRNA untranslated regions on expression. Plant Physiol. 106: 929–939.

Gill, K.S., Gill, B.S., Endo, T.R., and Boyko, E.V. (1996) Identification and high-density mapping of gene-rich regions in chromosome group 5 of wheat. Genetics 143: 1001–1012.

Goff, S.A., Klein, T.M., Roth, B.A., Fromm, M.E., Cone, K.C., Radicella, J.P., and Chandler, V.L. (1990) Transactivation of anthocyanin biosynthetic genes following transfer of *B* regulatory genes into maize tissues. EMBO J. 9: 2517–2522.

Goldsbrough, A.P., Lastrella, C.N., and Yoder, J.I. (1993) Transposition mediated re-positioning and subsequent elimination of marker genes from transgenic tomato. Bio/Technology 11: 1286–1292.

Golovkin, M.V., Abraham, M., Morocz, S., Bottka, S., Feher, A., and Dudits, D. (1993) Production of transgenic maize plants by direct DNA uptake into embryogenic protoplasts. Plant Sci. 90: 41–52.

Goodall, G.J., and Filipowicz, W. (1989) The AU-rich sequences present in the introns of plant nuclear pre-mRNAs are required for splicing. Cell 58: 473–483.

Goodall, G.J., and Filipowicz, W. (1991) Different effects of intron nucleotide composition and secondary structure on pre-mRNA splicing in monocot and dicot plants. EMBO J. 10:

2635–2644.

Gordon-Kamm. W.J., Spencer, T.M., Mangano, M.L., Adams, T.R., Daines. R.J., Start, W.G., O'Brien, J.V., Chambers, S.A., Adams, Jr. W.R., Willetts, N.G., Rice, T.B., Mackey, C.J., Krueger, R.W., Kausch, A.P., and Lemaux, P.G. (1990) Transformation of maize cells and regeneration of fertile transgenic plants. Plant Cell 2: 603–618.

Graham, M.W., and Larkin, P.J. (1995) Adenine methylation at *dam* sites increases transient gene expression in plant cells. Transgen. Res. 4: 324–331.

Hänsch, R., Koprek, T., Mendel, R.R., and Schulze, J. (1995) An improved protocol for eliminating endogenous $\beta$-glucuronidase background in barley. Plant Sci. 105: 63–69.

Haseloff, J., Siemering, K.R., Prasher, D.C., and Hodge, S. (1997) Removal of a cryptic intron and subcellular localization of green fluorescent protein are required to mark transgenic *Arabidopsis* plants brightly. Proc. Nat. Acad. Sci. USA 94: 2122–2127.

Haughn, G.W., and Somerville, C. (1986) Sulfonylurea-resistant mutants of *Arabidopsis thaliana*. Mol. Gen. Genet. 204: 430–434.

Hauptmann, R.M., Ozias-Akins, P., Vasil, V., Tabaeizadeh, Z., Rogers, S.G., Horsch, R.B., Vasil, I.K., and Fraley, R.T. (1987) Transient expression of electroporated DNA in monocotyledonous and dicotyledonous species. Plant Cell Rep. 6: 265–270.

Hauptmann, R.M., Vasil, V., Ozias-Akins, P., Tabaeizadeh, Z., Rogers, S.G., Fraley, R.T., Horsch, R.B., and Vasil, I.K. (1988) Evaluation of selectable markers for obtaining stable transformants in the Gramineae. Plant Physiol. 86: 602–606.

Hayakawa, T., Zhu, Y., Itoh, K., Kimura, Y., Izawa, T., Shimamoto, K., and Toriyama, S. (1992) Genetically engineered rice resistant to rice stripe virus, an insect-transmitted virus. Proc. Nat. Acad. Sci. USA 89: 9865–9869.

Heim, R., Prasher, D.C., and Tsien, R.Y. (1994) Wavelength mutations and posttranslational autoxidation of green fluorescent protein. Proc. Nat. Acad. Sci. USA 91: 12501–12504.

Hensgens, L.A.M. de Bakker, E.P.H.M., van Os-Ruygrok, E.P., Rueb, S., van de Mark, F., van der Maas, H., van der Veen, S., Kooman-Gersmann, M., Hart, L., and Schilperoort, R.A. (1993) Transient and stable expression of *gusA* fusions with rice genes in rice, barley and perennial ryegrass. Plant Mol. Biol. 23: 643–669.

Hensgens, L.A.M., Fornerod, M.W.J., Rueb, S., Winkler, A.A., van der Veen, S., and Schilperoort, R.A. (1992) Translation controls the expression level of a chimaeric reporter gene. Plant Mol. Biol. 20: 921–938.

Hiei, Y., Ohta, S., Komari, T., and Kumashiro, T. (1994) Efficient transformation of rice (*Oryza sativa* L.) mediated by *Agrobacterium* and sequence analysis of the boundaries of the T-DNA. Plant J. 6: 271–282.

Hoekema, A., Kastelein, R.A., Vasser, M., and De Boer, H.A. (1987) Codon replacement in the *PGK1* gene of *Saccharomyces cerevisiae* : experimental approach to study the role of biased codon usage in gene expression. Mol. Cell Biol. 7: 2914–2924.

Howard, E.A., Walker, J.C., Dennis, E.S., and Peacock, W.J. (1987) Regulated expression of an alcohol dehydrogenase 1 chimeric gene introduced into maize protoplasts. Planta 170: 535–540.

Hu, C., Chee, P.P., Chesney, R.H., Zhou, J.H., Miller, P.D., and O'Brien, W.T. (1990) Intrinsic GUS-like activities in seed plants. Plant Cell Rep. 9: 1–5.

Huang, M.T.F., and Gorman, C.M. (1990) Intervening sequences increase efficiency of RNA 3′ processing and accumulation of cytoplasmic RNA. Nuc. Acids Res. 18: 937–947.

Hunt, A.G., Chu, N.M., Odell, J.T., Nagy, F., and Chua, N.H. (1987) Plant cells do not properly recognize animal gene polyadenylation signals. Plant Mol. Biol. 8: 23–35.

Ingelbrecht, I.L.W., Herman, L.M.F., Dekeyser, R.A., van Montagu, M.C., and Depicker, A.G. (1989) Different 3′ end regions strongly influence the level of gene expression in plant cells. Plant Cell 1: 671–680.

Ishida, Y., Saito, H., Ohta, S., Hiei, Y., Komari, T., and Kumashiro, T. (1996) High efficiency transformation of maize (*Zea mays* L.) mediated by *Agrobacterium tumefaciens*. Nature Biotechnology 14: 745–750.

Itoh, K., Nakajima, M., and Shimamoto, K. (1997) Silencing of waxy genes in rice containing *Wx* transgenes. Mol. Gen. Genet. 255: 351–358.

Jefferson, R.A. (1987) Assaying chimeric genes in plants: The GUS gene fusion system. Plant Mol. Biol. Rep. 5: 387–405.

Jefferson, R.A., Burgess, S.M., and Hirsh, D. (1986) $\beta$-glucuronidase from *Escherichia coli* as a gene-fusion marker. Proc. Nat. Acad. Sci. USA 83: 8447–8451.

Jefferson, R.A., Kavanagh, T.A., and Bevan, M.W. (1987) GUS fusions: $\beta$-glucuronidase as a sensitive and versatile gene fusion marker in higher plants. EMBO J. 6: 3901–3907.

Jensen, L.G., Olsen, O., Kops, O., Wolf, N., Thomsen, K.K., and von Wettstein, D. (1996) Transgenic barley expressing a protein-engineered, thermostable (1,3-1,4)-$\beta$-glucanase during germination. Proc. Nat. Acad. Sci. USA 93: 3487–3491.

Kalla, R., Shimamoto, K., Potter, R., Nielsen, P.S., Linnestad, C., and Olsen, O. (1994) The promoter of the barley aleurone-specific gene encoding a putative 7 kDa lipid transfer protein confers aleurone cell-specific expression in transgenic rice. Plant J. 6: 849–860.

Kartha, K.K., Chibbar, R.N., Georges, F., Leung, N., Caswell, K., Kendall, E., and Qureshi, J. (1989) Transient expression of chloramphenicol acetyltransferase (CAT) gene in barley cell cultures and immature embryos through microprojectile bombardment. Plant Cell Rep. 8: 429–432.

Keith, B., and Chua, N. (1986) Monocot and dicot pre-mRNAs are processed with different efficiencies in transgenic tobacco. EMBO J. 5: 2419–2425.

Kilby, N.J., Snaith, M.R., and Murray, J.A.H. (1993) Site-specific recombinases: tools for genome engineering. Trends Genet. 9: 413–421.

Kleffe, J., Hermann, K., Vahrson, W., Wittig, B., and Brendel, V. (1996) Logitlinear models for the prediction of splice sites in plant pre-mRNA sequences. Nuc. Acids Res. 24: 4709–4718.

Klein, T.M., Fromm, M., Weissinger, A., Tomes, D., Schaaf, S., Sletten, M., and Sanford, J.C. (1988) Transfer of foreign genes into intact maize cells with high-velocity microprojectiles. Proc. Nat. Acad. Sci. USA 85: 4305–4309.

Klein, T.M., Roth, B.A., and Fromm, M.E. (1989) Regulation of anthocyanin biosynthetic genes introduced into intact maize tissues by microprojectiles. Proc. Nat. Acad. Sci. USA 86: 6681–6685.

Komari, T., Hiei, Y., Saito, Y., Murai, N., and Kumashiro, T. (1996) Vectors carrying two separate T-DNAs for co-transformation of higher plants mediated by *Agrobacterium tumefaciens* and segregation of transformants free from selection markers. Plant J. 10: 165–174.

Kosugi, S., Ohashi, Y., Nakajima, K., and Arai, Y. (1990) An improved assay for $\beta$-glucuronidase in transformed cells: methanol almost completely suppresses a putative endogenous $\beta$-glucuronidase activity. Plant Sci. 70: 133–140.

Kozak, M. (1989) Context effects and inefficient initiation at non-AUG codons in eucaryotic cell-free translation systems. Mol. Cell Biol. 9: 5073–5080.

Kozak, M. (1991) Structural features in eukaryotic mRNAs that modulate the initiation of translation. J. Biol. Chem. 266: 19867–19870.

Koziel, M.G., Beland, G.L., Bowman, C., Carozzi, N.B., Crenshaw, R., Crossland, L., Dawson, J., Desai, N., Hill, M., Kadwell, S., Launis, K., Lewis, K., Maddox, D., McPherson, K., Meghji, M.R., Merlin, E., Rhodes, R., Warren, G.W., Wright, M., and Evola, S.V. (1993) Field performance of elite transgenic maize plants expressing an insecticidal protein derived from *Bacillus thuringiensis*. Bio/Technology 11: 194–200.

Kyozuka, J., Fujimoto, H., Izawa, T., and Shimamoto, K. (1991) Anaerobic induction and tissue-specific expression of maize *Adh1* promoter in transgenic rice plants and their progeny. Mol. Gen. Genet. 228: 40–48.

Kyozuka, J., McElroy, D., Hayakawa, T., Xie, Y., Wu, R., and Shimamoto, K. (1993) Light-regulated and cell-specific expression of tomato *rbcS-gusA* and rice *rbcS-gusA* fusion genes in transgenic rice. Plant Physiol. 102: 991–1000.

Kyozuka, J., Olive, M., Peacock, W.J., Dennis, E.S., and Shimamoto, K. (1994) Promoter elements required for developmental expression of the maize *Adh1* gene in transgenic rice. Plant Cell 6: 799–810.

Lamppa, G., Nagy, F., and Chua, N. (1985) Light-regulated and organ-specific expression of a wheat Cab gene in transgenic tobacco. Nature 316: 750–752.

Last, D.I., Brettell, R.I.S., Chamberlain, D.A., Chaudhury, A.M., Larkin, P.J., Marsh, E.L., Peacock, W.J., and Dennis, E.S. (1991) pEmu: an improved promoter for gene expression in cereal cells. Theor. Appl. Genet. 81: 581–588.

Lee, B., Murdoch, K., Topping, J., Kreis, M., and Jones, M.G.K. (1989) Transient gene expression in aleurone protoplasts isolated from developing caryopses of barley and wheat. Plant Mol. Biol. 13: 21–29.

Leisy, D.J., Hnilo, J., Zhao, Y., and Okita, T.W. (1989) Expression of a rice glutelin promoter in transgenic tobacco. Plant Mol. Biol. 14: 41–50.

Li, Z., Hayashimoto, A., and Murai, N. (1992) A sulfonylurea herbicide resistance gene from *Arabidopsis thaliana* as a new selectable marker for production of fertile transgenic rice plants. Plant Physiol. 100: 662–668.

Li, Z., Upadhyaya, N.M., Meena, S., Gibbs, A.J., and Waterhouse, P.M. (1997) Comparison of promoters and selectable marker genes for use in Indica rice transformation. Mol. Breeding 3: 1–14.

Liu, H., and Filipowicz, W. (1996) Mapping of branchpoint nucleotides in mutant pre-mRNAs expressed in plant cells. Plant J. 9: 381–389.

Lloyd, A.M., Walbot, V., and Davis, R.W. (1992) *Arabidopsis* and *Nicotiana* anthocyanin production activated by maize regulators *R* and *C1*. Science 258: 1773–1775.

Lou, H., McCullough, A.J., and Schuler, M.A. (1993) Expression of maize *Adh1* intron mutants in tobacco nuclei. Plant J. 3: 393–403.

Ludwig, S.R., and Wessler, S.R. (1990) Maize *R* gene family: tissue-specific helix-loop-helix proteins. Cell 62: 849–851.

Luehrsen, K.R., and Walbot, V. (1991) Intron enhancement of gene expression and the splicing efficiency of introns in maize cells. Mol. Gen. Genet. 225: 81–93.

Luehrsen, K.R., and Walbot, V. (1994a) Addition of A- and U-rich sequence increases the splicing efficiency of a deleted form of a maize intron. Plant Mol. Biol. 24: 449–463.

Luehrsen, K.R., and Walbot, V. (1994b) Intron creation and polyadenylation in maize are directed by AU-rich RNA. Genes Dev. 8: 1117–1130.

Luehrsen, K.R., and Walbot, V. (1994c) The impact of AUG start codon context on maize gene expression *in vivo*. Plant Cell Rep. 13: 454–458.

Lund, G., Ciceri, P., and Viotti, A. (1995) Maternal-specific demethylation and expression of specific alleles of zein genes in the endosperm of *Zea mays* L. Plant J. 8: 571–581.

Lusardi, M.C., Neuhaus-Url, G., Potrykus, I., and Neuhaus, G. (1994) An approach towards genetically engineered cell fate mapping in maize using the *Lc* gene as a visible marker: transactivation capacity of *Lc* vectors in differentiated maize cells and microinjection of *Lc* vectors into somatic embryos and shoot apical meristems. Plant J. 5: 571–582.

Lütcke, H.A., Chow, K.C., Mickel, F.S., Moss, K.A., Kern, H.F., and Scheele, G.A. (1987) Selection of AUG initiation codons differs in plants and animals. EMBO J. 6: 43–48.

Lyznik, L.A., Hirayama, L., Rao, K.V., Abad, A., and Hodges, T.K. (1995) Heat-inducible expression of *FLP* gene in maize cells. Plant J. 8: 177–186.

Lyznik, L.A., Rao, K.V., and Hodges, T.K. (1996) FLP-mediated recombination of *FRT* sites in the maize genome. Nuc. Acids Res. 24: 3784–3789.

Maas, C., Laufs, J., Grant, S., Korfhage, C., and Werr, W. (1991) The combination of a novel stimulatory element in the first exon of the maize *Shrunken-1* gene with the following intron 1 enhances reporter gene expression up to 1000-fold. Plant Mol. Biol. 16: 199–207.

Marcotte, Jr. W.R., Russell, S.H., and Quatrano, R.S. (1989) Abscisic acid-responsive sequences from the Em gene of wheat. Plant Cell 1: 969–976.

Marris, C., Gallois, P., Copley, J., and Kreis, M. (1988) The 5′ flanking region of a barley B hordein gene controls tissue and developmental specific CAT expression in tobacco plants. Plant Mol. Biol. 10: 359–366.

Marrs, K.A., Alfenito, M.R., Lloyd, A.M., and Walbot, V. (1995) A glutathione S-transferase involved in vacuolar transfer encoded by the maize gene *Bronze-2*. Nature 375: 397–400.

Marrs, K.A., and Walbot, V. (1997) Expression and RNA splicing of the maize glutathione *S*-transferase *Bronze-2* gene is regulated by cadmium and other stresses. Plant Physiol. 113: 93–102.

Mascarenhas, D., Mettler, I.J., Pierce, D.A., and Lowe, H.W. (1990) Intron-mediated enhancement of heterologous gene expression in maize. Plant Mol. Biol. 15: 913–920.

Matassi, G., Montero, L.M., Salinas, J., and Bernardi, G. (1989) The isochore organization and the compositional distribution of homologous coding sequences in the nuclear genome of plants. Nuc. Acids Res. 17: 5273–5290.

Matsuki, R., Onodera, H., Yamauchi, T., and Uchimiya, H. (1989) Tissue-specific expression of the *rolC* promoter of the Ri plasmid in transgenic rice plants. Mol. Gen. Genet. 220: 12–16.

Matsuoka, M., Kyozuka, J., Shimamoto, K., and Kano-Murakami, Y. (1994) The promoters of two carboxylases in a $C_4$ plant (maize) direct cell-specific, light-regulated expression in a $C_3$ plant (rice). Plant J. 6: 311–319.

Matsuoka, M., and Sanada, Y. (1991) Expression of photosynthetic genes from the $C_4$ plant, maize, in tobacco. Mol. Gen. Genet. 225: 411–419.

Matsuoka, M., Tada, Y., Fujimura, T., and Kano-Murakami, Y. (1993) Tissue-specific light-regulated expression directed by the promoter of a $C_4$ gene, maize pyruvate, orthophosphate dikinase, in a $C_3$ plant, rice. Proc. Nat. Acad. Sci. USA 90: 9586–9590.

Matzke, M.A., Primig, M., Trnovsky, J., and Matzke, A.J.M. (1989) Reversible methylation and inactivation of marker genes in sequentially transformed tobacco plants. EMBO J. 8: 643–649.

McElroy, D., Blowers, A.D., Jenes, B., and Wu, R. (1991) Construction of expression vectors based on the rice actin 1 (*Act1*) 5′ region for use in monocot transformation. Mol. Gen. Genet. 231: 150–160.

McElroy, D., Zhang, W., Cao, J., and Wu, R. (1990) Isolation of an efficient actin promoter for use in rice transformation. Plant Cell 2: 163–171.

McElroy, W.D., and DeLuca, M. (1978) Chemistry of firefly luminescence. In: Herring, P.J. (ed.) Bioluminescence in Action, pp. 109–127. Academic Press, Inc., London.

Meijer, E.G.M., Schilperoort, R.A., Rueb, S., van Os-Ruygrok, P.E., and Hensgens, L.A.M. (1991) Transgenic rice cell lines and plants: expression of transferred chimeric genes. Plant Mol. Biol. 16: 807–820.

Millar, A.J., Short, S.R., Chua, N., and Kay, S.A. (1992) A novel circadian phenotype based on firefly luciferase expression in transgenic plants. Plant Cell 4: 1075–1087.

Mittelsten Scheid, O., Jakovleva, L., Afsar, K., Maluszynska, J., and Paszkowski, J. (1996) A change of ploidy can modify epigenetic silencing. Proc. Nat. Acad. Sci. USA 93: 7114–7119.

Molina, A., Diaz, I., Vasil, I.K., Carbonero, P., and Garcia-Olmedo, F. (1996) Two cold-inducible genes encoding lipid transfer protein LTP4 from barley show differential responses to bacterial pathogens. Mol. Gen. Genet. 252: 162–168.

Murray, E.E., Lotzer, J., and Eberle, M. (1989) Codon usage in plant genes. Nuc. Acids Res. 17: 477–493.

Nayak, P., Basu, D., Das, S., Basu, A., Ghosh, D., Ramakrishnan, N.A., Ghosh, M., and Sen, S.K. (1997) Transgenic elite *indica* rice plants expressing CryIAc δ-endotoxin of *Bacillus thuringiensis* are resistant against yellow stem borer (*Scirpophaga incertulas*). Proc. Nat. Acad. Sci. USA 94: 2111–2116.

Nehra, N.S., Chibbar, R.N., Leung, N., Caswell, K., Mallard, C., Steinhauer, L., Båga, M., and Kartha, K.K. (1994) Self-fertile transgenic wheat plants regenerated from isolated scutellar tissues following microprojectile bombardment with two distinct gene constructs.

Plant J. 5: 285–297.
Nelson, T., and Langdale, J.A. (1992) Developmental genetics of $C_4$ photosynthesis. Annu. Rev. Plant Physiol. Plant Mol. Biol. 43: 25–47.
Neuhuber, F., Park, Y., Matzke, A.J.M., and Matzke, M.A. (1994) Susceptibility of transgene loci to homology-dependent gene silencing. Mol. Gen. Genet. 244: 230–241.
Oard, J.H., Paige, D., and Dvorak, J. (1989) Chimeric gene expression using maize intron in cultured cells of breadwheat. Plant Cell Rep. 8: 156–160.
Odell, J.T., Nagy, F., and Chua, N. (1985) Identification of DNA sequences required for activity of the cauliflower mosaic virus 35S promoter. Nature 313: 810–812.
Ogawa, Y., Tsuruoka, T., Inouye, S., and Niida, T. (1973) Studies on a new antibiotic SF-1293. Sci. Rep. Meiji Seika 13: 42–48.
Ohme-Takagi, M., Taylor, C.B., Newman, T.C., and Green, P.J. (1993) The effect of sequences with high AU content on mRNA stability in tobacco. Proc. Nat. Acad. Sci. USA 90: 11811–11815.
Olsen, O., Wang, X., and von Wettstein, D. (1993) Sodium azide mutagenesis: preferential generation of A◦T → G◦C transitions in the barley *Ant18* gene. Proc. Nat. Acad. Sci. USA 90: 8043–8047.
Omirulleh, S., Abraham, M., Golovkin, M., Stefanov, I., Karabaev, M.K., Mustardy, L., Morocz, S., and Dudits, D. (1993) Activity of a chimeric promoter with the doubled CaMV 35S enhancer element in protoplast-derived cells and transgenic plants in maize. Plant Mol. Biol. 21: 415–428.
Ortiz, J.P.A., Reggiardo, M.I., Ravizzini, R.A., Altabe, S.G., Cervigni, G.D.L., Spitteler, M.A., Morata, M.M., Elias, F.E., and Vallejos, R.H. (1996) Hygromycin resistance as an efficient selectable marker for wheat stable transformation. Plant Cell Rep. 15: 877–881.
Padgette, S.R., Re, D.B., Gasser, C.S., Eichholtz, D.A., Frazier, R.B., Hironaka, C.M., Levine, E.B., Shah, D.M., Fraley, R.T., and Kishore, G.M. (1991) Site-directed mutagenesis of a conserved region of the 5-enolpyruvylshikimate-3-phosphate synthase active site. J. Biol. Chem. 266: 22364–22369.
Pang, S., DeBoer, D.L., Wan, Y., Ye, G., Layton, J.G., Neher, M.K., Armstrong, C.L., Fry, J.E., Hinchee, M.A.W., and Fromm, M.E. (1996) An improved green fluorescent protein gene as a vital marker in plants. Plant Physiol. 112: 893–900.
Park, Y., Papp, I., Moscone, E.A., Iglesias, V.A., Vaucheret, H., Matzke, A.J.M., and Matzke, M.A. (1996) Gene silencing mediated by promoter homology occurs at the level of transcription and results in meiotically heritable alterations in methylation and gene activity. Plant J. 9: 183–194.
Peng, J., Wen, F., Lister, R.L., and Hodges, T.K. (1995) Inheritance of *gusA* and *neo* genes in transgenic rice. Plant Mol. Biol. 27: 91–104.
Perlak, F.J., Deaton, R.W., Armstrong, T.A., Fuchs, R.L., Sims, S.R., Greenplate, J.T., and Fischhoff. D.A. (1990) Insect resistant cotton plants. Bio/Technology 8: 939–943.
Peterhans, A., Datta, S.K., Datta, K., Goodall, G.J., Potrykus, I., and Paszkowski, J. (1990) Recognition efficiency of *Dicotyledoneae*-specific promoter and RNA processing signals in rice. Mol. Gen. Genet. 222: 361–368.
Phillips, R.L., Kaeppler, S.M., and Olhoft, P. (1994) Genetic instability of plant tissue cultures: Breakdown of normal controls. Proc. Nat. Acad. Sci. USA 91: 5222–5226.
Pierce, D.A., Mettler, I.J., Lachmansingh, A.R., Pomeroy, L.M., Weck, E.A., and Mascarenhas, D. (1987) Effect of 35S leader modifications on promoter activity. In: UCLA Symp. Mol. Cell. Biol., New Series 62: 301–310. Alan R. Liss, New York.
Poljak, L., Seum, C., Mattioni, T., and Laemmli, U.K. (1994) SARs stimulate but do not confer position independent gene expression. Nuc. Acids Res. 22: 4386–4394.
Pröls, F., and Meyer, P. (1992) The methylation patterns of chromosomal integration regions influence gene activity of transferred DNA in *Petunia hybrida*. Plant J. 2: 465–475.
Putterill, J.J., and Gardner, R.C. (1989) Initiation of translation of the $\beta$-glucuronidase reporter gene at internal AUG codons in plant cells. Plant Sci. 62: 199–205.

Rao, R.N., Allen, N.E., Hobbs, Jr. J.N., Alborn, Jr. W.E., Kirst, H.A., and Paschal, J.W. (1983) Genetic and enzymatic basis of Hygromycin B resistance in *Escherichia coli.* Antimicrob. Agents Chemother. 24: 689–695.

Rathus, C., Bower, R., and Birch, R.G. (1993) Effects of promoter, intron and enhancer elements on transient gene expression in sugar-cane and carrot protoplasts. Plant Mol. Biol. 23: 613–618.

Ray, T.B. (1984) Site of action of chlorsulfuron. Inhibition of valine and isoleucine biosynthesis in plants. Plant Physiol. 75: 827–831.

Register, J.C., Peterson, D.J., Bell, P.J., Bullock, W.P., Evans, I.J., Frame, B., Greenland, A.J., Higgs, N.S., Jepson, I., Jiao, S., Lewnau, C.J., Sillick, J.M., and Wilson, H.M. (1994) Structure and function of selectable and non-selectable transgenes in maize after introduction by particle bombardment. Plant Mol. Biol. 25: 951–961.

Reichel, C., Mathur, J., Eckes, P., Langenkemper, K., Koncz, C., Schell, J., Reiss, B., and Maas, C. (1996) Enhanced green fluorescence by the expression of an *Aequorea victoria* green fluorescent protein mutant in mono- and dicotyledonous plant cells. Proc. Nat. Acad. Sci. USA 93: 5888–5893.

Rhodes, C.A., Pierce, D.A., Mettler, I.J., Mascarenhas, D., and Detmer, J.J. (1988) Genetically transformed maize plants from protoplasts. Science 240: 204–207.

Richards, E.J. (1997) DNA methylation and plant development. Trends Genet. 13: 319–323.

Ritala, A., Aspegren, K., Kurten, U., Salmenkallio-Marttila, M., Mannonen, L., Hannus, R., Kauppinen, V., Teeri, T.H., and Enari, T. (1994) Fertile transgenic barley by particle bombardment of immature embryos. Plant Mol. Biol. 24: 317–325.

Rogers, J.C., and Rogers, S.W. (1995) Comparison of the effects of $N^6$-methyldeoxyadenosine and $N^5$-methyldeoxycytosine on transcription from nuclear gene promoters in barley. Plant J. 7: 221–233.

Rogers, S.W., and Rogers, J.C. (1992) The importance of DNA methylation for stability of foreign DNA in barley. Plant Mol. Biol. 18: 945–961.

Ronchi, A., Petroni, K., and Tonelli, C. (1995) The reduced expression of endogenous duplications (REED) in the maize *R* gene family is mediated by DNA methylation. EMBO J. 14: 5318–5328.

Rothnie, H.M. (1996) Plant mRNA 3′-end formation. Plant Mol. Biol. 32: 43–61.

Rouwendal, G.J.A., Mendes, O., Wolbert, E.J.H., and de Boer, A.D. (1997) Enhanced expression in tobacco of the gene encoding green fluorescent protein by modification of its codon usage. Plant Mol. Biol. 33: 989–999.

Russell, D.A., and Fromm, M.E. (1997) Tissue-specific expression in transgenic maize of four endosperm promoters from maize and rice. Transgen. Res. 6: 157–168.

Schernthaner, J.P., Matzke, M.A., and Matzke, A.J.M. (1988) Endosperm-specific activity of a zein gene promoter in transgenic tobacco plants. EMBO J. 7: 1249–1255.

Schledzewski, K., and Mendel, R.R. (1994) Quantitative transient gene expression: comparison of the promoters for maize polyubiquitin1, rice actin1, maize-derived *Emu* and CaMV 35S in cells of barley, maize and tobacco. Transgen. Res. 3: 249–255.

Schmitt, F., Oakeley, E.J., and Jost, J.P. (1997) Antibiotics induce genome-wide hypermethylation in cultured *Nicotiana tabacum* plants. J. Biol. Chem. 272: 1534–1540.

Schneider, M., Ow, D.W., and Howell, S.H. (1990) The *in vivo* pattern of firefly luciferase expression in transgenic plants. Plant Mol. Biol. 14: 935–947.

Schöffl, F., Schröder, G., Kliem, M., and Rieping, M. (1993) An SAR sequence containing 395 bp DNA fragment mediates enhanced, gene-dosage-correlated expression of a chimaeric heat shock gene in transgenic tobacco plants. Transgen. Res. 2: 93–100.

Seed, B., and Sheen, J. (1988) A simple phase-extraction assay for chloramphenicol acetyltransferase activity. Gene 67: 271–277.

Shaner, D.L., Anderson, P.C., and Stidham, M.A. (1984) Imidazolinones. Potent inhibitors of acetohydroxyacid synthase. Plant Physiol. 76: 545–546.

Sharp, P.M., and Li, W. (1986) Codon usage in regulatory genes in *Escherichia coli* does not

reflect selection for 'rare' codons. Nuc. Acids Res. 14: 7737–7749.
Sheen, J., Hwang, S., Niwa, Y., Kobayashi, H., and Galbraith, D.W. (1995) Green-fluorescent protein as a new vital marker in plant cells. Plant J. 8: 777–784.
Shimamoto, K., Terada, R., Izawa, T., and Fujimoto, H. (1989) Fertile transgenic rice plants regenerated from transformed protoplasts. Nature 338: 274–276.
Shimomura, O., Johnson, F.H., and Saiga, Y. (1962) Extraction, purification and properties of a bioluminescent protein from the luminous hydromedusan, Aequorea. J. Cell Comp. Physiol. 59: 223–240.
Simpson, C.G., Clark, G., Davidson, D., Smith, P., and Brown, J.W.S. (1996) Mutation of putative branchpoint consensus sequences in plant introns reduces splicing efficiency. Plant J. 9: 369–380.
Simpson, C.G., Simpson, G.G., Clark, G., Leader, D.J., Vaux, P., Guerineau, F., Waugh, R., and Brown, J.W.S. (1992) Splicing of plant pre-mRNAs. Proc. Royal Soc. Edinburgh 99B: 31–50.
Somers, D.A., Rines, H.W., Gu, W., Kaeppler, H.F., and Bushnell, W.R. (1992) Fertile, transgenic oat plants. Bio/Technology 10: 1589–1594.
Sonenberg, N. (1994) mRNA translation: influence of the 5′ and 3′ untranslated regions. Curr. Opin. Gen. Dev. 4: 310–315.
Sørensen, M.B., Müller, M., Skerritt, J., and Simpson, D. (1996) Hordein promoter methylation and transcriptional activity in wild-type and mutant barley endosperm. Mol. Gen. Genet. 250: 750–760.
Spencer, T.M., O'Brian, J.V., Start, W.G., Adams, T.R., Gordon-Kamm, W.J., and Lemaux, P.G. (1992) Segregation of transgenes in maize. Plant Mol. Biol. 18: 201–210.
Spiker, S., and Thompson, W.F. (1996) Nuclear matrix attachment regions and transgene expression in plants. Plant Physiol. 110: 15–21.
Srivastava, V., Vasil, V., and Vasil, I.K. (1996) Molecular characterization of the fate of transgenes in transformed wheat (*Triticum aestivum* L.). Theor. Appl. Genet. 92: 1031–1037.
Stam, M., Mol, J.N.M., and Kooter, J.M. (1997) The silence of genes in transgenic plants. Ann. Bot. 79: 3–12.
Stark-Lorenzen, P., Nelke, B., Hänβler, G., Mühlbach, H.P., and Thomzik, J.E. (1997) Transfer of a grapevine stilbene synthase gene to rice (*Oryza sativa* L.). Plant Cell Rep. 16: 668–673.
Tachibana, K., Watanabe, T., Sekizawa, Y., and Takematsu, T. (1986) Accumulation of ammonia in plants treated with Bialaphos. J. Pesticide Sci. 11: 33–37.
Tada, Y., Sakamoto, M., Matsuoka, M., and Fujimura, T. (1991) Expression of a monocot LHCP promoter in transgenic rice. EMBO J. 10: 1803–1808.
Takaiwa, F., Oono, K., and Kato, A. (1991) Analysis of the 5′ flanking region responsible for the endosperm-specific expression of a rice glutelin chimeric gene in transgenic tobacco. Plant Mol. Biol. 16: 49–58.
Takimoto, I., Christensen, A.H., Quail, P.H., Uchimiya, H., and Toki, S. (1994) Non-systemic expression of a stress-responsive maize polyubiquitin gene (Ubi-1) in transgenic rice plants. Plant Mol. Biol. 26: 1007–1012.
Tanaka, A., Mita, S., Ohta, S., Kyozuka, J., Shimamoto, K., and Nakamura, K. (1990) Enhancement of foreign gene expression by a dicot intron in rice but not in tobacco is correlated with an increased level of mRNA and an efficient splicing of the intron. Nuc. Acids Res. 18: 6767–6770.
Taylor, M.G., Vasil, V., and Vasil, I.K. (1993) Enhanced GUS gene expression in cereal/grass cell suspensions and immature embryos using the maize ubiquitin-based plasmid AHC25. Plant Cell Rep. 12: 491–495.
Tenover, F.C., Gilbert, T., and O'Hara, P. (1989) Nucleotide sequence of a novel kanamycin resistance gene, *aphA-7* from *Campylobacter jejuni* and comparison to other kanamycin phosphotransferase genes. Plasmid 22: 52–58.

Terada, R., Nakayama, T., Iwabuchi, M., and Shimamoto, K. (1993) A wheat histone H3 promoter confers cell division-dependent and -independent expression of the *gusA* gene in transgenic rice plants. Plant J. 3: 241–252.

Terada, R., and Shimamoto, K. (1990) Expression of CaMV35S-GUS gene in transgenic rice plants. Mol. Gen. Genet. 220: 389–392.

Thomas, M.S., and Flavell, R.B. (1990) Identification of an enhancer element for the endosperm-specific expression of high molecular weight glutenin. Plant Cell 2: 1171–1180.

Thornburg, R.W., An, G., Cleveland, T.E., Johnson, R., and Ryan, C.A. (1987) Wound-inducible expression of a potato inhibitor II-chloramphenicol acetyltransferase gene fusion in transgenic tobacco plants. Proc. Nat. Acad. Sci. USA 84: 744–748.

Tingay, S., McElroy, D., Kalla, R., Fieg, S., Wang, M., Thornton, S., and Brettell, R. (1997) *Agrobacterium tumefaciens*-mediated barley transformation. Plant J. 11: 1369–1376.

Toki, S., Takamatsu, S., Nojiri, C., Ooba, S., Anzai, H., Iwata, M., Christensen, A.H., Quail, P.H., and Uchimiya, H. (1992) Expression of a maize ubiquitin gene promoter-*bar* chimeric gene in transgenic rice plants. Plant Physiol. 100: 1503–1507.

Torbert, K.A., Rines, H.W., and Somers, D.A. (1995) Use of paromomycin as a selective agent for oat transformation. Plant Cell Rep. 14: 635–640.

Ueng, P., Galili, G., Sapanara, V., Goldsbrough, P.B., Dube, P., Beachy, R.N., and Larkins, B.A. (1988) Expression of a maize storage protein gene in petunia plants is not restricted to seeds. Plant Physiol. 86: 1281–1285.

Vasil, I.K. (1994) Molecular improvement of cereals. Plant Mol. Biol. 25: 925–937.

Vasil, I.K. (1996) Phosphinothricin-resistant crops. In: Duke, S.O. (ed.), Herbicide-Resistant Crops, pp. 85–91. CRC Press Inc.

Vasil, I.K., and Anderson, O.D. (1997) Genetic engineering of wheat gluten. Trends Plant Sci. 2: 292–297.

Vasil, V., Brown, S.M., Re, D., Fromm, M.E., and Vasil, I.K. (1991) Stably transformed callus lines from microprojectile bombardment of cell suspension cultures of wheat. Bio/Technology 9: 743–747.

Vasil, V., Castillo, A.M., Fromm, M.E., and Vasil, I.K. (1992) Herbicide resistant fertile transgenic wheat plants obtained by microprojectile bombardment of regenerable embryogenic callus. Bio/Technology 10: 667–674.

Vasil, V., Clancy, M., Ferl, R.J., Vasil, I.K., and Hannah, L.C. (1989) Increased gene expression by the first intron of maize *Shrunken-1* locus in grass species. Plant Physiol. 91: 1575–1579.

Vasil, V., Srivastava, V., Castillo, A.M., Fromm, M.E., and Vasil, I.K. (1993) Rapid production of transgenic wheat plants by direct bombardment of cultured immature embryos. Bio/Technology 11: 1553–1558.

Vellanoweth, R.L., and Okita, T.W. (1993) Regulation of expression of wheat and rice seed storage protein genes. In: Verma, D.P.S. (ed.), Control of Plant Gene Expression, pp. 377–392. CRC Press Inc.

Verdaguer, B., de Kochko, A., Beachy, R.N., and Fauquet, C. (1996) Isolation and expression in transgenic tobacco and rice plants, of the cassava vein mosaic virus (CVMV) promoter. Plant Mol. Biol. 31: 1129–1139.

Viotti, A., Balducci, C., and Weil, J.H. (1978) Adaptation of the tRNA population of maize endosperm for zein synthesis. Biochim. Biophys. Acta. 517: 125–132.

Viret, J., Mabrouk, Y., and Bogorad, L. (1994) Transcriptional photoregulation of cell-type-preferred expression of maize *rbcS*-m3: 3′ and 5′ sequences are involved. Proc. Nat. Acad. Sci. USA 91: 8577–8581.

Walker, J.C., Howard, E.A., Dennis, E.S., and Peacock, W.J. (1987) DNA sequences required for anaerobic expression of the maize alcohol dehydrogenase 1 gene. Proc. Nat. Acad. Sci. USA 84: 6624–6628.

Walters, D.A., Vetsch, C.S., Potts, D.E., and Lundqvist, R.C. (1992) Transformation and inheritance of a hygromycin phosphotransferase gene in maize plants. Plant Mol. Biol. 18: 189–200.

Wan, Y., and Lemaux, P.G. (1994) Generation of large numbers of independently transformed fertile barley plants. Plant Physiol. 104: 37–48.

Watakabe, A., Tanaka, K., and Shimura, Y. (1993) The role of exon sequences in splice site selection. Genes Dev. 7: 407–418.

Weeks, J.T., Anderson, O.D., and Blechl, A.E. (1993) Rapid production of multiple independent lines of fertile transgenic wheat (*Triticum aestivum*). Plant Physiol. 102: 1077–1084.

Weir, B.J., Lai, K.J., Caswell, K., Leung, N., Rossnagel, B.G., Båga, M., Kartha, K.K., and Chibbar, R.N. (1996) Transformation of spring barley using the enhanced regeneration system and microprojectile bombardment. In: Slinkard, A., Scoles, G., and Rossnagel, B. (eds), Proc. V Intern. Oat Conf. and VII Intern. Barley Genet. Symp. 2: 440–442. Univ. Extension Press, Univ. of Saskatchewan, Saskatoon, Canada.

White, O., Soderlund, C., Shanmugan, P., and Fields, C. (1992) Information contents and dinucleotide compositions of plant intron sequences vary with evolutionary origin. Plant Mol. Biol. 19: 1057–1064.

Wilmink, A., van de Ven, B.C.E., and Dons, J.J.M. (1995) Activity of constitutive promoters in various species from the Liliaceae. Plant Mol. Biol. 28: 949–955.

Wu, L., Ueda, T., and Messing, J. (1994) Sequence and spatial requirements for the tissue- and species-independent 3′-end processing mechanism of plant mRNA. Mol. Cell Biol. 14: 6829–6838.

Xu, D., Lei, M., and Wu, R. (1995) Expression of the rice *Osgrp1* promoter-*Gus* reporter gene is specifically associated with cell elongation/expansion and differentiation. Plant Mol. Biol. 28: 455–471.

Xu, D., McElroy, D., Thornburg, R.W., and Wu, R. (1993) Systemic induction of a potato *pin2* promoter by wounding, methyl jasmonate, and abscisic acid in transgenic rice plants. Plant Mol. Biol. 22: 573–588.

Xu, Y., Zhu, Q., Panbangred, W., Shirasu, K., and Lamb, C. (1996) Regulation, expression and function of a new basic chitinase gene in rice (*Oryza sativa* L.) Plant Mol. Biol. 30: 387–401.

Yamamoto, N., Tada, Y., and Fujimura, T. (1994) The promoter of a pine photosynthetic gene allows expression of a $\beta$-glucuronidase reporter gene in transgenic rice plants in a light-independent but tissue-specific manner. Plant Cell Physiol. 35: 773–778.

Yoder, J.A., and Bestor, T.H. (1996) Genetic analysis of genomic methylation patterns in plants and mammals. Biol. Chem. 377: 605–610.

Yoder, J.I., and Goldsbrough, A.P. (1994) Transformation systems for generating marker-free transgenic plants. Bio/Technology 12: 263–267.

Yokoi, S., Tsuchiya, T., Toriyama, K., and Hinata, K. (1997) Tapetum-specific expression of the *Osg6B* promoter-$\beta$-glucuronidase gene in transgenic rice. Plant Cell Rep. 16: 363–367.

Zhang, H.M., Yang, H., Rech, E.L., Golds, T.J., Davis, A.S., Mulligan, B.J., Cocking, E.C., and Davey, M.R. (1988) Transgenic rice plants produced by electroporation-mediated plasmid uptake into protoplasts. Plant Cell Rep. 7: 379–384.

Zhang, W., McElroy, D., and Wu, R. (1991) Analysis of rice *Act1* 5′ region activity in transgenic rice plants. Plant Cell 3: 1155–1165.

Zhang, W., and Wu, R. (1988) Efficient regeneration of transgenic plants from rice protoplasts and correctly regulated expression of the foreign gene in the plants. Theor. Appl. Genet. 76: 835–840.

Zheng, Z., Kawagoe, Y., Xiao, S., Li, Z., Okita, T., Hau, T.L., Lin, A., and Murai, N. (1993) 5′ distal and proximal *cis*-acting regulator elements are required for developmental control of a rice seed storage protein glutelin gene. Plant J. 4: 357–366.

Zhong, H., Zhang, S., Warkentin, D., Sun, B., Wu, T., Wu, R., and Sticklen, M.B. (1996) Analysis of the functional activity of the 1.4-kb 5′-region of the rice actin 1 gene in stable transgenic plants of maize (*Zea mays* L.) Plant Sci. 116: 73–84.

Zhou, H., Arrowsmith, J.W., Fromm, M.E., Hironaka, C.M., Taylor, M.L., Rodriguez, D., Pajeau, M.E., Brown, S.M., Santino, C.G., and Fry, J.E . (1995) Glyphosate-tolerant CP4 and GOX genes as a selectable marker in wheat transformation. Plant Cell Rep. 15: 159–163.

Zimny, J., Becker, D., Brettschneider, R., and Lörz, H. (1995) Fertile, transgenic *Triticale* (x *Triticosecale* Wittmack). Mol. Breeding 1: 155–164.

# 6. Transgenic Cereals: *Triticum aestivum* (wheat)

INDRA K. VASIL and VIMLA VASIL
*Laboratory of Plant Cell and Molecular Biology, 1143 Fifield Hall,*
*University of Florida, Gainesville, FL 32611-0690, USA. E-mail: ikv@gnv.ifas.ufl.edu*

## Introduction

Wheat is the most widely cultivated and important food crop in the world. Even in the developing countries of Asia, where rice has historically been the dominant and preferred crop, wheat is well on the way to becoming the number one food crop. It is the staple crop for about 35% of the human population, accounting for about 20% of the caloric intake. With global production at more than 584 million metric tons, wheat accounts for the largest share (28.5%) of the cereals market (FAO, 1996). The popularity of wheat is based largely on its high nutritive value (>10% protein, 2.4% lipids, and 79% carbohydrates), and the versatility of its use in the production of a wide range of food products, such as the many kinds of breads, pastas, cookies, etc.

The foundation crops first domesticated during the Neolithic age more than 10,000 years ago included primitive forms of wheat. The modern hexaploid bread wheat (*Triticum aestivum*) evolved later and became abundant about 8,000 years ago. The 25 or so species of the genus *Triticum* are divided into three groups – diploid, tetraploid and hexaploid – according to their chromosome number. The diploid einkorn wheat with the AA genome, *T. monococcum*, is of no economic significance, and is grown only occasionally as animal feed. The allotetraploid emmer wheat with the AABB genome, *T. turgidum* var. *durum*, grows best in warmer climates, and is prized for making pasta. The allohexaploid common or bread wheat, *T. aestivum*, is grown in cool climates with moderate rainfall, such as in North America, Europe, China, India and Australia. It has the genome constitution of AABBDD ($2n = 6\times = 42$), formed through hybridization of *T. urartu* (AA) with an unknown diploid B genome (possibly *Aegilops speltoides*), and subsequent hybridization with a diploid D genome, *T. tauschii*. The AA, BB and DD genomes of wheat are closely related, and its 21 chromosomes have been classified into seven homoeologous groups, each composed of three functionally similar chromosomes. The polyploid nature of the wheat genome makes it very suitable for the incorporation of alien genes. *T.*

*I.K. Vasil (ed.), Molecular Improvement of Cereal Crops, 133 – 147*

*aestivum* and *T. turgidum* var. *durum* account for most of the commercial production and uses of wheat.

Pests, pathogens and weeds cause enormous losses in the productivity of wheat. In spite of modern crop protection measures, actual losses of 12.4%, 9.3% and 12.3% (total losses of 34%) are caused by pathogens, pests and weeds, respectively (Oerke et al., 1994). Without crop protection the potential losses are even higher (16.7% by pathogens, 11.3% by pests, and 23.9% by weeds, for a total of 51.9%). Consequently, the development of pest and pathogen resistance has been a major objective of wheat breeding during the past several decades. Parallel efforts have been made to develop selective herbicides to control weeds in wheat fields.

Enormous increases in the productivity of wheat were obtained through innovative breeding methods and strategies of the Green Revolution (Borlaug and Dowswell, 1988, Vasil, 1990). For example, during 1965–1985, three- to four-fold increase in wheat production was achieved in India and Mexico, respectively. The same technologies have enabled China and India to become the two largest producers of wheat in the world (FAO, 1996). At the same time, however, a variety of biotic and abiotic factors have caused a decline in the global rate of increase in productivity. The following review describes how genetic transformation, combined with high efficiency plant regeneration from cultured cells, can complement and supplement wheat breeding by introducing defined genes that can protect the crop from losses caused by pests, pathogens and weeds, and can improve seed quality by manipulating genes involved in the regulation of starch and protein synthesis.

## Plant Regeneration in vitro

The earliest tissue cultures of wheat consisted primarily of proliferating masses of root meristems and primordia forming terminally differentiated cells and many aberrant roots (Mascarenhas et al., 1975, Cure and Mott, 1978, O'Hara and Street, 1978, Ozias-Akins and Vasil, 1983c). Regeneration of plants, which was rather infrequent and unpredictable, was attributed to the de-repression of shoot primordia present in the initial explant or the formation of de novo shoot meristems (Shimada et al., 1969, Dudits et al., 1975, Chin and Scott, 1977).

A variety of explants have been used in attempts to establish regenerable tissue cultures of wheat, including whole seed (caryopsis), mature and immature embryos, isolated scutellum, immature inflorescence, immature leaf, mesocotyl, apical meristem, coleoptilar-node, and root. The best and most consistent results have been obtained when immature embryos (Shimada, 1978, Shimada and Yamada, 1979, Ozias-Akins and Vasil, 1982) have been cultured on Murashige and Skoog's (1962) medium containing twice the concentration of inorganic salts, 2,4-dichlorophenoxyacetic acid, casein hydrolysate and glutamine (Ozias-Akins and Vasil, 1983a, Redway et al., 1990a). Plant regeneration has also been obtained from cultured immature inflorescences (Dudits et al., 1975, Chin and Scott, 1977, Ozias-Akins and Vasil, 1982) and isolated scutella (Bommineni and Jauhar, 1996, Maës et al., 1996).

Cultured immature embryos of wheat produce two types of callus. A friable callus that arises from the embryo axis, and a compact and nodular callus formed by the proliferation of the epithelial and sub-epithelial layers of the scutellum (Ozias-Akins and Vasil, 1982). Cells of the friable callus rapidly become terminally differentiated, but at times form organized root meristems which develop into roots. Histological and scanning electron microscopic examination of the compact and regenerable callus carried out by Ozias-Akins and Vasil (1982, 1983b) revealed the formation of somatic embryos which developed into normal and fertile plants. Regenerable embryogenic callus arises from the epiblast, instead of the scutellum, when the immature embryos are cultured such that the scutellum is in contact with the medium (Ozias-Akins and Vasil, 1983b); cereal immature embryos are normally cultured with the embryo axis in contact with the medium and the scutellum exposed.

Regenerable cell suspension cultures have been established from embryogenic callus tissues of wheat (Redway et al., 1990b). Protoplasts isolated from such suspensions have been successfully cultured to regenerate plants (Vasil et al., 1990, Chang et al., 1991, Qiao et al., 1992, He et al., 1992, Ahmed and Sagi, 1993, Yang et al., 1993). Nevertheless, the establishment and maintenance of embryogenic suspension cultures and regeneration of plants from protoplasts of wheat is still difficult and far from routine. For this reason, but particularly because more reliable methods of wheat transformation have become available (see below), interest in the culture of wheat protoplasts has declined considerably.

Plant regeneration in wheat is primarily through somatic embryogenesis and has been obtained in a wide variety of elite lines as well as commercial cultivars (Ozias-Akins and Vasil, 1982, 1983a, Maddock et al., 1983, Redway et al., 1990a, Fennell et al., 1996, Machii et al., 1998). The physiological and developmental state of the explant, the nutrient medium, and the genotype have all been described as critical factors in the establishment of regenerable embryogenic cultures (Sears and Deckard, 1982, Carman et al., 1988, He et al., 1989, Redway et al., 1990a). There has been considerable discussion about the genetic control of regeneration ability in wheat and suggestions have been made about the location of these traits on different chromosomes (Galiba et al., 1986, Felsenberg et al., 1987, Kaleikau et al., 1989, Henry et al., 1994, Ben Amer et al., 1995, 1997). The fact that the in vitro response of immature embryos is closely related to their physiological and developmental state strongly suggests that the genetic control is likely exerted through the regulation of the activity of genes involved in the synthesis and metabolism of plant growth regulators. This view is further supported by the marked influence of the polar transport of auxin on the development of wheat zygotic embryos (Fischer et al., 1997). Studies with other gramineous species have documented the important role of endogenous plant growth substances in tissue culture response (Rajasekaran et al., 1987a,b, Wenck et al., 1988).

The existence of epigenetic as well as stable genetic variation has been reported in plants recovered from tissue cultures of wheat (Karp and Maddock, 1984, Henry et al., 1996). However, there is considerable controversy about the usefulness of such variation for plant improvement as in spite of extensive attempts by many workers there are as yet no clear examples of any major new commercial varieties of wheat developed on the basis of tissue culture-induced variation (Qureshi et al., 1992). Many studies have suggested that the extent of variation observed may be a direct consequence of the genotype used (Karp and Bright, 1985, Breiman et al., 1987, Chowdhury et al., 1994).

## Methods of Transformation

Direct DNA delivery by osmotic or electric shock was used to demonstrate the transient expression of reporter genes in protoplasts of *Triticum monococcum* (Ou-Lee et al., 1986, Hauptmann et al., 1987, Oard et al., 1989). Protoplast-derived stably transformed callus lines of *T. monococcum* as well as *T. aestivum* were also obtained but no transgenic plants were recovered as the cell lines used for transformation in these studies were non-regenerable (Lörz et al., 1985, Hauptmann et al., 1988, Marsan et al., 1993, Zhou et al., 1993, Chamberlain et al., 1994, Müller et al., 1996). There is only a single report of the recovery of transgenic plants from wheat protoplasts, but fertility of the plants and transmission of the introduced genes to progeny were not demonstrated (He et al., 1994).

The direct delivery of DNA by microprojectile bombardment (the biolistic method, see chapter by Klein and Jones, this volume) is thus far the most reliable and satisfactory method for the production of fertile transgenic wheat plants (Vasil et al., 1992, 1993, Weeks et al., 1993, Becker et al., 1994, Nehra et al., 1994, Zhou et al., 1995, Altpeter et al., 1996a,b, Blechl and Anderson, 1996, Karunaratne et al., 1996, Ortiz et al., 1996, Pang et al., 1996, Takumi and Shimada, 1996, Barro et al., 1997, De Block et al., 1997). It allows the efficient delivery of plasmid DNA directly into the highly regenerable scutellar cells of the immature embryo, and the recovery of transgenic plants in as little as 8–9 weeks after bombardment, and production of R3 homozygous seed in less than one year (Altpeter et al., 1996a). The bombardment of immature embryos for the production of transgenic wheat plants, with an average transformation efficiency of about 2%, has become fairly routine, and is successfully practiced in over 20 academic and industry laboratories.

Efficient methods recently developed for *Agrobacterium tumefaciens*-mediated transformation of rice, maize and barley (see chapter by Komari and Kubo, this volume), have now been successfully adapted for wheat. Cheng et al. (1997) have described the production of a number of normal and fertile transgenic wheat plants following incubation of immature embryos and embryogenic calli with *A. tumefaciens.*

There are three reports of the production of transgenic wheat by the pollen tube method (Hess et al. 1990, Zeng et al. 1994, Zilberstein et al. 1994). Each is based on ambiguous and incomplete evidence and has never been independently substantiated.

Most of the transgenic wheat plants studied so far contained an average of 2–5 copies of the transgene, irrespective of the method of transformation used (see also chapter by Vasil, this volume). Many plants with single copy insertions were also found. There was no clear correlation between copy number and gene expression.

## Reporter Genes

In the early experiments, transient expression of the chloramphenicol acetyl transferase (CAT) gene was used to evaluate the efficiency of DNA delivery and expression (Hauptmann et al., 1987). This cumbersome method was soon abandoned in favor of the bacterial *uidA* gene encoding $\beta$-glucuronidase (GUS), which has been used extensively in wheat transformation (Vasil et al., 1991, 1992, 1993, Weeks et al., 1993, Becker et al., 1994, Nehra et al., 1994). However, because of the destructive nature of CAT and GUS assays, there has been some interest in using non-destructive reporter genes. One such gene, the synthetic green fluorescent protein (GFP) gene (*pgpf*) has been used by Pang et al., (1996) for the visual identification of transgenic wheat plants by fluorescence microscopy. The synthetic gene provided a 150-fold increase in fluorescence over the wild-type *gfp*. Also, transient as well as stable expression of the firefly luciferase gene has been demonstrated in bombarded scutella of wheat embryos (Lonsdale et al., 1998). These techniques allow the study of gene expression in tissue explants from the time of bombardment to plant regeneration, and can be useful in improving transformation efficiencies by providing an insight into various factors involved in stable transformation. Nevertheless, at the present time the *uidA* gene is still the most widely used reporter gene in wheat transformation.

## Selectable Marker Genes

Stably transformed callus lines were first obtained by the direct delivery of the neomycin phosphotransferase II (NPTII) gene into protoplasts of *Triticum monococcum* (Lörz et al., 1985) and cell suspension cultures of wheat (Vasil et al., 1991), followed by selection on kanamycin containing media. However, many gramineous species, including wheat, are naturally resistant to kanamycin (Hauptmann et al., 1988). Therefore, antibiotics like geneticin or G418 have been used for the selection of transformants expressing the NPTII gene (Nehra et al., 1994, Cheng et al., 1997). The hygromycin phosphotransferase (HPT) gene, *hph*, which confers resistance to hygromycin, has been used extensively for the selection of rice and forage grass transformants. It has been used also for the transformation of wheat in a few cases (Hauptmann et al., 1988, Ortiz et al., 1996), but has not proven to be entirely satisfactory because of the inhibitory effect of hygromycin on plant regeneration. Other selectable marker genes used for wheat transformation are 5-enolpyruvylshikimate phosphate synthase (EPSPS) and glyphosate oxidoreductase (GOX), which confer resistance to glyphosate, the active ingredient of the non-selective herbicide Roundup® (Vasil et al., 1991, Zhou et al., 1995). However, the most efficient selection of wheat transformants has been obtained with the *bar* gene, which codes for phosphinothricin acetyl transferase (PAT), and confers resistance to phosphinothricin (Vasil, 1996) , the active ingredient of the non-selective herbicide Basta® (Vasil et al., 1992, 1993, Weeks et al., 1993, Becker et al., 1994, Nehra et al., 1994, Altpeter et al., 1996a, Takumi and Shimada, 1996).

## Promoters

The 35S promoter of the cauliflower mosaic virus (CaMV35S) is the most widely used constitutive promoter in plant transformation. However, the levels of gene expression obtained in gramineous cells with this promoter are upto 100-fold less than in dicotyledonous species (Fromm et al., 1985, Hauptmann et al., 1987). A number of other strategies and promoters have, therefore, been used to increase the level of gene expression. These include the duplication of the 35S promoter sequence (Vasil et al., 1991, Pang et al., 1996), monocot promoters, and the introduction of an intron in the transcriptional unit of the reporter gene (Callis et al., 1987, Vasil et al., 1989, Clancy et al., 1994, Karunaratne et al., 1996). The promoter of the maize ubiquitin 1 (*Ubi-1*) gene (Taylor et al., 1993, Christensen and Quail, 1996) has been widely used in wheat transformation as it consistently provides high levels of gene expression (Vasil et al., 1993, Weeks et al., 1993). The plasmids used contain the promoter, exon and the first intron of *Ubi-1* (Christensen and Quail, 1996). The promoter and intron 1 of the *Adh1* gene of maize (Vasil et al., 1991, 1992), and the promoter and intron 1 of the rice *Act1* gene (Nehra et al., 1994, Takumi and Shimada, 1996), have also been used to obtain high levels of gene expression in transgenic wheat.

Gene expression regulated by constitutive promoters does not discriminate between different tissues in the plant. Constitutive expression of genes may not always be desirable or required, such as in those instances where expression of the transgene is needed only in the endosperm for manipulation of starch and/or protein quality and quantity. The use of tissue specific promoters in wheat so far has been limited. In three instances the promoters of high molecular weight glutenin sub-unit (HMW-GS) genes have been used to express HMW-GS proteins in the endosperm (Altpeter et al., 1996b, Blechl and Anderson, 1996, Barro et al., 1997). In addition, a tapetum-specific promoter was used by De Block et al. (1997) to express the *barnase* gene which causes pollen abortion resulting in male sterility.

## Agronomically Important Transgenes used

*Herbicide Resistance.* Even with the use of herbicides, productivity losses in wheat attributed to weeds amount to 12.3% (Oerke et al., 1994). Weeds commonly found in wheat fields are known to have developed a high degree of resistance to the commonly used herbicides (Gressel, 1998). Therefore, phosphinothricin (Basta®) and glyphosate (Roundup®) have been suggested as suitable herbicides for weed control in wheat fields (Gressel, 1998). The *bar* gene conferring resistance to the herbicide phosphinothricin is the most common gene that has been introduced into wheat (Vasil et al., 1992, 1993, Weeks et al., 1993, Becker et al., 1994, Nehra et al., 1994, Altpeter et al., 1996a, Vasil, 1996). Roundup® resistant transgenic wheat plants expressing the EPSPS (CP4) and GOX genes have also been obtained (Zhou et al., 1995). Basta® and Roundup® resistance have the potential to provide substantial relief from yield losses caused by weeds in wheat fields.

*Bread-Making Quality.* The unique quality of wheat to make leavened bread is related to the presence of prolamins, a novel group of storage proteins, in its endosperm (Payne, 1987, Galili, 1997, Vasil and Anderson, 1997). The gliadins and glutenins comprising the wheat prolamins form a continuous proteinaceous network, the gluten, when wheat flour is mixed with water. The elasticity/strength and extensibility/viscosity of the dough, important for bread making, are related to the types and quantity of glutenins and gliadins present. However, biochemical and genetic studies have identified a specific group of gluten proteins, the high-molecular-weight glutenin subunits (HMW-GS), to be the most important determinants of bread-making quality. Two tightly linked pairs of HMW-GS genes are present on the 1A, 1B and 1D chromosomes of hexaploid wheat. Each wheat cultivar, therefore, contains six HMW-GS genes, but only three, four or five HMW-GS are present as some of the genes are silent. HMW-GS represent up to 10% of the total seed protein, as each HMW-GS accounts for about 2% of the total extractable protein.

The close linkage of the HMW-GS genes makes it difficult to manipulate them by traditional breeding methods. A number of HMW-GS genes have been identified and cloned to study their effect on bread-making qualities of wheat. Among the HMW-GS known to impart good bread-making qualities are 1Ax1, 1Dx5 and 1Dy10 (Vasil and Anderson, 1997). Transformation of HMW-GS 1Ax1 and/or Dx 5 genes, or a hybrid Dy10:Dx5 gene construct, into wheat resulted in the accumulation of the transgenic HMW-GS in the endosperm at levels similar to or higher than the endogenous HMW-GS (Blechl and Anderson, 1996, Altpeter et al., 1996b, Barro et al., 1997). The amount of transgenic HMW-GS produced remained consistent over several generations, and resulted in improved functional properties (Altpeter et al., 1996b, Barro et al., 1997, Vasil et al., unpublished results).

*Insect Resistance.* Although many strategies and genes have been used to confer insect resistance to transgenic plants (Jouanin et al., 1998, Schuler et al., 1998), their use in wheat has so far been rather limited. A barley trypsin inhibitor CMe (BTI-CMe) gene was introduced into wheat by the biolistic bombardment of immature embryos (Altpeter et al., 1998). The BTI-CMe was correctly transcribed and translated in the transgenic plants, amounting to 1.1% of the total extracted protein in wheat seeds. BTI-CMe caused a significant reduction in the survival of the Angoumois Grain Moth (*Sitotroga cerealella*) which causes considerable damage to wheat during storage. The presence of the BTI-CMe in transgenic leaves did not have any significant effect on leaf feeding insects.

*Pathogen Resistance.* Plant pathogens are a major cause of losses in wheat productivity. Unfortunately, in spite of many attempts virus or fungal resistance has thus far not been successfully engineered in wheat. A gene encoding the barley yellow mosaic virus (BaYMV) coat protein was introduced into wheat (Karunaratne et al., 1996). The presence of BaYMV coat protein was detected in the transgenic plants. Although BaYMV does not infect wheat, it was suggested that the transgenic plants might be protected against wheat spindle streak virus because of serological and amino acid sequence homologies between the coat proteins of the two viruses. No supporting data were presented.

*Male Sterility.* Production of hybrid wheat has been one of the important objectives of many wheat breeding programs (Pickett, 1993). Complete male sterility has not been attained through the cytoplasmic (CMS) and nuclear (NMS) male-sterile systems and chemical agents that have been used to obtain hybrid wheat. A tapetum-specific promoter driving the *barnase* gene, that had been used earlier to induce premature dissolution of the anther tapetum causing complete abortion of pollen development in tobacco and rapeseed (Mariani et al., 1990), was successfully used by De Block et al. (1997) to induce nuclear male sterility in wheat. In such a system, the ribonuclease-inhibitor *barstar* gene (Mariani et al., 1992) is used to restore fertility to male-sterile plants.

## Stability of Transgenes

Instability of transgenes (gene silencing and excision) has been described in a number plant species (Finnegan and McElroy, 1994, Flavell, 1994, Matzke and Matzke, 1995). Fortunately, this does not appear to be a serious problem in wheat, where the stability of integration and expression of transgenes is common (Srivastava et al., 1996, Altpeter et al., 1996b, see also chapter by Barcelo et al., this volume), as also is the case in maize (see chapter by Gordon-Kamm et al., this volume) and rice (see chapter by Datta, this volume). On the other hand, considerable variability is reported in barley (see chapter by Lemaux et al., this volume). Such marked differences between species may well be related to inherent genomic instability and characteristics.

## Future Needs and Directions

Transformation of wheat, first achieved in 1992 (Vasil et al., 1992), has now become fairly routine in many laboratories. As described earlier, a number of important genes have been introduced and stably expressed in wheat. Some of the transgenic plants, especially those containing herbicide and HMW-GS genes, are being field-tested and are likely to be integrated soon into wheat improvement programs and cultivated commercially.

The bread-making quality of wheat depends on the quality and quantity of the two gluten proteins, glutenin and gliadin. Integration and expression of HMW-GS genes has been shown to increase the amount of HMW-GS in wheat endosperm (Blechl and Anderson, 1996, Altpeter et al., 1996b, Barro et al., 1997). Vasil and Anderson (1997) have identified a number of targets for the engineering of gluten, including changing the pattern of disulfide bonds, altering repeat domain length, extending the glutenin macropolymer, etc.

Starch is the major component of wheat endosperm. Genes regulating starch synthesis have been isolated and used to increase starch content in plants (Stark et al., 1992). In addition to being a major source of calories, starch also has multiple industrial uses. In countries such as Canada, which can produce far more wheat than they need or export, there is growing interest in manipulating wheat starch quality by altering amylose/amylopectin ratios in order to make it more suitable for industrial uses. This may be accomplished by the transformation of starch branching enzyme (SBE) and/or other starch synthesis genes for tissue specific expression in the wheat endosperm (Shimada et al., 1993, Visser and Jacobsen, 1993, Wasserman et al., 1995). As a first step in that direction, Nair et al. (1997) have isolated and characterized SBEII cDNA from wheat.

Engineering of wheat for pathogen and pest resistance is slow, and needs to be accelerated. There is evidence from several unpublished reports that some of the resistance genes used successfully in other crops may not provide the desired resistance in wheat. The reasons for this are not fully understood. It may be useful, therefore, to clone some of the better-characterized endogenous genes that have been described in wheat (Van Ginkel and Rajaram, 1993, McIntosh et al., 1995), as was done recently in rice (Song et al., 1995). The wheat genome is large ($1.6 \times 10^{10}$ bp) and consists of 80% repetitive sequences (Smith and Flavell, 1975). Map-based cloning of genes in wheat is therefore difficult and challenging. In this regard, the high degree of synteny found in cereal genomes may be a valuable asset (see chapter by Bennetzen, this volume). Progress in this direction is being made, as shown by the recent cloning of a receptor-like kinase gene in wheat (Feuillet et al., 1997).

## References

Ahmed, K.Z., and Sagi, F. (1993) Culture of and fertile plant regeneration from regenerable embryogenic cell-derived protoplasts of wheat (*Triticum aestivum* L.). Plant Cell Rep. 12: 175–179.

Altpeter, F., Diaz, I., McAuslane, H., Gaddour, K., Carbonero, P., and Vasil, I.K. (1998) Increased insect resistance in transgenic wheat stably expressing trypsin inhibitor CMe. Abstract (p. 35), IX Internat. Cong. Plant Tiss. Cult. Jerusalem. (*see also* Altpeter et al. 1999. Mol. Breed. In Press).

Altpeter, F., Vasil, V., Srivastava, V., Stöger, E., and Vasil, I.K. (1996a) Accelerated production of transgenic wheat (*Triticum aestivum* L.) plants. Plant Cell Rep. 16: 12–17.

Altpeter, F., Vasil, V., Srivastava, V., and Vasil, I.K. (1996b) Integration and expression of the high-molecular-weight glutenin subunit 1Ax1 gene into wheat. Nature Biotechnology 14: 1155–1159.

Barro, F., Rooke, L., Békés, F., Gras, P., Tatham, A.S., Fido, R., Lazzeri, P.A., Shewry, P.R., and Barcelo, P. (1997) Transformation of wheat with high molecular weight subunit genes results in improved functional properties. Nature Biotechnology 15: 1295–1299.

Becker, D., Brettschneider, R., and Lörz, H. (1994) Fertile transgenic wheat from microprojectile bombardment of scutellar tissue. Plant J. 5: 299–307.

Ben Amer, I.M., Worland, A.J., and Börner, A. (1995) Chromosomal location of genes affecting tissue culture response in wheat. Plant Breed. 114: 84–85.

Ben Amer, I.M., Korzun, V., Worland, A.J., and Börner, A. (1997) Genetic mapping of QTL controlling tissue-culture response on chromosome 2B of wheat (*Triticum aestivum* L.) in relation to major genes and RFLP markers. Theor. Appl. Genet. 94: 1047–1052.

Blechl, A.E., and Anderson, O.D. (1966) Expression of a novel high molecular weight glutenin subunit gene in transgenic wheat. Nature Biotechnology 14: 875–879.

Bommineni, V.R., and Jauhar, P.P. (1996) Regeneration of plantlets through isolated scutellum culture of durum wheat. Plant Sci. 116: 197–203.

Borlaug, N.E., and Dowswell, C.R. (1988) World revolution in agriculture. In '1988 Book of the Year', pp. 5–14. Encyclopedia Britannica, Chicago.

Breiman, A., Felsenberg, T., and Galun, E. (1987) *Nor* loci analysis in progenies of plants regenerated from the scutellar callus of bread-wheat. Theor. Appl. Genet. 73: 827–831.

Callis, J., Fromm, M.E., and Walbot, V. (1987) Introns increase gene expression in cultured maize cells. Genes Dev. 1: 1183–1200.

Carman, J.G., Jefferson, N.E., and Campbell, W.F. (1988) Induction of embryogenic *Triticum aestivum* calli. I. Quantification of genotype and culture medium effect. Plant Cell Tiss. Org. Cult. 12: 83–95.

Carnes, M.G., and Wright, M.S. (1988) Endogenous hormone levels of immature corn kernels of A188, Missouri-17, and DeKalb XL-12. Plant Sci. 57: 195–203.

Chamberlain, D.A., Brettell, R.I.S., Last, D.I., Witrzens, B., McElroy, D., Dolferus, R., and Dennis, E.S. (1994) The use of Emu promoter with antibiotic and herbicide resistance genes for the selection of transgenic wheat callus and rice plants. Australian J. Plant Physiol. 21: 95–112.

Chang, Y.F., Wang, W.C., Warfield, C.Y., Nguyen, H.T., and Wong, J.R. (1991) Plant regeneration from protoplasts isolated from long-term cell cultures of wheat (*Triticum aestivum* L.). Plant Cell Rep. 9: 611–614.

Cheng, M., Fry, J.E., Pang, S., Zhou, H., Hironaka, C.M., Duncan, D.R., Conner, T.W., and Wan, Y. (1997) Genetic transformation of wheat mediated by *Agrobacterium tumefaciens.* Plant Physiol. 115: 971–980.

Chin, J.C., and Scott, K.J. (1977) Studies on the formation of roots and shoots in wheat callus cultures. Ann. Bot. 41: 473–481.

Chowdhury, M.K.U., Vasil, V., and Vasil, I.K. (1994) Molecular analysis of plants regenerated from embryogenic cultures of wheat (*Triticum aestivum* L.). Theor. Appl. Genet. 87: 821–828.

Christensen, A.H., and Quail, P.H. (1996) Ubiquitin promoter-based vectors for high level expression of selectable and/or screenable marker genes in monocotyledonous plants. Transgen. Res. 5: 213–218.

Clancy, M., Vasil, V., Hannah, L.C., and Vasil, I.K. (1994) Maize *Shrunken-1* intron and exon regions increase gene expression in maize protoplasts. Plant Sci. 98: 151–161.

Cure, M.W., and Mott, R.L. (1978) A comparative anatomical study of organogenesis in cultured tissues of maize, wheat and oats. Physiol. Plant. 42: 91–96.

De Block, M., Debrouwer, D., and Moens, T. (1997) The development of a nuclear male sterility system in wheat. Expression of the *barnase* gene under the control of tapetum specific promoters. Theor. Appl. Genet. 95: 125–131.

Dudits, D., Nemet, G., and Haydu, Z. (1975) Study of callus growth and organ formation in wheat (*Triticum aestivum*) tissue cultures. Canadian J. Bot. 53: 957–963.

FAO (1996) FAO Production Yearbook. FAO, Rome.

Felsenberg, T., Feldman, M., and Galun, E. (1987) Aneuploid and alloplasmic lines as tools for the study of nuclear and cytoplasmic control of culture ability and regeneration of scutellar calli from common wheat. Theor. Appl. Genet. 74: 802–810.

Fennell, S., Bohorova, N., van Ginkel, M., Crossa, J., and Hoisington, D. (1996) Plant regeneration from immature embryos of 48 elite CIMMYT bread wheats. Theor. Appl. Genet. 92: 163–169.

Feuillet, C., Schachermayr, G., and Keller, B. (1997) Molecular cloning of a new receptor-like kinase gene encoded at the *Lr10* disease resistance locus of wheat. Plant J. 11: 45–52.

Finnegan, J., and McElroy, D. (1994) Transgenic inactivation: plants fight back! Bio/Technology 12: 883–888.

Fischer, C., Speth, V., Fleig-Eberenz, S., and Neuhaus, G. (1997) Induction of zygotic proembryos in wheat: influence of auxin polar transport. Plant Cell 9: 1767–1780.

Flavell, R.B. (1994) Inactivation of gene expression in plants as a consequence of novel sequence duplications. Proc. Nat. Acad. Sci. USA 91: 3490–3496.

Fromm, M.E., Taylor, L.P., and Walbot, V. (1985) Expression of genes transferred into monocot and dicot plant cells by electroporation. Proc. Nat. Acad. Sci. USA 82: 5824–5828.

Galiba, G., Kovacs, G., and Sutka, J. (1986) Substitution analysis of plant regeneration from callus culture in wheat. Plant Breed. 97: 261–263.

Galili, G. (1997) The prolamin storage proteins of wheat and its relatives. In: Larkins, B.A., and Vasil, I.K. (eds), Cellular and Molecular Biology of Plant Seed Development, pp. 221–256. Kluwer, Dordrecht.

Gressel, J. (1998) Biotechnology of weed control. In: Altman, A. (ed), Agricultural Biotechnology, pp. 295–325. Marcel Dekker, New York.

Hauptmann, R.M., Ozias-Akins, P., Vasil, V., Tabaeizadeh, Z., Rogers, S.G., Horsch, R.B., Vasil, I.K., and Fraley, R.T. (1987) Transient expression of electroporated DNA in monocotyledonous and dicotyledonous species. Plant Cell Rep. 6: 265–270.

Hauptmann, R.M., Vasil, V., Ozias-Akins, P., Tabaeizadeh, Z., Rogers, S.G., Fraley, R.T., Horsch, R.B., and Vasil, I.K. (1988) Evaluation of selectable markers for obtaining stable transformants in the Gramineae. Plant Physiol. 86: 602–606.

He, D.G., Mouradov, A., Yang, Y.M., Mouradova, E., and Scott, K.J. (1994) Transformation of wheat (*Triticum aestivum* L.) through electroporation of protoplasts. Plant Cell Rep. 14: 192–196.

He, D.G., Yang, Y.M., and Scott, K.J. (1989) The effect of macroelements in the induction of embryogenic callus from immature embryos of wheat (*Triticum aestivum* L.). Plant Sci. 64: 251–258.

He, D.G., Yang, Y.M., and Scott, K.J. (1992) Plant regeneration from protoplasts of wheat (*Triticum aestivum* cv. Hartog). Plant Cell Rep. 11: 16–19.

Henry, Y., Marcotte, J.L., and De Buyser, J. (1994) Chromosomal location of genes controlling short-term and long-term somatic embryogenesis in wheat revealed by immature embryo culture of aneuploid lines. Theor. Appl. Genet. 89: 344–350.

Henry, Y., Marcotte, J.L., and De Buyser, J. (1996) The effect of aneuploidy on karyotype abnormalities in wheat plants regenerated from short- and long-term somatic embryogenesis. Plant Sci. 114: 101–109.

Hess, D., Dressler, K., and Nimmrichter, R. (1990) Transformation experiments by pipetting *Agrobacterium* into the spikelets of wheat (*Triticum aestivum* L.). Plant Sci. 72: 233–244.

Jouanin, L., Bonadé-Bottino, M., Girard, C., Morrot, G., and Giband, M. (1998) Transgenic plants for insect resistance. Plant Sci. 131: 1–11.

Kaleikau, E.K., Sears, R.G., and Gill, B.S. (1989) Control of tissue culture response in wheat (*Triticum aestivum* L.). Theor. Appl. Genet. 78: 783–787.

Karp, A., and Bright, S.W.J. (1985) On the causes and origins of somaclonal variation. Oxford Surv. Plant Mol. Cell. Biol. 2: 199–234.

Karp, A., and Maddock, S.E. (1984) Chromosome variation in wheat plants regenerated from cultured immature embryos. Theor. Appl. Genet. 67: 249–255.

Karunaratne, S., Sohn, A., Mouradov, A., Scott, J., Steinbiss, H.-H., and Scott, K.J. (1996) Transformation of wheat with the gene encoding the coat protein of barley yellow mosaic virus. Australian J. Plant Physiol. 23: 429–435.

Lonsdale, D.M., Lindup, S., Moisan, L.J., and Harvey, A.J. (1998) Using firefly luciferase to identify the transition from transient to stable expression in bombarded wheat scutellar tissue. Physiol. Plant. 102: 447–453.

Lörz, H., Baker, B., and Schell, J. (1985) Gene transfer to cereal cells mediated by protoplast transformation. Mol. Gen. Genet. 199: 178–182.

Machii, H., Mizuno, H., Hirabayashi, T., Li, H., and Hagio, T. (1998) Screening wheat genotypes for high callus induction and regeneration capability from anther and immature embryo cultures. Plant Cell Tiss. Org. Cult. 53: 67–74.

Maddock, S.E., Lancaster, V.A., Risiott, R., and Franklin, J. (1983) Plant regeneration from cultured immature embryos and inflorescences of 25 cultivars of wheat (*Triticum aestivum*). J. Exp. Bot. 34: 915–926.

Maës, O.C., Chibbar, R.N., Caswell, K., Leung, N., and Kartha, K.K. (1996) Somatic embryogenesis from isolated scutella of wheat: effects of physical, physiological and genetic factors. Plant Sci. 121: 75–84.

Mariani, C., De Beuckeleer, M., Truettner, J., Leemans, J., and Goldberg, R.B. (1990) Induction of male sterility in plants by a chimaeric ribonuclease gene. Nature 347: 737–741.

Mariani, C., Gossele, V., De Beuckeleer, M., De Block, M., Goldberg, R.B., De Greef, W., and Leemans, J. (1992) A chimeric ribonuclease-inhibitor gene restores fertility to male-sterile plants. Nature 357: 384–387..

Marsan, P.A., Lupotto, E., Locatelli, F., Qiao, Y.-M., and Cattaneo, M. (1993) Analysis of stable events of transformation in wheat via PEG-mediated DNA uptake into protoplasts. Plant Sci. 93: 85–94.

Mascarenhas, A.F., Pathak, M., Hendre, R.R., and Jagannathan, V. (1975) Tissue cultures of maize, wheat, rice and sorghum. I. Initiation of viable callus and root cultures. Indian J. Exp. Biol. 13: 103–107.

Matzke, M.A., and Matzke, A.J.M. (1995) How and why do plants inactivate homologous (trans)genes? Plant Physiol. 107: 679–685.

McIntosh, R.A., Wellings, C.R., and Park, R.F. (1995) Wheat Rusts: An Atlas of Resistance Genes. Kluwer, Dordrecht.

Müller, E., Lörz, H., and Lütticke, S. (1996) Variability of transgene expression in clonal cell lines of wheat. Plant Sci. 114: 71–82.

Murashige, T., and Skoog, F. (1962) A revised medium for rapid growth and bioassays with tobacco tissue cultures. Physiol. Plant. 15: 473–497.

Nair, R.B., Båga, M., Scoles, G.J., Kartha, K.K., and Chibbar, R.N. (1997) Isolation, characterization and expression analysis of a starch branching enzyme II cDNA from wheat. Plant Sci. 122: 153–163.

Nehra, N.S., Chibbar, R.N., Leung, N., Caswell, K., Mallard, C., Steinhauer, L., Baga, M., and Kartha, K.K. (1994) Self-fertile transgenic wheat plants regenerated from isolated scutellar tissues following microprojectile bombardment with two distinct gene constructs. Plant J. 5: 285–297.

Oard, J.H., Paige, D., and Dvorak, J. (1989) Chimeric gene expression using maize intron in cultured cells of bread wheat. Plant Cell Rep. 8: 156–160.

Oerke, E.-C., Dehne, H.-W., Schöbeck, F., and Weber, A. (1994) Crop Production and Crop Protection. Elsevier, Amsterdam.

O'Hara, J.F., and Street, H.E. (1978) Wheat callus culture: the initiation, growth and organogenesis of callus derived from various explant sources. Ann. Bot. 42: 1029–1038.

Ortiz, J.P.A., Reggiardo, M.I., Ravizzini, R.A., Altabe, S.G., Cervigni, G.D.L., Spitteler, M.A., Morata, M.M., Elias, F.E., and Vallejos, R.H. (1996) Hygromycin resistance as an efficient selectable marker for wheat stable transformation. Plant Cell Rep. 15: 877–881.

Ou-Lee, T.-M., Turgeon, R., and Wu, R. (1986) Expression of a foreign gene linked to either a plant-virus or a *Drosophila* promoter, after electroporation of protoplasts of rice, wheat and sorghum. Proc. Nat. Acad. Sci. USA 83: 6815–6819.

Ozias-Akins, P., and Vasil, I.K. (1982) Plant regeneration from cultured immature embryos and inflorescences of *Triticum aestivum* L. (wheat): evidence for somatic embryogenesis. Protoplasma 110: 95–105.

Ozias-Akins, P., and Vasil, I.K. (1983a) Improved efficiency and normalization of somatic embryogenesis in *Triticum aestivum* (wheat). Protoplasma 117: 40–44.

Ozias-Akins, P., and Vasil, I.K. (1983b) Proliferation of and plant regeneration from the epiblast of *Triticum aestivum* (wheat; Gramineae) embryos. American J. Bot. 70: 1092–1097.

Ozias-Akins, P., and Vasil, I.K. (1983c) Callus induction and growth from the mature embryo of *Triticum aestivum* (wheat). Protoplasma 115: 104–113.

Pang, S.-Z., DeBoer, D.L., Wan, Y., Ye, G., Layton, J.G., Neher, M.K., Armstrong, C.L., Fry, J.E., Hinchee, M.A.W., and Fromm, M.E. (1996) An improved green fluorescent protein gene as a vital marker in plants. Plant Physiol. 112: 893–900.

Payne, P.I. (1987) Genetics of wheat storage proteins and the effect of allelic variation on bread-making quality. Ann. Rev. Plant Physiol. 38: 141–153.

Pickett, A.A. (1993) Hybrid Wheat: Results and Problems. Paul Parey, Berlin.

Qiao, Y.M., Cattaneo, M., Locatelli, F., and Lupotto, E. (1992) Plant regeneration from long term suspension culture-derived protoplasts of hexaploid wheat (*Triticum aestivum* L.). Plant Cell Rep. 11: 262–265.

Qureshi, J.A., Hucl, P., and Kartha, K.K. (1992) Is somaclonal variation a reliable tool for spring wheat improvement? Euphytica 60: 221–228.

Rajasekaran, K., Hein, M.B., and Vasil, I.K. (1987a) Endogenous abscisic acid and indole-3-acetic acid and somatic embryogenesis in cultured leaf explants of *Pennisetum purpureum* Schum. Plant Physiol. 84: 47–51.

Rajasekaran, K., Hein, M.B., Davis, G.C., Carnes, M.G., and Vasil, I.K. (1987b) Endogenous growth regulators in leaves and tissue cultures of *Pennisetum purpureum* Schum. J. Plant Physiol. 130: 13–25.

Redway, F.A., Vasil, V., Lu, D., and Vasil, I.K. (1990a) Identification of callus types for long-term maintenance and regeneration from commercial cultivars of wheat (*Triticum aestivum* L.) Theor. Appl. Genet. 79: 609–617.

Redway, F.A., Vasil, V., and Vasil, I.K. (1990b) Characterization and regeneration of wheat (*Triticum aestivum* L.) embryogenic cell suspension cultures. Plant Cell Rep. 8: 714–717.

Schuler, T.H., Poppy, G.M., Kerry, B.R., and Denholm, I. (1998) Insect-resistant transgenic plants. Trends Biotech. 16: 168–175.

Sears, R.G., and Deckard, E.L. (1982) Tissue culture variability in wheat: callus induction and plant regeneration. Crop Sci. 22: 546–550.

Shimada, H., Tada, Y., Kawasaki, T., and Fujimura, F. (1993) Anti-sense regulation of the rice waxy gene expression using a PCR-amplified fragment of the rice genome reduces the amylose content in grain starch. Theor. Appl. Genet. 86: 665–672.

Shimada, T. (1978) Plant regeneration from the callus induced from wheat embryo. Japanese J. Genet. 53: 371–374.

Shimada, T., Sasakuma, T., and Tsunewaki, K. (1969) In vitro culture of wheat tissues. I. Callus formation, organ redifferentiation and single cell culture. Canadian J. Genet. Cytol. 11: 294–304.

Shimada, T., and Yamada, Y. (1979) Wheat plants regenerated from embryo cell cultures. Japanese J. Genet. 54: 379–385.

Smith, D.B., and Flavell, R.B. (1975) Characterization of the wheat genome by renaturation kinetics. Chromosoma 50: 223–242.

Song, W.Y., Wang, G.L., Chen, L.L., Kim, H.S., Pi, L.Y., Holsten, T., Gardner, J., Wang, B., Zhai, W.X., Chu, L.H., Fauquet, C.M., and Ronald, P. (1995) A receptor kinase-like protein encoded by the rice disease resistance gene, *Xa21*. Science 270: 1804–1806.

Srivastava, V., Vasil, V., and Vasil, I.K. (1996) Molecular characterization of the fate of transgenes in transformed wheat. Theor. Appl. Genet. 92: 1031–1037.

Stark, D.M., Timmerman, K.P., Barry, G.F., Preiss, J., and Kishore, G.M. (1992) Regulation of the amount of starch in plant tissues by ADP glucose pyrophosphorylase. Science 258: 287–292.

Takumi, S., and Shimada, T. (1996) Production of transgenic wheat through particle bombardment of scutellar tissues: frequency is influenced by culture duration. J. Plant Physiol. 149: 418–423.
Taylor, M.G., Vasil, V., and Vasil, I.K. (1993) Enhanced GUS gene expression in cereal/grass cell suspensions and immature embryos using the maize ubiquitin-based plasmid pAHC25. Plant Cell Rep. 12: 491–495.
Van Ginkel, M., and Rajaram, S. (1993) Breeding for durable resistance to diseases in wheat: an international perspective. In: Jacobs, T., and Parlevliet, J.E. (eds), Durability of Disease Resistance, pp. 259–272. Kluwer, Dordrecht.
Vasil, I.K. (1990) The realities and challenges of plant biotechnology. Bio/Technology 8: 296–301.
Vasil, I.K. (1996) Phosphinothricin-resistant crops. In: Duke, S.O. (ed), Herbicide-Resistant Crops, pp. 85–91. CRC Press, Boca Raton.
Vasil, I.K., and Anderson, O.D. (1997) Genetic engineering of wheat gluten. Trends Plant Sci. 2: 292–297.
Vasil, V., Brown, S.M., Re, D., Fromm, M.E., and Vasil, I.K. (1991) Stably transformed callus lines from microprojectile bombardment of cell suspension cultures of wheat. Bio/Technology 9: 743–747.
Vasil, V., Castillo, A.M., Fromm, M.E., and Vasil, I.K. (1992) Herbicide resistant fertile transgenic wheat plants obtained by microprojectile bombardment of regenerable embryogenic callus. Bio/Technology 10: 667–674.
Vasil, V., Clancy, M., Ferl, R.J., Vasil, I.K., and Hannah, L.C. (1989) Increased gene expression by the first intron of maize *Shrunken-1* locus in grass species. Plant Physiol. 91: 1575–1579.
Vasil, V., Redway, F., and Vasil, I.K. (1990) Regeneration of plants from embryogenic suspension culture protoplasts of wheat (*Triticum aestivum* L.). Bio/Technology 8: 429–434.
Vasil, V., Srivastava, V., Castillo, A.M., Fromm, M.E., and Vasil, I.K. (1993) Rapid production of transgenic wheat plants by direct bombardment of cultured immature embryos. Bio/Technology 11: 1553–1558.
Visser, R.G.F., and Jacobsen, E. (1993) Towards modifying plants for altered starch content and composition. Trends Biotech. 11: 63–68.
Wasserman, B.P., Harn, C., Mu-Forster, C., and Huang, R. (1995) Progress towards genetically modified starches. Cereals Food World 40: 810–817.
Weeks, J.T., Anderson, O.D., and Blechl, A.E. (1993) Rapid production of multiple independent lines of fertile transgenic wheat (*Triticum aestivum*). Plant Physiol. 102: 1077–1084.
Wenck, A., Conger, B.V., Trigiano, R., and Sams, C.L. (1988) Inhibition of somatic embryogenesis in orchardgrass by endogenous cytokinins. Plant Physiol. 88: 990–992.
Yang, Y.M., He, D.D., and Scott, K.J. (1993) Plant regeneration from protoplasts of durum wheat (*Triticum durum* Desf. cv. D6962). Plant Cell Rep. 12: 320–323.
Zhou, H., Arrowsmith, J.W., Fromm, M.E., Hironaka, C.M., Taylor, M.L., Rodriguez, D., Pajeau, M.E., Brown, S.M., Santino, C.G., and Fry, J.E. (1995) Glyphosate-tolerant CP4 and GOX genes as selectable markers in wheat transformation. Plant Cell Rep. 15: 159–163.
Zhou, H., Stiff, C.M., and Konzak, C.F. (1993) Stably transformed callus of wheat by electroporation-induced direct gene transfer. Plant Cell Rep. 12: 612–616
Zilberstein, A., Schuster, S., Flaishman, M., Pnini-Cohen, S., Koncz, C., Maas, C., Schell, J., and Eyal, J. (1994) Stable transformation of spring wheat cultivars. Abs. 2013, 4th Internat. Cong. Plant Mol. Biol., Amsterdam.

# 7. Transgenic Cereals : *Oryza sativa* (rice)

SWAPAN K. DATTA
*Plant Breeding, Genetics and Biochemistry Division, International Rice Research Institute, P.O. Box 933, 1099 Manila, Philippines, E-mail : SDATTA@IRRI.CGIAR.ORG*

Abstract. Rice is one of the most important food crops of the world, feeding nearly 3 billion people. In order to meet the needs of the population expected around the year 2020, it is estimated that 70% more rice will need to be produced on less land and water with a smaller agricultural workforce. The rice genome is as dynamic as of other crop plants, but needs to be more amenable to the addition of new genes for further improvement of quantitative and qualitative traits. Genetic engineering is a novel method of plant breeding which is expected to play an important role in rice improvement. Several agronomic traits that could not be manipulated by conventional breeding have been incorporated into rice by gene technology, such as resistance to pests, sheath blight, bacterial blight and tolerance to abiotic stress. Field evaluation of rice transformed with a herbicide resistance gene showed excellent results. Large-scale cultivation of transgenic rice is expected soon.

## 1. Introduction

### *1.1 Economic importance, origin and history of rice*

Rice is one of the most important food crops in the world. Indica type rice varieties feed about 3 billion people, predominantly in developing countries. *Oryza sativa* (cultivated rice) originated in the humid tropics of South and Southeast Asia. *Oryza glaberrima* (red rice) originated in Niger basin several millions years ago and is still chiefly grown in the middle Niger delta. For more than 10,000 years rice evolved into various ecotypes and land races under the influence of natural and farmer selection.

For about 8,000 years, rice has been closely associated with human beings. Besides food, it has influenced religions, cultures and lifestyles, mainly in Asia and some other parts of the world. A Balinese (Indonesian) legend has it that lord 'Vishnu', the male god of fertility and water, made mother earth to give birth to rice and asked the lord of heaven, 'Indra', to teach people the art of growing rice. Although, the exact place and time of the origin of rice will perhaps never be known, scientists believe that the distant ancestors of rice originated in the humid tropics of Gondwanaland some 135 million years ago; the Indian plate separated from it 75–90

*I.K. Vasil (ed.), Molecular Improvement of Cereal Crops, 149 – 187*

million years ago. Based on the study of extensive collection of rice accessions, the famed Russian geneticist N.I. Vavilov proposed that rice originated from its wild progenitors in 'The Hindustan center of origin' (Vavilov, 1951). This area extends from eastern India (Assam) to present day Myanmar, Thailand, Malaysia and Indonesia. However, a high concentration of wild forms favor the origin site to be in southern China. The oldest records of domesticated rice come from the Yangtze delta area, where non-Chinese people first began to cultivate it about 7,000 years ago. The rice found there has been radiocarbon dated to 5005 BC. It corresponds to the Indica type of *Oryza sativa* and differs distinctly from wild types found there. The history of rice in India goes back to the time of Harappa culture in 1800 BC. Rice is not mentioned in the Rig-Veda, but is reported in Atharva Veda, which is thought to have been written between 1000–800 BC.

From India, rice moved West, first to Persia and then to Egypt where it was first recorded by Theophrastus. However, there are no records of rice either in ancient Egyptian writings and inscriptions or in the Bible. Use of rice was spread abroad mainly after Alexander's expeditions to different countries, including Persia (334 BC), and Northwest India (327–325 BC). Around 1000 AD, rice was brought to Spain by the Moors, reached the Camargue in France and the plain of Po in Italy. In the 18th century, it was finally recognized as an important food crop carried by traders across the Alpine passes into the towns of Central and Northern Europe. It reached Central and South America during the course of Spanish and Portuguese colonization. In 1646, it was grown for the first time in North America in the James river region of Virginia.

Two distinctly eco-geographical races evolved from the early *Oryza sativa* through wide dispersal, geographic isolation and farmers selection. The indica type spread from tropical Asia to Africa, Western Central Asia and Latin America about 400 years ago. Japonica types evolved from Indica dispersed to China, Korea, Japan and more recently to North America and Australia.

*O. glaberrima* originated in the central Niger Delta from Senegal and Gambia to Nigeria and spread to West Africa about 3,000 to 3,500 years ago. During the past three centuries, *O. sativa* became more popular due to higher yield and better grain quality. International trade and migration of people has transported rice in all directions and it is cultivated now on every continent. During the early Neolithic era in Asia, much like wheat, rice was direct-seeded, dependent on rainfall for water, and mostly grown in forest clearings. Under domestication mostly in Asia, over a period of about 10,000 years, land races evolved, making rice the most genetically diverse cereal. Today, about 20 species and over 100,000 varieties of rice are spread in more than 115 countries of the world. Rice grows from 50°N to 40°S of equator, from 2,500 m in the hills to below sea level, from alkaline to acidic soils, from dry uplands to 6 meters of deepwater. All these ecotypes evolved under domestication giving rice enormous genetic variability (Figure 1a).

Modern rice breeding has greatly improved yield of rice during the past 25 years, from 257 million tons in 1966 to 520 million tons in 1990. In order to meet the needs of the projected world population in 2020, 70% more rice must be produced (Hossain and Fischer, 1995) on less land and water and with fewer agricultural workers.

### *1.2 Principles, species and areas of cultivation*

Two species (*O. sativa* and *O. glaberrima*) of rice are important sources of human nutrition. *O. sativa* is grown worldwide and *O. glaberrima* is grown in parts of West Africa. Wild *Oryza* species have 4 complexes (1. *O. schlechteri*, tetraploid; *O. brachyantha*, 2. *O. ridleyi*, 3. *O meyeriana*, and 4. *O. officinalis*) with diverse genomes as CC, BBCC, BB, CCDD, EE. All the species in this complex are perennial. The *O. sativa* complexes consist of the wild and weedy relatives of the two rice cultigens. The perennial *O. rufipogon* is widely distributed in South and Southeast Asia, Southeast China and Oceania. *O. nivara,* an annual wild form, is found in Deccan Plateau and Indo-Gangetic plain of India and in many parts of Southeast Asia. *O glaberrima* varieties can be divided into two ecotypes: deepwater and upland. About 45% of the land planted to rice in Africa belongs to the upland culture. Farmers and traders have helped in ecological diversification of *O. sativa*, which involved hybridization, differentiation, and selection cycles using ancestral forms of the cultigen.

Within broad geographic regions, two major eco-geographic races were developed as a result of isolation and selection: (1) Indica, adapted to the tropics, and, (2) Japonica, adapted to the temperate regions and tropical uplands. Further subclassification has been made based on hydrologic-edaptic-cultural-seasonal regime (Figure 1b, Rice Almanac, 1997).

Rice occupies one-tenth of the arable land, but in most Asian countries, it occupies one-third or more of the total planted area. Since World War II, world rice area, yield, and production have increased considerably. During the last four decades, rice has became ever more important as a source of human food, as indicated by the increasing portion of the total arable land occupied by the crop. However, with accelerating loss of productive rice land to rising sea levels, salinization, erosion and urbanization, rice supply cannot meet with the demand. To meet future needs, yield increases of 3% per year – which have never been achieved with any major food crop – will be needed.

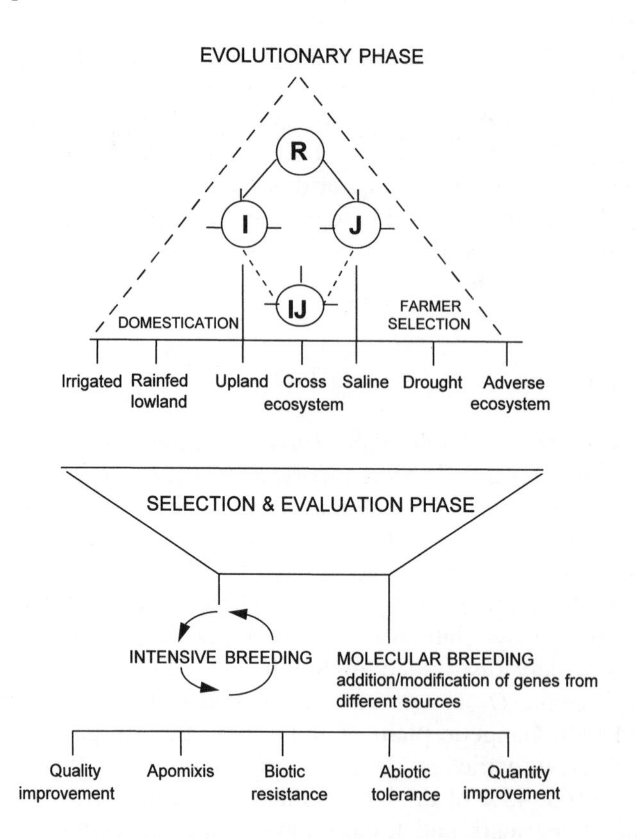


*Fig. 1a* Schematic presentation of diversification, selection and improvement of rice (R = rice, I = indica, J = japonica, IJ = indica-japonica hybrid)

## *1.3 Food value and market*

Rice provides 23% of global human per capita energy and 16% per capita protein. Rice protein ranks high in nutritional quality among cereals, though protein content is modest. Unmilled rice (brown rice) provides 4.3 to 18.2% protein, averaging 9.5% based on 17,587 cultivars in the IRRI germplasm (Rice Almanac, 1997). Rice also provides minerals, vitamins, and fiber. Milling removes roughly 80% of the thiamine from brown rice. For the majority of Asians who eat rice, the total intake is 2,531 calories per person per day, with 35% coming from rice, which is considerably high. However, breeding efforts to increase protein have so far been largely unsuccessful because of the considerable effects of environment and complex inheritance properties in the triploid rice endosperm tissue.

Only 4–5% of world rice enters the global export market. Most countries rely on domestic production to feed their populations. Asia produces 92.2%

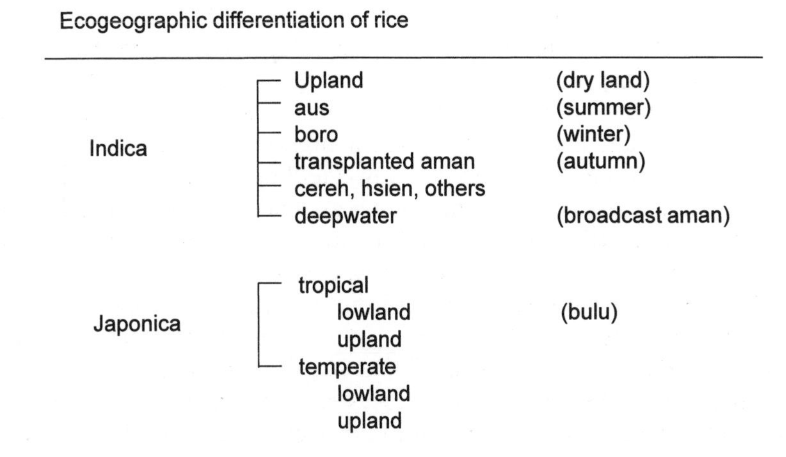


*Fig. 1b* Grouping of Asian rice cultivars by ecogeographical race, hydrologic-edaphic cultural regime and crop season (adapted from Rice Almanac 1997).

of world's rice, India and China together produce 58%. Any shortfall in major rice-growing countries could be a disaster for food security. Besides being a basic food in many Asian countries, glutinous rice is served in sweetened form for snacks, desserts or special foods for religious and ceremonial occasions. Alcoholic beverages (rice beer, sake, Wangtsiu) made from rice are used as popular drinks in many countries. In Europe and North America, rice is developing a new market as a staple and as a gourmet food.

## 2 Important problems of rice breeding and improvement

Rice is grown in four major ecosystems: (1) irrigated, (2) rainfed lowland, (3) upland, and (4) flood-prone. The irrigated rice ecosystem accounts for 55% of total rice area and contributes 76% of global rice production. This ecosystem is gradually losing land to urbanization and industrialization. The remaining 34% of rice comes from unfavorable rice growing areas (45%): rainfed lowland (20%), upland (12%) and flood prone areas (8%). Another 5% comes from favorable rainfed where high-yielding semidwarf varieties and improved technology have been adopted.

### *2.1 Irrigated rice ecosystems*

Irrigated rice is grown in bunded, puddled fields with assured irrigation for one or more cropping per year. Depending on rainfall variability, irrigated ecosystem is again subdivided into (1) irrigated wet season and (2) irrigated dry season. Worldwide, about 79 million ha of rice are grown under irrigated condition providing a yield of 3–9 t/ha. Improved rice cultivars developed at IRRI and in National programs for irrigated rice lands are of short duration, responsive to fertilization, moderately resistant to biotic stresses and with some tolerance for adverse soils. Egypt, China, Japan, Indonesia, Vietnam and Korea represent high-yielding areas. Cambodia, Eastern India, Nepal, Pakistan are low-yielding areas and the rest are medium-yielding areas. The main problems encountered in growing rice under irrigated conditions are the following: (1) an average on-farm yield plateau of about 6 t/ha in some countries, (2) yield instability due to pests, (3) difficulty of sustaining high yield due to intensive cropping, and (4) intensive use of agrochemicals with potential adverse effects on human health and environment (Khush, 1993). Despite the problems mentioned, research accomplishments for this ecosystem have been significant. These include: (1) semidwarf varieties with high yield potential (10–11 t/ha), (2) short growth duration (100–110 d), (3) greater yield stability, etc. Rice breeders are now developing plant types for favorable environments with yield potential at least 20–25% higher than the existing improved lines (Khush 1993, 1995). Newer technologies and improved rice plants are needed to improve yield with low inputs, reduced dependence on chemicals, and increased resistance to biotic and abiotic stresses.

### 2.2 *Rainfed lowland rice ecosystem*

Rainfed lowlands are characterized by lack of water control, with untimely floods and drought being potential problems. Thirty-six million ha are rainfed (about 25% of world rice areas). There has been no meaningful improvement in production in these areas owing to adverse climate, poor soils, and lack of suitable modern technologies. Often farmers have to grow a second crop, such as legumes, wheat, maize or vegetables. Most rainfed lowland farmers are poor and must cope with unstable yields and financial risks. They use traditional varieties with less inputs. New breeding materials for submergence tolerance, late transplanting, and resistance to lodging may help to improve the yield potential. Given proper technologies, the use of improved traditional rice varieties will assure higher yields from the current yield of 2–3 t/ha to 4–5 t/ha.

### 2.3 *Upland rice ecosystems*

Out of 148 million ha of world rice area in 1991, about 19 million ha were planted to upland rice. Upland rice is grown in Asia (12 m), Africa (2.5 m) and Latin America (3.3 m). Upland rice is grown in rainfed areas with naturally well-drained soils and no surface water accumulation. Major constraints are: (1) poor acidic and degraded soil, (2) weeds, (3) severe blast disease, (4) erratic rainfall and unspecified periods of drought stress. Weeds are the number one enemy causing yield loss and poor quality. The average yield under upland condition is 1 t/ha. Potential improvement is possible with 3–4 t/ha yield (Khush, 1993).

### *2.4 Flood-prone ecosystem*

Around 11.4 million ha of rice lands, mostly in South and Southeast Asia, are subject to uncontrolled flooding. Rice yields are low and extremely variable because of soil problems and unpredictable combination of drought and flood. Although average yields are only about 1.5 t/ha, these areas support more than 100 million people. Most rice varieties survive complete submergence for only 3 or 4 d but tolerant varieties can survive for about 12 d. Irrigated rices are grown in flood prone areas during non-flood periods, if irrigation is available. These are called boro rices in Bangladesh and India. Farmers are now using second crops (wheat, oil seeds, potato etc.) after the floods. Post-flood and dry season cropping are now widely practiced in Bangladesh, India, Myanmar, and Vietnam. Deepwater rices yield higher than floating rice, up to 4 t/ha, with high profit because of low cost of production. Submergence-tolerant varieties are preferred as these can be grown in more favorable areas and produce relatively high yields if flash floods are not severe (Setter et al., 1996).

### *2.5 Hybrid rice*

In addition to improved varieties and New plant type (potential yield of 15 t/ha), the use of hybrid rice may provide the target requirement of 15 t/ha. Hybrid rice has been grown in China since 1976, and has a yield advantage of 15–20% on the average over conventional high-yielding varieties (Virmani et al., 1993). Transgenic CMS and maintainer lines offer new possibilities for the improvement of hybrid rice (Alam et al., 1998b).

### *2.6 Molecular breeding*

Many potentially useful genes for rice have been identified, cloned and characterized. Gene transfer techniques and *in vitro* plant regeneration systems have been developed (Figure 1c). Tools for molecular characterization of transgenic plants (Southern, Enzyme assay, Northern and Western analysis) are available (Figures 2–10). Schematic presentation of gene transfer (*X*a-21 for bacterial blight resistance) by conventional breeding and transformation (by direct gene transfer) is shown in Figure 11. Each of these aspects is discussed in the following sections.

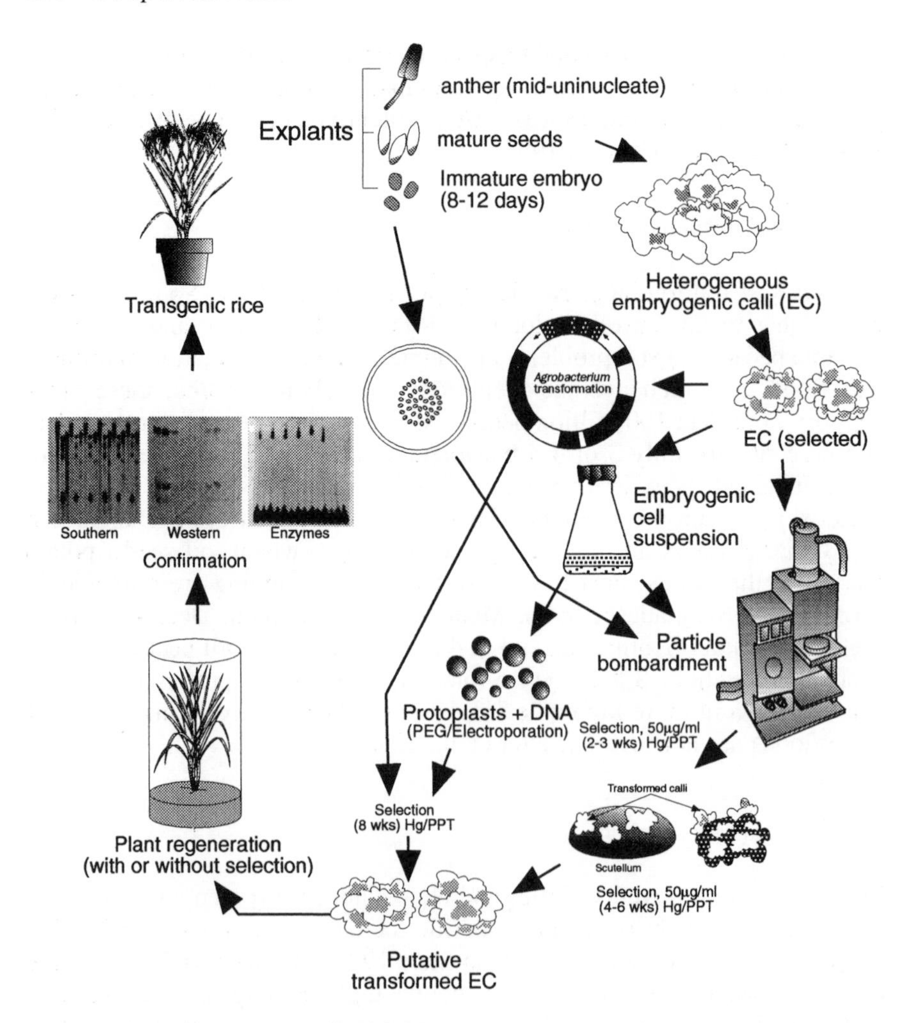


*Fig. 1c* Schematic presentation of protocol for production of fertile transgenic rice plants using biolistic, protoplast and *Agrobacterium* systems

## 3. Materials and Methods

### *3.1 Materials used*

Taipei 309 (japonica type) was initially used as a model for rice transformation, followed by Nipponbare, Yamabiko, Yamahousi, Norin 14, Zhonghua 6, IRRI-New Plant Type etc. (all japonica types) (Tables 1, 2). As indica type is more difficult to transform, only a few varieties of indica rice have been transformed (cv. Chinsurah boro II, Tepi boro, IR43, IR54, IR58, IR64, IR72, Basmati). Mediterranean rice, and US and European cultivars have also been transformed (Tables 1, 2).

Haploid culture (microspore/anther culture derived embryogenic cell suspension or callus) was also used for rice transformation to obtain homozygous transgenic plants (Datta et al., 1990, Chaïr et al., 1994, Tables 1, 2).

## 3.2 *Tissue Culture media*

Several nutrient media (see review by Vasil, 1987) have been used for the culture of rice cells and tissues, such as MS (Murashige and Skoog, 1962),

TABLE 1
Transformation of rice by protoplast method (1988–1996)[1]

| Year | Cultivar[2] | Genes | Fertility[3] | References |
|---|---|---|---|---|
| 1988 | Japonica | *npt II* | Planlets | Toriyama et al. |
| 1988 | Japonica | *gus* | S | Zhang and Wu |
| 1988 | Japonica | *kanamycin, nptII* | F | Zhang et al. |
| 1989 | Japonica | *hph* | F | Shimamoto et al. |
| 1990 | Japonica | *npt II* | F | Peterhans et al. |
| 1990 | Japonica | *hph* | F | Hayashimoto et al. |
| 1990 | Japonica | *CaMV35S, gus* (promoter) | T | Terada and Shimamoto |
| 1990 | Indica | *hph* | F | Datta et al. |
| 1991 | Japonica | *gus* | F | Battraw and Hall |
| 1991 | Japonica | *npt II* | F | Davey et al. |
| 1992 | Indica | *bar* | S | Datta et al.. |
| 1992 | Indica | *npt II* | F | Peng et al. |
| 1992 | Japonica | *rrsv-cp* | F | Hayakawa et al. |
| 1993 | Japonica | *crylA(b)* | F | Fujimoto et al. |
| 1993 | Japonica | *bar* | F | Goto et al. |
| 1993 | Japonica | *bar* | F | Rathore et al. |
| 1995 | CB II | *chitinase* | F | Lin et al. |
| 1995 | Japonica | *Oryza cystatin* | F | Hosoyama et al. |
| 1995 | Japonica | *cp Ti* | F | Xu et al. |
| 1995 | J & I | *nptII, gus* | F | Peng et al. |
| 1995 | Indica | *hph* | F | Alam et al. |
| 1996 | NPT | *hph + gus* | F | Alam et al. |
| 1996 | Japonica | *bar* | F | Chair et al. |
| 1996 | Japonica | *Em* | T | Rock & Quantrano |
| 1996 | Japonica | *corn cystatin* | F | Irie et al. |
| 1996 | Japonica | *Promoter* | T | Mitsuhara et al. |
| 1996 | J & I | *Bt, Chi II* | F | Datta et al. |

[1]Several Indica and Japonica cultivars have been transformed with a number of genes; selected references have been chosen;
[2]J = japonica, I = indica; NPT = IRRI New plant type; CBII = Chinsurah boro II
[3]S = sterile; F = fertile; T = transient gene expression

TABLE 2
Transformation of rice by biolistic method (1991–1997)

| Year | Cultivar | Genes | Fertility | References |
|---|---|---|---|---|
| 1991 | Indica and Japonica | *bar* | F | Christou et al. |
| 1992 | Japonica | *bar* | F | Cao et al. |
| 1993 | Indica and Japonica | *hph* and *gus* | F | Li et al. |
| 1995 | Japonica | *bar* | F | Shimada et al. |
| 1996 | Indica | *TWa, LEA3* | F | Jain et al. |
| 1996 | Indica | *hph* and *gus* | F | Sivamani et al. |
| 1996 | Indica | *hph* and *gus* | F | Zhang et al. |
| 1996 | IRRI-NPT | *hph* and *gus* | F | Alam et al. |
| 1996 | Indica | *crylA(b)* | F | Wünn et al. |
| 1996 | Japonica | *Xa-21* | F | Wang et al. |
| 1996b | Indica and Japonica | *crylA(b)* | F | Datta et al. |
| 1997 | Indica | *crylA(c)* | F | Nayak et al. |
| 1997 | Japonica | *Phytoene* | F | Burkhardt et al. |
| 1997 | Japonica | *crylA(b)* | F | Wu et al. |
| 1998a | Indica and Japonica | *chitinase* | F | Datta et al. |

N6 (Chu et al., 1975), AA (Muller and Graffe, 1978), $R_2$ (Ohira et al., 1973, Fujimura, 1985). Protocols used for genetic transformation include polyethylene glycol treatment or electroporation for protoplasts, and the biolistic or *Agrobacterium* methods for cells, tissues and immature/mature embryos (Datta et al., 1990, Christou et al., 1991, Hiei et al., 1994).

### *3.3 Plasmid constructs, gene expression and selectable marker genes*

Expression level of marker gene is important for transformation efficiency the stronger the gene the higher the possibility of efficient selection. Expression of the transgene depends on several factors: (1) efficient vector, suitable promoter, (2) position in the plant genome, (3) copy number, (4) 3′ noncoding sequences, codon frequency, and (5) the gene product itself, and unknown factors (Perlak et al., 1990, Maas et al., 1992). The promoter determines the expression pattern of transgene in plants. Two types of promoters – constitutive and – tissue specific – have been used (Table 3). The constitutive promoter directs expression in all tissues independent of developmental or environmental signals. The CaMV35S promoter is a strong, constitutive promoter commonly used in many transformation studies (Benfey and Chua, 1990) and extensively used in rice (Figure 2, Table 3), wheat (Vasil et al., 1992) and maize (Gordon-Kamm et al., 1990). McElroy et al. (1991) indicated that the expression of this promoter is weak

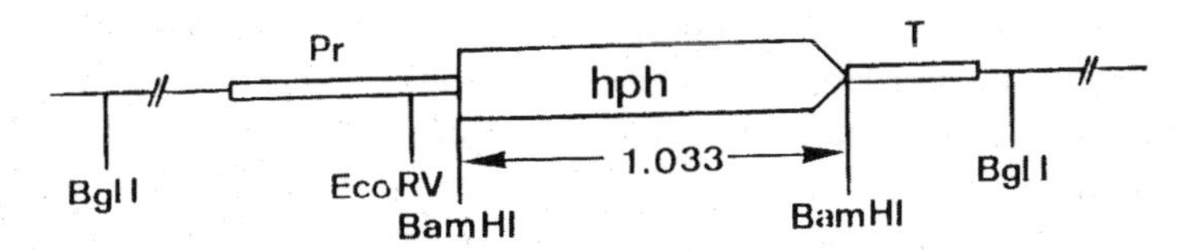


*Fig. 2* Part of plasmid pGL2 containing the bacterial gene for hygromycin phosphotransferase (*hph*) under the control of the expression signals of the 35S transcript of CaMV. For transformation experiments, plasmid pGL2 was cut with*BglI* releasing the fragment shown. The fragment contains a single *Eco*RV site within the promoter region (from Datta et al., 1990)

TABLE 3
Various promoters used in rice transformation

| Promoter | Source | Function | References |
|---|---|---|---|
| 35SP | CaMV | constitutive | Terada and Shimamoto 1990 |
| 35SP | CaMV | constitutive | Datta et al. 1990 |
| 35SP | CaMV | constitutive | Battraw and Hall 1991 |
| 35SP | CaMV | constitutive | Mitsuhara et al. 1996 |
| Emu | maize | constitutive | Chaimberlain et al. 1994 |
| Ubiquitine | maize | constitutive | Toki et al. 1992 |
| Ubiquitine | maize | constitutive | Cornejo et al. 1993 |
| Actin-1 | rice | constitutive | McElroy et al. 1990 |
| Actin-1 | rice | constitutive | Zhang et al. 1991 |
| Actin-1 | rice | constitutive | Wu et al. 1997 |
| PEPc | maize | green tissue | Matsuoka et al. 1994 |
| PEPc | maize | green tissue | Datta et al. 1996b |
| Pith | maize | pith tissue | Datta et al. 1998b |
| Pin2 | potato | wound inducible | Duan et al. 1996 |
| NOD | legume | selective expressioin | Reddy et al. 1998 |
| rolC | R1 plasmid | vascular and embryogenic | Matsuki et al. 1989 |
| Adh1 | maize | constitutive | Kyozuka et al. 1991 |
| LHCP | rice | leaves, stems, floral organs | Tada et al. 1991 |
| Histone H3 | wheat | meristematic tissue | Terada et al. 1993\ |
| rbc-S | rice | mesophyll | Kyozuka et al. 1993 |
| RTBV | RTBV | phloem tissue | Yin and Beachy 1995 |
| Osg 6B | rice | tapetum | Yokoi et al. 1997 |

in rice and maize as compared to tobacco. However, our experience with this promoter suggests that it is a very strong promoter for gene expression in rice but there may be variability in expression. Therefore, it is important to obtain a good number of independent transformants (based on Southern blot analysis, Figures. 3, 5, 6) and select the one which is desirable and suitable for target gene expression.

Gene expression in target tissues is becoming more popular because of its potential impact in crop improvement (Datta et al., 1997). Rice is now used as a model plant for gene expression and promoter study in monocots. Studies on plant promoters in transgenic rice, particularly the synthetic ones, will help in understanding the functional interactions between defined elements and DNA binding factors (Katagiri and Chua, 1992).

Introns have been shown to enhance gene expression in maize (Callis et al., 1987, Vasil et al., 1989, Luehrsen and Walbot, 1991, Walbot and Gallie, 1991, Clancy et al., 1994) as well as rice (Peterhans et al., 1990, McElroy et al., 1991). It is evident that a promoter in combination with a monocot intron between, can enhance the expression of the gene in rice and other monocot species.

Use of selective agents, particularly antibiotics, is often critical in selecting transformants. For example, kanamycin which is most widely used in dicot transformation is not suitable and effective in monocots (Potrykus et al., 1985, Hauptmann et al., 1988, Dekeyser et al., 1989, Peng et al., 1992) and in rice (Shimamoto, personal communication; our unpublished data). Nearly 70% of untransformed rice calli grow even in the presence of 100 mg/l kanamycin (Dekeyser et al., 1989). Often plants regenerated through kanamycin selection produce albino plants. However, G418 is slightly better than kanamycin for selection as well as green plant regeneration.

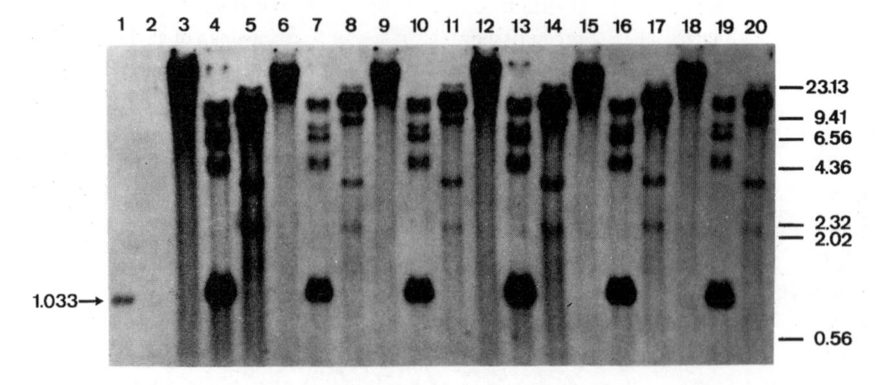


*Fig. 3* Southern data are presented for one representative primary transgenicplant (lanes 3-5), and five offspring of this plant (lanes 6-8, 9-11, 12-14, 15-17,18-20). Lane 1 represents a 3 copy reconstruction of the transforming plasmidcut with *Bam*HI, lane 2 contains DNA from control (untransformed) rice plant, also cut with *Bam*HI. Samples are arranged in groups of three each. The first lane of these triplets (3, 6, 9, 12, 15, 18) contains undigested DNA, the secondlane (4, 7, 10, 13, 16, 19) contains DNA restricted with *BamHI* (releasing a 1.033kb fragment characteristic of the protein-coding region of the gene), the thirdlane (5, 8, 11, 14, 17, 20) contains DNA restricted with *EcoRV* (yielding borderfragments between transforming DNA and host DNA). The size markers in kilobase pairs are derived from lambda DNA digested with *HindIII* (from Datta et al., 1990)

Aminoglycoside (antibiotics) inhibit protein synthesis in prokaryotic cells. Kanamycin, gentamycin, and its derivatives geneticin (G418), neomycin and paromomycin bind to the 30S ribosomal sub-unit thus inhibiting the initiation of translation and finally protein synthesis (Wilmink and Dons, 1993). The most visible effect of these on plants is chlorosis or bleaching of the leaves, caused by lack or inhibition of chlorophyll synthesis. Hygromycin can be inactivated by phosphorylation of a hydroxyl group by the enzyme APH (4′), also known as hygromycin phosphotransferase (HPT). The *hph-aphIV* gene which codes for this enzyme was isolated from *E.coli* and is so far the best selectable marker gene for rice transformation, particularly in the protoplast system (Shimamoto et al., 1989, Datta et al., 1990, Takano et al., 1997). HPT assay can be easily used for screening and confirmation of transgenic rice (Figure 4, Datta et al., 1990).

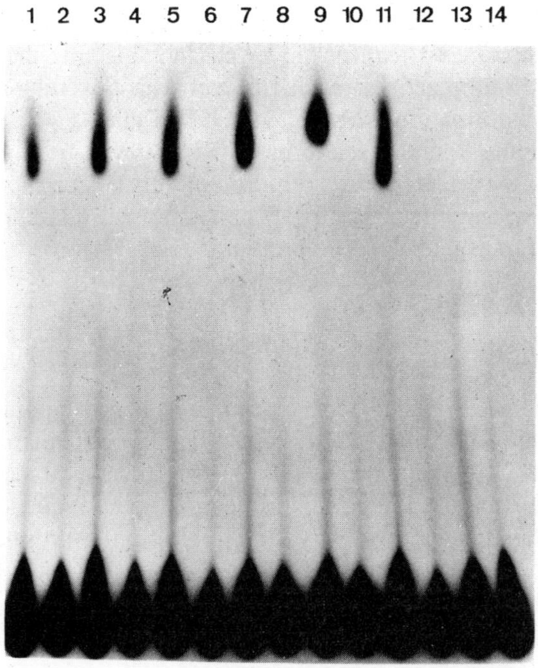

*Fig. 4* Enzyme assay for hygromycin phosphotransferase of primary transgenic and offspring plants. The plants assayed are identical to those used in the Southern blot analysis shown in Figure 3. The autoradiogram of the TLC plate shows labeled ATP at the start in all lanes and labeled hygromycin B in lanes 1, 3, 5, 7, 9, 11. Even numbers represent enzyme reactions carried out without Hygromycin B, odd numbers represent reactions containing Hygromycin. Lanes 1, 2 contain crude protein extract derived from the primary transgenic plant; lanes 3-12 contain extracts derived from five offspring of the same primary plant; lanes 13 and 14 contain extracts isolated from an untransformed control rice plant (from Datta et al., 1990)

Genes conferring resistance to the herbicide phosphinothricin (PPT) are frequently used for rice and monocot transformation. Resistance to the herbicide is conferred by phosphinothricin N-acetyl transferase (PAT) which inactivates PPT by acetylation using acetyl coenzyme as a cofactor (Figure 7, Datta et al., 1992).

Two similar genes encode PAT: *bar* isolated from *Streptomyces hygroscopicus* and *S. viridochromogenes* (Vasil, 1996). PPT is synthesized chemically (Basta®, Hoechst AG, Germany) or by fermentation of *S. hygroscopicus* (Herbiace, a product of Meiji Seika Ltd., Japan). For rice protoplast transformation, 5–25 mg/l PPT should be used at day 14 after protoplast isolation. Different rice cultivars show varying degrees of natural resistance to PPT, which is the same for all selectable agents (kanamycin, G418, hygromycin). In the biolistic method, 25 mg/l PPT is usually used, a very high concentration as compared to maize (Fromm et al., 1990) wheat (Vasil et al., 1993) or barley (Lazzeri et al., 1991, Wan and Lemaux, 1994).

The effective concentration of the selective agent and the timing of selection pressure are critical for the success of selecting the transformants and need to be optimized for each rice cultivar. Several selective agents used in rice transformation are summarized in Table 4.

Northern (Figure 8) and Western blots (Figures 9, 10) are useful molecular techniques to detect RNA and protein levels corresponding to the transgenes. Bioassay data for pest or pathogen resistance should be correlated with transgene-specific protein levels.

### *3.4 Techniques used in rice transformation*

Transgenic rice plants have been obtained in major japonica and indica cultivars including 'Basmati', IRRI-New Plant Type (Christou et al., 1991, Shimamoto et al., 1992, Datta et al., 1992, 1996b). Several methods of transformation have been used, e.g., microinjection, macroinjection, laser beam technique, pollen tube pathway, dry seed imbibition, cell/tissue electroporation. However, the best and most unambiguous results have been obtained from the protoplast, biolistic and *Agrobacterium* methods (Figure 1c).

## 4. Results and Discussion

### *4.1 Protoplast system*

The first report of plant regeneration from protoplasts of gramineous species (Vasil and Vasil, 1980) opened new possibilities of genetic transformation in cereals and grasses (Potrykus et al., 1985, Lörz et al., 1985, reviews: Gasser and Fraley, 1989, Potrykus, 1991, Hinche et al., 1994, Vasil, 1994).

TABLE 4
Use of various selective agents in rice transformation

| Selective agents | Concentration (mg/l) | Resistance gene | Resistance enzyme | Effect | Resistance mechanism | References |
|---|---|---|---|---|---|---|
| **1. Antibiotics** | | | | | | |
| Hygromycin | 25.0[1]<br>50.0[2]<br>75.0[3] | *hph (aph IV)* | APH (4′) | Protein synthesis | Phosphorylation of hygromycin | Shimamoto et al. 1989<br>Datta et al. 1990 |
| Kanamycin | 100.0 | *aph A2* | APH (3′) II (NPT II) | Protein synthesis | Phosphorylation of kanamycin | Zhang et al. 1988<br>Battraw and Hall, 1990 |
| G418 | 25.0 | *aph A2* | APH (3′) II (NPT II) | Protein synthesis | Phosphorylation of gentamycin | Peterhans et al. 1990 |
| **2. Herbicides** | | | | | | |
| Basta, (PPT, | 5–25 | *bar, pat* | PAT | amino acid synthesis | Acetylation of PP | Cao et al. 1992<br>Datta et al. 1992<br>Rathore et al. 1993<br>Christou et al. 1991 |
| **3. Methotrexate** | | *dhfr* | DHFR | nucleotide synthesis | inhibiting the enzyme DHFR | Meijir et al. 1991<br>Hauptmann et all. 1988 |

[1]protoplast, [2]biolistic, [3]*Agrobacterium* method

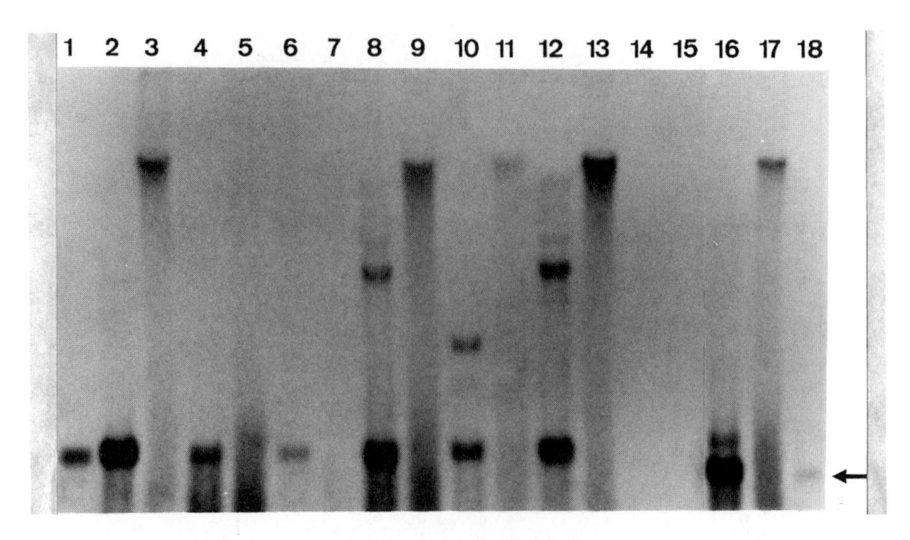


*Fig. 5* Southern blot analysis of genomic DNA probed with *nptII* specific probe of kanamycin resistant rice cell lines transformed with pAH (lanes 2-13) and pHP28 (lanes 16 and 17). Lane 1, 10 pg of *HindIII*-digested pAH plasmid DNA; lanes 2 and 3, *HindIII* digested and undigested DNA, respectively (the same for all clones) of clone 2; lanes 4 and 5, DNA of clone 8; lanes 6 and 7, DNA of clone 12; lanes 8 and 9, 12 and 13, DNA of clone 41; lanes 10 and 11, DNA of clone 42; lanes 14 and 15, DNA of kanamycin sensitive control clone; lanes 16 and 17, kanamycin resistant clone transformed with pHP28; lane 18, 2 pg of pHP28 DNA digested with *HindIII* (from Peterhans et al., 1990)

The first report of rice transformation was based on regeneration of plants from protoplasts (Toriyama et al., 1988). Since then several laboratories have reported transgenic rice recovered from protoplasts either by polyethylene glycol treatment or by electroporation. Indica rice, particularly IRRI varieties and breeding lines, are recalcitrant to *in vitro* culture and limited success has been reported. The single cell origin, non-chimeral nature and genetic fidelity of plants derived from somatic embryos are very attractive features for any genetic transformation system (Vasil, 1994). Early success was based the following:

1) Establishment of embryogenic cell suspensions (ECS), the key source of regenerable protoplasts.
2) Use of suitable plant transformation vectors with selectable marker genes.
3) Continuous efforts on improvement of tissue culture protocols by different laboratories, e.g., use of 2,4-D, maltose, nurse culture, PEG-MW 6000, osmotic adjustment, etc.

ECS are the only source of regenerable protoplasts in the Gramineae (Vasil, 1987). Establishment of ECS in japonica rice is easier as compared to indica rice. Only a few papers have shown the potential use of such cultures in

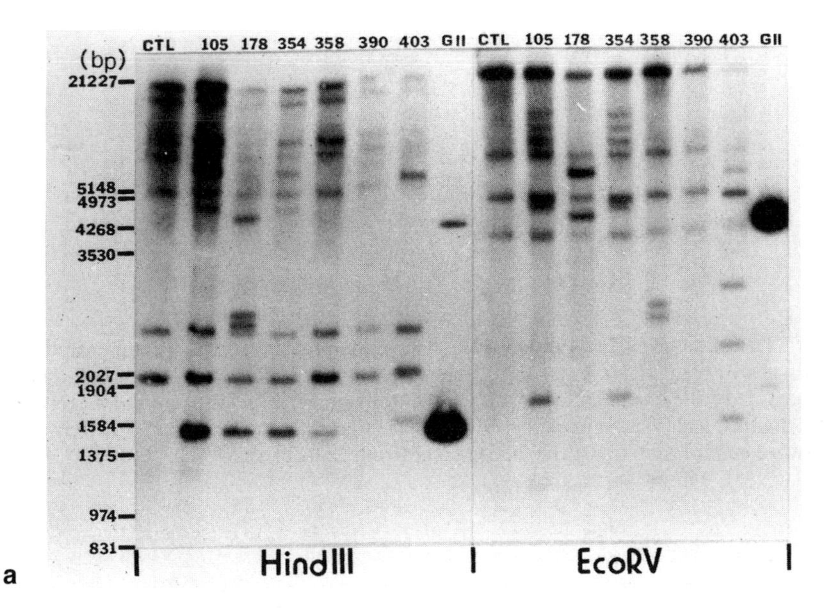


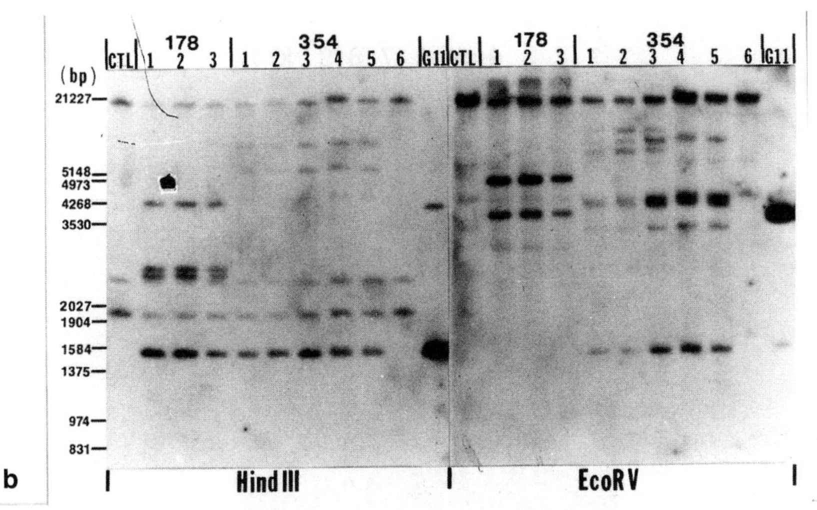


*Fig. 6* Southern blot analysis of the $T_0$ (Panel A) and T1 (Panel B) transgenic rice plants. Total DNA was extracted from $T_0$ and $T_1$ plants by the CTAB method. Aliquots (15 ($\mu$g) of the DNA were digested with *Hind*III (H) or *EcoRV* (EV) and subjected to Southern blot analysis using $^{32}$P-labeled 1.1 kb *Sac*I-*Hind*III or the 557 bp-long ScaI-*Hind*III fragment of chitinase gene, *chi11* as the hybridization probe. Estimates of copy number of the chitinase transgenes were obtained by counting the total number of bands in the *Eco*RV digest of DNA from each transformant and subtracting the number found in control. CTL = untransformed control; G11 = pGL2 (CaMV-*chi11*) DNA. The migration positions and sizes (in base pairs, bp) of markers are indicated on the left (from Lin et. al., 1995)

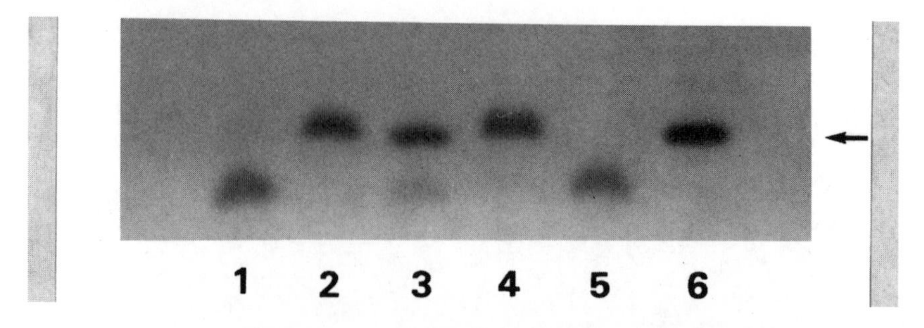


*Fig. 7* Detection of PAT activity by TLC. The crude protein extracts were prepared from leaf tissues of plants growing in the greenhouse (survived after Basta spraying). Lane 1, represents L-PPT $^{14}$C standard; lanes 2, 3 and 4 from three transgenic plants show acetylated PAT activities; lane 5, protein extract from control rice plant showing no PAT activity; lane 6, positive control from transgenic maize plant (from Datta et al., 1992)

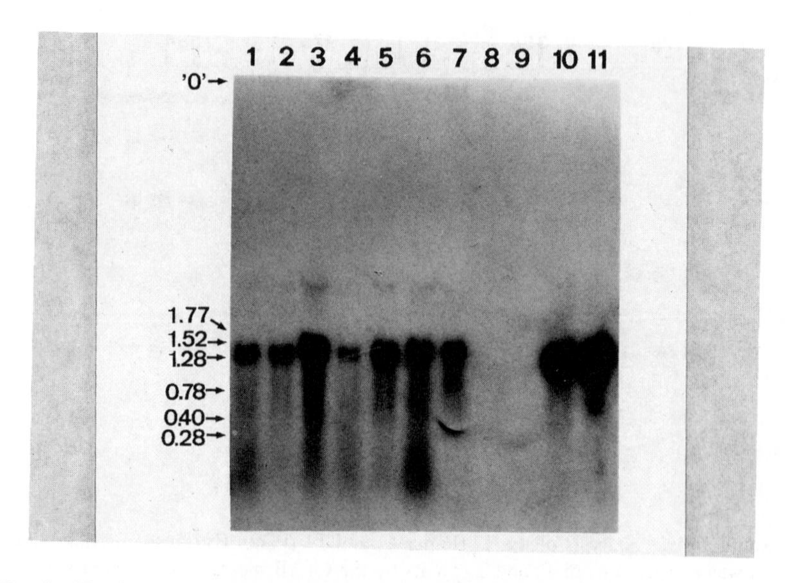


*Fig. 8* Northern blot analysis of RNA isolated from kanamycin resistant rice and tobacco clones. RNA was probed with the *nptII* specific probe. Lane 1, RNA of rice clone 2; lane 2, rice clone 8; lane 3, rice clone 12; lanes 4 and 6, rice clone 41; lane 5, rice clone 42; lane 7, kanamycin resistant rice cell line transformed with pHP28; lane 8, non-transformed rice control; lane 9, non-transformed tobacco control; lane 10, tobacco clone transformed with pAH; lane 11, tobacco clone transformed with pHP28. Ten micrograms of total RNA was used with the exception of lane 4 where 2 ($\mu$g of RNA was loaded. The start of the gel is marked '0'. The *numbers* next to the arrows indicate RNA size markers in kb

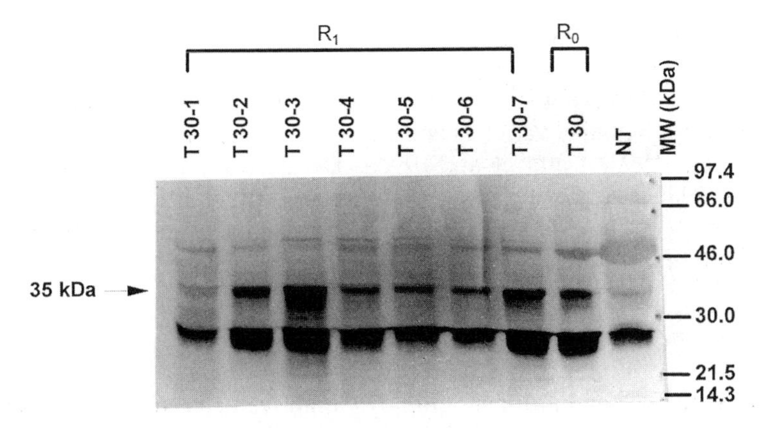


*Fig. 9* Western blot analysis of transgenic rice showing presence of chitinase protein (NT = non-transformed)

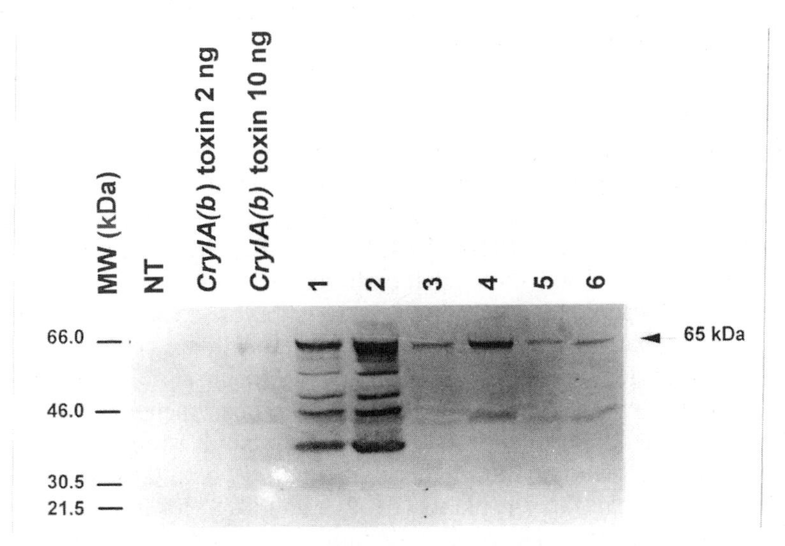


*Fig. 10* Western blot analysis of transgenic rice showing the presence of CryIA(b) proteins (NT = non-transformed)

traditional indica and IRRI developed indica varieties (Datta et al., 1990, 1992, Peng et al., 1992, Alam et al., 1996). Selected papers on rice transformation using protoplast system have been summarized in Table 1. Ingo Potrykus and co-workers attempted to regenerate cereal mesophyll protoplasts for decades without success. However, a report of plant regeneration from rice mesophyll protoplasts Gupta and Pattanayak (1993) created some optimism that did not last long as the work could not be repeated by others.

### *4.2 Agrobacterium system*

Genetic transformation of rice by *Agrobacterium tumefaciens* has been confirmed for japonica varieties and extended to recalcitrant indica varieties (Hiei et al., 1994, Datta et al., 1996a, Dong et al., 1996, Aldemita and Hodges, 1996). However, this system needs further improvement for its utilization in indica rice transformation (for details see chapter by Komari and Kubo, this volume).

### *4.3 Biolistic system*

Microprojectile bombardment employs high velocity metal particles to deliver biologically active DNA into plant cells and whole plants can be recovered from the transformed cells through selection. Genotype independent transformation may be carried out with this method as there is no biological limitation to DNA delivery.

#### *4.3.1 Explants*

Immature embryos (IE), ECS, and IE-derived primary and secondary calli (EC) have been reported as suitable explants for DNA delivery (Table 2). IE may be better for those genotypes where the material is available throughout the year, as is the case at IRRI. ILTAB (Roger Beachy's group) experience with japonica rice T309 shows that embyrogenic calli are more efficient in producing transgenic plants than immature embryos (25% vs. 3%). Further, it was found that embryogenic suspensions and calli are equally useful for obtaining transgenic calli (41 ± 3 vs 45 ± 9%, Fauquet et al., 1996). The size and age of the embryogenic suspensions for bombardment are critical. The suspension should be subcultured for 7d before bombardment, and the size of the explants should not be smaller than 2 mm for optimal transformation. At IRRI, we use all three explants (IE, EC, and ECS) for biolistic transformation of indica rice. However, our biolistic transformation involves IE (70%) and to some extent EC (> 25%). IE of all genotypes are available at IRRI throughout the year. Donor plants grown under 10,000 lux and at 25–31°C ± 2°C during the day and 21 ± 2°C at night time are most suitable.

#### *4.3.2 Osmotic treatment*

Usually, gold particle bombardment is detrimental to cells (Christou et al., 1991). However wounded cells/tissue can be protected to some extent by keeping them under high osmotic pressure before and after bombardment. An osmotic treatment (30 g mannitol $l^{-1}$, 30 g sorbitol $l^{-1}$) of cell suspensions for 4 h before and 16–20 h after bombardment enhances transformation efficiency in maize (Vain et al., 1993). Several groups found a positive correlation of high osmotic pressure with frequency of transformation. The osmotic pressure can be adjusted with either mannitol or sorbitol, or both.

### *4.4 Selection and frequency*

From work with protoplast transformation, it was found that hygromycin is an excellent selectable agent for both japonica (Shimamoto et al., 1989) and indica (Datta et al., 1990) rice when the *hph* gene driven by CaMV35S promoter is used as selectable marker. It works well also with the biolistic method (Li et al., 1993, Sivamani et al., 1996, Zhang et al., 1996). However, several groups have used PPT for selection and their results clearly show that the *bar* gene is also an effective selectable marker for rice transformation (Christou et al., 1991, Datta et al., 1992, Cao et al., 1992). Frequency of transformation is a relative term and different laboratories have expressed the data in different ways. Efficiency of transformation may be defined in terms of the number of $Hg^+$ or $PPT^+$ Southern-analyzed plants regenerated from 100 explants (IE or IE-derived units of calli).

## 5 Critical factors for producing transgenic rice

Several reviews and commentaries (Potrykus, 1991, Schell, 1991, Shimamoto, 1992, Hinche et al., 1994, McElroy and Bretell, 1994, Vasil 1994) have highlighted the critical factors for the transformation of plants. I would like to emphasize the following based on our own experience in the laboratory with rice transformation:

1. Knowledge of rice genotype and its response in tissue culture is important.
2. Induction and maintenance of embryogenic calli is a prerequisite for regeneration of plants.
3. Embryogenic cell suspensions are an absolutely requirement to obtain rice plants if the protoplast system is used.
4. The lesser the time in selection, the better for normal plant regeneration.
5. Suitable vector, selectable marker genes and promoters are required.
6. Use of control cultures in selection media up to plant regeneration is helpful to monitor the process and selection of transformants. This will help avoid escapes.
7. Freshly-isolated immature embryos from donor plants grown in optimum conditions are best.
8. Newly-developed immature embryo-derived embryogenic calli are more responsive in transformation and regeneration.
9. A minimum amount of high quality plasmid DNA should be used.

10. A minimum time of tissue culture phase before transferring the plants to the greenhouse is recommended.
11. Handling of transgenic plants in the greenhouse (under optimized conditions) is very critical to obtain normal fertile plants.

## 6. Useful genes introduced

Remarkable progress has been made in the production of transgenic rice since the initial reports of Shimamoto et al., (1989, japonica rice), Datta et al., (1990, indica rice) and Christou et al., (1991, indica/japonica rice). The introduction of useful genes, promoters and selective agents used in rice transformation is summarized in Tables 5 and 6.

## 7. Fate of introduced DNA and stability of transgenes

Biolistic and protoplast methods have been routinely used in rice transformation (Tables 1, 2) and several reports on the inheritance of transgenes are available, including a detailed study of plants transformed by the *Agrobacterium* system (Hiei et al., 1994). The results show that the integration of genes into the rice genome appears to be random. The precise mechanism of DNA integration remains elusive. Once integrated, foreign genes appear to be susceptible to most of the controlling mechanisms known for normal plant genes, although the flanking DNA is likely to play some role in expression (Saul and Potrykus, 1990). Illegitimate recombination is believed to be the mechanism of DNA integration in plant cells, but the role of DNA sequences and topological genome structure around the integration site has not been elucidated. In a recent paper, an extensive study of integration sites in transgenic rice has been reported. The study indicates that the transgene resides in a region with inverted structure and a large duplication of rice genome over 2 kb (Takano et al., 1997). Inheritance of transgenes generally follows Mendelian rules with some exceptions (Goto et al., 1993, Bao et al., 1996, Tu et al., 1998a). Integration patterns range from simple single copy to complex multiple (ca 50) copies (Peterhans et al., 1990, Datta et al., 1996b, Oard et al., 1996).

### *7.1 Carrier DNA*

Carrier DNA (mostly calf thymus and salmon sperm DNA) helps in achieving higher transformation frequency in tobacco (Shillito et al., 1985). With or without carrier DNA, we found simple as well as multiple integration bands in transgenic rice (Datta et al., 1998b). It is reported in animal cell transfection that carrier DNA may form a complex with the transgene which is more prone to rearrangement than plasmid DNA alone

TABLE 5
Genes of agronomic value introduced in rice

| Rice variety | Method used | Genes transferred | Transformation efficiency | Trait conferred | Reference |
|---|---|---|---|---|---|
| Indica/Japonica | Biolistic | *bar/gus* | H | Resistance to herbicide | Christou et al. 1991 |
| IR72 | Protoplast (PEG) | *bar* | H | Resistance to herbicide | Datta et al. 1992 |
| Japonica | Protoplast (electroporation) | *CP-stripe* virus | M | Resistance to stripe virus | Hayakawa et al. 1992 |
| Japonica | Protoplast (electroporation) | Bt crylA(b) | M | Resistance to insect | Fujimoto et al. 1993 |
| Indica | Protoplast (PEG) | Chitinase *chi11* | H | Resistance to sheath blight (SB) | Lin et al. 1995 |
| Japonica | Biolistic | *Xa-21* | H | BLB resistance | Song et al. 1995 |
| IR58 | Gene flow gun | Bt *crylA(b)* | P | Resistance to insect | Wünn et al. 1996 |
| NPT/IRRI | Protoplast/Biolistic | Bt/*chi11* | H | SB resistance | Alam et al. 1996 |
| Japonica | Biolistic/protoplast | *pinII* | M | Resistance to insect | Duan et al. 1996 |
| Indica | Biolistic/protoplast | Bt | H | Stem borer resistance | Datta et al. 1996 |
| Japonica | Biolistic | *HVA1* | M | Osmoprotectant | Xu et al. 1996 |
| Indica | Biolistic | *crylA(c)* | P | Stemborer resistance | Nayak et al. 1997 |
| Indica/Japonica | Biolistic/protoplast | *adh/pdc* | M | Submergence tolerance | Quimio et al. 1997 |
| Indica | Biolistic | Bt | M | Stem borer resistance | Tu et al. 1998a |
| Indica/Japonica | Biolistic/protoplast | Bt | H | Stem borer resistance | Datta et al. 1998b |
| Indica | Biolistic | *Xa-21* | M | BLB resistance | Tu et al. 1998b |

H – more than 100 independent transgenic plants; M – more than 10 independent transgenic plants; P – less than 10 independent transgenic plants

TABLE 6
Status of transgenic rice research at IRRI

| Cultivars | Method[1] | Genes | Plants in greenhouse (total number) | Southern+ Enzyme assay | Fertility status (%) | Transgenic lines selected[2] |
|---|---|---|---|---|---|---|
| IR72 | B | *Bt*, CH, *Xa-21* | 32 | 20 | 60 | 12 |
| IR64 | B, P | *Bt*, CH | 20 | 3 | 20 | 1 |
| CB II | P | *Bt*, CH, ST | 1800 | 160 | 80 | 18 |
| IR58 | P, B | *Bt*, CH, ST, others | 210 | 2 | 60 | 1 |
| IR51500 | P, B, A | *Bt*, CH, ST | 60 | 3 | 40 | 2 |
| Basmati 370 | P, B | *Bt*, CH, others | 72 | 16 | 20 | 2 |
| Basmati 122 | P, B, A | *Bt*, CH, ST | 66 | 18 | 30 | 4 |
| New Plant Type | B, P | *Bt*, CH,others | 120 | 66 | 30 | 4 |
| ML for hybrid rice | B, P | *Bt*, CH, others | 140 | 22 | 60 | 12 |
| MH63 | B | *Bt*, CH | 62 | 16 | 70 | 11 |
| Vaidehi | B | *Bt*, CH | 26 | 6 | 80 | 2 |
| T309 | B, P, A | *Bt*, CH, others | 1370 | 125 | 90 | 16 |

[1]B = biolistic; P = protoplast-mediated; A = *Agrobacterium* (this method was only used for model transformation system); *Bt* = encoding resistance for stemborer; CH = chitinase gene for sheath blight resistance; ST = gene for submergence tolerance
[2]Plants chosen based on Souther, Western and bioassay data

(Perucho et al., 1980). Carrier DNA was also found integrated in the genome of transformants without any known functions (Peerbolte et al., 1985).

### 7.2 *DNA concentration*

Lower concentration of the gene used in direct gene transfer resulted in a relatively simple integration pattern although higher concentration enhanced transformation frequency (unpublished data). For protoplast transformation, 4-10 $\mu$g DNA per $1.5 \times 10^6$ protoplasts and for biolistic transformation 10 $\mu$g DNA (for 6 shots) are optimum.

### 7.3 *Gene silencing*

Gene silencing has been reported in many plants and has emerged as a topic of interest (Matzke and Matzke, 1995). We found a few cases of gene silencing (unpublished data), but a perusal of the literature shows that it is not a major problem in rice transformation (Arencibia et al., 1998).

### *7.4 Co-transformation and co-integration*

The frequency of co-transformation varies based on the method used. The biolistic method results in higher co-transformation (up to 60%) as compared to protoplasts ( up to 40%). Unlike the protoplast system, the non-selectable DNA was more frequently found linked with the selectable marker showing similar pattern of integration in subsequent progeny. We have studied up to five generations of transgenic rice containing *hph* and *gus* genes, and upto three generations for chitinase and *Bt* gene. We did not notice any abnormality of the function of the transgene except that the $T_0$ transgenic rice plants were usually short in height, early-maturing and with sterility problems. Transgenic plants in $T_1$ and subsequent generations showed similar phenotypes as that of the seed-derived wild type plants. Selectable marker gene and gene of interest were used in 1:4 ratio for biolistic and 1:2 for protoplast system.

## 8. Case study of transgenic rice

During the past few years a large number of transgenic rice plants expressing many agronomically important genes have been obtained (Table 5). Selected examples from these are discussed below.

### *8.1 Transgenic rice with* Xa-21 *conferring bacterial blight resistance*

Bacterial blight (BB) caused by *Xanthomonas* pv. oryzae (*Xoo*) is one of the most destructive diseases of rice throughout the world. Rice yield losses caused by BB in some areas of Asia can be as high as 50%. The use of resistant cultivars is the most economical and effective method to control this disease (Ogawa, 1993).

A dominant gene for resistance to BB, designated *Xa-21*, was transferred from a wild species, *O. longistaminata*, to the cultivated variety IR24 (Khush et al., 1990). The resulting line with *Xa-21* is described as IRBB21. *Xa-21* confers resistance to all the known races of *Xoo* in India and Philippines (Ikeda et al., 1990). The molecular structure of *Xa-21* represents an uncharacteristic class of plant disease resistance genes. From its deduced amino acid sequence, the gene was found to be translated into a receptor kinase-like protein carrying leucine-rich repeats (LRR) in the putative extracellular domain, a single pass transmembrane domain, and a serine theonine kinase (STK) intracellular domain (Song et al., 1995). Further, *Xa-21* supports a role for cellular signaling in plant disease resistance (Song et al., 1995).

*Xa-21* has been transferred to susceptible japonica rice, T309 and showed resistance to BB (Wang et al., 1996). As T309 is not a commercial variety, we introduced the gene in elite breeding cultivars, eg. IR72, MH63, IR51500 etc. Molecular analysis of transgenic plants revealed the presence of a 3.8 kb *Eco*RV-digested DNA fragment corresponding to most of the *Xa-21* coding region and its complete intron sequence, indicating the integration of *Xa-21* in the rice genome (Figure 12, Tu et al., 1998b). Transgenic plants were challenged with two prevalent races (4 and 6) of *Xanthomonas oryzae*. $T_0$ and $T_1$ plants positive for transgene were found to be resistant to bacterial blight. We also observed that the level of resistance

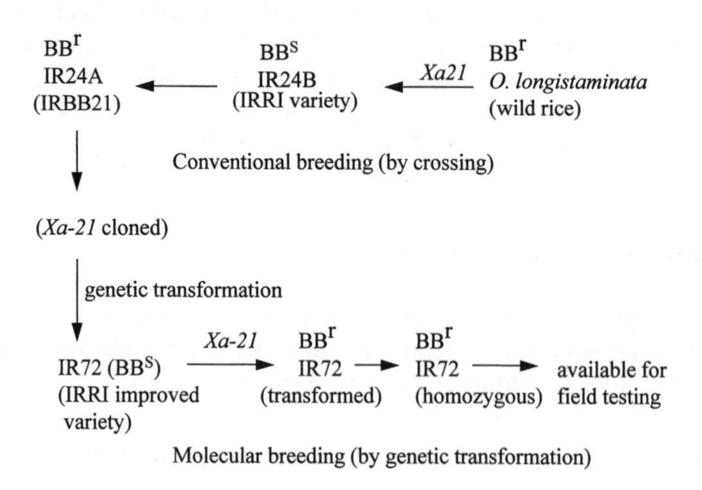


*Fig. 11* Schematic presentation of transfer of *Xa*-21 from original source to cultivated rice by conventional and molecular breeding

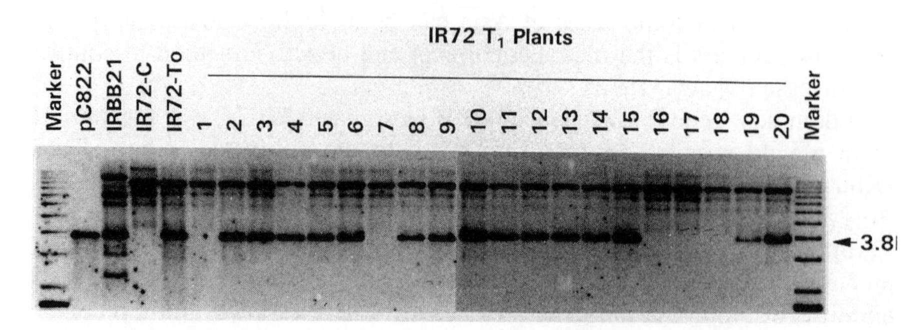


*Fig. 12* Southern analysis of transgenic IR72 $T_1$ plants. A total of 5 $\mu$g plant genomic DNA and 30 pg of plasmid DNA were digested with EcoRV and hybridized with the same enzyme-digested plasmid DNA fragment. The arrow indicates the expected 3.8 kb hybridizing band, which appeared in the plasmid, Xa21-donor line, and positive transgenic $T_0$ and $T_1$ plants. IRBB21, IR72-C, IR72-$T_0$ and IR72-$T_1$ plants represent *Xa-21* introgression line, non-transgenic control plant, transgenic primary and progeny plants, respectively. pC822, the plasmid containing *Xa-21* used to generate transgenic line T103. Markers in the first and last lane were labeled by a 1 kb ladder (from Tu et al., 1998b)

to race 4 of *Xoo* was higher due to pyramiding of *Xa-21* in addition to *Xa-4* already present in IR72. This is a very efficient way to improve BB resistance of rice without genetic dragging and requiring less that 2 years.

### *8.2 Transgenic rice with* BT *genes conferring insect pest resistance*

Stem borer damage is a serious problem in rice, causing estimated losses of 10–30% of the total yield. *Bacillus thuringiensis* (*Bt*), the common soil bacterium, produces crystals containing insecticidal proteins. These toxins kill insects by binding to and creating pores in the midgut membranes. Bt toxins are highly specific and therefore are not toxic to the beneficial insects, birds and mammals including humans (Koziel et al., 1993). Transgenic rice with *Bt* genes has been obtained by several groups (see Table and review, Kratïger, 1997). A few important points are elaborated here:

1. The amount of Bt protein in transgenic rice may not have significant effect on larval mortality. It is important to select a right *Bt* rice plant which shows 100% insect mortality irrespective of Bt protein contents. We have observed some transgenic rice lines with low Bt protein content conferring 100% insect mortality (Datta et al., 1998b).
2. Use of combined *Bt* genes with different receptor binding sites may provide durable resistance.
3. Use of tissue-specific promoters driving *Bt* genes might help better management of Bt resistance and reduce phenotypic cost (Datta et al., 1998b).
4. Special cultivars like DWR (deep water rice), maintainer lines for hybrid rice, and basmati rice with *Bt* genes provide great advantage to the breeders as other alternatives are less attractive (Alam et al., 1998a, b, Datta et al., 1998b).

### *8.3 Transgenic rice for micronutrient improvement*

In developing countries, the challenge of micronutrient deficiencies in diet is great. Iodine, Vitamin A and iron deficiencies are serious concerns in Asia and Africa. Rice breeders at IRRI have undertaken research on enriching rice genetically with iron and other micronutrients (D. Senadhira, personal communication). The study shows that some cultivars, particularly aromatic rice, have greater uptake of iron. A group of Japanese scientists have shown improvement in iron content by the introduction of *ferritin* gene (Yoshihara et al., 1997).

Several groups (USA, Germany, Switzerland in collaboration with IRRI) are working on the incorporation of provitamin A in rice supported by the Rockefeller Foundation. It is estimated that over 124 million children suffer from Vitamin A deficiency (Humphrey et al., 1992). Improved vitamin A nutrition could prevent 1–2 million deaths of children annually. In maize and sorghum endosperm cells can produce and accumulate carotenoids, source of vitamin A (Buckner et al., 1991). Rice in its milled form, as consumed, does not contain any provitamin A ($\beta$-carotene).

Four key enzymes, phytoene synthase, phytoene desaturase, $\delta$-carotene desaturase and cycopene cyclase are necessary for caroterroid biosynthesis. $\beta$-carotene is synthesized from the general $C_{20}$ isoprenoid precursor geranyl geranyl diphosphate by the above mentioned four enzymes. The genes encoding for these enzymes are available from both higher plants (Fray and Grierson, 1993; Linden et al., 1993; Bartley and Scolnik, 1995) and bacteria (Armstrong, 1994).

The ETH-group has produced transngeic model rice (T309) with cDNA coding for phytoene synthase from daffodil (*Narcissus pseudonarcissus*) under the control of either a constitutive or an endosperm specific promoter. They reported that the daffodil enzyme is active in transgenic rice as measured by the in vivo accumulation of phytoene in rice endosperm, a critical step towards the goal of provitamin synthesis in rice (Burkhardt et al., 1997). However, it will of great interest to see the introduction and expression of the remaining genes encoding the other enzymes required to produce $\beta$-carotene in rice. It is also essential to have 2 $\mu$g $\beta$-carotene per gram dry seed weight, to provide the daily requirements of provitamin A for children.

## 9. Field trials of transgenic rice

There is only limited information available on field trials of transgenic rice (Schuh et al., 1993, Oard et al., 1996). The first report deals with the phenotypic changes of transgenic rice under field conditions. The other evaluated two rice cultivars, 'Koshihikari' and 'Gulfmont', containing the herbicide resistance gene *bar*. Field tests of transgenic rice were conducted at the LSU Agricultural Center Rice Research Station over two consecutive years to evaluate agronomic performance of 11 independently transformed lines. $R_2$ and $R_3$ transgenic lines were drill-seeded in 2 m rows in a plot size measuring 3.6 m × 1.42 m. Normal agronomic and fertilization practices were followed to promote optimal plant growth. Weeds were controlled by using propanil at the rate of 3 kg/ha. Glufosinate treatment was used when the plants were at 3–4 leaf stage. All non-transgenic plants were killed within 7 days after 1:12 and 2:24 kg/ha glufosinate application. However, transgenic 'Koshihikari' and 'Gulfmont' survived glufosinate treatment and produced normal seeds. This is the first field demonstration of performance of herbicide resistance in rice (Oard et al., 1996). Several transgenic rice varieties with *Bt*, *Xa-21*, and chitinase genes produced at IRRI (Table 6) are now available for field testing. Our preliminary field experiments conducted in China showed that transgenic IR72 developed at IRRI were resistant to bacterial blight and insect-pests (unpublished data).

## 10. Biodiversity and transgenic rice

Rice genome is extremely diverse in nature (see section 2, Figure 1a, b). With the advancement of improved varieties, particularly in irrigated rice ecosystem, the number of traditional varieties being planted has declined in recent years. The genetic uniformity of improved rices has replaced the diversity of locally adapted varieties developed by farmers over generations (Jackson, 1995). On the other hand, farmers harvest on average 5–7 tons of rice from modern varieties compared to 1–3 tons from traditional varieties. However, introduction of any foreign gene in rice provides a new genetic make-up for the rice genome. The introduced character is often completely new and was not present in rice genome, e.g., sheath blight resistance. Transfer and expression of *TLP-D34, RC7* or *chi11* genes in rice showed enhanced expression of sheath blight resistance (Lin et al., 1995, Datta et al., 1998a). Similarly, the introduction of *Bt* (Fujimoto et al., 1993, Datta et al., 1998b, Nayak et al., 1997), and *Xa-21* genes (Figure 11), or genes for tolerance to water stress (Quimio et al., 1997, Xu et al., 1996a) will widen biodiversity and enrich the rice genome.

## 11. Environmental concerns and risk assessment

Some gene products with fungicidal and insecticidal mode of action have been reported as food allergens. Three classes of multigene families that encode both proteins and allergens involved in plant defense (Lectins, PR-proteins, RIPs, $\alpha$-amylase/trypsin inhibitors) have been identified. Some of them are heat stable and resistant to digestive proteolysis. So after ingestion, they may reach the blood supply and trigger allergic reactions. Therefore, all the transgenic plants with such classes of proteins should be carefully monitored for their allergenic potential. However, there is so far no report of such allergens in transgenic rice with PR-proteins. Some high-value transgenic rice, e.g. Basmati, needs special attention, as unintended changes (e.g. quality of aroma, grain quality) may become an additional matter of concern. Only a few plants from a transgene population will behave in the expected way without insertional mutagenesis, copy number effects, somaclonal variation, and pleiotropic effects (Franck-Oberaspach and Keller, 1997, our unpublished work).

Selectable marker genes (e.g. *nptII* or *hph*) are normally driven by constitutive promoters. Such genes are important during the process of selection and regeneration of transgenic plants, but offer no benefit to the plant during growth and development. They have generated public debate and concerns. Transformation strategies are now available which may help to eliminate the unwanted DNA sequence from the transgenic plants. On some rare occasions we have obtained transgenic *Bt* rice without a functioning *hph* gene (unpublished data). The presence of *Bt* toxin protein in food plants, particularly rice, is also a concern to the public. A large amount of the *Bt* protein may be ingested in the several Asian countries where rice is eaten three times a day. Although it is known that CryIA(b) or CryIA(c) proteins are toxic to lepidopteran insects only, and not to other insects or human beings, concerns remain about dose effects for prolonged periods of time. More scientific data with time and field performance of *Bt* rice may help in improving public perception. The use of tissue specific promoter for *Bt* protein expression in target tissue might minimize public concerns and provide environment friendly rice plants (Datta et al., 1998b).

Commonly used breeding procedures may systematically reduce genetic diversity. The genetic gap between elite gene pools and unimproved pools/ germplasm collection is growing larger with each breeding cycle. As a result, introgression of genetic diversity from unimproved germplasm has become more difficult and has most often been limited to genes for pest resistance (Rasmussion and Philipps, 1997). There are new concerns about genetic engineering promoting monoculture and reducing diversity. A very common concern popularly known as horizontal gene flow of herbicide resistance genes from transgenic rice plants to the related wild species (e.g. red rice) may cause the permanent production of weeds. Cultivated rice is a self-pollinating plant. However, outcrossing or cross pollination does occur at a low level under natural conditions. *O. sativa* is able to interbreed naturally with only seven of 22 wild species of the genus *Oryza*. These seven species, like cultivated rice, possess what is called the AA genome as their genetic material. Spontaneous hybridization between cultivated rice and wild rice is common in many Asian countries but has not caused the disappearance of the wild varieties, which have been a component of Asian rice communities for thousands of years. It is unlikely that a single gene (through genetic engineering) change would disturb this balance significantly, but this possibility can be investigated through studies on factors controlling the population dynamics of wild varieties in the natural environment. The pollen of rice has very limited viability outside the flower. Experience with hybrid rice suggests that pollen remains viable only over a limited distance. Systematic study, careful selection, and proper evaluation following the standard biosafety guidelines should be followed to avoid known or unknown risks associated with transgenic rice (OECD 1993). Public awareness and acceptance of new products are also equally important. I believe that scientists working in this area are capable of judging the possible risks and can bring safe products to the people.

## 12. Future prospects and conclusions

Gene transfer into rice is now possible beyond the laboratory scale. Several agronomically important genes have been introduced in rice by protoplast as well as biolistic methods and work with *Agrobacterium* system is in progress. Availability of efficient transformation systems and codon optimized transgenes driven by suitable promoters provide an excellent opportunity to widen the biodiversity of rice, including the improvement of traditional varieties. It should be possible to transfer into rice genes of interest available from rice or other sources. Besides being a staple food, genetically altered rice may also be useful for industrial products.

Conventional breeding requires 8–10 years in producing a variety because extensive backcrossing to restore the desirable character combinations is required. In contrast, genetic engineering may require only two years to produce an improved variety. The present intensive knowledge-based information technology will bring molecular and conventional breeding together for rice improvement to feed more people. It is now time to give a serious thought regarding the rights of the small farmers of the developing countries to ensure the rights to share the benefits of biotechnology (Borlaug, 1997).

## Acknowledgments

The Rockefeller Foundation network on rice biotechnology and BMZ/GTZ, (Germany) grants to Asian Rice Biotechnology are gratefully acknowledged. Thanks are due to Drs. K. Fischer, G.S. Khush, P.S. Teng, J. Bennett, T. Mew, B. Courtois, JK Ladha, D.S. Brar and M. Hossain of IRRI for their valuable discussion and support. I thank my collaborators and co-workers in Tissue Culture Laboratory: K. Datta, L. Torrizo, M.F. Alam, J. Tu, C. Wu, L.D. Hou, N. Oliva, A. Vasquez, E. Abrigo, M. Alejar, R. Garcia and M. Viray for their valuable contributions and help. Special thanks go to Carolyn Dedolph for her reading and comments on the manuscript.

## References

Alam, M.F., Datta, K., Abrigo, E., Vasquez, A., Senadhira, D., and Datta, S.K. (1998a) Production of transgenic deepwater indica rice plants expressing a synthetic *Bacillus thuringiensis cryIA(b)* gene with enhanced resistance to yellow stem borer. Plant Sci. 135: 25–30.

Alam, M.F., Datta, K., Abrigo, E., Oliva, N., Tu, J., Virmani, S.S., and Datta, S.K. (1998b) Transgenic insect resistant maintainer line (IR68899B) for improvement of hybrid rice. Plant Cell Rep. (in press).

Alam, M.F., Oliva, N.P., Zapata., F.J., and Datta, S.K. (1995) Fertile transgenic indica rice plants by PEG-mediated protoplast transformation. J. Genet. Breed 49: 303–308.

Alam, M.F., Datta, K., Vasquez, A.R., Oliva, N., Khush, G.S., and Datta, S.K. (1996) Production of fertile transgenic New Plant Type rice using protoplast and biolistic systems. Rice Genet. Newslet. 13: 139–141.

Aldemita, R.R., and Hodges, T.K. (1996). *Agrobacterium tumefaciens*-mediated transformation of indica and japonica rice varieties. Planta. 199: 612–617.

Arencibia, A., Gentinetta, E., Cuzzoni, E., Castglione, S., Kohli, A., Vain, P., Leech, M., Christou, P., and Sala, F. (1998) Molecular analysis of the genome of transgenic rice (*Oryza sativa* L.) plants produced via particle bomdardment or intact cell electroporation. Mol. Breed 4: 99–109.

Armstrong, G.A. (1994) Eubacteria show their true colors: genetics of carotenoid pigment biosythesis from microbes to plants. J. Bacteriol. 176:4795–4802.

Bao, P.H., Granata, S., Castiglione, S., Wang, G., Giordani, C., Cuzzoni, E., Damiani, G., Bandi, C., Datta, S.K., Datta, K., Potrykus, I., Callegarin, A., and Sala, F. (1996) Evidence for genomic changes in transgenic rice (*Oryza sativa* L.) recovered from protoplasts. Transgen. Res. 5: 97–103.

Bartley, G.E., and Scolnik, P.A. (1995) Plant carotenoids-pigments for photoprotection, visual attraction, and human health. Plant Cell 7: 1027–1038.

Battraw, M.J., and Hall, T.C. (1990) Histochemical analysis of CaMV promoter ($\beta$-glucuronidase gene expression in transgenic rice plants. Plant Mol. Biol. 15: 527–538.

Benfey, P.N., and Chua, N.H., (1990) The cauliflower mosaic virus 35S promoter: combinatorial regulation of transcription in plants. Science 250: 959–966.

Borlaug, N. (1997) Feeding a world of 10 billion people: the miracle ahead. Lecture presented at DeMontfort University, May 6, 1977, pp. 1–13, Leicester, U.K.

Buckner, B., Kelson, T.L., and Robertson, D.S. (1991) Cloning of the $\gamma$ locus of maize, a gene involved in the biosynthesis of carotenoids. Plant Cell 2: 867– 876.

Burkhardt, P., Beyer, P., Wünn, J., Klöti, A., Armstrong, G.A., Schledz, M., Lintig, J.V., and Potrykus, I. (1997) Transgenic rice (*Oryza sativa*) endosperm expressing daffodil (*Narcissus pseudonarcissus*) phytoene synthase accumulatesphytoene, a key intermediate of provitamin A biosynthesis. Plant J. 11(5): 1071–1078.

Callis, J., Fromm, M., and Walbot, V. (1987) Introns increase gene expression in cultured maize cells. Gene. Dev. 1: 1183–1200.

Cao, J., Duan, X., McElroy, D., and Wu R. (1992) Regeneration of herbicide resistant transgenic rice plants following microprojectile-mediated transformation of suspension culture cells. Plant Cell Rep. 11: 586–591.

Chaïr, H., Legavre, T., and Guiderdoni, E. (1996) Transformation of haploid, microspore-derived cell suspension protoplasts of rice (*Oryza sativa* L.). Plant Cell Rep. 15: 766–770.

Chamberlain, D.A., Bretell, RIS., Last, D.I., Witrzens, B., McElroy, D., Dolferus, R., and Dennis, E.S. (1994) The use of the Emu promoter with antibiotic and herbicide resistance genes for the selection of transgenic wheat callus and rice plants. Aust. J. Plant Physiol. 21: 95–112.

Christou, P., Ford, T.F., and Kofron, M. (1991) Production of transgenic rice (*Oryza sativa*) plants from agronomically important Indica and Japonica varieties via electric discharge particle acceleration of exogenous DNA into immature zygotic embryos. Bio/Technology 9: 957–962.

Chu, C.C., Wang, C.C., Sun, S.S., Hsu, C., Yin, K.C., Chu, C.Y., and Bi, F.Y. (1975) Establishment of an efficient medium for anther culture of rice through comparative experiments on the nitrogen sources. Sci. Sin. 18: 659–668.

Clancy, M., Vasil, V., Hannah, L.C., and Vasil, I.K. (1994) Maize Shrunken-1 intron and exon regions increase gene expression in maize protoplasts. Plant Sci. 98: 151–161.

Cornejo, M.J., Luth, D., Blankenship, K.M., Anderson, O.F., and Blechl, A.E. (1993) Activity of maize ubiquitin promoter in transgenic rice. Plant Mol. Biol. 23: 567–581.

Datta, K., Velazahan, R., Oliva, N., Ona, I., Mew, T., Khush, G.S., Muthukrishnan, S., Datta, S.K. (1998a) Over expression of cloned rice thaumatin-like protein (PR-5) gene in

transgenic rice plants enhances environmental friendly resistance to *Rhizoctonia solani* causing sheath blight disease. Theor. Appl. Genet. (in press).
Datta, K., Vasquez, A., Tu, J., Torrizo, L., Alam, M.F., Oliva, N., Abrigo, E., Khush, G.S., and Datta. S.K. (1998b) Constitutive and tissue-specific differential expression of *cryIA(b)* gene in transgenic rice plants conferring enhanced resistance to rice insect pests. Theor. Appl. Genet. 97: 20–30.
Datta, K., Oliva, N., Torrizo, L., Abrigo, E., Khush, G.S., and Datta, S.K. (1996a). Genetic transformation of indica and japonica rice by *Agrobacterium tumefaciens*. Rice Genet. Newslet. 13: 136–139.
Datta, K., Torrizo, L., Oliva, N., Alam, M.F., Wu C., Abrigo, E., Vasquez, A., Tu, J., Quimio, C., Alejar, M., Nicola, Z., Khush, G.S., and Datta, S.K. (1996b) Production of transgenic rice by protoplast, biolistic and *Agrobacterium* systems. Proc. Vth Internat. Symp. Rice Mol. Biol. pp. 159–167, Yi Hsiem Pub. Co., Taipei, Taiwan.
Datta, S.K. (1995) Polyethylene-glycol-mediated direct gene transfer to indica rice protoplasts and regeneration of transgenic plants. In: Potrykus, I., and Spangenberg, G. (eds), Gene Transfer to Plants, pp. 66–74. Springer, Heidelberg.
Datta, S.K., Datta, K., Soltanifar, N., Donn, G., and Potrykus, I. (1992) Herbicide-resistant indica rice plants from IRRI breeding line IR72 after PEG mediated transformation of protoplasts. Plant Mol. Biol. 20: 619–629.
Datta, S.K., Peterhans, A., Datta, K., and Potrykus, I. (1990) Genetically engineered fertile Indica-rice recovered from protoplasts. Bio/Technology 8: 736–740.
Davey, M.R., Kothari, S.L., Zhang, H., Rech, E.L., Cocking, E.C., and Lynch, P.T. (1991) Transgenic rice: characterization of protoplast-derived plants and their seed progeny. J. Exp. Bot. 42: 1159–1169.
Dekeyser, R., Claes, R., Marichal, M., Van Montagu, M., and Caplan, A. (1989) Evaluation of selectable markers for rice transformation. Plant Physiol. 90: 217–223.
Dong, J., Teng, W., Buchholz, W.G., and Hall, T.C. (1996) *Agrobacterium* mediated transformation of Javanica rice. Mol. Breed 2: 267–276.
Duan, X., Li, X., Xue, Q., Abo-El-Saad, M., Xu, D., and Wu, R. (1996) Transgenic rice plants harboring an introduced potato proteinase inhibitor II gene are insect resistant. Nature Biotechnology 14: 494–498
Fauquet, C.M., Zhang, S., Chen, L., Marmey, P., deKochko, A., and Beachy, R.N. (1996) Biolistic transformation of rice: now efficient and routine for japonica and indica rices. Rice Genetics III. In: Khush, G.S. (ed.), Proc 3rd Intern. Rice Genet. Symp. International Rice Research Institute, Manila, Philippines.
Fraley, R., and Schell, J. (1991) Plant biotechnology. Curr. Opin. Biotech. 2: 145–146.
Franck-Oberaspach, and Keller, B. (1997) Consequences of classical and biotechnological resistance breeding for food toxicology and allergenicity. Plant Breed. 116: 1–17.
Fray, R.G., and Grierson, D. (1993) Identification and genetic analysis of normal and mutant phytoene synthase of tomato by sequencing, complementation and co-suppression. Plant Mol. Biol. 22: 589–602.
Fromm, M., Morrish, F., Armstrong, C., Williams, R., Thomas, J., and Klein, T.M. (1990) Inheritance and expression of chimeric genes in the progeny of transgenic maize plants. Bio/Technology 8: 833–844.
Fujimoto, H., Itoh, K., Yamamoto, M., Kyozuka, J., and Shimamoto, K. (1993) Insect resistant rice generated by introduction of a modified $\delta$-endotoxin gene of *Bacillus thuringiensis*. Bio/Technology 11: 1151–1155.
Fujimura, T., Sakurai, M., Akagi, H., Negishi, and Hirose, A. (1985) Regeneration of rice plants from protoplasts. Plant Tissue Cult. Lett. 2: 74–75.
Gasser, C.S., and Fraley, R.T. (1989) Genetically engineering plants for crop improvement. Science 244: 1293–1299.
Gordon-Kamm, W.J., Spencer, T.M., Mangano, M.L., Adams, T.R., Daines, R.J., Start, W.G., O'Brien, J.V., Chambers, A.S., Adams, W.R., Willetts, N.G., Rice, T.B., Mackey, C.J.,

Krueger, R.W., Kausch, A.P., and Lemaux, P.G. (1990) Transformation of maize cells and regeneration of fertile transgenic plants. Plant Cell. 2: 603–618.
Goto, F., Toki, S., and Uchimiya, H. (1993) Inheritance of a co-transferred foreign gene in the progenies of transgenic rice plants. Transg. Res. 2: 300–305.
Gupta, H.S., and Pattanayak, A. (1993) Plant regeneration from mesophyll protoplasts of rice (*Oryza sativa* L.). Bio/Technology 11: 90–94.
Hauptmann, R.M., Vasil, V., Ozias-Akins, P., Tabaeizadeh, Z., Rogers, S.G., Fraley, R.T., Horsch, R.B., and Vasil, I.K. (1988) Evaluation of selectable markers for obtaining stable transformants in the gramineae. Plant Physiol. 86: 602–606.
Hayakawa, T., Zhu, Y., Itoh, K., Kimura, Y., Izawa, T., Shimamoto, K., and Toriyama, S. (1992) Genetically engineered rice resistant to rice stripe virus, an insect-transmitted virus. Proc. Natl. Acad. Sci. USA. 89: 9865–9869.
Hayashimoto, A., Li, Z., and Murai, N. (1990) A polyethylene glycol mediated protoplast transformation system for production of fertile transgenic rice plants. Plant Physiol. 93: 857–863.
Hiei, Y., Ohta, S., Komari, T., and Kumashiro, T. (1994) Efficient transformation of rice (*Oryza sativa* L.) mediated by *Agrobacterium* and sequence analysis of the boundaries of the T-DNA. Plant J. 6: 271–282.
Hinche, M.W.D., Corbin, D.R., Armstrong, C.L., Fry, J.E., Sato, S.S., Deboer, D.L., Petersen, W.L., Armstrong, T.A., Connor-Ward, T.V., Layton, J.G., and Horsch, R.B. (1994) Plant Transformation. In: Vasil, I.K., and Thorpe, T.A. (eds), Plant Cell and Tissue Culture, pp. 231–270. Kluwer Academic Publishers, The Netherlands.
Hosoyama, H., Irie, K., Abe, K., and Arai, S. (1995) Introduction of a chimeric gene encoding an oryzacystatin-$\beta$-glucorinidase fusion protein into rice protoplasts and regeneration of transforrmed plants. Plant Cell Rep. 15: 174–177.
Hossain, M., and Fischer, K. (1995) Rice research for food security and sustainable agricultural development in Asia: achievements and future challenges. Geo. Journal 35: 286–298.
Humphrey, J.H., West, K.P. Jr., and Sommer, A. (1992) Vitamin A deficiency and attributable mortality among under-5-year-olds. Bull. World Health Org. 70: 225–232.
Ikeda, R., Khush, G.S., and Tabien, R.E. (1990) A new resistance gene to bacterial blight derived from *O. longistaminata*. Japanese J. Breed 40 (suppl 1): 280–281.
Irie, K., Hosoyama, H., Takeuchi, T., Iwabuchi, K., Watanabe, H., Abe, M., Abe, K., and Arai, S. (1996) Transgenic rice established to express corn cystatin exhibits strong inhibitor activity against insect gut proteinases. Plant Mol. Biol. 30: 149–157.
Jackson, M.T. (1995) Protecting the heritage of rice biodiversity. Geo. Journal 35: 267–274.
Jain, R.K., Jain, S., Wang, B., and Wu, R. (1996) Optimization of biolistic method for transient gene expression and production of agronomically useful transgenic basmati rice plants. Plant Cell Rep. 15: 963–968.
Katagiri, F., and Chua, N.H. (1992) Plant transcription factors: present knowledge and future challenges. Trends Genet. 8: 22–27.
Khush, G.S. (1995) Modern varieties: their real contribution to food supply and equity. Geo. Journal 35: 275–284.
Khush, G.S. (1993) Varietal needs for different environments and breeding strategies. In: Muralidharan, K., and Siddiq, E.A. (eds), New Frontier in Rice Research, pp. 68–75. Directorate of Rice Research, Hyderabad, India.
Khush, G.S., Bacalangco, E., and Ogawa, T. (1990) A new gene for resistance to bacterial blight from *O. longistaminata*. Rice Genet. Newsl. 7: 121–122.
Khush, G.S. and Toenniessen, G.H. (1991) Rice Biotechnology. CAB International, Wallingford, UK in association with the International Rice Research Institute, Manila, Philippines.
Koziel, M.G., Beland, G.L., Bowman, C., Carozzi, N.B., Crenshaw, R., Crossland, L., Dawson, J., Desai, N., Hill, M., Kadwell, S., Launis, K., Lewis, K., Maddox, D.,

McPherson, K., Meghji, M.R., Merlin, E., Rhodes, R., Warren, G.W., Wrights, M., and Evola, S.V. (1993) Field performance of elite transgenic maize plants expressing an insecticidal protein derived from *Bacillus thuringiensis*. Biotechnology 11: 194–200.

Krattiger, A.F. (1997) Insect Resistance in Crops: A case study of *Bacillus thuringiensis (Bt)* and its transfer to developing countries. ISAAA Briefs No. 2., pp. 42. ISAAA, Ithaca, NY.

Kyozuka, J., McElroy, D., Hayakawa, T., Xie, Y., Wu, R., and Shimamoto, K. (1993) Light regulated and cell-specific expression of tomato *rbcS-gusA* fusion genes in transgenic rice. Plant Physiol. 102: 991–1000.

Kyozuka, J., Fujimoto, H., Izawata, T., and Shimamoto, K. (1991) Anaerobic induction and tissue-specific expression of maize *Adh 1* promoter in transgenic rice plants and their progeny. Mol. Gen. Genet. 228: 40–48.

Lazzeri, P.A., Brettschneider, R., Lührs, R., and Lörz, H. (1991) Stable transformation of barley via PEG-induced direct DNA uptake into protoplasts. Theor. Appl. Genet. 81: 437–444.

Li, L., Rongda, Q., Kochko, de A., Fauquet, C.M., and Beachy, R.N. (1993) An improved rice transformation system using the biolistic method. Plant Cell Rep. 12: 250–255.

Lin, W., Anuratha, C.S., Datta, K., Potrykus, I., Muthukrishnan, S., and Datta, S.K. (1995) Genetic engineering of rice for resistance to sheath blight. Bio/Technology 13: 686–691.

Linden, H., Vioque, A., and Sandmann, G. (1993) Isolation of a carotenoid biosynthesis gene coding for *β*-carotene desaturase from *Anabena* PCC 7120 by heterologous complementation. FEMS Microbiol. Lett. 106: 99–104.

Lörz, H., Baker, B., and Schell, J. (1985) Gene transfer to cereal cells mediated by protoplast transformation. Mol. Gen. Genet. 199: 178–182.

Luehrsen, K.R., and Walbot, V. (1991) Intron enhancement of gene expression and the splicing efficiency of introns in maize cells. Mol. Gen. Genet. 225: 81–93.

Maas, C., Schell, J., and Steinbiβ, H.H. (1992) Applications of an optimized monocot expression vector in studying transient gene expression and stable transformation of barley. Physiol. Plant 85: 367–373.

Matsuki, R., Onodera, H., Yamauchi, T., and Uchimiya, H. (1989) Tissue specific expression of the *rolC* promoter of the $R_1$ plasmid in transgenic rice plants. Mol. Gen. Genet. 220: 12–16.

Matsuoka, M., Kyozuka, J., Shimamoto, K., and Kano-Murakami, Y. (1994) The promoter of two carboxylase in a $C_4$ plant (maize) direct cell-specific, light-regulated expression in a $C_3$ plant (rice). Plant J. 6: 311–319.

Matzke, M.A., and Matzke, A.J.M. (1995) How and why do plants inactivate homologous (trans) genes? Plant Physiol. 107: 679–685.

McElroy, D., Blowers, A.D., Jenes, B., and Wu, R. (1991) Construction of expression vectors based on the rice actin 1 (*Act1*) 5' region for use in monocot transformation. Mol. Gen. Genet. 231: 150–160.

McElroy, D., Zhang, W., and Wu, R. (1990) Isolation of an efficient actin promoter for use in rice transformation. Plant Cell. 2: 163–171.

Meijir, E.G.M., Schilperoort, R.A., Rueb, S., Os-Ruygrot, P.E., and Hensgens, L.A.M. Transgenic rice cell lines and plant expression of transferred chimeric genes. Plant Mol. Biol. 16: 807–820.

Mitsuhara, I., Ugaki, M., Hirochika, H., Ohshima, M., Murakami, T., Gotoh, Y., Katayose, Y., Nakamura, S., Honkura, R., Nishimiya, S., Ueno, K., Mochizuki, A., Tanimoto, H., Tsugawa, Y., Otsuki., and Ohashi, Y. (1996) Efficient promoter cassettes for enhanced expression of foreign genes in dicotyledonous and monocotyledonous plants Plant Cell Physiol. 37: 49–59.

Müller, A.J., and Graffe, R. (1978) Isolation and characterization of cell lines of *Nicotiana tabacum* lacking nitrate reductase. Mol. Gen. Genet. 161: 67–76.

Murashige, T., and Skoog, F. (1962) A revised medium for rapid growth and bioassays with tobacco tissue cultures. Physiol. Plant 15: 473–497.

Murray, E.E., Rochelaeau, T., Eberle, M., Stock, C., Sekar, V., and Adang, M.J. (1991) Analysis of unstable RNA transcripts of insecticidal crystal protein genes of *Bacillus thuringiensis* in transgenic plants and electroporated protoplasts. Plant Mol. Biol. 16: 1035–1050.

Nayak, P., Basu, D., Das, S., Basu, A., Ghosh, D., Ramakrishna, N.A., Ghosh, M., and Sen, S.K. (1997) Transgenic elite indica rice plants expressing *CryIAc* δ–endotoxin of *Bacillus thuringiensis* are resistant against yellow stem borer (*Scirpophaga incertulas*). Proc. Nat. Acad. Sci. USA 94: 2111–2116.

Oard, J.H., Linscombe, S.D., Braverman, M.P., Jodari, F,. Blouin, D.C., Leech, M., Kohli, A., Vain, P., Cooley, J.C., and Christou, P. (1996) Development, field evaluation, and agronomic performance of transgenic herbicide resistant rice. Mol. Breed 2: 359–368.

OECD. (1993) Safety evaluation of foods derived by modern biotechnology. OECD, Paris.

Ogawa, T. (1993) Methods and strategy for monitoring race distribution and identification of resistance genes to bacterial leaf blight (*Xanthomonas campestris pv. oryzae*) in rice. JARQ 27: 71–80.

Ohira, K., Ojima, K., and Fujiwara, A. (1973) Studies on the nutrition of rice cell culture. 1. A simple, defined medium for rapid growth in suspension culture. Plant Cell Physiol. 14: 1113–1121.

Peerbolte, R., Krens, F.A., Mans, R.W.M., Floor, M., Hoge, J.H.C., Wullems, G.J., and Schiperoort, R.A. (1985) Transformation of plant protoplasts with DNA: co-transformation of non-selected calf thymus DNA and meiotic segregation of transforming DNA sequences. Plant Mol. Biol. 5: 235–246.

Peng, J., Fujiang, W., Lister, R.L., and Hodges, T.K. (1995) Inheritance of *gusA* and neo genes in transgenic rice. Plant Mol. Biol. 27: 91–104.

Peng, J., Kononowicz, H., and Hodges, T.K. (1992) Transgenic indica rice plants. Theor. Appl. Genet. 83: 855–863.

Perlak, F.J., Fuchas, R.L., Dean, D.A., McPherson, S.L., and Fischhoff, D.A. (1991) Modification of the coding sequence enhances plant expression of insect control protein genes. Proc. Nat. Acad. Sci. USA 88: 3324–3328.

Perlak, F.J., Deaton, R.W., Armstrong, T.A., Fuchs, R.L., Sims, S., Greenplate, J.T., and Fischhoff, D.A. (1990) Insect resistant cotton plants. Bio/Technology 8: 939–943.

Perucho, M., Hanahan, D., and Wigler, M. (1980) Genetic and physical linkage of exogenous sequences in transformed cells. Cell. 22: 309–317.

Peterhans, A., Datta, S.K., Datta, K., Goodall, G.J., Potrykus, I., and Paszkowski, J. Recognition efficiency of *Dicotyledoneae*-specific promoter and RNA processing signals in rice. Mol. Gen. Genet. 22: 361–368.

Potrykus, I., (1991) Gene transfer to plants assessment of published approaches and results. Ann. Rev. Plant Physiol. Plant Mol. Biol. 42: 205–225.

Potrykus, I., Saul, M.W., Petruska, J., Paszkowski, J., and Shillito, R.D. (1985) Direct gene transfer to cells of graminaceous monocot. Mol. Gen. Genet. 199: 183–188.

Quimio, C., Torrizo, L., Abrigo, E., Oliva, N., Ella, E., Setter, T., Ellis, M., Ito, O., Datta, K., and Datta, S.K. (1997) Transgenic rice for submergence tolerance. Philippine J. Crop Sci. Supp. 1, 22: 1.

Rasmusson, D.C., and Phillips, R.L. (1997) Plant Breeding progress and genetic diversity from De Novo variation and elevated epistasis. Crop Sciences 37(2): 303–310.

Rathore, K.S., Chowdhurry, V.K., and Hodges, T.K. (1993) Use of *bar* as selectable marker gene for the production of herbicide resistant rice plants from protoplasts. Plant Mol. Biol. 21: 871–884.

Reddy, P.M., Ladha, J.K., Ramos, M.C., Maillet, F., Hernandez, R.J., Torrizo, L.B., Oliva, N.P., Datta, S.K. and Datta, K. (1998) Rhizobial lipochitooligosaccharide nodulation factors activate expression of the legume early nodulin gene *ENOD12* in rice. The Plant J. 14: 693–702.

Rice Almanac (1997) 2nd edition, International Rice Research Institute, Philippines, Centro Internacional de Agricultura Tropical, Colombia, and West Africa Rice Development Association, Côte d'Ivoire.

Rock, C.D., and Quatrano, R.S. (1996) Lanthanide ions are atagonists of transient gene expression in rice protoplasts and act in synergy with ABA to increase *Em* gene expression. Plant Cell Reports 15: 371–376.

Saul, M.W., and Potrykus, I. (1990) Direct gene transfer to protoplasts: fate of the transferred genes. Developmental Genetics 11: 176–181.

Schell, J., (1987) Transgenic plants as tools to study the molecular organization of plant genes. Science 237: 1176–1183.

Schuh, W., Nelson, M.R., Bigelow, D.M., Orum, T.V., Orth, C.E., Lynch, P.T., Eyles, P.S., Blackhall, N.W., Jones, J., Cocking, E.C., and Davey, M.R. (1993) The phenotypic characterization of $R_2$ generation transgenic rice plants under field conditions. Plant Sci. 89: 69–79.

Setter, T.L., Ellis, M., Laureles, E.V., Ella, E.S., Senadhira, D., Mishra, S.B., Sarkarung, S., and Datta, S.K. (1996) Physiology and genetics of submergence tolerance in rice. Annals of Botany 79: 67–77.

Shillito, R.D., Saul, M.W., Pazkowsi, J., Mueller, M., and Potrykus, I. (1985) High efficiency direct gene transfer to plants. Bio/Technology 3: 1099–1103.

Shimada, T., Gürel, F., and Takumi, S. (1995) Simple and rapid production of transgenic rice plants by particle bombardment. Ishikawa Agr. Coll. 4: 1–8.

Shimamoto, K., (1992) Genetic manipulation of rice: from protoplasts to transgenic plants. Jpn. Journal Genet. 67: 273–290.

Shimamoto, K., Terada, R., Izawa, T., and Fujimoto, H. (1989) Fertile transgenic rice plants regenerated from transformed protoplast. Nature 337: 274–276.

Sivamani, E., Shen, P., Oplaka, N., Beachy, R.N., and Fauquet, C.M. (1996) Selection of large quantities of embryogenic subcultured calli from indica rice seeds for production of fertile transgenic plants using the biolistic method. Plant Cell Reports 15: 322–327.

Song, W.Y., Wang, G.L., Chen, L.L., Kim, H.S., Pi, L.Y., Holsten, T., Gardner, J., Weng, B., Zhai, W.X., Chu, L.H., Fauquet, C.M., and Ronald, P. (1995) A receptor kinase-like protein encoded by the rice disease resistance gene, *Xa-21*. Science 270: 1804–1806.

Stewart, C.N., Adang, J.M., All. J.N., Raymer, P.L., Ramachandran, S., and Parrott, W.A. (1996) Insect control and dosage effects in transgenic canola containing a synthetic *Bacillus thuringiensis cryIAc* gene. Plant Physiol. 112: 115–120.

Strizhov, N., Keller, M., Mathur, J., Koncz-Kalman, Z., Bosch, D., Prudovsky, E., Schell. J., Sneh, B., Koncz, C., and Zilberstein, A. (1996) A synthetic *cryIA(c)* gene, encoding a *Bacillus thuringiensis* δ-endotoxin, confers *Spodoptera* resistance in alfalfa and tobacco. Proc. Natl. Acad. Sci. 13: 15012–15017.

Tada, Y., Sakamoto, M., Matsuoka, M., and Fujimura, T. (1991) Expression of a monocot LHCP promoter in transgenic rice. The EMBO Journal 10(7): 1803–1808.

Takano, M., Egawa, H., Ikeda, J.E., and Wakasa, K. (1997) The structures of integration sites in transgenic rice. The Plant Journal 11: 353–361.

Terada, R., Nakayama, T., Iwabuchi, M., and Shimamoto, K. (1993) A wheat histone H3 promoter confers cell division dependent and independent expression of the *gus* A gene in transgenic rice plants. The Plant Journal 3(2): 241–252.

Terada, R., and Shimamoto, K. (1990) Expression of CaMV35S-GUS gene in transgenic rice plants. Mol. Gen. Genet. 220: 389–392.

Toki, S., Takamatsu, S., Nojiri, C., Ooba, S., Anzai, H., Iwata, M., Christensen, A.H., Quail, P.H., and Uchimiya, H. (1992) Expression of a maize ubiquitin gene promoter-bar chimeric gene in transgenic rice plants. Plant Physiol. 11: 1503–1507.

Toriyama, K., Arimoto, Y., Uchimiya, H., and Hinata, K. (1988) Transgenic rice plants after direct gene transfer into protoplasts. Bio/Technology 6: 1072–1074.

Tu, J., Datta, K., Alam, M.F., Khush, G.S., and Datta, S.K. (1998a) Expression and function of a hybrid *Bt* toxin gene in transgenic rice conferring resistance to insect pests. Plant Biotechnology (Japan), in press.

Tu, J., Ona, I., Zhang, Q., Mew, T.W., Khush, G.S., and Datta, S.K. (1998b) Transgenic rice variety IR72 with *Xa21* is resistance to bacterial blight. Theor. Appl. Genet. 97: 31–36.

Vain, P., McMullen, M.D., and Finer, J.J. (1993) Osmotic treatment enhances particle bombardment-mediated transient and stable transformation of maize. Plant Cell Reports. 12: 84–88.

Vasil, I.K. (1987) Developing cell and tissue culture systems for the improvement of cereal and crop plants. J. Plant Physiol. 128: 193–218.

Vasil, I.K. (1994) Molecular improvement of cereals. Plant Mol. Biol. 25: 925–937.

Vasil, I.K. (1996) Phosphinothricin-resistant crops. In: Duke, S.O. (ed.), Herbicide-Resistant Crops, pp. 85–91. CRC Press, Boca Raton, Florida.

Vasil, I.K., Castillo, A.M., Fromm, M.E., and Vasil, I.K. (1992) Herbicide resistant fertile transgenic wheat plants obtained by microprojectile bombardment of regenerable embryogenic callus. Bio/Technology 10: 667–674.

Vasil, V., Clancy, M, Ferl, R.J., Vasil, I.K., and Hannah, L.C. (1989) Increased gene expression by the first intron of maize shrunken-1 locus in grass species. Plant Physiol. 91: 1575–1579.

Vasil, V., Srivastava, V., Castillo, A.M., Fromm, M.E., and Vasil, I.K. (1993) Rapid production of transgenic wheat plants by direct bombardment of cultured immature embryos. Bio/Technology 11: 1553–1558.

Vasil, V., and Vasil, I.K. (1980) Isolation and culture of cereal protoplasts. II. Embryogenesis and plantlet formation from protoplasts of *Pennisetum americanum.* Theor. Appl. Genet. 56: 97–99.

Vavilov, N.I. (1951) The origin of variation, immunity and cultivated plants. Chronica Botanica. 13: 1–1364, Waltham, Mass USA (Translation from Russian).

Virmani, S.S., Prasad, M.N., and Kumar, I. (1993) Breaking yield barrier of rice through exploitation of heterosis. In: Muralidharan, K., and Siddiq, E.A. (eds) New Frontiers in Rice Research (pp. 76–85). Directorate of Rice Research, Hyderabad, India.

Walbot, V., and Gallie, D. (1991). Gene expression in rice. Rice biotechnology. CAB International, Wallingford, UK in association with the International Rice Research Institute, Manila, Philippines 6: 225–251.

Wan, Y., and Lemaux, P.G. (1994) Generation of large numbers of independently transformed fertile barley plants. Plant Physiol. 104: 37–48.

Wang, G.L., Song, W.Y., Ruan, D.L., Sideris, S., and Ronald, P.C. (1996) The cloned gene, *Xa21* confers resistance to multiple *Xanthomonas oryzae pv. oryzae* isolates in transgenic plants. Mol. Plant-Microbe Interact. 9: 850–855.

Wilmink, A., and Dons, J.M. (1993) Selective agents and marker genes for use in transformation of monocotyledonous plants. Plant Molecular Biology Reporter. 11(2): 165–185.

Wünn, J., Klöti, A., Burkhardt, P.K., Ghosh, Biswas, G.C, Launis, K., Iglesia, V.A., and Potrykus, I. (1996) Transgenic indica rice breeding line IR58 expressing a synthetic *cryIA(b)* gene from *Bacilllus thuringiensis* provides effective insect pest control. Bio/Technology 14: 171–176.

Xu, D., Duan, X., Wang, B., Hong, B., David, T.H., and Wu, R. (1996a) Expression of a late embryogenesis abundant protein gene, HVA1 from barley confers tolerance to water deficit and salt stress in transgenic rice. Plant Physiol. 110: 249–257.

Xu, D., Xue, Q., McElroy, D., Mawal, Y., Hilder, V.A., and Wu R (1996b) Constitutive expression of a cowpea trypsin inhibitor gene *CpTi,* in transgenic rice plants confers resistance to two major rice insects pests. Molecular Breeding 2: 167–173.

Yin, Y., and Beachy, R. (1995) The regulatory regions of the rice tungro bacilli form virus promoter and interacting nuclear factor in rice (*Oryza sativa L.*). Plant J. 7: 969–980.

Yokoi, S., Tsuchiya, T., Toriyama, K., and Hinata, K. (1997) Tapetum-specific expression of the Osg6B promoter $\beta$-glucuronidase gene in transgenic rice. Plant Cell Reports 16: 363–367.

Yoshida, S., Forno, D.A., Cock, J.H., and Gomez, K.A. (1976) Laboratory manual for physiological studies of rice. The International Rice Research Institute. Los Baños, Philippines.

Yoshihara, T., Naoki, S., Goto, F., Toki, S., and Takaiwa, F. (1997) Accumulation of iron in rice seed by transferring the soybean iron storage protein *ferritin* gene. Abstract 303, 5th Internat. Congr.. Plant Mol. Biol. Singapore.

Zhang, S., Chen, L., Qu, R., Marmey, P., Beachy, R.N., and Fauquet, C.M. (1996) Regeneration of fertile transgenic indica (group I) rice plants following microprojectile transformation of embryogenic cell suspension culture cells. Plant Cell Reports 15: 465–469.

Zhang, W., McElroy, D., and Wu, R. (1991) Analysis of rice *Act5'* region activity in transgenic plants. Plant Cell 3: 1155–1165.

Zhang, H.M., Yang, H., Rech, E.L., Golds, T.J., Davis, A.S., Mulligan, B.J., Cocking, E.C., and Davey, M.R. (1988) Transgenic rice plants produced by electroporation mediated plasmid uptake into protoplast. Plant Cell Reports 7: 379–384.

Zhang, W., and Wu, R. (1988) Efficient regeneration of transgenic plants from rice protoplasts and correctly regulated expression of the foreign gene in the plants. Theor. Appl. Genet. 76: 835–840.

Zheng, Z., Sumi, K., Tanaka, K., and Murai, N. (1995) The bean seed storage protein *β*-phaseolin is synthesized, processed, and accumulated in the vacuolar type-II protein bodies of transgenic rice endosperm. Plant Physiol. 109: 777–786.

# 8. Transgenic Cereals – *Zea mays* (maize)

W.J. GORDON-KAMM, C.L. BASZCZYNSKI, W.B. BRUCE and D.T. TOMES
*Pioneer Hi-Bred International, Inc., 7300 N.W. 62nd Ave., P.O. Box 1004, Johnston, IA 50131, USA*

ABSTRACT. Genetic transformation of maize is routine in several genotypes despite the many difficulties encountered in developing reliable transformation techniques in this major cereal species. Aspects of maize tissue culture, including the target explant, subsequent rapid *in vitro* proliferation and dependable regeneration from competent cells were prerequisite developments for gene delivery into maize. Recovery of transgenic, fertile maize required high levels of gene expression and identification of new selectable markers, along with DNA delivery into competent maize cells. DNA delivery by particle bombardment, *Agrobacterium*, electroporation and silica fiber methods have been the most carefully documented, each of which can now be used for gene transfer into maize. Promoters such as those from the CaMV 35S or ubiquitin genes, together with various introns have been widely used to achieve high expression levels, while the herbicide resistance gene, *bar*, has served as an important selectable marker for numerous studies in maize transformation. Although tissue culture cells were instrumental in the development of maize transformation, the direct use of explants such as the immature embryo and/or meristems has found favor in more recent applications. Gene delivery in maize has shifted from emphasis on technology development to evaluation of gene expression with various transgenes, some of which are already in large-scale commercial development (e.g. insect and herbicide resistance). Maize transformation is increasingly being used to address more sophisticated aspects of gene regulation, plant development and physiology. The stability of transgene expression in primary transgenic plants and subsequent generations is of obvious academic and commercial importance. The isolation of promoters with a variety of expression profiles that are tissue-specific and/or temporally regulated will become more important as trait modification strategies evolve. Technologies such as site-directed integration, homologous recombination, 'chimeraplasty', and others will likely become routine in higher plants such as maize as this research area, now in its infancy, continues to develop. These technologies have the potential to aid our understanding of gene regulation, and to more directly make changes in endogenous gene sequences or to permit targeting of new genes (or regulatory elements) into precise genomic locations. With an assortment of accompanying genetic tools such as reverse genetic methods, mapping, genome-scale analysis and gene expression information, maize transformation has evolved into an important tool for both basic and applied studies in plants.

## I. Introduction

Transformation in maize is only about 10 years old, and as a result, the proposition that new genetic technologies can complement established

*I.K. Vasil (ed.), Molecular Improvement of Cereal Crops, 189–253*

maize breeding techniques is a relatively recent idea. Even so, there has been rapid progress in the field of maize biotechnology, to the point that many of our current research initiatives are well beyond what we would have imagined in the early days of maize transformation. Early progress was relatively modest, but the substantial private and public research devoted to maize allowed innovative approaches to transformation and molecular biology problems that were not possible in many species.

Maize, wheat, and rice clearly represent the most important cereal crops, although their relative importance varies in different parts of the world. Over the period between 1950 and 1980, the increase in maize production world-wide outpaced both wheat and rice. Despite a temporary downswing in the early to mid-1980s (due to both environmental and political factors) world production has risen steadily from around 145 million tons in 1950 to nearly 500 million tons by 1990. Increases in yield and harvested area have been the predominant contributors to enhanced world production; with yield playing the major role in industrialized countries and harvested area expansion being most important in developing countries (see Dowswell et al., 1996 for review).

At present, 20 countries account for over 90% of the world's maize production and 80% of the total maize area, with the United States and China accounting for 42% and 19% of the world maize crop, respectively. Uses for maize are led by feed for livestock, followed by direct human consumption, industrially processed foods or non-food products, and finally as seed. With its many current and potential new uses, the world maize demand will undoubtedly grow. Cellular and molecular biology in conjunction with plant breeding and improved production practices, will play a critical role in helping to meet this demand for large and small-scale farmers in both industrialized and developing countries. As breeding and biotechnological methods are melded together into one technology, we will see continued improvements in genetic resistance to abiotic stress (clearly the biggest detractor from overall productivity), as well as to insects and disease. Other agronomic traits, and new traits targeted toward specialty markets will also have a major impact.

The importance and future potential of maize as a major world crop has fueled the long, intensive effort at obtaining gene transfer technology in maize. The history of this effort provides an excellent example of how scientific problem solving can be brought to bear on such applied objectives. The challenges that confronted this effort included i) developing the basic transformation methods, ii) improving transformation as a practical, useful technology, iii) regulating transgene expression in a predictable, consistent fashion, and iv) improving our understanding of the maize genome in order to design the most appropriate improvement strategies. The knowledge gained in these areas will assist both in defining future technology efforts, and in trait modification to sustain and enhance productivity to meet food production needs. The following sections recall the historical development of transformation technology and important advances that have led to understanding the complex genetic and physiological structure of maize.

## II. Maize Transformation

Maize transformation research began in the shadow of the success of petunia and tobacco transformation in the late 1970s. Gene transfer was accomplished either by direct DNA uptake in protoplasts (Negrutiu et al., 1987) or by direct DNA transfer using *Agrobacterium* (Fraley et al., 1984; De Block et al., 1984; Hernalsteens et al., 1984). Methodology that seemed at the time both elegant and simple in these model dicots eluded researchers in maize transformation for almost 10 years. Gene transfer in maize required the simultaneous development of three independent research areas; i) identification of competent recipient cells by maize cell biologists for DNA uptake *and* recovery of fertile plants, ii) DNA transfer into competent cells, and iii) identification of control elements and selectable markers with efficiency in maize analogous to those already available in dicots. The long wait for maize transformants required the underlying development of these disparate research disciplines before they could be successfully integrated into a gene transfer method, and continued refinement has finally produced the techniques currently in place.

The dichotomous research goals of wide germplasm accessibility and large scale testing of different gene combinations has driven both public and private research in maize transformation. Plant breeders saw obvious advantages of making rapid use of transgenes in *their* favorite elite genotypes (often a very inefficient process), while the imprecision of gene integration and expression in maize has necessitated large scale testing of putative control elements and structural genes. On a practical level, most gene testing in maize has been accomplished in a handful of amenable genotypes: A188 or A188 × B73 derivatives (Armstrong, 1994; Armstrong and Green, 1985; Ishida et al., 1996), H99, Pa91 and Fr16 (Wan et al., 1995) and a small number of elite inbreds (Koziel et al., 1993; Aves et al., 1992). Transgene expression and characterization has made tremendous progress with commercial products for insect resistance, herbicide resistance, male sterility, and disease resistance that are either available to consumers or in advanced product development cycles (see discussion below). The number of field test permits in maize accounts for over 50% of the testing done in the United States which reflect in part the diversity of gene testing now being conducted (APHIS Biotechnology and Scientific Services Newsletter, updated December 5, 1997).

*Callus-based Methods*

*In-vitro* culture technology for tobacco in the mid-1970s developed rapidly due to a number of important technical accomplishments, including ease of culture establishment, plant hormonal combinations that allowed protoplast isolation and culture, and recovery of plants that retained seed fertility after prolonged culture (Gamborg et al., 1981). In contrast, the first reports of cultures of maize inbred A188 that would regenerate plants at *any* age were in 1975 (Green and Phillips, 1975). Subsequent investigation in maize and other cereals indicated that progressive loss of regeneration capacity occurred as a function of prolonged culture, and among regenerated plants changes in plant phenotype and loss of fertility were observed (Scowcroft et al., 1985; Scowcroft and Larkin, 1988). The success of gene transfer by direct DNA uptake in maize protoplasts thus immediately faced serious technical difficulties because of the extended period required for culture initiation, adaptation, protoplast isolation and recovery (Vasil and Vasil, 1987; Rhodes et al., 1988a). The other major gene transfer option developed in petunia and tobacco, the elegant leaf disk *Agrobacterium* procedure (De Block et al., 1984; Fraley et al., 1984) was thought to be inaccessible due to the limited host range of *Agrobacterium.*

DNA delivery methods first developed in tobacco and petunia were rapidly adopted by cell biologists in cereals. Howard et al. (1985) and Fromm et al. (1985, 1986) demonstrated transient gene expression of the *cat* gene using electroporation as the DNA delivery vehicle. Polyethylene glycol (PEG) and electroporation methods were used to obtain transgenic wheat, rye and grass transgenic cell lines (Potrykus et al., 1985; Shillito et al., 1985). Meanwhile in the mid-1980s, John Sanford and colleagues (Sanford et al., 1987) demonstrated that high velocity, DNA-coated tungsten particles of approximately 4 $\mu$m diameter could be delivered into intact onion cells without loss of viability. Later, Klein et al. (1988a) demonstrated transient expression of the *cat* gene in both embryogenic and non-embryogenic maize cultures using microprojectile bombardment. *Agrobacterium*-mediated DNA delivery in cereals during this time frame was limited to agroinfection methodology in which viral symptoms could be propagated throughout maize plants by using *Agrobacterium* to deliver viral DNA (Hohn et al., 1987). Paradoxically, the most efficient DNA delivery and expression was demonstrated in protoplast based systems with their attendant problems, while the inefficient DNA delivery methodologies of microinjection, micro-projectile bombardment, and *Agrobacterium* resulted in low efficiency transformation for intact cells that demonstrated more robust regeneration and fertility.

During this period, improvements in gene expression and marker systems for monocots were also being made. Gene expression was enhanced by incorporation of introns from maize genes into the control regions of constitutive promoters such as CaMV 35S (Callis et al., 1987; Planckaert and Walbot, 1989; Vasil et al., 1989; Mascarenhas et al., 1990). This was the first demonstration that monocots might have different requirements for high gene expression compared to dicots. Transient gene expression studies using adapted cell cultures such as Black Mexican Sweet (BMS), electroporation, and the chloramphenicol acetyltransferase gene (*cat*) whose expression could be accurately quantified at very low levels of expression were instrumental in the rapid development of promoters specifically enhanced for use in maize and other cereals (Fromm et al., 1985, 1986). Despite the recovery of transgenic rice plants in 1988 (Zhang and Wu, 1988; Toriyama et al., 1988) by the direct delivery of DNA into protoplasts, and recovery of several non-regenerable transgenic cell lines of maize using the same antibiotic resistance genes shown to function in tobacco and petunia, the efficacy of available markers remained a problem for the recovery of transgenic monocot plants.

During the late 1980s, intense research efforts were being directed toward maize transformation. Beginning then, consensus slowly developed regarding the criteria necessary to demonstrate integrative transformation in maize and other cereals (and plants in general). These 'proofs' of integrative transformation were the object of several reviews by Vasil (1988) and Potyrkus (1989, 1990a,b, 1991). The essential elements of maize transformation in which the transforming DNA is integrated into the genome of the host (termed integrative transformation) are:

- Correlation between growth on selective medium and physical presence of the transforming DNA.
- Demonstration of the physical presence of transforming DNA by Southern analysis in high molecular weight plant DNA and/or demonstration of the protein product of the transgene.
- Clear evidence to discriminate between genuine and false positives indicated by phenotypic data (growth on selective conditions).
- Correlation of phenotypic and physical evidence in transgenic parents and their sexual offspring including genetic and molecular analysis of offspring populations.

The final 'proof' of physical evidence of transmission to sexual progeny required years to obtain the combination of adequate number of competent plant cells, DNA delivery into cells which could ultimately give rise to sexual progeny, efficient regulatory elements to drive expression, and selectable markers that function in maize.

A non-regenerable but rapidly growing suspension culture, Black Mexican Sweet (BMS; see Sheridan, 1975, 1982) allowed for the development of several essential elements necessary for maize transformation. DNA delivery and gene expression in maize was first demonstrated in BMS cells for both direct DNA uptake (Fromm et al., 1986; Antonelli and Stadler, 1990) and microprojectile bombardment to deliver DNA into intact BMS cells by Klein et al. (1988a,b). The first transgenic maize callus was obtained from BMS cells (Fromm et al., 1986). This result clearly demonstrated that gene expression in maize was distinct from earlier information obtained in dicot species (see Table 1). Transgenic suspension and callus cultures of maize were obtained by direct DNA delivery to protoplasts derived from regenerable suspension cultures, some of which also gave rise to non-fertile plants (Rhodes et al. 1988b). Both electroporation (Fromm et al., 1986; Callis et al., 1987; Huang and Dennis, 1989) and polyethylene glycol-mediated DNA delivery (Armstrong, et al., 1990; Shillito, et. al. 1985) were effective methods for delivery into protoplasts of maize (see Table 1) and other monocots. Soon after the demonstration of DNA delivery into intact cells (Klein et al., 1988c) by microprojectile bombardment, transgenic callus of BMS was obtained following selection for kanamycin resistance (D. Tomes et. al., personal communication.). During this phase of maize transformation research, the primary limitation was the concomitant ability to obtain large numbers of competent cells for both DNA integration and to produce fertile, regenerated plants (Vasil, 1988; Potrykus, 1989). Microprojectile bombardment into intact maize cells (Klein et. al., 1988a,b, 1989a,b), either from intact plant tissue or cultures, dramatically increased the potential for recovery of fertile transgenic offspring from maize.

More reliable selectable marker genes such as acetolactate synthase (*als*; see Fromm et al., 1990) and *bar* (Gordon-Kamm et al., 1990), combined with DNA delivery by microprojectile bombardment into intact embryogenic suspension cells gave the first reported transgenic maize seed. Earlier attempts using aminoglycosides for selection gave sporadic results, and further complicated the recovery of transgenic maize seed (Gordon-Kamm et al., 1989). Direct DNA delivery into protoplasts also gave rise to transgenic maize seed when maize hybrids and inbreds were developed specifically for their ability to yield large numbers of protoplasts which retained the ability to regenerate fertile plants (Donn et al., 1990; Morocz et al., 1990). Thus, all the components necessary for maize transformation came together, allowing the design of large-scale experiments to increase efficiency and to focus on expression of genes important for basic and product development. While the use of maize protoplasts for production of transgenic plants has been limited, protoplast transformation *per se* has continued to have great value in such applications as testing structural genes and investigating aspects of gene expression, such as promoter analysis, site-directed recombination and intron splicing (see Table 1).

The *bar* (or *pat*) selectable marker gene(s), encoding phosphinothricin acetyltransferase, has been used most frequently (Table 1), although several other selectable marker genes such as hygromycin (Walters et. al., 1992), methotrexate (Golovkin et al., 1993), and kanamycin (D'Halluin et al., 1992; Omirulleh et al., 1993) have also been used to obtain transgenic seed.

TABLE 1
Maize transformation using protoplasts

| Purpose/protoplast source | DNA delivery method | Marker gene(s) | Observation | Reference |
|---|---|---|---|---|
| *Development of protoplast transformation methods* | | | | |
| Black Mexican Sweet (BMS) cell suspension | DEAE-dextran | CAT | Transient | Howard et al. 1985 |
| BMS cell suspension | electroporation | CAT | Transient | Fromm et al. 1985 |
| BMS cell suspension | electroporation | CAT | Transient | Fromm et al. 1986 |
| A188 × B73 suspension | electroporation | GUS, NPT-II | Transgenic plants | Rhodes et al. 1988b |
| BMS cell suspension | PEG | CAT | Transient | Maas et al. 1989 |
| Cell suspension | PEG | NPT-II, GUS | Transgenic tissue | Lyznik et al. 1989 |
| BMS cell suspension | PEG | NPT-II | Transgenic tissue | Armstrong et al. 1990 |
| BMS cell suspension | Cationic lipids | CAT, NPT-II | Transient | Antonelli and Stadler 1990 |
| Microspore-derived cell suspension | electroporation | NPT-II, GUS | Transgenic plants | Sukhapinda et al. 1993 |
| HE/89 suspension | PEG | NPT-II, PAT, GUS | Transgenic seed | Omirulleh et al. 1993 |
| HE/89 suspension | PEG | mDHFR | Transgenic seed | Golovkin et al. 1993 |
| Cell suspension; inbred (B73 related) | PEG | bar | Transgenic tissue | Kramer et al. 1993 |
| Cell suspension (Sen-60 & H99 × FR16) | PEG | GUS | Transient | Krautwig et al. 1994 |
| A69Y endosperm suspension | PEG | NPT-II, GUS | Transgenic tissue | Faranda et al. 1994 |
| Cell suspension; inbred (B73 related) | PEG | bar | Transient | Fleming et al. 1995 |
| *Testing structural genes in maize* | | | | |
| Leaf, root, stem | electroporation | Maize PPDK | Transient | Sheen 1991 |
| Leaf mesophyll from seedlings | PEG | U3sn RNA | Transient | Marshallsay et al. 1992 |
| Leaf mesophyll from seedlings | PEG | U5sn RNA | Transient | Leader et al. 1993 |
| Embryogenic callus | PEG | LC | Transient | Neuhaus-Url et al. 1994 |
| Leaf mesophyll | electroporation | GFP | Transient | Sheen et al. 1995 |
| *Regulation of transgene expression* | | | | |
| BMS & embryogenic callus (A188-derived F1 lines) | electroporation | luci | Transient | Planckaert and Walbot 1989 |
| BMS cell suspension | electroporation | CAT | Transient | Vasil et al. 1989 |
| BMS suspension | electroporation | CAT | Transient | Mascarenhas et al. 1990 |
| BMS cell suspension | PEG | NPT-II | Transient | Maas et al. 1990 |

TABLE 1 *continued*

TABLE 1 *(continued)*

| Purpose/protoplast source | DNA delivery method | Marker gene(s) | Observation | Reference |
|---|---|---|---|---|
| BMS cell suspension | electroporation | CAT, GUS | Transient | Olive et al. 1990 |
| BMS cell suspension | electroporation | GUS | Transient | Last et al. 1991 |
| BMS cell suspension | PEG | CAT | Transient | Maas et al. 1991 |
| Endosperm suspension culture | electroporation | CAT | Transient | Ueda and Messing 1991 |
| A636 & BMS cell suspensions | PEG | luci | Transient | Quayle et al. 1991 |
| Cell suspension (Pioneer Hybrid P3377) | PEG or electroporation | CAT | Transient | Fennell & Hauptmann 1992 |
| BMS cell suspension | electroporation | luci/R, C1, Bz2 | Transient | Bodeau & Walbot 1992 |
| BMS cell suspension | electroporation | CAT | Transient | Christensen et al. 1992 |
| Cell suspension (DeKalb XL82) | electroporation | CAT | Transient | Fox et al. 1992 |
| *Regulation of transgene expression* | | | | |
| Cell suspension (inbred P3377) | PEG | GUS | Transient | Lu and Ferl 1992 |
| Immature endosperm | electroporation | GUS | Transient | Giovinazzo et al. 1992 |
| A636 endosperm suspension culture | PEG | luci | Transient | Quayle and Feix 1992 |
| BMS cell suspension | electroporation | GUS | Transient | Rigau et al. 1993 |
| Oh43 aleurone & endosperm; BMS cell suspension | PEG/or electroporation | luci | Transient | Gallie & Young 1994 |
| Callus | electroporation | GUS | Transient | Kao et al. 1996 |
| BMS cell suspension | electroporation | GUS & *hsp* promoters | Transient | Marrs & Sinibaldi 1997 |
| *Exploring molecular mechanisms* | | | | |
| Mesophyll from seedlings | PEG | Intron splicing | Transient | Goodall & Filipowicz 1990 |
| BMS cell suspension | electroporation | luci, modified introns | Transient | Luehrsen & Walbot 1992 |
| Cell suspension (A188 × BMS) | PEG | GUS | Transient | Lyznik et al. 1991a,b |
| Cell suspension (A188 × BMS) | PEG | GUS | Transient | Lyznik et al. 1993 |
| Leaf mesophyll | electroporation | GUS, luci | Transient | Russell et al. 1993 |
| BMS cell suspension | electroporation | luci, modified Bz2 intron | Transient | Carle-Urioste et al. 1997 |

Widespread use of microprojectile bombardment for DNA delivery has made possible direct use of type II maize callus (Aves et al., 1992; Weymann et al., 1993; Dennehey et al., 1994; Zhang et al., 1996; Pareddy et al., 1997), type I callus (Wan et al., 1995) and immature embryos (Koziel et al., 1993; Songstad et al., 1996). Recently, DNA delivery by *Agrobacterium* has been used to transform A188 and A188-derivatives which expands the methods that can be used for maize transformation (Ishida et al., 1996; see also Komari and Kubo, this volume).

### *Alternatives to callus-based transformation methods*

Maize transformation based on tissue culture procedures, such as callus culture, are restrictive because of the narrow range of responding genotypes and the high level of expertise required to recover transgenic plants and seed. For this reason, the search has continued for transformation strategies aimed at avoiding tissue culture. Pollen-mediated transformation methods have been reported in maize (De Wet et al., 1986; Ohta et al., 1986; Hess, 1987; Langridge et al., 1992) but most lacked careful physical proof of the presence of transforming DNA or expression of the transgene product (see Potrykus, 1991). Among other explants, young actively growing tissue has received attention including, for example, pieces of germinating seedlings (Cao et al., 1990) and immature inflorescences (Dupuis and Pace, 1993; Pareddy et al., 1997; Lusardi et al., 1994). A central theme in many of these reports is the targeting of meristem cells. In some studies, the meristem is exposed by physical removal of sheathing leaves from germinating seedlings (see for example, Gould et al., 1991). Others have simply bissected germinating embryos (Cao et al., 1990) to prepare a similar meristem target for bombardment. To date, none of these methods that target meristems in the mature embryo or germinating seedling have produced transgenic plants that meet stringent criteria for transmission of the new genes and the desired transgenic phenotype to progeny. Others have looked for developmental stages and tissues where the meristem is already exposed for DNA introduction. For example, the exposed meristems found on the surface of the immature tassel or ear (Dupuis and Pace, 1993; Pareddy et al., 1997) have been shown to provide numerous meristems for DNA targeting. In both reports, transient GUS expression showed that multiple meristems per inflorescence could receive DNA, but transgenic plants and progeny have yet to be reported. Even so, the immature inflorescence remains a viable target; sorghum transgenics have recently been produced and documented from this explant (Casas et al., 1997a,b, see also chapter by Castillo and Casas, this volume).

### *Targeting the Meristem of Coleoptilar-stage Zygotic Embryos*

The value of transforming the apical meristem has been recognized for many years in the plant biotechnology arena. As with other crops, attempts have been made to transform the maize apical meristem (Cao et al., 1990; Gould et al., 1991). Many of these studies started by targeting the meristem in imbibed or germinating seeds, which contained mature embryos with well-developed shoot apices covered by sheathing leaves. In order to obtain germline transmission from meristems of mature embryos, at least one of the four apical initials in the L2 layer would need to be targeted. The chimeric sectors produced under these conditions would be unlikely (as predicted by fate mapping) to transmit the transgenes to progeny (Cao et al., 1990).

One potential solution is to use earlier-staged embryos in which the exposed meristem is comprised of fewer cells. Accordingly, the apical dome of the zygotic embryo in early developmental stages has also been used; successful delivery of DNA into the meristem of coleoptilar-stage embryos has been demonstrated (Bowen, 1993). Using this meristem as a target for particle delivery of marker genes into maize, transgenic sectors have been produced at high frequencies; 4–45% depending on the genotype (see Lowe et al., 1995). Sectored plants have been produced in all maize inbred lines tested, suggesting that this method of meristem transformation may be useful in diverse maize genotypes. Although reducing the size of targeted meristems increased sector frequency and size in the chimeric $T_0$ plant, this alone was insufficient to produce transgenic seed. Thus, transgenic sector size must be even larger to increase the probability of gene transmission to progeny.

As one method of increasing transgenic sector size, multiple shoot proliferation on high cytokinin-containing medium was tested. Meristems of coleoptilar-staged embryos were bombarded with marker genes (either *nptII+ uidA, or aadA + uidA*) and the embryos were germinated. After initial germination, the small shoots were moved onto multiple shoot-induction medium including the appropriate selective antibiotic (kanamycin or streptomycin, respectively). Using this method, transgenic shoots have been recovered from a sweetcorn hybrid (Honey N'Pearl) and a proprietary Pioneer inbred (Lowe et al., 1995). For both examples, Mendelian inheritance and stable transgene expression through multiple generations have been observed. These experiments demonstrated that cells in the apical meristem of maize are capable of integrating introduced DNA, and that progeny expressing transgenes can be obtained from chimeric shoot meristems with appropriate manipulation. Unfortunately, multiple shoot proliferation can be laborious with many genotypes (and not all inbreds respond similarly to exogenously applied hormones), so alternative routes were explored.

One of the salient features of meristem bombardment is the production of chimeric plants. Researchers working with crops such as soybean, (Christou and McCabe, 1992), cotton (McCabe and Martinell, 1993) and sunflower (Bidney et al., 1992) have been able to exploit the branching growth pattern and predictable nature of transgenic sectors in these species. Our research to recover transgenic progeny suggested that some degree of transgenic sector reorganization was essential for maize, which required simple, more rapid methods of enlarging sectors in the meristem. Various treatments were tested with the objectives of i) enlarging transgenic sectors, ii) increasing their persistence and iii) increasing inheritance of transgene expression. These include the use of earlier-stage embryos as explants, physical disruption of the meristem, and the application of non-lethal selection regimes that were compatible with continued meristem development. While the use of earlier-stage embryos (i.e. transitional stage) as targets for meristem transformation will produce larger-sectored plants, this alone did not result in germline transmission. Delivering DNA to even earlier-stage embryos such as late-proembryos theoretically should result in larger sector size and increase the likelihood of germline transmission. This in fact was attempted using microinjection in *Brassica* by Neuhaus et al. (1987). However, despite this potential advantage we decided to emphasize coleoptilar-stage embryos simply because of the larger size and increased ease of handling.

Physical disruption of the apical meristem at the coleoptilar stage produced an interesting result. Piercing the center of the meristem with a 5 $\mu$m diameter glass needle arrested the apical dome and stimulated the formation of 3–6 new meristems that radiated out from the pre-existing dome within 3–4 days (see Figure 1*). Sectioning of Feulgen-stained embryos showed that the newly-formed meristems appeared below the coleoptilar ring in the scutellar tissue. When needle puncturing was applied after particle-mediated transformation of the apical meristem, plants with large sectors were produced, but again, no germline transmission was observed (data not shown). However, we found that a brief selective treatment with streptomycin during germination resulted in transmission of transgenes to progeny from single chimeric plants (i.e. derived from directly-germinated individual meristems).

To accomplish this, we used a non-lethal selective agent, streptomycin, at 100 mg/l for 2–4 weeks during the early stages of germination. Plants with large, persistent streptomycin-resistant, GUS+ sectors were produced. Additionally, use of streptomycin has permitted unambiguous identification of transgenic sectors (i.e. dark-green leaf phenotype on an albino background) from which plants can be readily recovered (see Figure 2). Recent experiments using a single protocol have produced sectors in $T_0$ plants that persist into the tassel. Experiments using this protocol have demonstrated transmission of transgenes and expression of these genes in progeny for five Pioneer inbreds, spanning multiple heterotic groups, and for the Hi-II genotype (Lowe et al., 1997; Ross et al, personal communication).

---

*Colour figures 1, 2, 3, 7, 9 and 10 will be found on pages 217–224.

Based on these observations, it appears that mild forms of meristem reorganization may be sufficient to shift the transgenic sector into a lineage that will contribute to the reproductive tissues. Why is this important, when callus-based transformation is so routine? For a number of years, there has been growing speculation that monocot meristem cells might have an intrinsic block that prevented transgene integration (see articles by Potrykus, 1989, 1990a,b, 1991). Generating individual chimeric, transgenic maize plants that produced transgene-expressing progeny unambiguously validates the potential of this approach. Also, for many species and or varieties, *in vitro* culture methods remain onerous, laborious or even impossible. Targeting the meristem and proceeding through a normal developmental pathway should greatly increase the range of germplasm that can be transformed.

## New Tools for Maize Transformation

### *Selectable Markers*

Even though maize transformation methods have been available for ten years, some of the first selective agents reported for maize remain the most widely used. The *bar* or *pat* genes, which confer bialaphos (or PPT) resistance have been widely used for the recovery of transgenic maize (Table 2). While nearly all of these reports use callus-based selection methods, a recent report described the efficacy of this marker for selection during shoot multiplication (Zhong et al., 1996a). Other established selective agents successfully used in maize include chlorsulfuron (Fromm et al., 1990), and antibiotics such as kanamycin (D'Halluin et al., 1992; Omirulleh et al., 1993; Lowe et al., 1995), streptomycin (Lowe et al., 1995), hygromycin (Walters et al., 1992) and methotrexate (Golovkin et al., 1993).

It is noteworthy that one of the first successful selection systems reported for maize worked so well and has been used so widely. The efficacy of glufosinate chemistry (selection using PPT, the PPT-derivative bialaphos, or commercial formulations such as BASTA in conjunction with the *bar* or *pat* genes) helped establish the basic methods for maize transformation, but also decreased the effort to explore other selection schemes. As a consequence, researchers in maize transformation turned their attention to other areas such as DNA delivery, gene expression and trait development. Recently however, the interest in developing additional selection systems has been rekindled, and alternatives will undoubtedly emerge.

A positive selection system relying on mannose utilization by mannose-isomerase expressing cells was recently reported by Evans et al. (1996). Although this system did not appear as efficient as *bar*/bialaphos selection in the preliminary report, it is still exciting to see the development of selection schemes that rely on providing a positive growth advantage to

TABLE 2
Maize Transformation for Whole Cells or Tissues

| Tissue/Cell source | Marker Gene(s) | Observation | Reference |
|---|---|---|---|
| *Particle delivery; methods developed in cell suspensions* | | | |
| Suspension Cells (B73, B73 × G35, BMS) | CAT | Transient | Klein et al. 1988a,b |
| Suspension Cells (BMS) | GUS, NPT-II | Transgenic tissue | Klein et al. 1989a |
| Suspension Cells (BMS) | GUS | Transient | Oard et al. 1990 |
| Suspension Cells (BMS) | GUS,bar | Transgenic tissue | Spencer et al. 1990 |
| Suspension Cells (A188 × B73, reciprocal and derived lines) | GUS, bar, luci, mutant-ALS | Transgenic plants & seed | Fromm et al. 1990 |
| Suspension Cells (A188 × B73) | GUS, bar | Transgenic plants & seed | Gordon-Kamm et al. 1990 |
| Suspension Cells (A188 × B73) | GUS, bar | Transgenic plants & seed | Spencer et al. 1992 |
| Suspension Cells (A188 × B73) | GUS, bar | Transgenic tissue | Vain et al. 1993 |
| Suspension Cells (R90 & R91; two sweet corn inbreds) | NPT-II | Transgenic plants & seed | Murry et al. 1993 |
| Suspension Cells (A188 × B73) | GUS, bar, NPT-II | Transgenic plants & seed | Register et al. 1994 |
| Suspension Cells (BMS) | B, C1, GUS, bar | Transgenic tissue | Rasmussen et al. 1994 |
| Suspension Cells (A188 × B73) | GUS, bar, B-peru | Transient | Kausch et al. 1995 |
| *Particle delivery; method development in callus* | | | |
| Type II callus (A188 × B73) | GUS, HPT | Transgenic plants & seed | Walters et al. 1992 |
| Type II callus (DeKalb inbred) | GUS, bar | Transgenic plants & seed | Aves et al. 1992 |
| Type II callus (Ciba inbred) | aphIV | Transgenic plants & seed | Weymann et al. 1993 |
| Type II callus (Hi-II; A188 × B73 derivative) | GUS, bar | Transgenic plants & seed | Dennehey et al. 1994 |
| Type II callus (Hi-II) | GUS, bar | Transgenic plants & seed | Zhang et al. 1996 |
| Type II callus | GUS, bar | Transgenic tissue | Pareddy et al. 1997 |
| Type I callus (H99 × FR16, H99 × Pa91; doubled haploids) | GUS, bar | Transgenic plants & seed | Wan et al. 1995 |
| *Particle delivery to immature embryos* | | | |
| Scutellum of immature embryo, (A188) | GUS | Transient | Klein et al. 1988b |
| Scutellum of immature embryo, aleurone (A1 & Bz1 mutants) | A1 or Bz1 genes | Transient (complement; produce anthocyanins) | Klein et al. 1989b |
| Scutellum of immature embryo, (and aleurone layer) | luci & Bz1 promoter | Bz1 promoter; deletion series | Roth et al. 1991 |
| Scutellum of immature embryos (Cargill R-160) | GUS | Transient | Reggiardo et al. 1991 |
| Scutellum of immature embryos (Ciba Geigy elite Lancaster) | GUS, bar, CryIA(b), bar | Transgenic plants & seed | Koziel et al. 1993 |
| Scutellum of immature embryo | CryIA(b), bar | Transgenic plants & seed | Hill et al. 1995 |
| Scutellum of immature embryos | GUS, ALS | Transgenic plants & seed | Songstad et al. 1996 |

TABLE 2 *continued*

TABLE 2 *(continued)*

| Tissue/Cell source | Marker Gene(s) | Observation | Reference |
|---|---|---|---|
| Scutellum of immature embryos | GUS | Transgenic plants & seed | Russell and Fromm 1997 |
| Scutellum of immature embryos (A188, H99, Pa91 & crosses) | bar | Transgenic plants & seed | Brettschneider et al., 1997 |
| *Particle delivery to meristems* | | | |
| Meristem, immature embryo (Honey 'N Pearl, PHI inbred) | GUS, NPT-II, aadA | Transgenic plants & seed | Lowe et al. 1995 |
| Exposed shoot apices in shoot culture (Honey 'N Pearl) | GUS, bar | Transgenic plants & seed | Zhong et al. 1996a |
| Immature inflorescence (Oh43) | GUS; C1, B-peru | Transient | Dupuis and Pace 1993 |
| Isolated microspores (DH99 × DH222) | GUS | Transient | Jardinaud et al. 1995 |
| *Particle delivery; testing genes and/or regulation* | | | |
| Type II callus (A188 × B73) | C1, B-peru, B-I | Transient | Goff et al. 1990 |
| Suspension Cells (Zm85) | GUS, bar | Transient | Taylor et al. 1993 |
| Suspension Cells (BMS) | GUS, ALS, glgC16 | Transgenic tissue | Russell et al. 1993 |
| Endosperm from opaque2 lines (inbred O2 or o2-null lines) | GUS, luci; opaque 2 promoter | Transient | Unger et al. 1993 |
| Endosperm 15 DAP (Inbred F-352) | GUS; alpha-coixin promoter | Transient | Yunes et al. 1994 |
| Leaf mesophyll cells | GFP | Transient | Chiu et al. 1996 |
| Leaf tissue | GFP | Transgenic plants | Pang et al. 1996 |
| Type II callus (Hi-II) | GUS; Caffeic O-methyltransferase promoter | Transient | Capellades et al. 1996 |
| Suspension Cells (BMS) | bar, LysC, DHPS | Transgenic tissue | Bittel et al. 1996 |
| Suspension Cells (endosperm) | CryIA (b & c) | Transgenic tissue | Sardana et al. 1996 |
| Suspension Cells (A188 × B73) | GUS; with introns | Transient | Vain et al. 1996 |
| Callus (A188 × B73) | bar, ALS, SacB | Transgenic plants & seed | Caimi et al. 1996. |
| Suspension Cells (inbred) | luci | Transient | Sainz et al. 1997 |
| Suspension Cells | GUS, luci | Transient | Schledzewski and Mendel 1994 |
| Scutellum of immature embryos | 24kDa alpha-zein | Transgenic plants & seed | Coleman et al. 1997 |
| Various tissues from whole-plant | GUS with Hrgp promoter | Transient | Menossi et al. 1997 |
| Endosperm | Modified (-zein) | Transient | Torrent et al. 1997 |
| Bisected kernels | GUS with CMd promoter | Transient | Grosset et al. 1997 |
| Callus | GUS with actin promoter | Transgenic plants | Zhong et al. 1996b |

TABLE 2 *continued*

TABLE 2 *(continued)*

| Tissue/Cell source | Marker Gene(s) | Observation | Reference |
|---|---|---|---|
| *Other delivery methods; tissue electroporation* | | | |
| Suspension Cells (hybrid P3377) and microspores | CAT | Transient | Fennell and Hauptmann 1992 |
| Suspension Cells (A188 × B73 and a DeKalb inbred) | GUS, bar | Transgenic plants & seed | Laursen et al. 1994 |
| Suspension Cells (BMS) | GUS, CAT, PAT | Transgenic tissue | Sabri et al. 1996 |
| Type II callus (both Hi-II and a B73 × A188 BC3 line) | GUS, bar, NPT-II | Transgenic plants & seed | Pescitelli and Sukhapinda 1995 |
| Immature zygotic embryos (H99 and Pa91) | NPT-II | Transgenic plants & seed | D'Halluin et al. 1992 |
| Immature zygotic embryos (Hi-II) | GUS, anthocyanin genes) | Transient | Songstad et al. 1993 |
| Immature zygotic embryos (Huanong Supersweet No. 42) | NPT-II | Transgenic plants | Xiayi et al. 1996 |
| *Silica-carbide fibers* | | | |
| Suspension Cells (BMS) | GUS | Transient | Kaeppler et al. 1990 |
| Suspension Cells (BMS) | GUS, bar | Transgenic tissue | Kaeppler et al. 1992 |
| Suspension Cells (A188xB73) | GUS, bar | Transgenic plants & seed | Frame et al. 1994 |
| *Agrobacterium* | | | |
| Agro-infection of stem, near apex | MSV-DNA | Viral symptoms observed in plant | Grimsley et al. 1988 |
| Exposed shoot tips | GUS, NPT-II | Southern for GUS and NPT-II in progeny | Gould et al. 1991 |
| Agro-infection at coleoptilar node | GUS | Expression in sectors of the plant | Shen and Hohn 1994 |
| Excised segments from freshly-germinated seedlings | GUS | Transient | Ritchie et al. 1993 |
| Immature embryos and type I callus (A188 inbred) | GUS, bar | Transgenic plants & seed | Ishida et al. 1996 |
| *Microinjection* | | | |
| Somatic embryos (B79) and shoot apical meristems (K55) | Lc (anthocyanin gene) | Transgenic sectors in plant | Lusardi et al. 1994 |
| Transgene assays in other tissues | | | |
| Aleurone | C1, B-peru, B-I | Transient | Goff et al. 1990 |

transgenic callus as opposed to conventional systems that rely on killing non-transgenic cells.

A potential method to avoid the use of inhibitory compounds to recover transgenics is by using visible marker genes. A number of these marker genes have been tested in maize, including genes that regulate anthocyanin production (Ludwig et al., 1990; Bowen, 1993), and genes such as *uidA* or luciferase whose encoded enzymatic functions are detectable, respectively, through colorimetric assays (see Bowen, 1993) or as luminescence (Fromm et al., 1990). These reporter genes were very important tools for gene expression studies and for the verification of transgenic events during selection. However, despite the initial anticipation, each class of reporter genes were of limited value when used as the sole means of identifying transgenic events.

The gene encoding the green fluorescent protein (GFP) first cloned from the jellyfish *Aequoria sp.* (Prasher et al., 1992) appears to hold promise as a visible marker. GFP has received great attention since the first report of its use as a transformation marker (Chalfie et al., 1994) and its expression in *Arabidopsis* (Haseloff et al., 1995). A number of reports demonstrate expression of GFP in maize, including transient expression in mesophyll cells (Chiu et al., 1996) and expression in chemically-selected transgenic plants (Pang et al., 1996). We have demonstrated its utility for the recovery of maize transformants in the absence of chemical selection. In these experiments, a GFP sequence optimized for maize codon usage (see Campbell et al., 1990) was utilized. A UBI::GFP::pinII cassette and CaMV 35s::bar::pinII were cotransformed using particle-bombardment-mediated delivery into immature embryos of the Hi-II genotype (see Songstad et al., 1996, for a generic protocol). After bombardment of 147 immature embryos per treatment, the embryos were placed onto either 3 mg/l bialaphos or onto medium containing no chemical selective agent. After 8 weeks, 12 of the 16 bialaphosresistant colonies recovered expressed the GFP transgene (see Figure 3a,b). Furthermore, 12 GFP-expressing colonies also were visually identified in the non-selected treatments. Of these, all expressed *bar* and exhibited BASTA™-resistance when the leaves were painted with a lanolin paste that contained BASTA™ (see Gordon-Kamm et al., 1990, for lanolin application method). Transformants have been reproducibly recovered through GFP-visualization in multiple experiments (data not shown). Additionally, no substantial differences in Southern hybridization patterns were observed between the bialaphos selected and the GFP-visualized transformants (Figure 4). This result showed that the transformed calli were not biased by the method of recovery. GFP-visualized callus exhibited similar regeneration as bialaphos-selected events.

Visual identification of transformants could have very positive implications, particularly with germplasm that is sensitive to the trauma associated with chemical selection. These issues still need to be addressed. However, these preliminary experiments clearly demonstrate the potential value of GFP-visualization. With an effort in this area similar to that expended on developing such selective agents as bialaphos, GFP could make a major contribution to further advances in maize transformation technology.

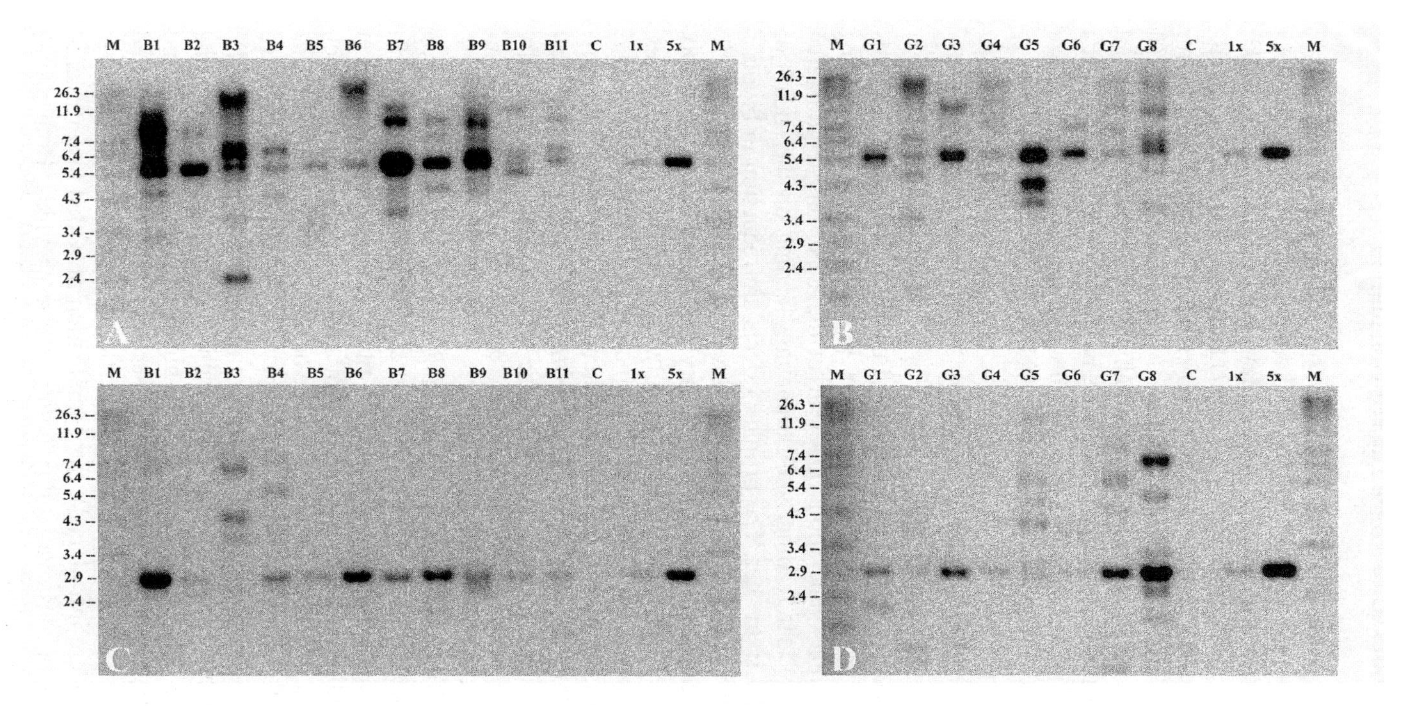


*Fig. 4.* Southern blots from transgenic embryogenic callus recovered through GFP visualization (with no chemical selection; Figures B & D) and bialaphos-selected transformants (Figures A & C). Genomic DNA was digested and probed with either the *pat* (A & B) or GFP (C & D) sequences; Genomic DNA was restricted with PvuII (cutting once in the *pat*-containing plasmid as in Figure A, or once in the GFPm-containing plasmid, Figure B), or with SphI + SpeI (cutting out the *pat* or GFPm structural genes, in Figures C & D, respectively). Hybridization patterns in A & B are variable (which is common between independently-derived transformants), showing a typical range from low copy number to complex integration events. In figures C & D, an intact insert approximately 2.9 kb in size was observed in all the transformants. Lane 'C' was genomic DNA extracted from non-transformed control callus. Lanes 1x and 5x were loaded with plasmid DNA at approximately 1 and 5 copy levels, respectively (M. Miller, G. Sandahl & W. Gordon-Kamm, unpublished results).

## III. Traits Introduced to Maize via Transformation

Over the last few years, the first wave of transgenic crops, including maize, have begun to emerge in the market. Several companies have reported field trials with transgenic maize over the past few years (Koziel et al., 1993; Armstrong et al., 1995; Williams et al., 1997) and some have begun releasing the first products with transgenes conferring European Corn Borer (ECB) resistance. The number and scope of transgenic maize products will expand rapidly in the next few years, especially as information gained from genomic-related projects in maize and rice is used to modify the plant's development and metabolism. At present, however, the new traits in maize are rather limited, and are briefly described below.

### *Insect Resistance*

In the United States, *Ostrina nubilalis*, the European corn borer (ECB) is among the most notorious of maize insect pests. Typically progressing through two to three generations per year in the U.S., this lepidopteran causes damage by leaf feeding, stalk tunneling, feeding on the sheath or collar, and finally in the ear. The primary effects are reduction in biomass and disruption of vascular flow as the plant develops, but increased susceptibility to microorganisms also causes secondary problems (Burkhardt, 1978).

Due to the widely accepted practice of applying Bt-toxins as insecticides, this single gene-encoded protein (originally cloned from *Bacillus thuringiensis*) was enthusiastically developed as one of the earlier potential crop traits. Within the last few years, *cryIIIA*-expressing potato, *cryI*-expressing cotton and *cryI*-expressing maize have become commercially available, yet for maize there are very few reports in the literature concerning Bt-expressing maize plants.

Similar to the first reports on insect resistant tobacco and tomato (Barton et al., 1987; Fischhoff et al., 1987; Vaeck et al., 1987), ECB resistance in maize was first conferred by expression of a truncated *cryIA(b)* (Koziel et al., 1993). In addition, this group used a synthetic structural gene in which the G + C content increased from 37% (in the original *Bacillus thuringiensis* gene) up to 65% while encoding the same amino acid sequence. Two transgenic lines were field tested in 1992 and 1993. The first, a CaMV 35S::*cryIA(b)*-expressing line showed variability within each event and appeared to vary between genotypes, with highest expression in pith and root and some expression in kernels. In the second, *cryIA(b)* was expressed by the PEP carboxylase and pollen-specific promoters (two separate structural genes were introduced); the combination provided lower overall expression levels but less variability among the transgenics. Both events conferred protection against repeated challenges with ECB.

In 1995, two additional reports were published on maize expressing *cryIA(b)*. In the first of these, a maize hybrid (Huobai × Lai1029) was used to produce numerous independent maize transformants expressing *cryIA(b)* (Wang et al., 1995). Based on feeding assays using corn borer larvae, variable gene expression and insect inhibition was observed between seven $T_0$ plants from different transgenic events. A pragmatic solution to the commonly observed problem of variable expression between transgenic events is to generate many transformants. This strategy is illustrated in a report by Armstrong et al. (1995). After particle-mediated delivery of a CaMV 35S::Adh1 intron::*cryIA* expression cassette into Hi-II germplasm and selection for transformants, 314 of 715 calli tested showed insecticidal activity. After regeneration and crossing to B73, the $T_1$ progeny from 89 events (of 173 tested) showed transgene segregation ratios consistent with Mendelian inheritance. Field tests in 1993 using this $T_1$ generation resulted in 34 of 77 lines tested showing reduced tunneling by second-brood ECB.

To be commercially useful, transgene expression must be predictable: expressed in the correct tissue(s), at the appropriate developmental stage(s), at an effective level and stable over numerous generations and in different environments. Armstrong et al. (1995) stressed that genome position effects can influence gene expression, but other criteria must also be evaluated. Expression levels across several generations must be monitored. In addition, for each transgenic event that appears to maintain the desired expression patterns, it must be empirically confirmed that the transgenic sequence or its expression have no deleterious impact on yield or other agronomic characteristics. In a recent report by Williams et al. (1997), Bt-conferred insect resistance was evaluated after 6 generations of back-crossing. Seven different hybrid lines were generated from a single CaMV 35S::*cryIA(b)*-expressing transformant. There was variability – likely due to genotype influence – between the seven lines tested. However all seven lines appeared to be more resistant to fall armyworm and Southwestern corn borer in field tests conducted in 1994, relative to similar test results in 1995 (although not stated, this reduction could have been due to reduced gene expression in successive generations, or be reflective of year to year variation in insect feeding pressure).

Although the results to date with transgenic insect resistant maize are encouraging, some cautions were raised as well. Variability of expression between transgenic events appears to be the most common observation (i.e. likely due to genome position effect) as well as additional variability within a given event depending on the genetic background (i.e. position effect by genotype interaction). Taking advantage of transgenes will require more exhaustive examination of expression patterns in different genetic back-grounds and over several generations.

### *Herbicide resistance*

A few of the well-characterized and widely-used herbicides also have been developed as transgenic selection systems. This has resulted in the development of herbicide resistant maize plants (Yoder and Goldsbrough, 1994). Glufosinate-resistant maize was one of the first to be released in field tests (APHIS Biotechnology and Scientific Services Newsletter, updated December 5, 1997). While it was one of the first selection systems successfully used in maize (Spencer et al., 1990; Gordon-Kamm et al., 1990; Fromm et al., 1990), it has continued to be a popular choice (for example, see Kaeppler et al., 1992; Kramer et al., 1993; Omirulleh et al., 1993; Frame et al., 1994; Register et al., 1994; Laursen et al., 1994; Dennehey et al., 1994; Pescitelli and Sukhapinda, 1995; and Zhong et al., 1996a). This selection system has worked for numerous cell types (Table 1) and various culture conditions.

While chlorsulfuron has been used for selection with the corresponding mutant acetolactate synthase (*als*) gene and fertile transgenic plants resistant to chlorsulfuron were generated (Fromm et al., 1990), expression in subsequent generations has not been reported. *In vitro* selected mutants for the sulfonylurea class of herbicides are herbicide resistant (see Somers and Anderson, 1994), and these same mutant genes have conferred resistance in transgenic plants (Haughn et al., 1988; Sathasivan et al., 1991). These *als* mutants, also referred to as AHAS mutants, can confer resistance to another class of herbicides called imidazolinones (see Chang and Duggleby, 1997). Glyphosate is the other widely used herbicide for which transgenic resistance maize is in field trials, although published data on these transgenics has yet to be presented.

### *Disease Resistance*

Resistance to viral and fungal disease resistance is another area that may be effectively impacted by transgenic maize. Numerous strategies to introduce viral resistance into crops are being pursued (Wilson, 1993). The most extensively exploited phenomenon is that of coat protein-mediated protection. In maize, the expression of maize dwarf mosaic virus (MDMV) strain B coat protein has been demonstrated to confer resistance against MDMV infections (Murry et al., 1993). Similarly, there are many transgenic approaches being tested for fungal resistance in crops (see Lamb et al., 1993 for review), although in maize there are no published transgenic studies to date. Generic strategies such as combined expression of chitinase and glucanase genes (Jongedijk et al., 1995), expression of ribosome-inactivating proteins (Stirpe et al., 1992; Leah et al., 1991), or engineering secondary metabolic pathways to increase anti-fungal compounds all represent potential approaches for enhancing fungal resistance in maize. However, for specific pathogens we may be able to turn to maize itself for solutions. Characterization and cloning of genes such as the maize *Hm-1*, which encodes a reductase that inactivates HC-toxin (produced by the maize pathogen *Cochliobolus carbonum*; see Meeley et al., 1992; Johal and Briggs, 1992) may provide opportunities to improve or alter pre-existing fungal resistance traits through enhanced or altered expression patterns. Alternatively, manipulating compounds involved in the normal defense mechanisms of plant responses to pathogens could lead to sources of transgenes or mechanisms to effectively control diseases.

### *Male Sterility*

Production methods for hybrid corn seed are both cost and resource intensive, mainly due to the requirement of emasculating or detasseling female rows. Systems based on molecular biology that would render the intended female inbred rows sterile (to produce the hybrid in the production field), but also permit restoration of fertility (to perpetuate the female lines), would be more convenient to produce and, in most cases, would increase grain yield.

Several strategies are being developed for male sterility in plants. Male sterility has been demonstrated by the pollen- or tapetum-specific expression of a cytotoxin that prevents pollen development. A restoration system (i.e. expression of a second gene that blocks the action of the cytotoxic gene) can then be used to regain fertility. For example, expression of the ribonuclease-encoding gene, barnase, during pollen development has been successfully used to confer sterility in *Brassica* and tobacco (Mariani et al., 1990; DeBlock and Debrouwer, 1993; Zhan et al., 1996). As anticipated based on the early characterization of this system (Hartely, 1988) fertility was restored in the *Brassica* system by the expression of a second gene, barstar, that effectively inhibits the cytotoxic action of barnase (Mariani et al., 1992). Although published details are lacking, it has been reported that the Barnase/barstar system has been successfully incorporated into maize (Leemans, 1992), and field tests have been conducted (APHIS Biotechnology and Scientific Services Newsletter, updated December 5, 1997).

Although effective, a caution has also been raised about this approach; the barnase/barstar system requires expression of a transgene for sterility, and if the expression control degenerates then fertility will be inadvertently restored. A different approach being developed by Albertsen et al. (1993) is to introduce a known fertility gene, such as Ms45, under the control of a chemical induction system into a genotype with non-functional fertility-gene alleles such as ms45/ms45 (therefore sterile). The result will be constitutive male sterility that does not rely on gene expression for maintenance of the trait (see Albertsen et al., 1993 for detailed description). However, adequate inducible expression of the Ms45 transgene is required to perpetuate the female inbred line in a parent seed production field. The rudimentary components for this system such as inducible promoters (Mett et al., 1993) and gene replacement strategies (Morrow and Kucherlapati, 1993) exist, but require further development for successful application at a commercial level.

*Industrial Protein Production*

In addition to improving agronomic performance through transgenics, numerous groups have recognized the potential value of using transgenic plants for the production of recombinant proteins (Krebbers et al., 1992; Verwoerd et al., 1995; Whitelam et al., 1993; Krebbers et al., 1994). A major advantage of this strategy is that well-established methods exist for harvesting, storage and processing our most important crop plants (Whitelam et al., 1993). In this arena, corn has been used to set the precedent; transgenic maize has been utilized for the commercial production of a high value protein, chicken egg white avidin (Hood et al., 1997). Chicken egg white avidin produced in transgenic maize is currently being marketed by a major chemical company. Recently, commercial-scale production of beta-glucuronidase (GUS) in maize has been accomplished as well (Witcher et al., 1998). With further refinements, the commercial benefit of using plants in this manner will likely expand beyond these first examples.

*Metabolic and Developmental Engineering*

Many of the first generation traits described above took advantage of the opportunity to introduce non-plant genes into corn. As the example describing constitutive male sterility exemplifies, there will be opportunity to directly manipulate maize metabolism and development. The biochemistry and molecular biology of many complex pathways are becoming increasingly well understood; a few examples include starch biosynthesis (Morell et al., 1995), lipids and oils (Browse and Somerville, 1991; Ohlrogge and Jaworski, 1997) and lignins (Campbell and Sederoff, 1996; Boudet and Grima-Pettenati, 1996; Dixon et al., 1996). Together with genome-related projects, the opportunities to engineer maize with increasing precision will be staggering. This will likely require gene targeting/replacement strategies and knockout techniques, some of which are described later in this chapter.

Cloning of maize genes will also afford us the opportunity to use transgenics to address developmental questions. One of the first examples of this in maize is from some recent transgenic experiments involving the maize knotted1 gene (Sinha et al., 1993) or *kn1*-like genes from other species

(Matsuoka et al., 1993; Chuck et al., 1996). In both tobacco and *Arabidopsis*, expression of kn1 (or *kn1*-like genes) resulted in ectopic meristem formation in leaf tissue (Sinha et al., 1993; Chuck et al., 1996), while constitutive expression in maize produced vein clearing knots and ligule displacement (R. Williams-Carrier; unpublished data). Constitutive expression of kn1 in barley produced a phenotype similar to the hooded mutant (caused by a duplicated intron in the knotted gene), in which reiterating clusters of meristems form along the awn (Williams-Carrier et al., 1997). These studies may help us gain insights into how meristem pattern formation occurs in the monocot inflorescence.

## IV. Regulation of Transgene Expression

To a large extent, the first generation of maize transgenic products evolved by taking advantage of the long-standing efforts in the molecular and biochemical characterization of bacterial genes. This will certainly expand to include the use of genes from many other organisms. However, practical application of maize transgenics will require rigorous control of gene expression. Now that a variety of methods for DNA delivery are available, the focus has shifted to expression analysis of transgenes and understanding of maize gene structure and function (see Table 2 for recent examples). Appropriate transgene expression will, of course, be dictated by the specific trait being modified, but will require precise control of expression level, tissue and/or cell specificity, and stability across generations.

There are many examples to date of regulatory regions that are useful for effecting desired levels and, in a subset of examples, correct spatial and temporal specificity of gene expression. These sequences include promoters, transcriptional and translational leaders, 3′ untranslated regions, motifs with enhancing or attenuating properties, and intracellular or extracellular targeting sequences.

As objectives in maize transgenic research have become more sophisticated, a corresponding diverse arsenal of regulatory elements, and a greater need to elucidate the functional expectations of any given regulatory element are required. Thus, a better understanding of tissue specificity, temporal regulation, and stable expression across generations must be considered before choosing a promoter (or any combination of gene regulatory elements) for transgenic experiments. Two examples of evaluating regulatory regions for suitability in transgene expression are provided below. These include a detailed analysis of the Ubiquitin1 promoter from maize (Christensen et al., 1992) and the isolation and characterization of the promoter from the maize DnaJ gene. These examples will demonstrate a usual course taken for promoter analysis, yet also show the inherent problems in expression controls and reproducibility of expression patterns.

These data support the need for confirmatory approaches to validate such an analysis. Some of these limitations are related to the basic problem of non-targeted gene transformation resulting in variable genome positions and/or DNA modifications grossly affecting expression level and patterns.

*Maize Ubiquitin 1 Promoter*

The ubiquitin gene product can be found in all eukaryotes and in relatively high abundance within most cell types. It has been shown to be involved in several cellular processes including protein turnover, histone modifications, stress-related responses and cell cycle control (Johnson et al., 1992; Li et al., 1993; Ciechanover and Schwartz, 1994; Genschik et al., 1994; Sudakin et al., 1995). The genes encoding the maize polyubiquitin, a 76 amino acid polypeptide repeated 6–7 times in a head to tail fashion, exist as a small multigene family (Christensen et al., 1992). Christensen et al. (1992) demonstrated that the maize *Ubi1* gene is highly active in leaves of non-stressed plants and inducible by heat treatments. Didierjean et al. (1996) found that a maize polyubiquitin gene, nearly identical to the maize *Ubi1* gene (except for a C-terminal amino acid change) also is induced by wounding, mercuric chloride and cold treatments, suggesting that the maize polyubiquitin gene family is involved in multiple-stress responses.

The sequence encompassing the maize ubiquitin 1 promoter, first exon and intron (designated UBI-1) have been used in driving reporter genes transiently in a variety of species (Taylor et al., 1993), including maize (Christensen et al., 1992; Graham and Larkin, 1995; Schledzewski and Mendel, 1994), rice (Bruce et al., 1989; Chair et al., 1996), wheat (Loeb and Reynolds, 1994; Martiensen, 1996) and barley (Rogers and Rogers, 1995; Schledzewski and Mendel, 1994), effectively establishing that this sequence can potentiate high and reproducible gene expression in monocot cells. Futhermore, the UBI-1 sequence was also shown to produce constitutive expression in transgenic rice (Toki et al., 1992; Uchimiya et al., 1993; Cornejo et al., 1993; Takimoto et al., 1994), wheat (Vasil et al., 1993, Weeks et al., 1993), rye (Castillo et al., 1994), and barley (Wan and Lemaux, 1994), yet it functions poorly in dicot species tested to date (Bruce et al., 1989; Christensen et al., 1992; Schledzewski and Mendel, 1994). In addition, UBI-1 also confers heat-shock and mechanical wounding responses in transgenic rice (Cornejo et al., 1993; Schledzewski and Mendel, 1994).

Despite the numerous reports on maize UBI-1 expression in various plant species , there has been only one report of transgenic maize with the maize UBI-1::GUS/UBI-1::bar genes, with little detail of the resulting expression levels and patterns (Wan et al., 1995). Thus, a more careful characterization of UBI-1 regulated expression patterns in maize was needed. In addition, a major concern in transgenic research is the observation of decreased or silenced gene expression in cases where the regulatory elements or transgenes are introduced into species containing identical sequences in the genome. Examples include the *cab140* promoter of Arabidopsis (Brusslan et al., 1993, 1995) and the CaMV 35S and 19S promoters (Vaucheret et al., 1993). It is possible that introduction of the maize Ubi1 promoter back into maize could result in similar silencing episodes not previously seen in heterologous transgenic monocot species.

Transgenic progeny of numerous events containing the –900 to +1 bp 5′-flanking region, first exon and intron of the maize Ubi1 gene, the GUS structural gene (uidA; see Jefferson, 1987) and a polyA 3′ region from the potato pinII gene (An et al., 1989) were analyzed for GUS expression patterns throughout the plant. The contribution of the Ubi1 5′ flanking sequence and the Ubi-intron on transient GUS expression as shown in Figure 5 was also evaluated by transient assays (Unger et al., 1993) using three day old maize seedlings. Following bombardment, seedlings were incubated for 20 hours at 26°C in the dark, and seedling tissue was homogenized in Eppendorf tubes using a Kontes pestle in 50 mM sodium phosphate (pH 7.0), 10 mM EDTA, 1 mM/DTT phosphate buffer. The tubes were centrifuged and the supernatant recovered and subjected to luciferase (Callis et al., 1987) and GUS assays (fluorometric assay as described by Rao and Flynn, 1990; or the GUS Light assay from Tropix, Inc., Bedford, MA).

As shown in Figure 5, either replacing the UBI-1 exon1 and intron with the tobacco mosaic virus omega' sequence alone (Ubi/no intron) or the TMV-omega' sequence plus the Adh1 intron (Ubi/adh1 intron) dramatically reduced the expression levels relative to the intact UBI-1 sequence (Ubi/ubi intron). In our hands, gene expression driven by the CaMV 35S promoter was not affected by removing the maize Adh1 intron or its replacement with the maize Ubi exon/intron sequence. This contradicts previously reported results using the Adh1 intron (Callis et al., 1987; Mascarenhas et al., 1990). The inability to enhance Ubi-based gene activity with the omega' or the omega'/Adh1 intron sequences suggests a selective interaction between the maize *Ubi1* promoter and its corresponding exon and intron sequences. The practice of combining particular promoters with downstream sequences for transgene expression is questionable given these results. Many reports in the past have demonstrated the modularity of certain intron sequences (eg. the maize Adh1 intron 1) with a variety of promoters that ultimately improve expression levels. Yet it is unknown if the phenomenon demonstrated in Figure 5 is Ubi1-specific or an example of a more generalized restrictive interaction between promoter and downstream sequences. The contribution of introns to gene expression may be more promoter-specific than first thought.

The use of both GUS and LUC (luciferase) driven by either Ubi1 or CaMV 35S promoters provided an opportunity to test the contribution of the coding sequence to gene expression levels. A comparison of enzyme activity

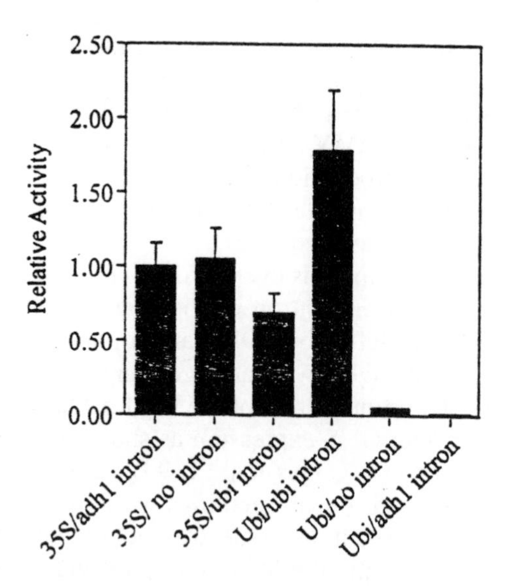


*Fig. 5.* Promoter and intron transient assays. Various constructs were bombarded into an embryogenic suspension line to assess the function of the maize *Ubi1* intron. '35S/Adh1 intron1' refers to the 35S::*Adh1* intron1-LUC-PinII transgene while ubi1/ubi intron refers to the UBI-1::LUC-PinII transgene. Part of the first exon and the whole intron of the UBI-1::LUC-PinII was removed in the 'ubi1/no intron' whereas this part was replaced with the maize *Adh1* intron1 sequences designated 'ubi1/adh1 intron'. The *Adh1* intron1 was removed in the '35S/no intron' construct. At least three independent bombardments were conducted and the data were normalized to the '35S/adh1 intron' construct. Error bars are the SE. (L. Simms & W. Bruce, unpublished results)

and rate of accumulation between the UBI-1 and the CaMV 35S constructs was conducted as shown in Figure 6. The highest rate of increase for all constructs was observed between 5 and 18 hours following bombardment. During this period, the rate of the maize Ubi-driven GUS or LUC activity was 3–4 fold greater than the CaMV 35S-driven constructs (Figure 6). Enzyme activity of the UBI-1-driven constructs reached levels approximately 3 to 4-fold greater at 18 hours and 15–17 fold greater at 42 hours than the CaMV 35S-driven constructs. These data demonstrate that the maize UBI1 sequence can potentiate higher levels of gene activity than the CaMV 35S-driven constructs independent of the reporter gene used.

UBI-1::GUS expression patterns were examined in stably transformed plants from Hi-II (Armstrong et al., 1991), embryogenic callus lines and a Pioneer elite inbred line. From 14 independent Hi-II events, six yielded transgenic plants with high levels of GUS expression, five produced intermediate to low levels and three produced plants with no detectable activity. Individual plants from two lines of the GUS-positive Hi-II events showed varied staining patterns whereas the remainder had homogeneous GUS staining patterns in all cell types evaluated. Eleven of the sixteen indepen-

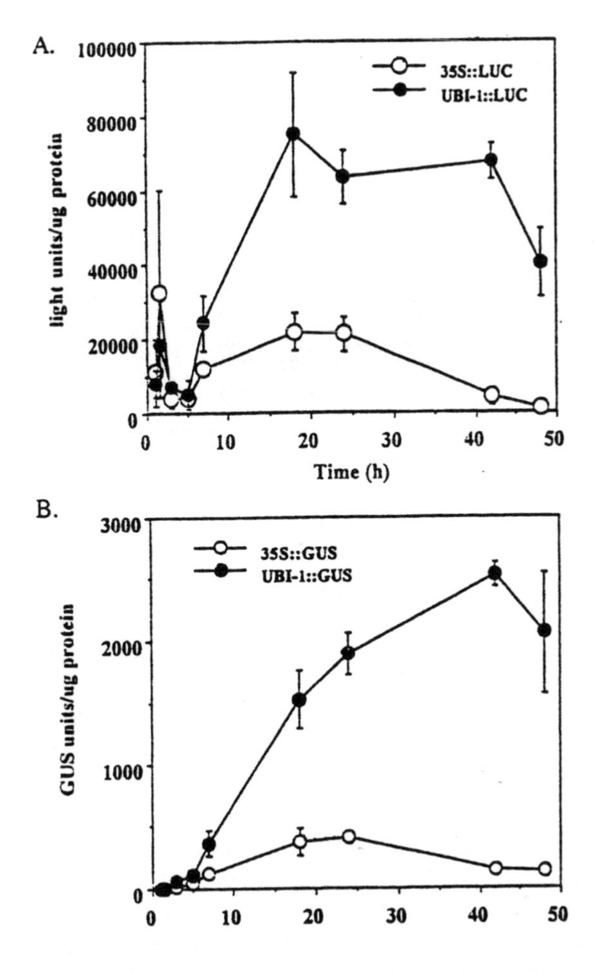


*Fig. 6.* A time course of transient activity comparing the maize *Ubi1* promoter against the CaMV 35S promoter in maize seedlings. Whole 3-day-old seedlings were bombarded with the designated LUC or GUS construct, incubated at 26°C for the times indicated before harvesting and assaying for enzyme activity. Each value is the mean of at least six repetitions with error bars as the SE. The UBI-1-driven constructs show a 3- to 4-fold higher rate of enzyme accumulation over the 35S-driven constructs regardless of the coding sequence during the 6–18 hours following bombardments. (L. Simms & W. Bruce, unpublished results)

dent elite inbred events produced plants expressing high levels of GUS whereas three events yielded low expressing plants and two events produced no expression.

Based on histochemical staining for GUS activity, virtually all tissues from mature plants derived from the Hi-II and the elite inbred transgenic lines examined resulted in expression (Figure 7). Expression was low in anthers although activity in developing and mature pollen was quite evident (Figure 7J). Little GUS activity was observed in the coleoptile, similar to the pattern of endogenous ubiquitin RNA based on *in situ* hybridization (see below). Transgenic plants from most of the expressing events showed similar staining patterns. Plants from several Hi-II and elite inbred transgenic events were carried out to the T3 generation; histochemical staining was typically uniform throughout the plants (data not shown). In contrast, enhanced CaMV 35S-driven GUS expression was lower than Ubi-driven expression, which is consistent with other comparisons made between CaMV 35S and Ubi1 promoters (Chair et al., 1996; Christensen et al., 1992; Cornejo et al., 1993; Schledzewski and Mendel 1994). Although quantitative GUS assays on various tissues from representative T1 plants with the Ubi promoter at the mature tassel stage revealed varied levels of specific activity within the assayed plant, there seemed to be a tendency toward higher specific activity in the root (Figure 8A).

GUS histochemical expression was compared to various biochemical assays since other studies have shown inconsistent results with *in situ* RNA hybridization and Northern analyses (Sieburth and Meyerowitz, 1997; Fu et al., 1995; Larkin et al., 1993). These examples suggest that for certain genes, expression is governed by sequences within the encoded gene in addition to their respective promoter elements. Thus, *in situ* RNA hybridization for ubiquitin on sections of non-transformed root and shoot tissues was conducted to localize cell-specific expression and confirm these observations with the transgenic GUS results. Ubiquitin RNA was readily detected throughout the meristematic region of the vegetative shoot with higher prevalence in the cells associated with the vascular strands (Figure 9A). The RNA signal diminished in the mesocotyl distal from the coleoptilar node yet was most notable where vascular strands were visible (Figure 9A, arrow). In roots, expression appeared highest in the root tip including root cap cells (Figure 9C) contrasting with results of Jackson et al. (1994) who detected very little ubiquitin mRNA in root cap cells. In the mature portion of the root, ubiquitin expression was diminished relative to the root tip yet it was detectable in the epidermis, pericycle, and phloem with very little signal in the cortex and endodermis (Figure 9C and F). These data correlated with the histochemical GUS staining pattern of the transgenic maize containing the UBI-1::GUS construct except that ubiquitin RNA hybridization was below detection in the mature tissue where GUS activity was readily detected, possibly due to the stability of the GUS gene product.

Gene silencing was not observed in subsequent generations for most transgenic events, but one of the high expressing transgenic elite inbred lines exhibited a progressive reduction in GUS activity, as seen in the fully developed kernels of the $T_1$ to $T_3$ generation (Figure 10A–C). This reduction in activity also was observed in other plant tissues in successive generations. Notably, this line showed little activity in the embryo whereas

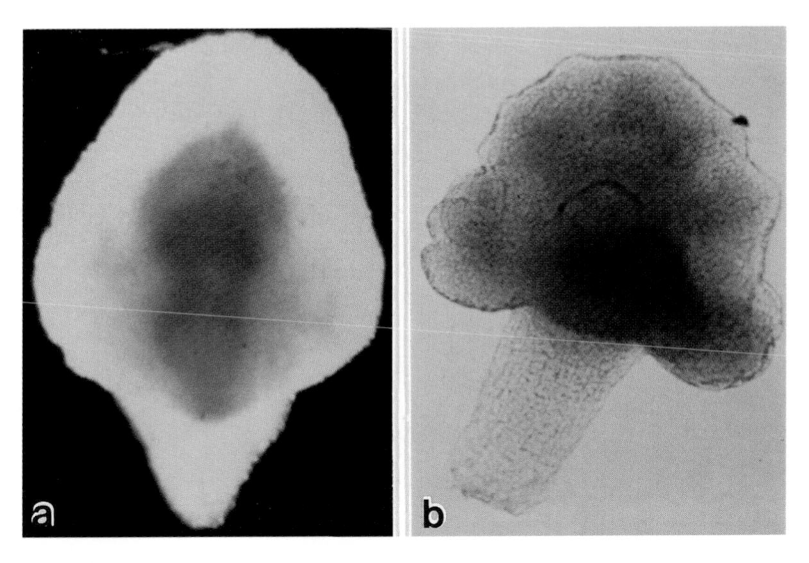


*Fig 1.* A coleoptilar-stage immature embryo freshly-isolated from an eight day-after-pollination kernel (A), showing a typical staining pattern for meristematically active cells (magenta) after Feulgen staining (Berlyn and Miksche, 1976; Chamberlin et al., 1993). If the exposed apical dome of the embryo was pierced by a glass needle (approximately 5 $\mu$m diameter), multiple meristems were observed to radiate away from the original meristematic dome (seen in the Feulgen staining pattern in B) and 3–6 new shoots could be seen forming within a 3–4 day period (M. Ross & W. Gordon-Kamm, unpublished data).

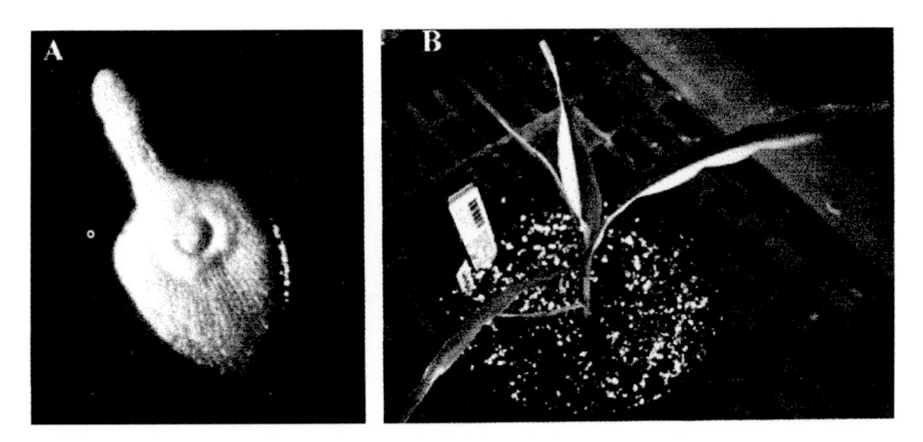


*Fig. 2.* The exposed meristem of coleoptilar-stage embryos (A) was bombarded and placed under streptomycin selection during germination. Plants that survived into the greenhouse were tested following a variation on the method of D'Halluin, et al. (1992); streptomycin was applied to the cut surface of a leaf and the next few leaves to emerge from the whorl clearly show chimeric transgenic resistance (B) (M. Ross, K. Lowe, G. Hoerster, L. Church, G. Sandahl, M. Miller, K. Field, D. Elfvin & W. Gordon-Kamm, unpublished data).

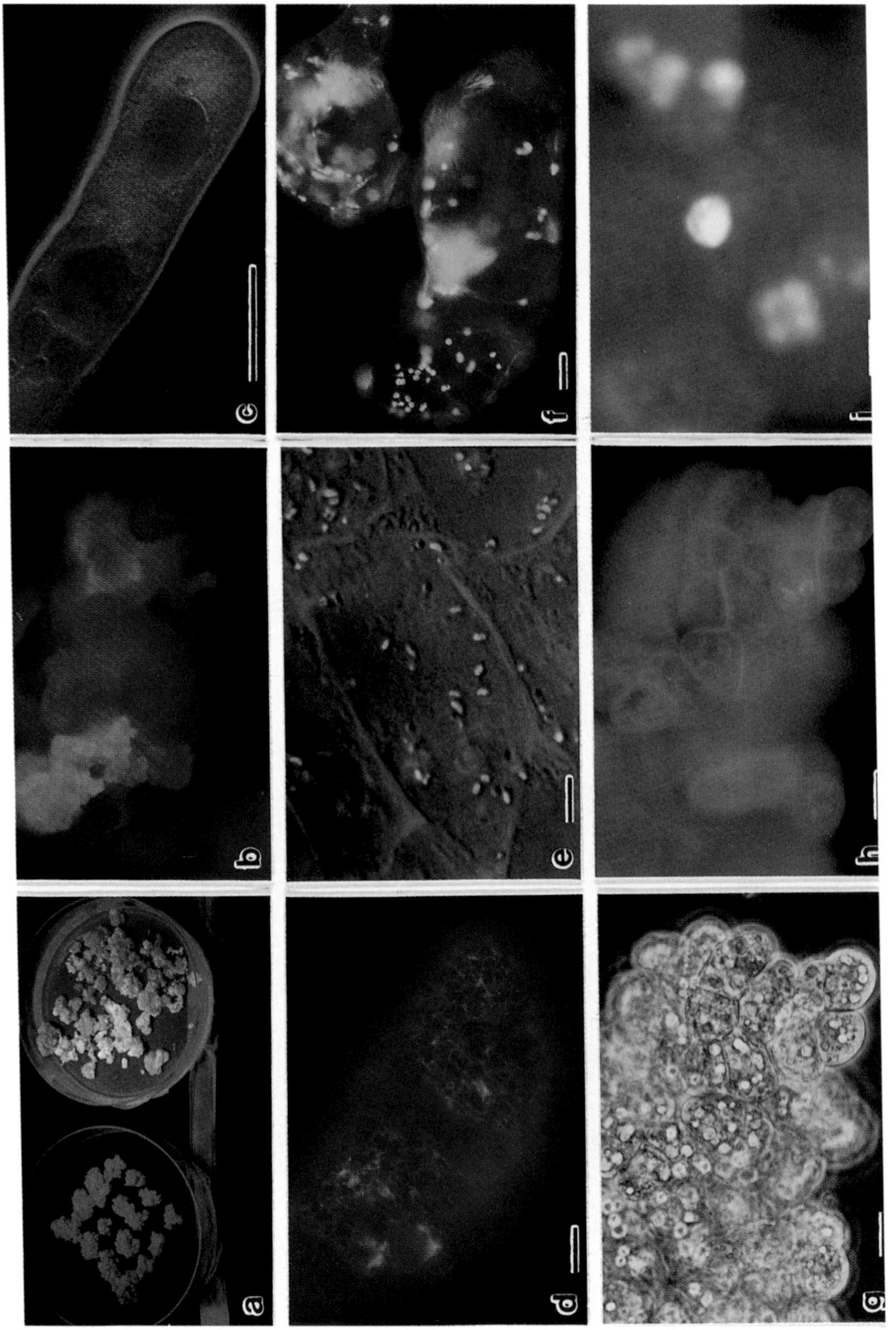

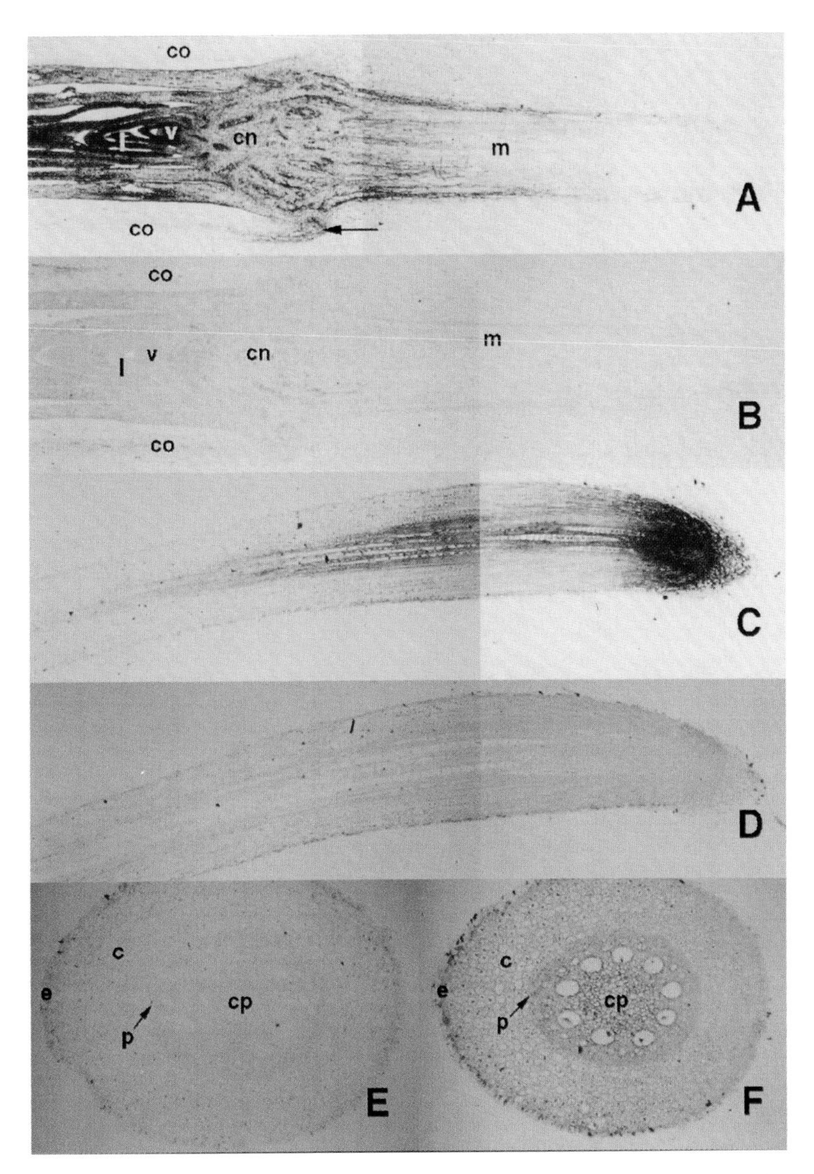


*Fig. 9.* *In situ* RNA hybridization with maize *Ubi1* cDNA probe. (A and B) Longitudinal section from the coleoptilar node of a 1-week-old seedling hybridized with the anti-sense and sense probes, respectively. Note the detection of polyubiquitin mRNA preferentially in the vascular cells of the coleoptilar node. (C and D) Longitudinal section of adventitious root from a 3-week-old plant hybridized with the antisense and sense probes, respectively. (E and F) Transverse section of adventitious root, 2 cm from the root tip beyond the elongation zone hybridized with sense and antisense probes, respectively. Note polyubiquitin expression is pronounced in the epidermal and vascular regions. (S. Hantke, unpublished data).

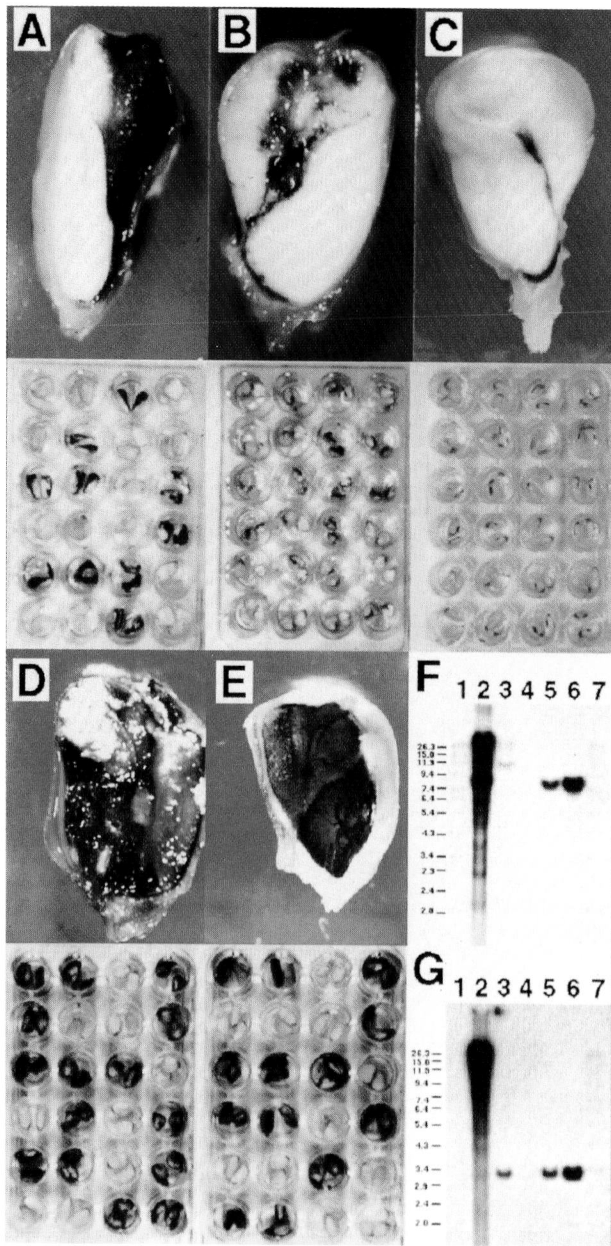


*Fig. 10.* A UBI-1::GUS 'gene quieting' event. T1 to T3 generation seeds (A–C) from a line that exhibited reduction in activity and T1 and T2 generation seeds (D and E) from an active line were bisected and stained histochemically for GUS activity. Each set of seeds were a product of a backcross with nearly a 1:1 expected segregation. Genomic DNA from the T2 seeds was digested with *Hind*III (F) or *BglII/Not*I (G) and probed with the complete GUS coding region. Lanes 1 and 7, molecular weight markers; lanes 2, the 'gene quieting' line; lane 3, the GUS active line; lane 4, control inbred; lane 5 and 6, one and five copy number reconstruction of the UBI-1::GUS plasmid (Z.-Y. Zhao, W. Gu & W. Bruce, unpublished results).

the active lines stained positively throughout the kernel. Decreased gene activity, referred to here as 'gene quieting' in successive generations has been previously reported (Kilby et al., 1992) and was suggested to involve DNA methylation. To test this, genomic DNA of both the 'quieted' line and the GUS-active lines were digested with the methylation sensitive enzymes HindIII (Huang et al., 1982; McClelland and Nelson, 1985) and NotI (Quiang et al., 1990) (NotI was used in a double enzyme digestion with BglII) and probed with the GUS coding sequence. In the transgenic maize lines, digestion of the integrated UBI-1::GUS gene with HindIII was expected to result in varied bands due to unique integration sites in the genome, whereas digestion with NotI/BglII was expected to produce a single band comprising the complete gene insert. Digestion of the genomic DNA from the 'gene quieting' line resulted in high molecular weight products (Figure 10F, G, lane 2) suggesting that modification of the DNA such as methylation prevented complete digestion (Quiang et al., 1990; Smulders et al., 1995). The genomic DNA from the 'active' line (i.e. GUS expressing, Figure 10D and E) was digested by the restriction enzymes producing a single band as expected (Figure 10F, G, lane 3). Taken together, these data indicate that a correlation exists between methylation and 'gene quieting'. It should be noted that the intense autoradiographic signal of the 'quieted' line (Figures 10F, G. lane 2) implies multiple integration copies of the transgene, possibly contributing a negative influence on expression as observed by others (Meyer and Saedler, 1996). Despite the few transformed maize lines exhibiting low gene activity or any evidence of gene silencing, the UBI-1::GUS transgene is generally very active in nearly all tissue types of transgenic maize throughout various developmental stages and over several generations studied to date.

This study demonstrated some of problems encountered with promoter analysis in transgenic maize, including 1) the potential of varied expression levels and patterns between transformed events (categorically referred to as genome position effects), 2) promoter/intron interactions and 3) gene silencing. With transgenic technology, there is an inherent expectation that parental levels of gene expression will be retained in the progeny across many generations. To realize this, progress is being made to circumvent such silencing phenomena and such methods will be discussed below.

*The ZmdJ1 Promoter (from a maize DnaJ-related gene)*

One of the first constitutive promoters described, CaMV 35S (Odell et al., 1985) gained rapid popularity for maize transgenic research and has been used extensively over the years (see Table 1). Analogous to the effect that the *bar* (or *pat*) gene(s) had on reducing the impetus to find new selectable markers, many groups took advantage of the strong, reasonably non-specific expression conferred by CaMV 35S and concentrated on other areas. Nonetheless, alternative promoters that confer general, non-specific expression in maize have been identified, including the actin promoter from rice (McElroy et al., 1990a and b) and the maize ubiquitin promoter (Christensen et al., 1992). In addition, there are other promoters that would be predicted to express in a wide variety of tissues based on the normal function of the associated genes in important, ongoing cellular processes. Examples include genes for histones and non-histone chromosomal proteins (Baumbach et al., 1987; Landsman et al., 1989), ribosomal proteins (Dudov and Perry, 1986; Vera et al., 1996) or other cellular 'housekeeping' processes (Zernik et al., 1990; Stapleton et al., 1993; Grula et al., 1995).

We have cloned a novel maize promoter that expresses in a tissue-general manner (Baszczynski et al., 1997). The cDNA for this gene has sequence homology to the DnaJ or DnaJ-related protein genes from bacteria (Bardwell et al., 1986; Anzola et al., 1992; Narberhaus et al., 1992; Van Asseldonk et al., 1992), yeast (Caplan and Douglas, 1991; Atencio and Yaffe, 1992) and other plant species (Bessoulle, 1993; Preisig-Muller and Kindl, 1993; Zhu et al., 1993). In bacteria, DnaJ-related proteins are involved in the process of chaperone-mediated protein folding (Langer et al., 1992). In bacteria and yeast they also play a role in other fundamental cellular processes such as DNA replication, translation and polypeptide translocation across intracellular membranes as well as cell viability at elevated temperatures (Liberek et al., 1988; Wickner, 1990; Caplan and Douglas, 1991; Liberek et al., 1991; Malki et alk 1991; Wickner et al., 1991; Caplan et al., 1992; Zhong and Arndt, 1993). At least one plant gene has been shown to complement yeast DnaJ mutations which is strong evidence for functional similarity (Zhu et al., 1993). Thus, DnaJ appears to be important in basic cellular functions (i.e. it qualifies as a housekeeping-type gene). The control regions (promoter) of this gene should be able to direct expression of associated genes in a wide variety of cell types. Northern analysis of maize total RNA from a broad range of maize tissues probed with the ZmdJ1 cDNA demonstrated expression in all tissue types tested. Strong expression was primarily observed in the whorl of developing plants, in both the pith and node tissues of the stalk, and in kernels (Baszczynski et al., 1997).

Functionality of the ZmdJ1 promoter was tested and confirmed using a ZmdJ1P::*uidA*::pinII construct. While expression in BMS callus was strong enough to produce a uniform, solid-blue histochemically detectable product indicative of GUS expression, the staining was not as intense as in callus transformed with CaMV 35S::*uidA* stained for a similar length of time. Although the ZmdJ1 promoter provided clear GUS expression in callus, it did not drive *pat* expression in callus to sufficient levels to recover transgenic BMS or Hi-II callus. This observation does not preclude the use of this promoter for other, potentially more robust selectable markers or for other applications. For example, expression of R and C, transcriptional activators of the maize anthocyanin pathway (Dooner et al., 1991; Bowen, 1993), under the control of the ZmdJ1 promoter resulted in effective transactivation of target genes or chimeric constructs (Ben Bowen, personal communication). Additionally, whole plant insect (ECB) bioassays on transgenic maize plants containing a Bt gene under the control of the ZmDJ1 promoter yielded several events with high ECB lethality scores and intermediate levels of BT protein as measured by ELISA (Table 3). These results emphasize that evaluation of promoters or other regulatory regions is best done in the context of their anticipated applications, to avoid discounting the potential value of a novel promoter.

At the plant level, ZmdJ1-driven GUS expression was generally strong in seedling tissues, although levels and distribution varied among events. Staining was particularly intense in the mature stalk pith and in ear sections, and less intense in the other tissues examined including the tassel, upper stalk, leaf, leaf collar, and roots. Within any one event, expression generally varied by less than 10-fold, although as anticipated, expression between independent events showed greater variability (Baszczynski et al., 1997). In at least two events, all tissues except the leaf collar exhibited intermediate to high expression levels, comparable to CaMV 35S-driven controls.

Expression levels in various plant parts were tested in the $T_0$ and $T_1$ generations. Overall GUS expression levels were reduced from the $T_0$ to $T_1$ generation, although several events consistently expressed well in most tissues. One event was consistently the strongest expressor in both the $T_0$ and $T_1$ generations, while another showed a marked reduction in $T_1$ expression, again suggesting that gene silencing or quieting occurred. For one

TABLE 3

Comparison of insect bioassay and ELISA scores for constructs with either the ZmdJ1 or the maize ubiquitin promoter driving expression of the *Bt* gene, as described in the text.

| Total number of events | Maize ZmDJ1 promoter | Maize Ubiquitin promoter |
|---|---|---|
| Positive for *Bt* by ELISA or PCR | 38 | 20 |
| Infested with ECB | 25 | 15 |
| Average ECB scores ≥ 6* | 4 (16%) | 7 (47%) |
| ELISA score range (pg/ug):** | | |
| 0–25 | 10 | 3 |
| 25–50 | 6 | 2 |
| 50–100 | 6 | 2 |
| 100–150 | 1 | 1 |
| >150 | – | 7 |

*Control susceptible checks had an average ECB score of 2.0, while control resistant checks had an average ECB score of 8.5.

**No ELISA data was available for 2 of the 25 infested ZmDJ1::*Bt* events

CaMV 35S::GUS event examined, the level of expression in some tissues (e.g. leaf, ear) was higher in $T_1$ than in $T_0$ plants, while other tissues showed a reduction or no change in expression levels between generations.

The copy number of the ZmdJ1 transgene ranged from single copy in some transgenic events to others with complex patterns, as with other studies using particle-mediated DNA delivery. One event with a complex pattern (i.e. > 5 copies) consistently showed a high level of expression in both the $T_0$ and $T_1$ generations. The remaining events, which spanned the spectrum from single to complex integration, generally had reduced expression levels in the $T_1$ generation. Since an endogenous promoter was introduced as part of the GUS expression construct, these events may have been predisposed to quieting or silencing, irrespective of the introduced transgene copy number. It remains to be seen whether use of homologous or heterologous promoters will similarly impact the stability of transgene expression across generations. This data will come from future studies using various promoter/species combinations.

Several general observations pertaining to transgene expression in plants and other specific conclusions can be drawn from the ZmdJ1 study. First, ZmdJ1 represents a functional and effective regulatory region for driving expression of genes in a variety of maize tissues. While the overall level of expression is clearly lower than that of the CaMV 35S promoter, it still is useful for driving transgene expression, particularly so in cases where lower or intermediate expression levels of a given gene product are desirable or where the gene product is potent or functionally efficacious at lower concentrations.

Reflective of more generalized observations in maize transgenics, the two above examples clearly re-emphasize the commonly observed themes of i) how integration position, achieved through a more or less random process using current transformation methods, influences transgene expression (the most obvious impact here was variability in expression between events), and ii) the impact of gene silencing (or quieting) on stability of transgene expression levels across generations (often a result of multiple copies of the introduced sequence, or homology between the introduced sequence and endogenous sequences).

Integration position effects and gene silencing are further illustrated in Figure 11, in which expression of the Gus marker gene under the control of five different promoters (including the ubiquitin and ZmdJ1 promoters described here) is examined in primary transgenics and their progeny. Several conclusions can be drawn from this data set. First, a given promoter usually demonstrates a characteristic level of GUS expression despite variability between events. Second, expression levels decline from $T_0$ to $T_1$ generations in at least some events of all promoters examined to date. Further, some events show gene silencing in the $T_1$ generation. There appeared to be no correlation between transgene copy number and down-regulation of

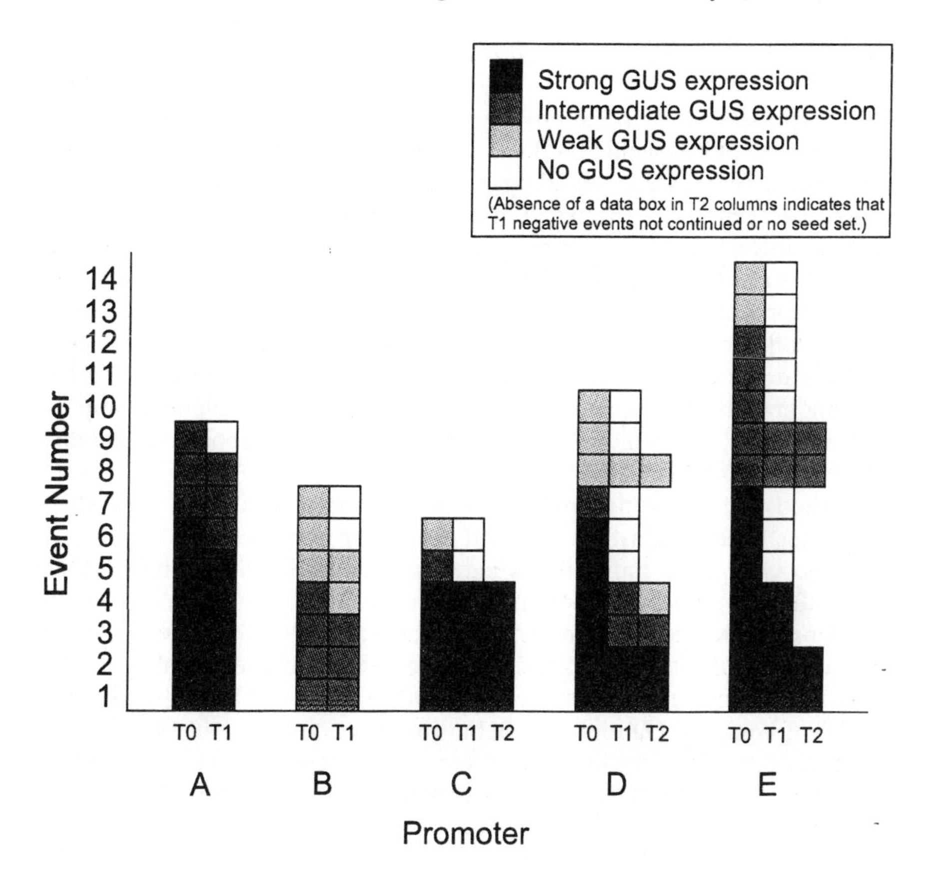


*Fig. 11.*

transgene expression in the $T_1$ and $T_2$ generations (data not shown). The phenomenon of transgene silencing has been observed by many groups (for example, van der Krol, 1990; Hobbs et al., 1990; Matzke and Matzke, 1993; Matzke et al., 1994a,b), and as the recent review by Maessen (1997) indicates, there are many distinct types of silencing which can in turn be mitigated by factors such as genome position, genetic background, environment and developmental stage. Further changes in gene expression over generations can arise, but the sources and cause of such changes are beyond the scope of this review. Because of these concerns, studies have been initiated with the aim of circumventing some of these problems including gene targeting approaches described later.

*Subcellular Targeting of Encoded Proteins*

Gene expression must be tightly regulated and predictable for commercial development of transgenic products. Numerous factors can influence how a transgene is expressed, including 5′ and 3′ regulatory elements, introns, codon bias, genome position and overall chromatin architecture. Stability of both the mRNA and encoded protein can also influence overall efficacy. In addition to absolute amount, sub-cellular location of specific transgene-encoded products also can be critical in determining efficacy. The influence of sub-cellular targeting on transgene-encoded protein efficacy has been well documented in plants (Scofield et al., 1994; O'Keefe et al., 1994; Chaumont et al., 1995), and is an important aspect of transgenic manipulation in plants, particularly as our strategies for metabolic engineering become more sophisticated.

Methods for assessing proper sub-cellular targeting usually depend on antibodies raised against the encoded protein, in conjunction with methods to assess changes in molecular weight as a result of cleavage upon transit across the membrane (i.e. Western analysis) or through cross-linking to sub-cellular membranes (for example, see Akita et al., 1997). In some cases such as nuclear localization, GUS has been used effectively to visualize protein entry into the nucleus (Restepo et al., 1990; Varagona et al., 1992; Shieh et al., 1993; Yang and Gabriel, 1995), but for many types of other sub-cellular compartments GUS localization has not been reported. As discussed by Grebenok et al. (1997) the GUS histochemical assay compromises cellular membrane integrity, and thus this assay typically cannot be performed on living cells and conclusions regarding sub-cellular targeting must be qualified. The recent discovery of GFP as a transgenic marker (Chalfie et al., 1994) and the various mutants of GFP (Delagrave et al., 1995; Heim and Tsien, 1996) have added a valuable tool for cell biology. Expression of GFP in transient assays has been used to evaluate sub-cellular targeting (Rizzuto et al., 1995; Grebenok et al., 1997). Similar to reports in other eukaryotic systems (Rizzuto et al., 1995), transgenic maize expressing GFP with various targeting sequences has been used to demonstrate sub-cellular targeting of GFP in endoplasmic reticulum, mitochondria, peroxisomes, plastids and the nucleus (Figure 3d–i). Stable transgenic maize that expressed GFP fused to a nuclear localization signal did not remain localized in the nucleus. This was likely due to the GFP protein being small enough to passively diffuse back into the cytosol after nuclear import (similar to the transient results reported by Grebenok et al. 1997). To test this hypothesis, the effective size of GFP was increased beyond the nuclear pore exclusion limit by the addition of a large, unrelated protein, the maize acetolactate synthase (ALS) to the carboxy-terminus of GFP. Stably-expressed GFP-ALS accumulated in the nucleus of callus cells, similar to the transient results of Grebenok et al. (1997).

## V. Gene Targeting (Targets and Challenges)

While production and recovery of successful transformation events in plants is practical, including creation of commercial products, the process to achieve results is more or less a random process. That is, creation of one set of transgenic events is not predictive of the recovery or 'performance' of successive events. Repeated use of identical regulatory elements, genes, and delivery methods will still produce sites, frequencies and patterns of integration that differ with each experiment. The availability of a predictive integration site or pattern would provide many benefits to the process of creating transgenic plants for both basic research and commercial applications. More specifically, the ability to deliver a gene selectively and repeatedly to a predefined chromosomal location in the genome would theoretically eliminate the 'position effect' impact, barring other more specific 'gene by chromosomal site' interactions. Secondly, the ability to selectively modify or manipulate existing DNA sequences at a predefined location in the genome would provide opportunities to study gene function, or make allelic modifications useful in developing plants with altered agronomic properties or improved quality traits. Research in the area of gene targeting thus specifically seeks to understand mechanisms and develop tools for achieving this predictive and controlled, directional targeted integration or gene modification capacity in the plant.

For plants and especially crop species like maize, overall benefits and potential value of gene targeting technologies are numerous and include:

a) providing a non-random gene delivery and integration process to increase predictability of performance of a transgenic event (in essence performance of different genes or promoters can be evaluated within the same genomic context);
b) reducing transgene integration complexity, thereby aiding in the process of breeding and trait introgression;
c) increasing the probability of long term expression stability by delivering genes to sites determined to better tolerate newly integrated DNA or gene modifications;
d) providing opportunities to target modifications to specific genomic locations, again to facilitate breeding, to link value-added traits to quantitative trait loci (QTL) of interest, and to maintain trait linkage over generations;
e) providing the ability to target modifications to specific genomic sequences, thereby permitting manipulation of endogenous gene structure or function for creating improved quality traits or altering biochemical pathways;
f) providing a tool to selectively 'knock-out' endogenous genes to study their cellular function;

g) reducing product development time and costs by allowing one to recover desired events more quickly and with less effort, which in turn would allow one to evaluate larger numbers of promoter and gene combinations for the same amount of resource expenditure;
h) reducing regulatory hurdles by being able to precisely document the sequences and genomic location of insertions, as well as being able to eliminate undesirable sequences such as selectable markers, antibiotic resistance genes, and so forth.

A range of gene targeting methodologies being developed and evaluated should bring these capabilities into the hands of researchers and companies working to understand the function and expression of both endogenous genes and introduced transgenes, and how to manipulate these processes towards the goal of developing better commercial plant products.

### *Promising Schemes and Approaches*

A few selected approaches will be discussed that, if developed for routine use, would be particularly attractive for the applications described.

1. Site-specific recombination: Limited success has been achieved to date in manipulating some of the site-specific recombination systems occurring in non-plant species for use in plants. A number of these site-specific recombination systems have been expressed in plants to date (reviewed by Odell and Russell, 1994). The most extensive plant studies in this area primarily have employed either the Cre-*lox* system of bacteriophage P1 (Dale and Ow, 1990; Odell et al., 1990, 1994; Albert et al., 1995) or the FLP/*Frt* system derived from *Saccharomyces cerevisiae* (Lyznik et al., 1993, 1995, 1996; Lloyd and Davis, 1994). Both of these systems have the advantage of functioning without the need for accessory or host factors for the basic recombination process; that is, Cre or FLP recombinases will recognize their corresponding target site (*lox* or *Frt*, respectively), and catalyze site-specific recombination leading to DNA sequence excision, inversion, integration or replacement, depending on the arrangement and composition of target sites in the DNA. These features make them particularly amenable and attractive for use in plants.

   The natural molecular mechanisms by which the recombinases in these two systems catalyze the recombination reactions have been extensively studied and are described elsewhere (Sternberg and Hamilton, 1981; Ambreski et al., 1983; Hoess et al., 1986; Sadowski, 1995). As a point of relevance for this discussion, the intra- or inter-molecular recombination by these proteins at site-specific target sequences is a reversible process. In the continued presence of active recombinase, integration and excision of a sequence at recombinase target sites proceeds to equilibrium; at low levels of substrate such as single or low copy integrated sequences flanked by recombinase target sites, excision would be the preferred state. As a result, the Cre/*lox* and FLP/*Frt* systems primarily have been used for DNA excision. For example, a DNA sequence (gene) flanked by target sites is introduced into host cells. Following subsequent expression of the corresponding recombinase, the intervening sequences are excised leading to gene attenuation or activation (Odell et al, 1994; Lyznik et al., 1996). The ability to create conditions which favor the forward reaction leading to stable integration continues to be the challenge to using site-specific recombination as a general approach for targeted integration of genes into genomic DNA. Recently, tools have been developed that will allow for the design of experiments to test schemes for site-specific gene integration. The use of non-identical recombinase target sites to flank DNA sequences reduces the efficiency of recombination leading to excision (Senecoff et al., 1988; Schlake and Bode, 1994; Albert et al., 1995; Baszczynski et al., unpublished). An additional strategy includes expression of the recombinase protein for a limited period of time for transient expression (Baubonis and Sauer, 1993) or under inducible control (Logie and Stewart, 1995). These strategies and others should lead to working site-specific recombination systems for more routine integration of DNA sequences into previously created genomic target sites.

While offering the potential for retargeting alternate or additional genes to previously created genomic target sites, site-specific recombination systems do require the initial creation of suitable target sites in genomic regions that provide stable expression of introduced genes and with minimal negative impact on agronomic performance of the transgenic plants. Generation and selection of useful targeting sites will be one of the challenges. In one recent study for example (Ow et al., 1997), it was found unexpectedly that independent transformants for a reporter gene site-specifically integrated at a genomic location using the Cre/*lox* system exhibited expression variability. A number of factors may be contributing to this observation and more extensive studies are being conducted to determine if the results are particular to that system. Related studies using the FLP/*Frt* system are also underway in our laboratories.

2. Homologous Recombination: Despite the potential benefits associated with site-directed integration, these strategies still require the initial introduction of a target sequence. The preferred situation would be one where a gene of interest could routinely be integrated into any selected genomic location, or where an endogenous gene could be modified to provide a new valuable phenotype, such as in knocking out or modifying biochemical pathway genes (e.g., for oil or protein modification). While examples exist of success with basic homologous recombination (HR) experiments in plant cells (reviewed in Puchta and Hohn, 1996), the efficiency of the process continues to be very low ($10^{-4}$ to $10^{-6}$) in both animal and plant systems. Recent developments with non-plant systems,

however, have provided promising results with respect to being able to achieve 'enhanced' levels of homologous recombination. Two particularly interesting technologies, Double D-loop and chimeraplasty are being evaluated for use in plants.

In the Double D-Loop (DDL) approach (Sena and Zarling, 1993; Revet et al., 1993), two complementary strands of a targeting DNA molecule are first complexed with recombinase protein (e.g. bacterial RecA). Following introduction of the DNA-protein complexes into cells, high rates (10 to 100-fold greater than for conventional methods) of homologous recombination have been reported at a corresponding chromosomal or extrachromosomal target site. It has been demonstrated that RecA protein can enhance the frequency of homologous recombination in plants, although the mechanism of the process has not been fully elucidated (Reiss et al., 1996). Continued studies regarding the roles of various proteins or factors in the general recombination process in plants will be important. Based on the limited success to date with conventional HR, approaches such as DDL merit further investigation.

Another technology for gene targeting in plants based on enhanced homologous recombination is 'chimeraplasty' (Yoon et al., 1996; Cole-Strauss et al., 1996). This strategy uses synthetic chimeric oligonucleotides that assume an intramolecular double-stranded configuration and which incorporate short stretches of 2'-O-methyl RNA residues flanking a DNA sequence with altered nucleotides designed to make specific nucleotide changes in a target DNA molecule. The method has been shown to make specifically targeted single base pair changes in mammalian cells at significantly higher frequencies over comparable DNA-only oligonucleotides or other conventional HR approaches (Yoon et al., 1996; Cole-Strauss et al., 1996). Moreover, recent progress has been made in adapting this approach to plants. At least two groups (including our own) have converted an endogenous acetohydroxyacid synthase (AHAS or *als*) gene to a herbicide resistant form (Kipp et al., personal communication; Zhu et al., in preparation). There are many potential applications of this technology, where changing a single or a few base pairs to alter gene function or expression could lead to modified or improved traits. More work is required to determine the efficiency of chimeraplasty in plants, in particular, how to effectively use the system for traits where selection for a newly created phenotype can not be easily determined in culture.

Thinking beyond existing approaches to gene targeting, the potential exists to combine various components of different gene targeting approaches that could lead to a more practical and usable system. Existing incompatibilities between some of the different approaches may be difficult to overcome. On the other hand some technologies can be combined easily, such as using *Agrobacterium* transformation to create transgenics with single or lower copy number of integrated *Frt* or *lox* sites for subsequent retargeting experiments.

While development, adaptation or evaluation of gene targeting technologies for routine use in plants represents a long term investment, the ability to selectively target insertion or modification of a gene at any selected genomic location, directly in a preferred inbred or variety, will provide significant value to breeding efforts utilizing advanced genetic pedigrees. In particular, such capabilities will greatly reduce backcrossing time and effort, as well as the associated performance penalty often associated with introgression of traits from events generated using easily-transformed maize lines like Hi-II (Armstrong et al., 1991). The anticipated result is increased expression stability and predictability leading to better performance. As such, the potential benefits of creating better performing transgenic plants as part of a total genetic package, in a more cost-effective manner, make the investment worthwhile.

## V. Understanding the Maize Genome

Recently maize research has followed in the footsteps of the human genome-related projects. A global view of gene organization, function and modification should uncover new insights into maize development and its interaction with the environment. Many groups have embarked on various programs that include mapping both single and multigenic traits, expressed sequence tag (EST) sequencing and functional genomic analyses. These programs, while providing new avenues for insight into biological events, will create an enormous amount of data requiring methods for sorting and extracting salient information.

The rapidly advancing field of genomics has resulted in the development of novel approaches encompassing genome-wide information gathering, including areas such as genome mapping, EST sequencing, mRNA profiling and protein characterization.

Variations on physical mapping approaches, such as gel-based RFLP-derived techniques, have arisen that were intended to provide scorable markers. More recent advances, such as chromosomal *in situ* hybridization or FISH and QTL mapping, have added to the available mapping technology and are improving the resolution of specific target sites while adding new information in the quest for genes that are related to desired traits (Quarrie, 1996). A common goal for mapping projects is to saturate the genome with as many genetic markers as possible that will benefit breeders and geneticists alike for genetic analysis of crop plants. Tools that have been recently developed for the pharmaceutical industry include Affymetrix's gene chip technology which provides the means of very high through-put analysis of bi-allelic markers. Affymetrix (Santa Clara, CA) recently described their '4L tiled array or re-sequencing gene chip' used for determining sequence polymorphisms present in the entire human mitochondrial genome (Chee et al., 1996). Affymetrix scientists have perfected the production of high density oligonucleotide arrays on very small 'chips' relying on photolithographic methods for oligonucleotide synthesis (McGall et al., 1996). In collaboration with Eric Lander's lab, Affymetrix is developing gene chips expected to have the capacity for simultaneous analysis of nearly 5000 single nucleotide polymorphic sequences (SNPs) per chip. Such polymorphisms can be detected with single-base resolution and unprecedented efficiency. This technology will provide a rapid, facile and efficient means of genotyping parents and progeny and likely help improve breeding outcomes immensely. Current markers used in mapping populations for maize number in the few hundred at best. Development of re-sequencing chips with all the attributes of reproducible polymorphism detection may drastically change current breeding strategies.

Recently, several groups have developed methods of profiling the steady state levels of mRNA of desired samples as a means to assess which genes are active at a particular time or developmental stage, in order to correlate this information to particular traits. Diverse methods have been reported that include hybridization-based, massive parallel approaches, as well as AFLP-like analyses, such as differential display and SAGE. Of the former group, the most notable again are the Affymetrix 'Gene Chips' (Hacia et al., 1996), as mentioned above. Current Gene Chips are capable of housing as many as 400,000 oligonucleotides (corresponding to some 65,000 gene sequences) on a 1 cm$^2$ glass chip. Samples of mRNA preparations are interrogated by hybridizing to these chips as fluorescently-labeled cDNA probes. Following relatively stringent wash conditions, the chips are fluorescently imaged using a laser-based scanner. It is possible to combine two to four different probes, each labeled with an unique fluorescent tag, into a hybridization reaction thus reducing error in hybridization between samples and increasing mRNA sampling throughput. Analyses of Affymetrix Gene Chips with 1500 maize EST sequences are in progress at Pioneer Hi-Bred. Similar approaches have been developed that rely on spotting DNA onto glass or nylon at lower densities. Despite the success of these methods, they are basically limited in that knowledge of the clone or gene sequence is necessary for their application.

Methods of profiling mRNA where there is no *a priori* knowledge of sequence information have been developed and are analogous to AFLP/SAGE. Recently, CuraGen Corp. (New Haven, CT) has employed a gel-based method of profiling mRNA samples. Their process has the capacity for profiling and identifying up to 95% of all the mRNA species in a given sample with sensitivity of detection to 1 per 125,000 copies, as well as the means of associating particular 'bands' to entries in sequence databases. Other companies also are currently developing technologies to determine the nature and complexity of steady-state mRNA samples.

Naturally, there may be examples where little correlation exists between mRNA levels and the desired phenotypic trait. In such cases, it may be more appropriate to analyze protein profiles in a genome-wide fashion. The name 'Proteomics' was coined to reflect this new area of research. Just as high through-put mapping and mRNA profiling analyses were developed, rapid improvements and novel approaches for protein analyses are becoming more prevalent and accessible. Many of the techniques under development at pharmaceutical companies are being targeted at disease-related genes, but plant research is expected to take advantage of this technology as well. As an example, companies have improved a high-through-put 2D protein gel approach that is robust and reproducible. Large Scale Biology (LSB; Rockville, MD) has the capacity to profile as many as 4000 protein spots on 2-D gels, collect these spots individually and microsequence the isolated proteins using one of various mass spectrometer-based (MS) methods. Still other companies have developed high through-put ligand-binding analyses coupled with MS that can determine not only the protein species within complex mixtures but also rudimentary protein modifications that occurred within the interrogated sample.

The future of genomic-related studies in maize undoubtedly will include mapping of genetic networks such as signal transduction pathways related to developmental or environmental responses and high through-put *in situ* analyses either with protein reporter genes such as GFP or other phytofluors (Murphey and Lagarias, 1997) or with cDNA probes. Such studies as genome-wide 'transcription-factor profiling' will be dependent on new tools like gene chips and microarrays as well as developing methodologies that improve the reproducibility and sensitivity of analysis of transcriptional control. These studies will continue to generate a comprehensive biological view of not only maize but of plants in general. The real limitation will be effective management of the immense volume of data that continues to accumulate, through the ever-increasing availability of innovative research tools.

## References

Abremski, K., Hoess, R., and Sternberg, N. (1983) Studies on the properties of P1 site-specific recombination: evidence for topologically unlinked products following recombination. Cell 32: 1301–1311.

Akita, M., Nielsen, E., and Keegstra, K. (1997) Identification of protein transport complexes in the chloroplastic envelope membranes via chemical cross-linking. J. Cell Biol. 136: 983–994.

Albert, H., Dale, E.C., Lee, E., and Ow, D.W. (1995) Site-specific integration of DNA into wild-type and mutant lox sites placed in the plant genome. Plant J. 7: 649–659.

Albertsen, M.C., Fox, T.W., and Trimnell, M.R. (1993) Cloning and utilizing a maize nuclear male sterility gene. Proc. 48th Ann. Corn. and Sorghum Industry Res. Conf. pp. 224–233.

An, G., Mitra, A., Choi, H.K., Costa, M.A., An, K., Thornburg, R.W., and Ryan, C.A. (1989) Functional analysis of the 3′ control region of the potato wound-inducible proteinase inhibitor II gene. Plant Cell 1: 115–122.

Antonelli, N.M., and Stadler, J. (1990) Genomic DNA can be used with cationic methods for highly efficient transformation of maize protoplasts. Theor. Appl. Genet. 80: 395–401.

Anzola, J., Luft, B.J., Gorgone, G., and Peltz, G. (1992) Characterization of a *Borrelia burgdorferi* dnaJ homolog. Infect. and Immun. 60: 4965–4968.

Armstrong, C.L. (1994) Regeneration of plants from somatic cell cultures: Applications for in vitro genetic manipulation. In: Freeling, M., and Walbot, V. (eds), The Maize Handbook, pp. 663–670. Springer-Verlag, New York.

Armstrong, C.L., and Green, C.E. (1985) Establishment and maintenance of friable, embryogenic maize callus and the involvement of L-proline. Planta. 164: 207–214.

Armstrong, C.L., Green, C.E., and Phillips, R.L. (1991) Development and availability of germplasm with high Type II culture formation response. Maize Genet. Coop. Newslett. 65: 92–93.

Armstrong, C.L., Parker, G.B., Pershing, J.C., Brown, S.M., Sanders, P.R., Duncan, D.R., Stone, T., Dean, D.A., Deboer, D.L., Hart, J., Howe, A.R., Morrish, F.M., Pajeau, M.E., Petersen, W.L., Reich, B.J., Rodriguez, R., Santino, C.G., Sato, S.J., Schuler, W., Sims, S.R., Stehling, S., Tarochione, L.J., and Fromm, M.E. (1995) Field evaluation of European corn borer control in progeny of 173 transgenic corn events expressing an insecticidal protein from *Bacillus thuringiensis*. Crop Sci. 35: 550–557.

Armstrong, C.L., Petersen, W.L., Buchholz, W.G., Bowen, B.A., and Sulc, S.L. (1990) Factors affecting PEG-mediated stable transformation of maize protoplasts. Plant Cell Rep. 9: 335–339.

Asseldonk, Mv., Simons, A., Visser, H., Vos, W.Md., and Simons, G. (1993) Cloning, nucleotide sequence, and regulatory analysis of the *Lactococcus lactis* dnaJ gene. J. Bacteriol. 175: 1637–1644.

Atencio, D.P., and Yaffe, M.P. (1992) MAS5, a yeast homolog of DnaJ involved in mitochondrial protein import. Mol. Cell Biol. 12: 283–291.

Aves, K., Genovesi, D., Willetts, N., Zachwieja, S., Mann, M., Spencer, T., Flick, C., and Gordon-Kamm, W. (1992) Transformation of an elite maize inbred through micro-projectile bombardment of regenerable embryogenic callus. In Vitro Cell Dev. Biol. 28: 74a.

Bardwell, J.C.A., Tilly, K., Craig, E., King, J., Zylicz, M., and Georgopoulos, C. (1986) The nucleotide sequence of the *Escherichia coli* K12 dnaJ+ gene. J. Biol. Chem. 261: 1782–1785.

Barton, K.A., Whiteley, H.R., and Yang, N. (1987) *Bacillus thuringiensis* delta-endotoxin expressed in transgenic *Nicotiana tabacum* provides resistance to lepidopteran insects. Plant Physiol. 85: 1103–1109.

Baszczynski, C.L., Barbour, E., Zeka, B.L., Maddock, S.E., and Swenson, J.L. (1997) Characterization of a genomic clone for a maize DnaJ-related gene, ZmdJ1, and expression analysis of its promoter in transgenic plants. Maydica. 42: 189–201.

Baubonis, W., and Sauer, B. (1993) Genomic targeting with purified Cre recombinase. Nuc. Acids Res. 21: 2025–2029.

Baumbach, L.L., Stein, G.S., and Stein, J.L. (1987) Regulation of human histone gene expression: transcriptional and posttranscriptional control in the coupling of histone messenger RNA stability with DNA replication. Biochemistry 26: 6178–6187.

Becker, T.W., Templeman, T.S., Viret, J.F., and Bogorad, L. (1992) The cab-m7 gene: a light-inducible, mesophyll-specific gene of maize. Plant Mol. Biol. 20: 49–60.

Berlyn, G.P., and Miksche, J.P. (1976) Botanical Microtechnique and Cytochemistry. The Iowa State University Press, Ames, IA.

Bessoule, J.J. (1993) Occurrence and sequence of a DnaJ protein in plant (*Allium porrum*) epidermal cells. FEBS Lett. 323: 51–54.

Bidney, D., Scelonge, C., Martich, J., Burrus, M., Sims, L., and Huffman, G. (1992) Microprojectile bombardment of plant tissues increases transformation frequency by *Agrobacterium tumefaciens*. Plant Mol. Biol. 18: 301–313.

Bittel, D.C., Shver, J.M., Somers, D.A., and Gengenbach, B.G. (1996) Lysine accumulation in maize cell cultures transformed with a lysine-insensitive form of maize dihydropicolinate synthase. Theor. Appl. Genet. 92: 70–77.

Block, Md., and Debrouwer, D. (1993) Engineered fertility control in transgenic *Brassica napus* L.: histochemical analysis of anther development. Planta. 189: 218–225.

Block, Md., Herrera-Estrella, L., Montagu, Mv., Schell, J., and Zambryski, P. (1984) Expression of foreign genes in regenerated plants and in their progeny. EMBO J. 3: 1681–1689.

Bodeau, J.P., and Walbot, V. (1992) Regulated transcription of the maize Bronze2 promoter in electroporated protoplasts requires the C1 and R gene products. Mol. Gen. Genet. 233: 379–387.

Boudet, A.M., and Grima-Pettenati, J. (1996) Lignin genetic engineering. Mol. Breed. 2: 25–39.

Bowen, B.A. (1993) Markers for plant gene transfer. In: Kung, S.-D., and Wu, R. (eds), Transgenic Plants, pp. 89–123, Academic Press, Inc., San Diego, CA.

Brettschneider, R., Becker, D., and Lörz, H. (1997) Efficient transformation of scutellar tissue of immature maize embryos. Theor. Appl. Genet. 94: 737–748.

Browse, J., and Somerville, C. (1991) Glycerolipid synthesis: biochemistry and regulation. Ann. Rev. Plant Physiol. Plant Mol. Biol. 41: 467–506.

Bruce, W.B., Christensen, A.H., Klein, T., Fromm, M., and Quail, P.H. (1989) Photoregulation of a phytochrome gene promoter from oat transferred into rice by particle bombardment. Proc. Nat. Acad. Sci. USA 86: 9692–9696.

Brusslan, J.A., Karlinneumann, G.A., Huang, L., and Tobin, E.M. (1993) An *Arabidopsis* mutant with a reduced level of cab140 RNA is a result of cosuppression. Plant Cell. 5: 667–677.

Brusslan, J.A., and Tobin, E.M. (1995) Isolation of new promoter-mediated co-suppressed lines in *Arabidopsis thaliana*. Plant Mol. Biol. 27: 809–813.

Burkhardt, C.C. (1978) Insect pests of corn. In: Pfadt, R.E. (ed.), Fundamentals of Applied Entomology, pp. 303–334. MacMillan Publishing Co. Inc, New York.

Caimi, P.G., McCole, L.M., Klein, T.M., and Kerr, P.S. (1996) Fructan accumulation and sucrose metabolism in transgenic maize endosperm expressing a *Bacillus amyloliquefaciens* SacB gene. Plant Physiol. 110: 355–363.

Callis, J., Fromm, M., and Walbot, V. (1987) Introns increase gene expression in cultured maize cells. Genes Dev. 1: 1183–1200.

Campbell, M.M., and Sederoff, R.R. (1996) Variation in lignin content and composition – mechanisms of control and implications for the genetic improvement of plants. Plant Physiol. 110: 3–13.

Campbell, W.H., and Gowri, G. (1990) Codon usage in higher plants, green algae, and cyanobacteria. Plant Physiol. 92: 1–11.

Cao, J., Wang, Y.C., Klein, T.M., Sanford, J., and Wu, R. (1990) Transformation of rice and maize using the biolistic process. In: Lamb, C.J., and Beachy, R.N. (eds), Plant Gene Transfer, pp. 21–33. UCLA Symp. Mol. Cell Biol., Vol. 129 Wiley-Liss Inc, New York.

Capellades, M., Torres, M.A., Bastisch, I., Stiefel, V., Vignols, F., Bruce, W.B., Peterson, D., Puigdomenech. P., and Rigau, J. (1996) The maize caffeic acid O-methyltransferase gene promoter is active in transgenic tobacco and maize plant tissues. Plant Mol. Biol. 31: 307–322.

Caplan, A.J., and Douglas, M.G. (1991) Characterization of YDJ1: A yeast homologue of the bacterial DnaJ protein. J. Cell Biol. 114: 609–621.

Caplan, A.J., Tsai, J., Casey, P.J., and Douglas, M.G. (1992) Farnesylation of YDJ1p is required for function at elevated growth temperatures in *Saccharomyces cerevisiae*. J. Biol. Chem. 267: 1880–1895.

Carle-Urioste, J.C., Brendel, V., and Walbot, V. (1997) A combinatorial role for exon, intron and splice site sequences in splicing in maize. Plant J. 11: 1253–1263.

Casas, A.M., Kononowicz, A.K., Haan, T.G., Zhang, L., Tomes, D.T., Bressan, R.A., and Hasegawa, P.M. (1997a) Transgenic sorghum plants obtained after microprojectile bombardment of immature inflorescences. In Vitro Cell Dev. Biol. 33P: 92–100.

Casas, A., Kononowicz, A., Zehr, U., Zhang, L., Haan, T., Tomes, D., Bressen, R., and Hasegawa, P. (1997b) Approaches to the genetic transformation of sorghum. In: Tsaftaris, A.S. (ed.), Genetics, Biotechnology and Breeding of Maize and Sorghum, pp. 88–93. Royal Society of Chemistry, Cambridge.

Castillo, A.M., Vasil, V., and Vasil, I.K., (1994) Rapid production of fertile transgenic plants of rye (*Secale cereale* L.) Bio/Technology 12: 1366–1371.

Chair, H., Legavre, T., and Guiderdoni, E. (1996) Transformation of haploid, microspore-derived cell suspension protoplasts of rice (*Oryza sativa* L.). Plant Cell Rep. 15: 766–770.

Chalfie, M., Tu, Y., Euskirchen, G., Ward, W.W., and Prasher, D.C. (1994) Green fluorescent protein as a marker for gene expression. Science 263: 802–805.

Chamberlin, M.A., Horner, H.T., and Palmer, R.G. (1993) Nuclear size and DNA content of the embryo and endosperm during their initial stages of development in *Glycine max* (Fabaceae). Amer. J. Bot. 80: 1209–1215.

Chang, A.K., and Duggleby, R.G. (1997) Expression, purification and characterization of *Arabidopsis thaliana* acetohydroxyacid synthase. Biochem. J. 327: 161–169.

Chaumont, F., Bernier, B., Buxant, R., Williams, M.E., Levings, C.S., and Boutry, M. (1995) Targeting the maize T-urf13 product into tobacco mitochondria confers methomyl sensitivity to mitochondrial respiration. Proc. Nat. Acad. Sci. USA 92: 1167–1171.

Chee, M., Yang, R., Hubbell, E., Berno, A., Huang, X.C., Stern, D., Winkler, J., Lockhart, D.J., Morris, M.S., and Fodor, S.P. (1996) Accessing genetic information with high-density DNA arrays. Science 274: 610–614.

Chiu, W., Niwa, Y., Zeng, W., Hirano, T., Kobayashi, H., and Sheen, J. (1996) Engineered GFP as a vital reporter in plants. Curr. Biol. 6: 325–330.

Christensen, A.H., Sharrock, R.A., and Quail, P.H. (1992) Maize polyubiquitin genes: Structure, thermal perturbation of expression and transcript splicing, and promoter activity following transfer to protoplasts by electroporation. Plant Mol. Biol. 18: 675–689.

Christou, P., and McCabe, D.E. (1992) Prediction of germ-line transformation events in chimeric R(0) transgenic soybean plantlets using tissue-specific expression patterns. Plant J. 2: 283–290.

Chuck, G., Lincoln, C., and Hake, S. (1996) KNAT1 induces lobed leaves with ectopic meristems when overexpressed in *Arabidopsis*. Plant Cell 8: 1277–1289.

Ciechanover, A., and Schwartz, A.L. (1994) The ubiquitin-mediated proteolytic pathway: mechanisms of recognition of the proteolytic substrate and involvement in the degradation of native cellular proteins. FASEB J. 8: 182–191.

Close, P.S. (1993) Cloning and molecular characterization of two nuclear genes for *Zea mays* mitochondrial chaperonin 60. PhD Thesis, pp. 218, Dept of Genetics, Iowa State University, Ames, IA.

Coleman, C.E., Clore, A.M., Ranch, J.P., Higgins, R., Lopes, M.A., and Larkins, B.A. (1997) Expression of a mutant alpha-zein creates the *floury2* phenotype in transgenic maize. Proc. Nat. Acad. Sci. USA 94: 7094–7097.

Cole-Strauss, A., Yoon, K., Xiang, Y., Byrne, B.C., Rice, M.C., Gryn, J., Holloman, W.K., and Kmiec, E.B. (1996) Correction of the mutation responsible for sickle cell anemia by an RNA–DNA oligonucleotide. Science 273: 1386–1389.

Cornejo, M.J., Luth, D., Blankenship, K.M., Anderson, O.D., and Blechl, A.E. (1993) Activity of a maize ubiquitin promoter in transgenic rice. Plant Mol. Biol. 23: 567–581.

Dale, E.C., and Ow, D.W. (1990) Intra- and intermolecular site-specific recombination in plant cells mediated by bacteriophage P1 recombinase. Gene 91: 79–85.

De Wet, J.M.J., De Wet, A.E., Brink, D.E., Hepburn, A.G., and Woods, J.A. (1986) Gametophyte transformation in maize (*Zea mays,* Gramineae). In: Mulcahy, D.L., Mulcahy, G.B., and Ottaviano, E. (eds) Biotechnology and Ecology of Pollen, pp. 59–64. Springer-Verlag, New York.

Delagrave, S., Hawtin, R.E., Silva, C.M., Yang, M.M., and Youvan, D.C. (1995) Red-shifted excitation mutants of the green fluorescent protein. Bio/Technology 13: 151–154.

Dennehey, B.K., Petersen, W.L., Ford-Santino, C., Pajeau, M., and Armstrong, C.L. (1994) Comparison of selective agents for use with the selectable marker gene bar in maize transformation. Plant Cell Tiss. Org. Cult. 36: 1–7.

D'Halluin, K., Bonne, E., Bossut, M., De Beuckeleer, M., and Leemans, J. (1992) Transgenic maize plants by tissue electroporation. Plant Cell. 4: 1495–1505.

Didierjean, L., Frendo, P., Nasser, W., Genot, G., Marivet, J., and Burkhard, G. (1996) Heavy-metal-responsive genes in maize: Identification and comparison of their expression upon various forms of abiotic stress. Planta. 199: 1–8.

Dixon, R.A., Lamb, C.J., Masoud, S., Sewalt, V.J.H., and Paiva, N.L. (1996) Metabolic engineering: prospects for crop improvement through the genetic manipulation of phenylpropanoid biosynthesis and defense responses – a review. Gene 179: 61–71.

Donn, G., Nilges, M., and Morocz, S. (1990) Stable transformation of maize with a chimaeric, modified phosphinothricin acetyltransferase gene from *Streptomyces viridochromogenes*. Abstracts VIIth International Congress on Plant Tissue and Cell Culture, Amsterdam, p. 53.

Dooner, H.K., Robbins, R.P., and Jorgensen, R.A. (1991) Genetic and developmental control of anthocyanin biosynthesis. Ann. Rev. Genet. 25: 173–199.

Dowswell, C.R., Paliwal, R.L., and Cantrell, R.P. (1996) Maize in the Third World. Westview Press, Boulder, CO.

Dudov, K.P., and Perry, R.P. (1986) Properties of a mouse ribosomal protein promoter. Proc. Nat. Acad. Sci. USA 83: 8545–8549.

Dupuis, I., and Pace, G.M. (1993) Gene transfer to maize male reproductive structure by particle bombardment of tassel primordia. Plant Cell Rep. 12: 607–611.

Evans, R., Wang, A., Hanten, J., Altendorf, P., and Mettler, I. (1996) A positive selection system for maize transformation. In Vitro Cell Dev. Biol. 32: 72a.

Faranda, S., Genga, A., Viotti, A., and Manzocchi, L.A. (1994) Stably transformed cell lines from protoplasts of maize endosperm suspension cultures. Plant Cell Tiss. Org. Cult. 37: 39–46.

Fennell, A., and Hauptmann, R. (1992) Electroporation and PEG delivery of DNA into maize microspores. Plant Cell Rep. 11: 567–570.

Fischhoff, D.A., Bowdish, K.S., Perlak, F.J., Marrone, P.G., McCormick, S.M., Niedermeyer, J.G., Dean, D.A., Kusano-Kretzmer, K., Mayer, E.J., Rochester, D.E., Rogers, S.G., and Fraley, R.T. (1987) Insect tolerant transgenic tomato plants. Bio/Technology 5: 807–813.

Fleming, G.H., Kramer, C.M., Le, T., and Shillito, R.D. (1995) Effect of DNA fragment size on transformation frequencies in tobacco (*Nicotiana tabacum*) and maize (*Zea mays*). Plant Sci. 110: 187–192.

Fontes, E.B.P., Shank, B.B., Wrobel, R.L., Moose, S.P., Obrian, G.R., Wurzel, E.T., and Boston, R.S. (1991) Characterization of an immunoglobin binding protein homolog in the maize *floury-2* endosperm mutant. Plant Cell 3: 483–496.

Fox, P.C., Vasil, V., Vasil, I.K., and Gurley, W.B. (1992) Multiple ocs-like elements required for efficient transcription of the mannopine synthase gene of T-DNA in maize protoplasts. Plant Mol. Biol. 20: 219–233.

Fraley, R.T., Horsch, R.B., Matzke, A., Chilton, M.D., Chilton, W.S., and Sanders, P.R. (1984) In vitro transformation of petunia cells by an improved method of co-cultivation with *Agrobacterium tumefaciens* strains. Plant Mol. Biol. 3: 371–378.

Frame, B.R., Drayton, P.R., Bagnall, S.V., Lewnau, C.J., Bullock, W.P., Wilson, H.M., Dunwell, J.M., Thompson, J.A., and Wang, K. (1994) Production of fertile transgenic maize plants by silicon carbide whisker-mediated transformation. Plant J. 6: 941–948.

Fromm, M.E., Morrish, F., Armstrong, C., Williams, R., Thomas, J., and Klein, T.M. (1990) Inheritance and expression of chimeric genes in the progeny of transgenic maize plants. Bio/Technology 8: 833–839.

Fromm, M., Taylor, L.P., and Walbot, V. (1985) Expression of genes transferred into monocot and dicot plant cells by electroporation. Proc. Nat. Acad. Sci. USA 82: 5824–5828.

Fromm, M.E., Taylor, L.P., and Walbot, V. (1986) Stable transformation of maize after gene transfer by electroporation. Nature 319: 791–793.

Fu, H., Kim, S.Y., and Park, W.D. (1995) High-level tuber expression and sucrose inducibility of a potato *Sus4* sucrose synthase gene require 5′ and 3′ flanking sequences and the leader intron. Plant Cell 7: 1387–1394.

Gallie, D.R., and Young, T.E. (1994) The regulation of gene expression in transformed maize aleurone and endosperm protoplasts. Analysis of promoter activity, intron enhancement, and mRNA untranslated regions on expression. Plant Physiol. 106: 929–939.

Gamborg, O.L., Shyluk, J.P., and Shahin, E.A. (1981) Isolation, fusion and culture of plant protoplasts. In: Thorpe, T.A. (ed.) Plant Tissue Culture: Methods and Applications in Agriculture, pp. 115–154. Academic Press, New York.

Genschik, P., Jamet, E., Philipps, G., Parmentier, Y., Gigot, C., and Fleck, J. (1994) Molecular characterization of a beta-type proteasome subunit from *Arabidopsis thaliana* co-expressed at a high level with an alpha-type proteasome subunit early in the cell cycle. Plant J. 6: 537–546.

Giovinazzo, G., Manzocchi, L.A., Bianchi, M.W., Coraggio, I., and Viotti, A. (1992) Functional analysis of the regulatory region of a zein gene in transiently transformed protoplasts. Plant Mol. Biol. 19: 257–263.

Goff, S.A., Klein, T.M., Roth, B.A., Fromm, M.E., Cone, K.C., Radicella, J.P., and Chandler, V.L. (1990) Transactivation of anthocyanin biosynthetic genes following transfer of B regulatory genes into maize tissues. EMBO J. 9: 2517–2522.

Golovkin, M.V., Abraham, M., Morocz, S., Bottka, S., Feher, A., and Dudits, D. (1993) Production of transgenic maize plants by direct DNA uptake into embryogenic protoplasts. Plant Sci. 90: 41–52.

Goodall, G.J., and Filipowicz, W. (1990) The minimum functional length of pre-mRNA introns in monocots and dicots. Plant Mol. Biol. 14: 727–733.

Gordon-Kamm, W.J., Adams, T.R., Adams, W.R., Chambers, S.A., Courreges, V.C., Daines, R.J., Mangano, M.L., O'Brien, J.V., Spencer, T.M., Start, W.G., Willetts, N.G., Kausch, A.P., Krueger, R.W., Lemaux, P.G., and Mackey, C.J. (1989) Stable transformation of embryogenic maize cultures by microprojectile bombardment. J. Cell Biochem. 13D: 259.

Gordon-Kamm, W.J., Spencer, M.T., Mangano, M.L., Adams, T.R., Daines, R.J., Start, W.G., O'Brien, J.V., Chambers, S.A., Adams, W.R., Jr., Willetts, N.G., Rice, T.B., Mackey, C.J., Krueger, R.W., Kausch, A.P., and Lemaux, PG. (1990) Transformation of maize cells and regeneration of fertile transgenic plants. Plant Cell 2: 603–618.

Gould, J., Devey, M., Hasegawa, O., Ulian, E.C., Peterson, G., and Smith, R.H. (1991) Transformation of *Zea mays* L. using *Agrobacterium tumefaciens* and the shoot apex. Plant Physiol. 95: 426–432.

Gould, S.J., Keller, G.A., and Subramani, S. (1988) Identification of peroxisomal targeting signals located at the carboxy terminus of four peroxisomal proteins. J. Cell Biol. 107: 897–905.

Graham, M.W., and Larkin, P.J. (1995) Adenine methylation at dam sites increases transient gene expression in plant cells. Transgen. Res. 4: 324–331.

Grebenok, R.J., Pierson, E., Lambert, G.M., Gong, F.C., Afonso, C.L., Haldemann-Cahill, R., Carrington, J.C., and Galbraith, D.W. (1997) Green-fluorescent protein fusions for efficient characterization of nuclear targeting. Plant J. 11: 573–586.

Green, C.E., and Phillips, R.L. (1975) Plant regeneration from tissue cultures of maize. Crop Sci. 15: 417–427.

Grimsley, N.H., Ramos, C., Hein, T., and Hohn, B. (1988) Meristematic tissues of maize plants are most susceptible to agroinfection with maize streak virus. Bio/Technology 6: 185–189.

Grosset, J., Alary, R., Gautier, M.-F., Menossi, M., Martinez-Izquierdo, J.A., and Joudrier, P. (1997) Characterization of a barley gene encoding for a $\alpha$-amylase inhibitor subunit (CMd protein) and analysis of its promoter in transgenic tobacco plants and in maize kernels by microprojectile bombardment. Plant Mol. Biol. 34: 331–338.

Grula, J.W., Hudspeth, R.L., Hobbs, S.L., and Anderson, D.M. (1995) Organization, inheritance and expression of acetohydroxyacid synthase genes in the cotton allotetraploid *Gossypium hirsutum*. Plant Mol. Biol. 28: 837–846.

Hacia, J.G., Brody, L.C., Chee, M.S., Fodor, S.P., and Collins, F.S. (1996) Detection of heterozygous mutations in BRCA1 using high density oligonucleotide arrays and two-colour flourescence analysis. Nature Genetics 14: 441–447.

Hartley, R.W. (1988) Barnase and barstar expression of its cloned inhibitor permits expression of a cloned ribonuclease. J. Mol. Biol. 202: 913–915.

Haseloff, J., and Amos, B. (1995) GFP in plants. Trends Genet. 11: 328–329.

Haughn, G.W., Smith, J., Mazur, B., and Somerville, C. (1988) Transformation with a mutant Arabidopsis acetolactate synthase gene renders tobacco resistant to sulfonylurea herbicides. Mol. Gen. Genet. 211: 266–271.

Heim, R., and Tsien, R.Y. (1996) Engineering green fluorescent protein for improved brightness, longer wavelengths and fluorescence resonance energy transfer. Curr. Biol. 6: 178–182.

Hernalsteens, J.P., Thia-Toong, L., Schell, J., and Montagu, Mv. (1984) An *Agrobacterium*-transformed cell culture from the monocot *Asparagus officinalis*. EMBO J. 3: 3039–3041.

Hess, D. (1987) Pollen-based techniques in genetic manipulation. Int. Rev. Cytol. 107: 367–395.

Hill, M., Launis, K., Bowman, C., McPherson, K., Dawson, J., Watkins, J., Koziel, M., and Wright, M.S. (1995) Biolistic introduction of a synthetic *Bt* gene into elite maize. *Euphytica* 85: 119–123.

Hobbs, S.L.A., Kpodar, P., and DeLong, C.M.O. (1990) The effects of T-DNA copy number, position and methylation on reporter gene expression in tobacco transformants. Plant Mol. Biol. 15: 851–864.

Hoess, R.H., Wierzbicki, A., and Abremski, K. (1986) The role of the loxP spacer region in P1 site-specific recombination. Nuc. Acids Res. 14: 2287–2300.

Hohn, B., Hohn, T., Boulton, M.I., Davies, J.W., and Grimsley, N. (1987) Agroinfection of *Zea mays* with maize streak virus DNA. In: Wettstein, D., and Chua, N.H. (eds), NATO Adv. Sci. Inst., Ser. A, Life Sci., pp. 459–468. Plenum Press, New York.

Hood, E.E., Witcher, D.R., Maddock, S., Meyer, T., Baszczynski, C., Bailey, M., Flynn, P., Register, J., Marshall, L., Bond, D., Kulisek, E., Kusnadi, A., Evangelista, R., Nikolov, Z., Wooge, C., Mehigh, R.J., Hernan, R., Kappel, W.K., Ritland, D., Li, C.P., and Howard, J.A. (1997) Commercial production of avidin from transgenic maize: characterization of transformant, production, processing, extraction and purification. Mol. Breed 3: 291–306.

Howard, E.A., Danna, K.J., Dennis, E.S., and Peacock, W.J. (1985) Transient expression in maize protoplasts. J. Cell Biochem. 35: 225–234.

Huang, L.-H., Farnet, C.M., Erhlich, K.C., and Erhlich, M. (1982) Digestion of highly modified bacteriophage DNA by restriction endonucleases. Nuc. Acid. Res. 10: 1579–1591.

Huang, Y.-W., and Dennis, E.S. (1989) Factors influencing stable transformation of maize protoplasts by electroporation. Plant Cell Tiss. Org. Cult. 18: 281–296.

Ishida, Y., Saito, H., Ohta, S., Hiei, Y., Komari, T., and Kumashiro, T. (1996) High efficiency transformation of maize (*Zea mays* L.) mediated by *Agrobacterium tumefaciens*. Nature Biotechnology 14: 745–750.

Jackson, D., Veit, B., and Hake, S. (1994) Expression of maize KNOTTED1 related homeobox genes in the shoot apical meristem predicts patterns of morphogenesis in the vegetative shoot. Development 120: 405–413.

Jardinaud, M.F., Souvre, A., Alibert, G., and Beckert, M. (1995) *uidA* gene transfer and expression in maize microspores using the biolistic method. Protoplasma 187: 138–143.

Jefferson, R.A. (1987) Assaying chimeric genes in plants: The GUS gene fusion system. Plant Mol. Biol. Rep. 5: 387–405.

Johal, G.S., and Briggs, S.P. (1992) Reductase activity encoded by the HM1 disease resistance gene in maize. Science 258: 985–987.

Johnson, E.S., Bartel, B., Seufert, W., and Varshavsky, A. (1992) Ubiquitin as a degradation signal. EMBO J. 11: 497–505.

Jongedijk, E., Tigelaar, H., Van Roekel, J.S.C., Bres-Vloemans, S.A., Dekker, I., Van Den Elzen, P.J.M., Cornelissen, B.J.C., and Melchers, L.S. (1995) Synergistic activity of chitinases and beta-1,3-glucanases enhances fungal resistance in transgenic tomato plants. Euphytica 85: 173–180.

Kaeppler, H.F., Gu, W., Somers, D.A., Rines, H.W., and Cockburn, A.F. (1990) Silicon carbide fiber-mediated DNA delivery into plant cells. Plant Cell Rep. 9: 415–418.

Kaeppler, H.F., Somers, D.A., Rines, H.W., and Cockburn, A.F. (1992) Silicon carbide fiber-mediated stable transformation of plant cells. Theor. Appl. Genet. 84: 560–566.

Kalderon, D., Roberts, B.L., Richardson, W.D., and Smith, A.E. (1984) A short amino acid sequence able to specify nuclear location. Cell 39: 499–509.

Kao, C.Y., Cocciolone, S.M., Vasil, I.K., and McCarty, D.R. (1996) Localization and interaction of the cis-acting elements for abscisic acid, VIVIPAROUS1, and light activation of the C1 gene of maize. Plant Cell 8: 1171–1179.

Kausch, A.P., Adams, T.R., Mangano, M., Zachwieja, S.J., Gordon-Kamm, W., Daines, R., Willets, N.G., Chambers, S.A., Adams, W., Jr., and Anderson, A. (1995) Effects of microprojectile bombardment on embryogenic suspension cell cultures of maize (*Zea mays* L.) used for genetic transformation. Planta 196: 501–509.

Kilby, N.J., Leyser, H.M.O., and Furner, I.J. (1992) Promoter methylation and progressive transgene inactivation in *Arabidopsis*. Plant Mol. Biol. 20: 103–112.

Klein, T.M., Fromm, M., Weissinger, A., Tomes, D., Schaaf, S., Sletten, M., and Sanford, J.C. (1988a) Transfer of foreign genes into intact maize cells with high-velocity microprojectiles. Proc. Nat. Acad. Sci. USA 85: 4305–4309.

Klein, T.M., Gradziel, T., Fromm, M.E., and Sanford, J.C. (1988b) Factors influencing gene delivery into *Zea mays* cells by high-velocity microprojectiles. Bio/Technology 6: 559–563.

Klein, T.M., Harper, E.C., Svab, Z., Sanford, J.C., Fromm, M.E., and Maliga, P. (1988c) Stable genetic transformation of intact *Nicotiana* cells by particle bombardment process. Proc. Nat. Acad. Sci. USA 85: 8502–8505.

Klein, T.M., Kornstein, L., Sanford, J.C., and Fromm, M.E. (1989a) Genetic transformation of maize cells by particle bombardment. Plant Physiol. 91: 440–444.

Klein, T.M., Roth, B.A., and Fromm, M.E. (1989b) Regulation of anthocyanin biosynthetic genes introduced into intact maize tissue by microprojectiles. Proc. Natl. Acad. Sci. USA 86: 6681–6685.

Koziel, M.G., Beland, G.L., Bowman, C., Carozzi, N.B., Crenshaw, R., Crossland, L., Dawson, J., Desai, N., Hill, M., Kadwell, S., Launis, K., Lewis, K., Maddox, D., McPherson, K., Meghji, M.R., Merlin, E., Rhodes, R., Warren, G.W., Wright, M., and Evola, S.V. (1993) Field performance of elite transgenic maize plants expressing an insecticidal protein derived from *Bacillus thuringiensis*. Bio/Technology 11: 194–200.

Kramer, C., DiMaio, J., Carswell, G.K., and Shillito, R.D. (1993) Selection of transformed protoplast-derived Zea mays colonies with phosphinothricin and a novel assay using the pH indicator chlorophenol red. Planta 190: 454–458.

Krautwig, B., Lazzeri, P.A., and Lorz, H. (1994) Influence of enzyme solution on protoplast culture and transient gene expression in maize (*Zea mays* L.). Plant Cell Tiss. Org. Cult. 39: 43–48.

Krebbers, E., Rompaey, Jv., and Vandekerchhove, J. (1992) Expression of modified seed storage proteins in transgenic plants. In: Hiatt, A. (ed.), Transgenic Plants: Fundamentals and Applications, pp. 37–60. M. Dekker, New York.

Krol, A.R.vd., Mur, L.A., Beld, M., Mol, J.N.M., and Stuitje, A.R. (1990) Flavonoid genes in petunia: addition of a limited number of gene copies may lead to a suppression of gene expression. Plant Cell 2: 291–299.

Lamb, C.J., Rvals, J.A., Ward, E.R., and Dixon, R.A. (1993) Emerging strategies for enhancing crop resistance to microbial pathogens. Curr. Plant Sci. Biotech. Agric. 15: 45–60.

Landsman, D., McBride, O.W., and Bustin, M. (1989) Human non-histone chromosomal protein HMG-17: identification, characterization, chromosome localization and RFLPs of a functional gene from the large multigene family. Nuc. Acids Res. 17: 2301–2314.

Langer, T., Lu, C., Echols, H., Flanagan, J., Hayer, M.K., and Hartl, F.U. (1992) Successive action of DnaK, DnaJ, and GroEL along the pathway of chaperone-mediated protein folding. Nature 36: 683–689.

Langridge, P., Brettschneider, R., Lazzeri, P., and Lorz, H. (1992) Transformation of cereals via *Agrobacterium* and the pollen pathway: a critical assessment. Plant J. 2: 631–638.

Larkin, J.C., Oppenheimer, D.G., Pollock, S., and Marks, M.D. (1993) *Arabidopsis* GLABROUS1 gene requires downstream sequences for function. Plant Cell 5: 1739–1748.

Last, D.I., Brettell, R.I.S., Chamberlain, D.A., Chaudhury, A.M., Larkin, P.J., Marsh, E.L., Peacock, W.J., and Dennis, E.S. (1991) pEMU: an improved promoter for gene expression in cereal cells. Theor. Appl. Genet. 81: 581–588.

Laursen, C.M., Krzyzek, R.A., Flick, C.E., Anderson, P.C., and Spencer, T.M. (1994) Production of fertile transgenic maize by electroporation of suspension culture cells. Plant Mol. Biol. 24: 51–61.

Leader, D., Connelly, S., Filipowicz, W., Waugh, R., and Brown, J.W.S. (1993) Differential expression of U5snRNA gene variants in maize (Zea mays) protoplasts. Plant Mol. Biol. 21: 133–143.

Leah, R., Tommerup, H., Svendsen, I., and Mundy, J. (1991) Biochemical and molecular characterization of three barley seed proteins with antifungal properties. J. Biol. Chem. 266: 1564–1573.

Leemans, J. (1992) Genetic engineering for fertility control. J. Cell Biochem. Suppl. 16F: 203.

Li, W., Nagaraja, S., Delcuve, G.P., Hendzel, M.J., and Davie, J.R. (1993) Effects of histone acetylation, ubiquitination and variants on nucleosome stability. Biochem. J. 296: 737–744.

Liberek, K., Georgopoulos, C., and Zylicz, M. (1988) Role of the *Escherichia coli* DnaK and DnaJ heat shock proteins in the initiation of bacteriophage DNA replication. Proc. Nat. Acad. Sci. USA 85: 6632–6636.

Liberek, K., Marszalek, J., Ang, D., Georgopoulos, C., and Zlicz, M. (1991) *Escherichia coli* DnaJ and GrpE heat shock proteins jointly stimulate ATPase activity of DnaK. Proc. Nat. Acad. Sci. USA 88: 2874–2878.

Lloyd, A.M., and Davis, R.W. (1994) Functional expression of the yeast FLP/FRT site-specific recombination system in *Nicotiana tabacum*. Mol. Gen. Genet. 242: 653–657.

Loeb, T.A., and Reynolds, T.L. (1994) Transient expression of the uidA gene in pollen embryoids of wheat following microprojectile bombardment. Plant Sci. 104: 81–91.

Logie, C., and Stewart, A.F. (1995) Ligand-regulated site-specific recombination. Proc. Nat. Acad. Sci. USA 92: 5940–5944.

Lowe, K., Bowen, B., Hoerster, G., Ross, M., Bond, D., Pierce, D., and Gordon-Kamm, B. (1995) Germline transformation of maize following manipulation of chimeric shoot meristems. Bio/Technology 13: 677–682.

Lowe, K., Ross, M., Sandahl, G., Miller, M., Hoerster, G., Church, L., Tagliani, L., Bond, D. and Gordon-Kamm, W. (1997) Transformation of the maize apical meristem: Transgenic sector reorganization and germline transmission. In: Tsaftaris, A.S. (ed.), Genetics, Biotechnology and Breeding of Maize and Sorghum, pp. 94–97, Royal Society of Chemistry, Cambridge.

Lu, G., and Ferl, R.J. (1992) Site-specific oligodeoxynucleotide binding to maize Adh1 gene promoter represses Adh1-GUS gene expression *in vivo*. Plant Mol. Biol. 19: 715–723.

Ludwig, S.R., Bowen, B., Beach, L., and Wessler, S.R. (1990) A regulatory gene as a novel visible marker for maize transformation. Science 247: 449–450.

Luehrsen, K.R., and Walbot, V. (1992) Insertion of non-intron sequence into maize introns interferes with splicing. Nuc. Acids Res. 20: 5181–5187.

Lusardi, M.C., Neuhaus-Uri, G., Potrykus, I., and Neuhaus, G. (1994) An approach towards genetically engineered cell fate mapping in maize using the Lc gene as a visible marker: Transactivation capacity of Lc vectors in differentiated maize cells and microinjection of Lc vectors into somatic embryos and shoot apical meristems. Plant J. 5: 571–582.

Lyznik, L.A., Hirayama, L., Rao, K.V., Abad, A., and Hodges, T.K. (1995) Heat-inducible expression of FLP gene in maize cells. Plant J. 8: 177–186.

Lyznik, L.A., McGee, J.D., Tung, P.Y., Bennetzen, J.L., and Hodges, T.K. (1991a) Homologous recombination between plasmid DNA molecules in maize protoplasts. Mol. Gen. Genet. 230: 209–218.

Lyznik, L.A., Peng, J.Y., and Hodges, T.K. (1991b) Simplified procedure for transient transformation of plant protoplasts using polyethylene glycol treatment. Biotechniques 10: 294–300.

Lyznik, L.A., Mitchell, J.C., Hirayama, L., and Hodges, T.K. (1993) Activity of yeast FLP recombinase in maize and rice protoplasts. Nuc. Acids Res. 21: 969–975.

Lyznik, L.A., Rao, K.V., and Hodges, T.K. (1996) FLP-mediated recombination of FRT sites in the maize genome. Nuc. Acids Res. 24: 3784–3789.

Lyznik, L.A., Ryan, R.D., Ritchie, S.W., and Hodges, T.K. (1989) Stable co-transformation of maize protoplasts with gusA and neo genes. Plant Mol. Biol. 13: 151–161.

Maas, C., Schaal, S., and Werr, W. (1990) A feedback control element near the transcription start site of the maize Shrunken gene determines promoter activity. EMBO J. 9: 3447–3452.

Maas, C., Laufs, J., Grant, S., Korfhage, C., and Werr, W. (1991) The combination of a novel stimulatory element in the first exon of the maize Shrunken-1 gene with the following intron 1 enhances reporter gene expression up to 1000-fold. Plant Mol. Biol. 16: 199–207.

Maas, C., and Werr, W. (1989) Mechanism and optimized conditions for PEG mediated DNA transfection into plant protoplasts. Plant Cell Rep. 8: 148–151.

Maessen, G.D.F. (1997) Genomic stability and stability of expression in genetically modified plants. Acta Bot. Neerl. 46: 3–24.

Malki, A., Hughes, P., and Kohiyama, M. (1991) In vitro roles of *Escherichia coli* DnaJ and DnaK heat shock proteins in the replication of oriC plasmids. Mol. Gen. Genet. 225: 420–446.

Mariani, C., de Beuckeleer, M., Truettner, J., Leemans, J., and Goldberg, R.B. (1990) Induction of male sterility in plants by a chimaeric ribonuclease gene. Nature 347: 737–741.

Mariani, C., Gossele, V., de Beuckeleer, M., de Block, M., Goldberg, R.B., de Greef, W., and Leemans, J. (1992) A chimaeric ribonuclease-inhibitor gene restores fertility to male sterile plants. Nature 357: 384–387.

Marrs, K.A., and Sinibaldi, R.M. (1997) Deletion analysis of the maize *hsp82*, *hsp81*, and *hsp17.9* promoters in maize and transgenic tobacco: contributions of individual heat shock elements and recognition by distinct protein factors during both heat shock and development. Maydica 42: 211–226.

Marshallsay, C., Connelly, S., and Filipowicz, W. (1992) Characterization of the U3 and U6 snRNA genes from wheat: U3 snRNA genes in monocot plants are transcribed by RNA polymerase III. Plant Mol. Biol. 19: 973–983.

Martiensen, R. (1996) Epigenetic phenomena: Paramutation and gene silencing in plants. Curr. Opin. Biol. 6: 810–813.

Mascarenhas, D., Mettler, I.J., Pierce, D.A., and Lowe, H.W. (1990) Intron-mediated enhancement of heterologous gene expression in maize. Plant Mol. Biol. 15: 913–920.

Matsuoka, M., Ichikawa, H., Saito, A., Tada, Y., Fujimura, T., and Kano-Murakami, Y. (1993) Expression of a rice homeobox gene causes altered morphology of transgenic plants. Plant Cell 5: 1039–1048.

Matzke, M.A., and Matzke, A.J.M. (1993) Genomic imprinting in plants: parental effects and trans-inactivating phenomena. Ann. Rev. Plant Physiol. Plant Mol. Biol. 44: 53–76.

Matzke, M.A., Mazke, A.J.M., and Mittelsten-Scheid, O. (1994a) Inactivation of repeated

genes – DNA–DNA interaction? In: Paszkowski, J. (ed.), Homologous Recombination and Gene Silencing in Plants, pp. 271–307. Kluwer, Dortrecht.
Matzke, A.J.M., Neuhuber, F., Park, Y.-.D, Ambros, P., and Matzke, M.A. (1994b) Homology-dependent gene silencing in transgenic plants: epistatic silencing loci contain multiple copies of methylated transgenes. Mol. Gen. Genet. 244: 219–229.
McCabe, D.E., and Martinell, B.J. (1993) Transformation of elite cotton cultivars via particle bombardment of meristems. Bio/Technology 11: 596–598.
McClelland, M., and Nelson, N. (1985) The effect of site specific methylation on restriction endonuclease digestion. Nuc. Acids Res. 13: R201–R207.
McElroy, D., Rothenberg, M., Reece, K.S., and Wu, R. (1990a) Characterization of the rice (*Oryza sativa*) actin gene family. Plant Mol. Biol. 15: 257–268.
McElroy, D., Zhang, W., Cao, J., and Wu, R. (1990b) Isolation of an efficient actin promoter for use in rice transformation. Plant Cell 2: 163–171.
McGall, G., Labadie, J., Brock, P., Wallraff, G., Nguyen, T., and Hinsberg, W. (1996) Light-directed synthesis of high-density oligonucleotide arrays using semiconductor photoresists. Proc. Nat. Acad. Sci. USA 93: 13555–13560.
Meeley, R.B., Johal, G.S., Briggs, S.P., and Walton, J.D. (1992) A biochemical phenotype for a disease resistance gene of maize. Plant Cell 4: 71–77.
Menossi, M., Martinez-Izquierdo, J.A., and Puigdomenech, P. (1997) Promoter tissue specific activity and ethylene control of the gene coding for the maize hydroxyproline-rich glycoprotein in maize cells transformed by particle bombardment. Plant Sci. 125: 189–200.
Mett, V.L., Lockhead, L.P., and Reynolds, P.H.S. (1993) Copper-controllable gene expression system for whole plants. Proc. Nat. Acad. Sci. USA 90: 4567–4571.
Meyer, P., and Saedler, H. (1996) Homology-dependent gene silencing in plants. Ann. Rev. Plant Physiol. Plant Molec. Biol. 47: 23–48.
Morell, M.K., Rahman, S., Abrahams, S.L., and Appels, R. (1995) The biochemistry and molecular biology of starch synthesis in cereals. Aust. J. Plant Physiol. 22: 647–660.
Morocz, S., Donn, G., Nemeth, J., and Dudits, D. (1990) An improved system to obtain fertile regenerants via maize protoplasts isolated from a highly embryogenic suspension culture. Theor. Appl. Genet. 80: 721–726.
Morrow, B., and Kucherlapati, R. (1993) Gene targeting in mammalian cells by homologous recombination. Curr. Opin. Biotech. 4: 577–582.
Murphy, J.T., and Lagarias, J.C. (1997) The phytofluors: a new class of fluorescent protein probes. Curr. Biol. 7: 870–876.
Murry, L.E., Elliott, L.G., Capitant, S.A., West, J.A., Hanson, K.K., Scarafia, L., Johnston, S., Deluca-Flaherty, C., Nichols, S., and Cunanan, D. (1993) Transgenic corn plants expressing MDMV strain-B coat protein are resistant to mixed infections of maize dwarf mosaic virus and maize chlorotic mottle virus. Bio/Technology 11: 1559–1564.
Narberhaus, F., Giebeler, K., and Bahl, H. (1992) Molecular characterization of the dnaK gene region of *Clostridium acetobutylicum*, including grpE, dnaJ, and a new heat shock gene. J. Bact. 174: 3290–3299.
Negrutiu, I., Shillito, R., Potrykus, I., Biasini, G., and Sala, F. (1987) Hybrid genes in the analysis of transformation conditions. I. Setting up a simple method for direct gene transfer in plant protoplasts. Plant Mol. Biol. 8: 363–373.
Neuhaus, G., Spangenberg, G., Mittelsten-Scheid, O., Scheid, O.M., and Schweiger, H.-G. (1987) Transgenic rapeseed plants obtained by the microinjection of DNA into microspore-derived embryoids. Theor. Appl. Genet. 75: 30–36.
Neuhaus-Url, G., Lusardi, M.C., Imoberdorf, R., and Neuhaus, G. (1994) Integrative and self-replicating Lc vectors and their transactivation capacity in maize callus protoplasts. Plant Cell Rep. 13: 564–569.
Oard, J.H., Paige, D.F., Simmonds, J.A., and Gradziel, T.M. (1990) Transient gene expression in maize, rice, and wheat cells using an airgun apparatus. Plant Physiol. 92: 334–339.

Odell, J., Caimi, P., Sauer, B., and Russell, S. (1990) Site-directed recombination in the genome of transgenic tobacco. Mol. Gen. Genet. 223: 369–378.

Odell, J.T., Hoopes, J.L., and Vermerris, W. (1994) Seed-specific gene activation mediated by the Cre/lox site-specific recombination system. Plant Physiol. 106: 447–458.

Odell, J.T., Nagy, F., and Chua, N.-H. (1985) Identification of DNA sequences required for activity of the cauliflower mosaic virus 35S promoter. Nature 313: 810–812.

Odell, J.T., and Russell, S.H. (1994) Use of site-specific recombination systems in plants. In: Paszkowski, J. (ed.), Homologous recombination and gene silencing in plants, pp. 219–270. Kluwer Acad. Publishers, Netherlands.

Ohlrogge, J.B., and Jaworski, J.G. (1997) Regulation of fatty acid synthesis. Ann. Rev. Plant Physiol. Plant Mol. Biol. 48: 109–136.

Ohta Y, (1986) High-efficiency genetic transformation of maize by a mixture of pollen and exogenous DNA. Proc. Nat. Acad. Sci. USA 83: 715–719.

O'Keefe, D.P., Tepperman, J.M., Dean, C., Leto, K.J., Erbes, D.L., and Odell, J.T. (1994) Plant expression of a bacterial cytochrome P450 that catalyzes activation of a sulfonylurea pro-herbicide. Plant Physiol. 105: 473–482.

Olive, M.R., Walker, J.C., Singh, K., Dennis, E.S., and Peacock, W.J. (1990) Functional properties of the anaerobic responsive element of the maize Adh1 gene. Plant Mol. Biol. 15: 593–604.

Omirulleh, S., Abraham, M., Golovkin, M., Stefanov, I., Karabaev, M.K., Mustardy, L., Morocz, S., and Dudits, D. (1993) Activity of a chimeric promoter with the doubled CaMV 35S enhancer element in protoplast-derived cells and transgenic plants in maize. Plant Mol. Biol. 21: 415–428.

Ow, D., Day, C., Holappa, L., Lee, E., Kobayashi, J., Albert, H., and Koshinsky, H. (1997) Expression of site-specific transgenes. Vth International Congress of Plant Molecular Biology, Abstract #210.

Pang, S.Z., DeBoer, D.L., Wan, Y., Ye, G., Layton, J.G., Neher, M.K., Armstrong, C.L., Fry, J.E., Hinchee, M.A.W., and Fromm, M.E. (1996) An improved green fluorescent protein gene as a vital marker in plants. Plant Physiol. 112: 893–900.

Pareddy, D., Petolino, J., Skokut, T., Hopkins, N., Miller, M., Welter, M., Smith, K., Clayton, D., Pescitelli, S., and Gould, A. (1997) Maize transformation via helium blasting. Maydica. 42: 143–154.

Pescitelli, S.M., and Sukhapinda, K. (1995) Stable transformation via electroporation into maize type II callus and regeneration of fertile transgenic plants. Plant Cell Rep. 14: 712–716.

Planckaert, F., and Walbot, V. (1989) Transient gene expression after electroporation of protoplasts derived from embryogenic maize callus. Plant Cell Rep. 8: 144–147.

Potrykus, I. (1989) Gene transfer to cereals: an assessment. Trends Biotechnol. 7: 269–272.

Potrykus, I. (1990a) Gene transfer to cereals: an assessment. Bio/Technology 8: 535–542.

Potrykus, I. (1990b) Gene transfer to plants: assessment and perspectives. Physiol. Plant 79: 125–134.

Potrykus, I. (1991) Gene transfer to plants: assessment of published approaches and results. Ann. Rev. Plant Physiol. Plant Mol. Biol. 42: 205–225.

Potrykus, I., Saul, M.W., Petruska, J., Paszkowski, J., and Shillito, R.D. (1985) Direct gene transfer to cells of a graminaceous monocot. Mol. Gen. Genet. 199: 183–188.

Prasher, D.C., Eckenrode, V.K., Ward, W.W., Prendergast, F.G., and Cormier, M.J. (1992) Primary structure of the *Aequorea victoria* green-fluorescent protein. Gene 111: 229–233.

Preisig-Muller, R., and Kindl, H. (1993) Plant dnaJ homologue: molecular cloning, bacterial expression, and expression analysis in tissues of cucumber seedlings. Arch. Biochem. Biophys. 305: 30–37.

Puchta, H., and Hohn, B. (1996) From centiMorgans to base pairs: Homologous recombination in plants. Trends Plant Sci. 1: 340–348.

Qiang, B.Q., McClelland, M., Poddar, S., Spokauskas, A., and Nelson, M. (1990) The apparent specificity of NotI 5′-GCGGCCGC-3′ is enhanced by *M. fnudii* or *M. bepi* methyltransferases 5′-methyl-CGCG-3′ cutting bacterial chromosomes into a few large pieces. Gene 88: 101–106.

Quarrie, S.A. (1996) New molecular tools to improve the efficiency of breeding for increased drought resistance. Plant Growth Regul. 20: 167–178.

Quayle, T., and Feix, G. (1992) Functional analysis of the –300 region of maize zein genes. Mol. Gen. Genet. 231: 369–374.

Quayle, T.J.A., Hetz, W., and Feix, G. (1991) Characterization of a maize endosperm culture expressing zein genes and its use in transient transformation assays. Plant Cell Rep. 9: 544–548.

Rao, A.G., and Flynn, P. (1990) A quantitative assay for beta-D-glucuronidase (GUS) using microtiter plates. Biotechniques 8: 38–40.

Rasmussen, J.L., Kikkert, J.R., Roy, M.K., and Sanford, J.C. (1994) Biolistic transformation of tobacco and maize suspension cells using bacterial cells as microprojectiles. Plant Cell Rep. 13: 212–217.

Reggiardo, M.I., Luis-Arana, J., Orsaria, L.M., Permingeat, H.R., Spitteler, M.A., and Vallejos, R.H. (1991) Transient transformation of maize tissues by microparticle bombardment. Plant Sci. 75: 237–244.

Register, J.C., Peterson, D.J., Bell, P.J., Bullock, W.P., Evans, I.J., Frame, B., Greenland, A.J., Higgs, N.S., Jepson, I., Jiao, S.P., Lewnau, C.J., Sillick, J.M., and Wilson, H.M. (1994) Structure and function of selectable and non-selectable transgenes in maize after introduction by particle bombardment. Plant Mol. Biol. 25: 951–961.

Reiss, B., Klemm, M., Kosak, H., and Schell, J. (1996) RecA protein stimulates homologous recombination in plants. Proc. Nat. Acad. Sci. USA 93: 3094–3098.

Restrepo, M.A., Freed, D.D., and Carrington, J.C. (1990) Nuclear transport of plant potyviral proteins. Plant Cell 10: 987–998.

Revet, B.M., Sena, E.P., and Zarling, D.A. (1993) Homologous DNA targeting with RecA protein-coated short DNA probes and electron microscope mapping on linear duplex molecules. J. Mol. Biol. 232: 779–791.

Rhodes, C.A., Lowe, K.S., and Ruby, K.L. (1988a) Plant regeneration from protoplasts isolated from embryogenic maize cell cultures. Bio/Technology 6: 56–60.

Rhodes, C.A., Pierce, D.A., Mettler, I.J., Mascarenhas, D., and Detmer, J.J. (1988b) Genetically transformed maize plants from protoplasts. Science 240: 204–207.

Rigau, J., Capellades, M., Montoliu, L., Torres, M.A., Romera, C., Martinez-Izquierdo, J.A., Tagu, D., and Puigdomenech, P. (1993) Analysis of maize alpha-tubulin gene promoter by transient expression and in transgenic tobacco plants. Plant Journal 4: 1043–1050.

Ritchie, S.W., Lui, C.-N., Sellmer, J.C., Kononowicz, H., Hodges, T.K., and Gelvin, S.B. (1993) *Agrobacterium tumefaciens*-mediated expression of gusA in maize tissues. Transgenic Res. 2: 252–265.

Rizzuto, R., Brini, M., Pizzo, P., Murgia, M., and Pozzan, T. (1995) Chimeric green fluorescent protein as a tool for visualizing subcellular organelles in living cells. Curr. Biol. 5: 635–642.

Rogers, J.C., and Rogers, S.W. (1995) Comparison of the effects of N-6-methyldeoxyadenosine and N-5-methyldeoxycytosine on transcription from nuclear gene promoters in barley. Plant J. 7: 221–233.

Roth, B.A., Goff, S.A., Klein, T.M., and Fromm, M.E. (1991) C1- and R-dependent expression of the maize Bz1 gene requires sequences with homology to mammalian myb and myc binding sites. Plant Cell 3: 317–325.

Russell, D.A., and Fromm, M.E. (1997) Tissue-specific expression in transgenic maize of four endosperm promoters from maize and rice. Transgenic Res. 6: 157–168.

Russell, D.A., DeBoer, D.L., Stark, D.M., Preiss, J., and Fromm, M.E. (1993) Plastid targeting of *E. coli* beta-glucuronidase and ADP-glucose pyrophosphorylase in maize (*Zea mays*

L.) cells. Plant Cell Rep. 13: 24–27.

Sabri, N., Pelissier, B., and Teissie, J. (1996) Transient and stable electrotransformations of intact Black Mexican Sweet maize cells are obtained after preplasmolysis. Plant Cell Rep. 15: 924–928.

Sadowski, P.D. (1995) The Flp recombinase of the 2-microns plasmid of *Saccharomyces cerevisiae*. Prog. Nuc. Acid Res. Mol. Biol. 51: 53–91.

Sainz, M.B., Grotewold, E., and Chandler, V.L. (1997) Evidence for direct activation of an anthocyanin promoter by the maize C1 protein and comparison of DNA binding by related Myb domain proteins. Plant Cell 9: 611–625.

Sanford, J.C., Klein, T.M., Wolf, E.D., and Allen, N. (1987) Delivery of substances into cells and tissues using a particle bombardment process. Part. Sci. Tech. 5: 27–37.

Sardana, R., Dukiandjiev, S., Giband, M., Cheng, X.Y., Cowan, K., Sauder, C., and Altosaar, I. (1996) Construction and rapid testing of synthetic and modified toxin gene sequences CryIA (b&c) by expression in maize endosperm culture. Plant Cell Rep. 15: 677–681.

Sathasivan, K., Haughn, G.W., and Murai, N. (1991) Molecular basis of imidazolinone herbicide resistance in *Arabidopsis thaliana* var Columbia. Plant Physiol. 97: 1044–1050.

Schlake, T., and Bode, J. (1994) Use of mutated FLP recognition target (FRT) sites for the exchange of expression cassettes at defined chromosomal loci. Biochemistry 33: 12746–12751.

Schledzewski, K., and Mendel, R.R. (1994) Quantitative transient gene expression – comparison of the promoters for maize polyubiquitin1, rice actin 1, maize-derived Emu and CaMV 35S in cells of barley, maize and tobacco. Transgenic Res. 3: 249–255.

Scofield, S.R., Jones, D.A., Harrison, K., and Jones, J.D.G. (1994) Chloroplast targeting of spectinomycin adenyltransferase provides a cell-autonomous marker for monitoring transposon excision in tomato and tobacco. Mol. Gen. Genet. 244: 189–196.

Scowcroft, W.R., and Larkin, P.J. (1988) Somaclonal variation. In: Gregory, B., Marsh, J. (eds), Applications of Plant Cell and Tissue Culture, pp. 21–35. Wiley, Chichester, Sussex, UK.

Scowcroft, W.R., Ryan, S.A., Brettel, R.I.S., and Larkin, P.J. (1985) Somaclonal variation in crop improvement. Biotechnology in international agricultural research, Proc. Inter-Center Sem. Internat. Agric. Res. Cent. (IARCs) and Biotech. pp. 99–109. International Rice Research Institute, Manila, Philippines.

Sena, E.P., and Zarling, D.A. (1993) Targeting in linear DNA duplexes with two complementary probe strands for hybrid stability. Nature Genetics 3: 365–371.

Senecoff, J.F., Rossmeissl, P.J., and Cox, M.M. (1988) DNA recognition by the FLP recombinase of the yeast 2 mu plasmid. A mutational analysis of the FLP binding site. J. Mol. Biol. 201: 405–21.

Sheen, J. (1991) Molecular mechanisms underlying the differential expression of maize pyruvate, orthophosphate dikinase genes. Plant Cell 3: 225–245.

Sheen, J., Hwang, S., Niwa, Y., Kobayashi, H., and Galbraith, D.M. (1995) Green-fluorescent protein as a new vital marker in plant cells. Plant J. 8: 777–784.

Shen, W.H., and Hohn, B. (1994) Amplification and expression of the beta-glucuronidase gene in maize plants by vectors based on maize streak virus. Plant J. 5: 227–236.

Sheridan, W.F. (1975) Growth of corn cells in culture. J. Cell Biol. 67: 396a.

Sheridan, W.F. (1982) Black Mexican Sweet corn: its use for tissue culture. In: Sheridan, W.F. (ed.) Maize for Biological Research, pp. 385–387, University of North Dakota Press, Grand Forks.

Shieh, M.W., Wessler, S.R., and Raikhel, N.V. (1993) Nuclear targeting of the maize-R protein requires two nuclear localization sequences. Plant Physiol. 101: 353–361.

Shillito, R.D., Saul, M.W., Paszkowski, J., Muller, M., and Potrykus, I. (1985) High efficiency direct gene transfer to plants. Bio/Technology 3: 1099–1103.

Sieburth, L.E., and Meyerowitz, E.M. (1997) Molecular dissection of the AGAMOUS control region shows that cis elements for spatial regulation are located intragenically.

Plant Cell 9: 355–365.
Sinha, N.R., Williams, R.E., and Hake, S. (1993) Overexpression of the maize homeo box gene, KNOTTED-1, causes a switch from determinate to indeterminate cell fates. Gene Dev. 7: 787–795.
Smulders, M.J.M., Rus-Kortekaas, W., and Vosman, B. (1995) Tissue culture-induced DNA methylation polymorphisms in repetitive DNA of tomato calli and regenerated plants. Theor. Appl. Genet. 91: 1257–1264.
Somers, D.A., and Anderson, P.C. (1994) In vitro selection for herbicide tolerance in maize. Biotech. Agric. 25: 293–313.
Songstad, D.D., Armstrong, C.L., Petersen, W.L., Hairston, B., and Hinchee, M.A.W. (1996) Production of transgenic maize plants and progeny by bombardment of Hi-II immature embryos. In Vitro Cell. Dev. Biol. 32P: 179–183.
Songstad, D.D., Halaka, F.G., DeBoer, D.L., Armstrong, C.L., Hinchee, M.A.W., Ford-Santino, C.G., Brown, S.M., Fromm, M.E., and Horsch, R.B. (1993) Transient expression of GUS and anthocyanin constructs in intact maize immature embryos following electroporation. Plant Cell Tis. Org. Cult. 33: 195–201.
Spencer, T.M., Gordon-Kamm, W.J., Daines, R.J., Start, W.G., and Lemaux, P.G. (1990) Bialaphos selection of stable transformants from maize cell culture. Theor. Appl. Genet. 79: 625–631.
Spencer, T.M., O'Brien, J.V., Start, W.G., Adams, T.R., Gordon-Kamm, W.J., and Lemaux, P.G. (1992) Segregation of transgenes in maize. Plant Mol. Biol. 18: 201–210.
Stapleton, G., Somma, M.P., and Lavia, P. (1993) Cell type-specific interactions of transcription factors with a housekeeping promoter *in vivo*. Nucl. Acids Res. 21: 2465–2471.
Sternberg, N., and Hamilton, D. (1981) Bacteriophage P1 site-specific recombination. I. Recombination between loxP sites. J. Mol. Biol. 150: 467–486.
Stirpe, F., Barbieri, L., Battelli, M.G., Soria, M., and Lappi, D.A. (1992) Ribosome-inactivating proteins from plants: present status and future prospects. Bio/Technology 10: 405–412.
Sudakin, V., Ganoth, D., Dahan, A., Heller, H., Hershko, J., Luca, F.C., Ruderman, J.V., and Hershko, A. (1995) The cyclosome, a large complex containing cyclin-selective ubiquitin ligase activity, targets cyclins for destruction at the end of mitosis. Mol. Biol. Cell 6: 185–197.
Sukhapinda, K., Kozuch, M.E., Rubin-Wilson, B., Ainley, W.M., and Merlo, D.J. (1993) Transformation of maize (*Zea mays* L.) protoplasts and regeneration of haploid transgenic plants. Plant Cell Rep. 13: 63–68.
Takimoto, I., Christensen, A.H., Quail, P.H., Uchimiya, H., and Toki, S. (1994) Non-systemic expression of a stress-responsive maize polyubiquitin gene (Ubi-1) in transgenic rice plants. Plant Mol. Biol. 26: 1007–1012.
Taylor, M.G., Vasil, V., and Vasil, I.K. (1993) Enhanced GUS gene expression in cereal/grass cell suspensions and immature embryos using the maize ubiquitin-based plasmid pAHC25. Plant Cell Rep. 12: 491–495.
Toki, S., Takamatsu, S., Nojiri, C., Ooba, S., Anzai, H., Iwata, M., Christensen, A.H., Quail, P.H., and Uchimya, H. (1992) Expression of a maize ubiquitin gene promoter-bar chimeric gene in transgenic rice plants. Plant Physiol. 100: 1503–1507.
Toriyama, K., Arimoto, Y., Uchimaya, H., and Hinata, K. (1988) Transgenic rice plants after direct gene transfer into protoplasts. Bio/Technology 6: 1072–1074.
Torrent, M., Alvarez, I., Geli, M.I., Dalcol, I., and Ludevid, D. (1997) Lysine-rich modified $\gamma$-zeins accumulate in protein bodies of transiently transformed maize endosperms. Plant Mol. Biol. 34: 139–149.
Uchimiya, H., Iwata, M., Nojiri, C., Samarajeewa, P.K., Takamatsu, S., Ooba, S., Anzai, H., Christensen, A.H., Quail, P.H., and Toki, S. (1993) Bialaphos treatment of transgenic rice plants expressing a bar gene prevents infection by the sheath blight pathogen (*Rhizoctonia solani*). Bio/Technology 11: 835–836.

Ueda, T., and Messing, J. (1991) A homologous expression system for cloned zein genes. Theor. Appl. Genet. 82: 93–100.

Unger, E., Parsons, R.L., Schmidt, R.J., Bowen, B., and Roth, B.A. (1993) Dominant negative mutants of opaque2 suppress transactivation of a 22-kD zein promoter by opaque2 in maize endosperm cells. Plant Cell. 5: 831–841.

Vaeck, M., Reynaerts, A., Hofte, H., Jansens, S., De Beuckeleer, M., Dean, C., Zabeau, M., Van Montagu, M., and Leemans, J. (1987) Transgenic plants protected from insect attack. Nature 328: 33–37.

Vain, P., McMullen, M.D., and Finer, J.J. (1993) Osmotic treatment enhances particle bombardment-mediated transient and stable transformation of maize. Plant Cell Rep. 12: 84–88.

Vain, P., Finer, K.R., Engler, D.E., Pratt. R.C., and Finer, J.J. (1996) Intron-mediated enhancement of gene expression in maize (*Zea mays* L.) and bluegrass (*Poa pratensis* L.). Plant Cell Rep. 15: 489–494.

Varagona, M.J., Schmidt, R.J., and Raikhel, N.V. (1992) Nuclear localization signal(s) required for nuclear targeting of the maize regulatory protein Opaque-2. Plant Cell 4: 1213–1227.

Vasil, I.K. (1988) The contributions and prospects of plant biotechnology – an assessment. In: Pais, M.S.S., Mavituna, F., Novais, J.M. (eds), NATO ASI Ser. H. Cell Biol., pp. 15–19. Springer-Verlag, Berlin.

Vasil, V., Srivastava, V., Castillo, A.M., Fromm, M.E., and Vasil, I.K. (1993) Rapid production of transgenic wheat plants by direct bombardment of cultured immature embryos. Bio/Technology 14: 1553–1558.

Vasil, V., Clancy, M., Ferl, R.J., Vasil, I.K., and Hannah, L.C. (1989) Increased gene expression by the first intron of maize shrunken-1 locus in grass species. Plant Physiol. 91: 1575–1579.

Vasil, V., and Vasil, I.K. (1987) Formation of callus and somatic embryos from protoplasts of a commercial hybrid of maize (*Zea mays* L.). Theor. Appl. Genet. 73: 793–798.

Vaucheret, H. (1993) Identification of a general silencer for 19S and 35S promoters in a transgenic tobacco plant – 90 bp of homology in the promoter sequence are sufficient for trans-inactivation. C. R. Acad. Sci. [III] 316: 1471–1483.

Vera, A., Hirose, T., and Sugiura, M. (1996) A ribosomal protein gene (rpl32) from tobacco chloroplast DNA is transcribed from alternative promoters: similarities in promoter region organization in plastid housekeeping genes. Mol. Gen. Genet. 251: 518–525.

Verwoerd, T.C., Van Paridon, P.A., Van Ooyen, A.J.J., Van Lent, J.W.M., Hoekema, A., and Pen, J. (1995) Stable accumulation of *Aspergillus niger* phytase in transgenic tobacco leaves. Plant Physiol. 109: 1199–1205.

Walters, D.A., Vetsch, C.S., Potts, D.E., and Lundquist, R.C. (1992) Transformation and inheritance of a hygromycin phosphotransferase gene in maize plants. Plant Mol. Biol. 18: 189–200.

Wan, Y., and Lemaux, P.G. (1994) Generation of large numbers of independently transformed fertile barley plants. Plant Physiol. 104: 37–48.

Wan, Y., Widholm, J.M., and Lemaux, P.G. (1995) Type I callus as a bombardment target for generating fertile transgenic maize (*Zea mays* L). Planta. 196: 7–14.

Wang, G., Du, T., Zhang, H., Xie, Y., Dai, J., Mi, J., Li, T., Tian, Y., Qiao, L., and Mang, K. (1995) Transfer of Bt-toxin protein gene into maize by high-velocity microprojectile bombardments and regeneration of transgenic plants. Science in China, Ser. B. Chem. Life Sci. Earth Sci. 38: 817–824.

Weeks, J.T., Anderson, O.D., and Blechl, A.E. (1993) Rapid production of multiple independent lines of fertile transgenic wheat (*Triticum aestivum*). Plant Physiol. 102: 1077–1084.

Weymann, K., Urban, K., Ellis, D.M., Novitzky, R., Dunder, E., Jayne, S., and Pace, G. (1993) Isolation of transgenic progeny of maize by embryo rescue under selective conditions. In Vitro Cell. Dev. Biol. 29P: 33–37.

Whitelam, G.C., Cockburn, B., Gandecha, A.R., and Owen, M.R.L. (1993) Heterologous

protein production in transgenic plants. Biotech. Genet. Eng. Rev. 11: 1–29.
Wickner, S., Hoskins, J., and McKenny, K. (1991) Function of DnaJ and DnaK as chaperones in origin-specific DNA binding by RepA. Nature 350: 165–167.
Wickner, S.H. (1990) Three *Escherichia coli* heat shock proteins are required for P1 plasmid DNA replication: Formation of an active complex between *E. coli* DnaJ and P1 initiator protein. Proc. Natl. Acad. Sci. USA 87: 2690–2694.
Williams, W.P., Sagers, J.B., Hanton, J.A., Davis, F.M., and Buckley, P.M. (1997) Transgenic corn evaluated for resistance to fall armyworm and southwestern corn borer. Crop Sci. 37: 957–962.
Williams-Carrier, R.E., Lie, Y.S., Hake, S., and Lemaux, P.G. (1997) Ectopic expression of the maize *kn1* gene phenocopies the Hooded mutant of barley. Development 124: 3737–3745.
Wilson, T.M.A. (1993) Strategies to protect crop plants against viruses – pathogen-derived resistance blossoms. Proc. Natl. Acad. Sci. USA 90: 3134–3141.
Witcher, D.R., Hood, E.E., Peterson, D., Bailey, M., Bond, D., Kusnadi, A., Evangelista, R., Nikolov, Z., Wooge, C., Mehigh, R., Kappel, W., Register, J., and Howard, J.A. (1998) Commercial production of $\beta$-glucuronidase (GUS): a model system for the production of proteins in plants. Mol. Breed. 4: 301–312.
Xiayi, K., Xiuwen, Z., Heping, S., and Baojian, L. (1996) Electroporation of immature maize zygotic embryos and regeneration of transgenic plants. Transgen. Res. 5: 219–221.
Yang, Y., and Gabriel, D.W. (1995) Xanthomonas avirulence/pathogenicity gene family encodes functional plant nuclear targeting signals. Mol. Plant Microb. Interact. 8: 627–631.
Yoder, J.I., and Goldsbrough, A.P. (1994) Transformation systems for generating marker-free transgenic plants. Bio/Technology 12: 263–267.
Yoon, K., Cole-Strauss, A., and Kmiec, E.B. (1996) Targeted gene correction of episomal DNA in mammalian cells mediated by a chimeric RNA-DNA oligonucleotide. Proc. Natl. Acad. Sci. USA 93: 2071–2076.
Yunes, J.A., Neto, G.C., Silva, M.Jd., Leite, A., Ottoboni, L.M.M., and Arruda, P. (1994) The transcriptional activator Opaque2 recognizes two different target sequences in the 22-kD-like alpha-prolamin genes. Plant Cell 6: 237–249.
Zernik, J., Thiede, M.A., Twarog, K., Stover, M.L., Rodan, G.A., Upholt, W.B., and Rowe, D.W. (1990) Cloning and analysis of the 5′ region of the rat bone/liver/kidney/placenta alkaline phosphatase gene. A dual-function promoter. Matrix 10: 38–47.
Zhan, X., Wu, H.M., and Cheung, A.Y. (1996) Nuclear male sterility induced by pollen-specific expression of a ribonuclease. Sex. Plant Reprod. 9: 35–43.
Zhang, S., Warkentin, D., Sun, B., Zhong, H., and Sticklen, M. (1996) Variation in the inheritance of expression among subclones for unselected (*uidA*) and selected (bar) transgenes in maize (*Zea mays* L.). Theor. Appl. Genet. 92: 752–761.
Zhang, W., and Wu, R. (1988) Efficient regeneration of transgenic plants from rice protoplasts and correctly regulated expression of the foreign gene in the plants. Theor. Appl. Genet. 76: 835–840.
Zhong, H., Sun, B., Warkentin, D., Zhang, S., Wu, R., Wu, T., and Sticklen, M.B. (1996a) The competence of maize shoot meristems for integrative transformation and inherited expression of transgenes. Plant Physiol. 110: 1097–1107.
Zhong, H., Zhang, S., Warkentin, D., Sun, B., Wu, T., Wu, R., and Sticklen, M.B. (1996b) Analysis of the functional activity of the 1.4-kb 5′-region of the rice actin 1 gene in stable transgenic plants of maize (*Zea mays* L.). Plant Sci. 116: 73–84.
Zhong, T., and Arndt, K.T. (1993) The yeast SIS1 protein, a DnaJ homolog, is required for the initiation of translation. Cell 73: 1175–1186.
Zhu, J.K., Shi, J., Bressan, R.A., and Hasegawa, P.M. (1993) Expression of an *Atriplex nummularia* gene encoding a protein homologous to the bacterial molecular chaperone DnaJ. Plant Cell 5: 341–349.

# 9. Transgenic Cereals: *Hordeum vulgare* L. (barley)

PEGGY G. LEMAUX[1], MYEONG-JE CHO[1,3], SHIBO ZHANG[1,3] and PHIL BREGITZER[2,3]

[1]*Department of Plant and Microbial Biology, University of California, Berkeley CA 94720, USA. E-mail: lemauxpg@nature.berkeley.edu;* [2]*USDA-ARS, National Small Grains Germplasm Research Center, P.O. Box 307, Aberdeen, ID 83210, USA (*[3]*Authors made equal contributions to publication)*

ABSTRACT. The development of barley as a crop dates to the earliest agricultural activities of humans, and it remains one of the major cereals grown for feed and food, and for the production of beer. In this century, an understanding and application of quantitative genetic theory has created a genetically elite crop that is divergent from its ancestors. Further improvements in barley cultivars will depend on continued access to useful allelic variability. Sexual hybridization will continue to play an important role in such improvement, but its utility is limited because potentially useful alleles are either linked to undesirable alleles or unavailable because of sexual incompatibilty. The advent of molecular genetics and non-sexual gene transfer offers exciting opportunities to bypass these limitations and to provide access to more diverse sources of genes. Recent developments have added barley to the list of major crops that are amenable to this type of genetic manipulation either through direct DNA transfer (bombardment) or mediated by *Agrobacterium tumefaciens*. However, significant problems remain, and include: 1) the lack of reproducible, efficient transformation systems for commercial germplasm; 2) the induction of stable genetic and epigenetic changes during the *in vitro* process; and 3) transgene and transgene expression instability. In this chapter, we will discuss and describe the first systems used for the genetic transformation of barley, introduce and describe the development of new systems for barley transformation, and comment on past and future uses of barley transformation as a tool for basic science and commercial application.

## I. Introduction

In the context of classical breeding, the application of genetic engineering methods for commercially important barley cultivars could play an increasingly important role in solving some fundamental challenges that face agriculture, natural resources and the environment. This could include elevating yields by improving agronomic traits, such as enhancing pest, stress and herbicide resistance. Improvements could also be made in the quality of the crop or in its feed or malting characteristics. These methods can also be used to create new value-added products in barley which are presently being made by other means, e.g. industrial enzymes, plastics, neutraceuticals. The advent

*I.K. Vasil (ed.), Molecular Improvement of Cereal Crops, 255 – 316*

of molecular technologies provides an important adjunct for classical breeding to widen the genepool available for breeders to accomplish these tasks.

The first transgenic plants expressing engineered foreign genes were reported in 1984 (for review Gasser and Fraley, 1989). Despite the promise of biotechnological methods, however, significant problems arose in the practical application of these methods to the study of crop species and their improvement, particularly in the cereals. Over time and with much effort, some of these challenges have been met. However, significant inefficiencies remain for introducing and stably expressing foreign genes in commercially important cereal species. These include:

- Lack of reproducible, efficient transformation systems for commercial germplasm
- Need for reliable schemes to identify transformants and remove selectable/ screenable genes
- Potential somatic mutation and stable epigenetic changes arising during *in vitro* culture
- Transgene and transgene expression instability

Because these inefficiencies substantially increase the cost of cereal transformation, this review will focus on work being done to facilitate its efficient and effective application to cultivar development. Although the focus will be on barley, many of the principles and strategies will be of broad applicability to other cereals and even dicotyledonous crop species.

## II. Economic Impact of Barley

Barley is among the most ancient of crops. Archaeological evidence indicates domestication occurred at least 8,000, and possibly as early as 17,000 years B.C. (Roemer, et al., 1953, Harlan, 1979). Barley is also notable among cultivated cereals for its variability in form. This includes variation in color, presence or absence of awns, presence or absence of hulls after threshing, and variation in the fertility of lateral florets (two- versus six-rowed forms). Barley also exhibits a broad range of adaptability, including the ability to grow at high altitudes and latitudes, under a wide range of day-length variation and under marginal conditions, such as high salinity and low rainfall.

Barley is a major crop both worldwide and domestically. Worldwide it ranks fourth in production behind wheat, rice, and maize (cited in Mann, 1997). Approximately 196 million tons of barley are produced annually, which represent 9% of total world cereal production (FAO Yearbook Production 1992). Forty percent is grown in Europe, 29% in the former Soviet Union, 13% in North America, and 11% in Asia (Schildbach, 1994). In the U.S., the primary uses for barley are feed and malt (47% and 32%, respectively), with lesser amounts used for export (17%), food products

(1%), and seed (3%) (personal communication, Scott Heisel, American Malting Barley Association). The direct value of barley as a raw agricultural commodity in the United States is an estimated $928 million for 1987 through 1996 (source, USDA Economic Research Service, http://www.econ.ag.gov/Briefing/FBE/CAR/BARLEY3.HTM). Besides the domestic value of this commodity, barley exports annually average (1990–1994) $178 million for the grain itself and its milled products, $38 million for malt and malt extracts, and $211 million for beer (growing at the rate of 17% per year) (personal communication, Scott Heisel, American Malting Barley Association).

The development of barley as a crop species has been accomplished by utilizing the sources of genetic variability, which can be easily accessed through traditional sexual hybridization within the primary genepool. The primary pool is limited to various barley cultivars, breeding lines, and exotic sources. Exotic sources include archaic or unadapted primitive cultivars and landraces of cultivated barley (*Hordeum vulgare* subsp. *vulgare*), and genotypes within the closely related subspecies, *Hordeum vulgare* subsp. *spontaneum*. The secondary genepool is represented by one member, *Hordeum bulbosum*. This species exists in both dipolid and tetraploid forms and possesses related genomes, which show high meiotic pairing with those of *H. vulgare*, but hybrids are highly sterile. All other *Hordeum* species are members of the tertiary genepool. Their genomes are different from that of *H. vulgare* and significant barriers exist to sexual transfer of potentially useful alleles to cultivated barley (von Bothmer et al., 1995).

The primary genepool has been the sole source of allelic variation in commercial cultivars. For classical breeding methods, crosses between elite breeding lines and cultivars have been and will continue to be the foundation of commercial barley germplasm enhancement. Exotic members of the primary genepool have functioned as valuable sources of specific characteristics, for example, sources of resistance to the Russian wheat aphid (Mornhinweg et al., 1995) and resistance to barley yellow dwarf virus (Schaller et al., 1963).

*Hordeum bulbosum* contains allelic variation of great potential value for cultivated barley; however, the low hybrid fertility has substantially impeded intergeneric flow. Some progress, using wide-species crosses, has been made, e.g. resistance to powdery mildew (Xu and Kasha, 1992; Pickering et al., 1995; Pickering et al., 1997) and barley mild mosaic virus (Michel, 1995) from *H. bulbosum* has been transferred to *H. vulgare.*

Despite the potential and demonstrated value of exotic members of the primary and secondary genepools, modern breeding efforts make little use of them. Technical difficulties severely limit the utility of *H. bulbosum,* and until recently there has been no opportunity to incorporate *H. bulbosum*-derived alleles into modern cultivars. Exotic members of the primary genepool have made many contributions to modern cultivars, but not without a price. Recovery of these alleles into commercially acceptable cultivars entails time-consuming cycles of hybridization and selection. This difficulty arises primarily from the success of modern plant breeding efforts, which have produced barley cultivars with considerable genetic divergence from even their recent *H. vulgare* ancestors. A number of distinct breeding populations have been developed, each of which is adapted to commercial production within a set of defined production and end-use requirements. Within each population, particular combinations of alleles at many critical loci have presumably been fixed in combinations to produce the desirable commercial characteristics; the desirable allelic combinations are what distinguish 'elite' germplasm from 'exotic' germplasm. Potentially useful alleles at individual loci within exotic genotypes will be linked to other loci, many of which likely will possess undesirable alleles.

The limitations of sexually compatible germplasm resources emphasize the potential importance of applying molecular technologies to the task of barley improvement. Knowledge of the physiology underlying particular characteristics and of the genes involved in the expression of these pathways will enable geneticists to identify new sources of allelic variability. The existence of an efficient, non-mutagenic technology for genetic transformation of barley is a requirement if such variability is to be readily accessible to barley breeders for cultivar development.

## III. History of Barley Transformation

The worldwide importance of barley for food, feed and malt justifies the development of methods to genetically engineer commercially important barley cultivars to improve their agronomic performance, pest resistance, yield, quality traits or to enhance alternative uses. Transformation of barley, as with other plant species, utilizes methodologies based on the delivery to, and integration and expression of defined foreign genes in plant cells grown *in vitro*. The utility of these approaches is dependent upon the ability to generate large numbers of independently and stably transformed, fertile, green plants which maintain the important characteristics of the starting germplasm.

### A. *Transient Gene Expression*

Since the inception of efforts to transform barley, laboratories have utilized a variety of approaches to introduce foreign genes. Since the ability to transform cereals with *Agrobacterium* was not possible until very recently (Chan et al., 1994; Hiei et al., 1994; Park et al., 1996; Ishida et al., 1996;

Cheng et al., 1997) and barley even more recently (Tingay et al., 1997), most work with barley has utilized direct DNA introduction methods. The first available methods for direct DNA introduction, electroporation and PEG-mediated transfer, required the use of protoplasts. Protoplast-based cultures are highly genotype-dependent, are generally not reproducibly generated or manipulated and often are not regenerable. Nonetheless, initial attempts at transformation, utilizing protoplasts from mesophyll, aleurone or endosperm tissue or cell suspensions (Junker et al., 1987; Lee et al., 1989, 1991; Lazzeri et al., 1991), were useful in demonstrating transient uptake and expression of exogenous DNA. However, stable transformation and the generation of fertile transgenic plants using protoplasts were demonstrated (Funatsuki et al., 1995; Salmenkallio-Marttila et al., 1995) after the successful use of other explants and methods for DNA introduction (see Section III.B). Transient expression was also demonstrated with other DNA introduction methods. Microparticle bombardment of suspension cells (Mendel et al., 1989; Ritala et al., 1993), immature zygotic embryos (IEs) (Kartha et al., 1989; Lee et al., 1991), callus tissue (Kartha et al., 1989), endosperm (Knudsen and Müller, 1991) and microspores (King and Kasha, 1994; Wan and Lemaux, 1994) resulted in transient gene expression of several introduced genes, including *uidA* (*gus*), *nptII* and *cat*; however, no fertile, stably transformed plants were obtained. In addition, microinjection and tissue electroporation have been used to introduce the *uidA* gene into coleoptilar cells and excised scutella of IEs, respectively, transient expression being demonstrated with histochemical staining (Toyoda et al., 1990; Hänsch et al., 1996). Microinjection has been reported to lead to transient expression of introduced genes in microinjected zygotic embryos of maize (Leduc et al., 1996). Protoplasts of the barley zygote gave rise to regenerated plantlets (Holm et al., 1994) and preliminary data indicate that DNA could be introduced into such protoplasts without affecting the subsequent regeneration potential of the cells (Holm et al., 1996). This protoplast system has the advantage of leading to very high integration frequencies as seen in mammalian systems and to the introduction of larger-sized DNA fragments than are currently possible with microparticle bombardment.

Additional reports by Zhang et al. (1995) using PEG-mediated DNA uptake into protoplasts and Ritala et al. (1993) and Stiff et al. (1995) using microparticle bombardment of suspension cells described stable expression in barley cells, but fertile, transgenic plants were not reported. Stable expression was reported in transgenic plants using macroinjection of DNA into barley floral tillers at the three- to six-leaf stage (Mendel et al., 1990) or five-leaf stage (Rogers and Rogers, 1992). Both groups reported first- or second-generation plant, which contained the introduced transgenes (*uidA*, *npt*II); however, either the transgene(s) was reported as unstable or the transmission of the transgenes in later generation plants was questionable.

### B. *Stable Transformation by Direct DNA Introduction Methods*

The first demonstrations of stable transformation of barley which resulted in fertile, stably transformed plants was in 1994. In that year, three groups reported success (Wan and Lemaux, 1994; Ritala et al., 1994; Jähne et al., 1994), using three different approaches and five different explants. Wan and Lemaux, (1994) reported success from three of four explants attempted, IEs, young callus from IEs and microspore-derived embryos. The demonstration of stable transformants using microspores, the other explant attempted by Wan and Lemaux (1994), was reported by Jähne, et al. (1994). All of these studies utilized cultivars that were amenable to *in vitro* culture but are currently not commercially important. Ritala et al., (1994) utilized the axis of IEs without *in vitro* culturing in a commercial Finnish variety. Since that time, other groups have succeeded in generating fertile, transformed barley plants and the methods have varied somewhat depending on genotype, explant, selectable/screenable markers used, DNA introduction methods and details of tissue selection and manipulation. Listed in Table 1 are the reports to date of successful production of fertile transgenic barley plants. These efforts will be discussed in detail below and conclusions drawn about shortcomings and prospects for each.

The first report of recovery of stably transformed fertile plants utilized microparticle bombardment (Wan and Lemaux, 1994) of the various explants. More organized tissues, like IEs, were utilized because of problems encountered in barley transformation which were more marked than those observed with other cereals. These include the rapid loss of regenerability of *in vitro*-cultured tissue and an increase in albinism with time. The introduction of DNA directly into organized scutellar cells at the inception of callus formation has enabled the selection of transgenic callus and the regeneration of plants before the regenerability of these tissues falls below a critical level. The failure to recover transgenic, green plants from systems based on established callus or suspension culture may be explained by the additional time required by these systems from culture initiation to plant regeneration.

Because IEs have been the most common target tissue used in published barley transformation procedures, certain elements of this procedure will be noted below to provide contrast for the discussion of other methods. Wan and Lemaux (1994) used IEs of the spring, two-rowed cultivar Golden Promise. Exogenous plasmid DNA, containing the herbicide resistance gene, *bar*, and the screenable *uidA* gene, both under the control of the maize ubiquitin promoter and ubiquitin intron, was introduced by microparticle bombardment (PDS-1000 He, Bio-Rad Inc.). For transformation of IEs, embryos (1.5 to 2.5 mm) were bisected longitudinally to prevent germination and plated on callus-induction medium (see Table 2) containing the auxin dicamba (2.5 mg/L) and placed in the dark. One day post-bombardment, tissue was moved to callus-induction medium with 5 mg/L bialaphos for 10

TABLE 1
Summary of successful barley transformations

| Authors | Transformation Method | Starting Material | Genotype | Selective Agent | TF[g] | ETF[h] | Remarks |
|---|---|---|---|---|---|---|---|
| Jähne et al. (1994) | Proj.[a] | Microspore | Igri | Basta | | $1 \times 10^{-7}$ | |
| Ritala et al. (1994) | Proj. | Embryonic axis side of IE[d] | Kymppi | Plantlets were screened for NPTII activity directly without selection | | 0.4% | Chimeric |
| Wan and Lemaux (1994) | Proj. | IE | GP[f] | Bialaphos | 7.9% | 4.4% | Albino |
| | | Callus | | | – | | |
| | | MDE[e] | Igri | | 0.3% | 0.1% | Albino |
| Funatsuki et al. (1995) | Proto. | Suspension culture | Igri | G418 | | | Albino |
| Hagio et al. (1995) | Proj. | IE | GP | Hygromycin B | | 0.6% | No albino |
| | | | Haruna Nijo | | | 0.2% | |
| | | | Dissa | | | 0.7% | |
| | | | New Golden | | | – | |
| Salmenkaillo-Marttila et al. (1995) | Proto.[b] | Microspore | Kymppi | Plants were screened for NPTII activity directly without selection | | $2.5 \times 10^{-6}$ | |
| Jensen et al. (1996) | Proj. | IE | GP | Bialaphos | 6.5% | | |
| Jørgenesen (1996) | Proj. | IE | GP | Bialaphos | 13.2% | 5.0–5.6% | Albino |

*Continued on next page*

TABLE 1
Continued

| Authors | Transformation Method | Starting Material | Genotype | Selective Agent | TF[g] | ETF[h] | Remarks |
|---|---|---|---|---|---|---|---|
| Koprek et al. (1996) | Proj.* | IE | GP | Bialaphos | | 1.4% | Dorina not transformed |
| | | | Dera | | | 0.3% | |
| | | | Corniche | | | 0.3%–1.1% | |
| | | | Salome | | | 0–1.6% | |
| | | | Femina | | | 0–0.6% | |
| | | | Dorina | | | 0% | |
| Lemaux et al. (1996) | Proj. | IE | GP | Bialaphos | 7.7% | 3.9% | Albino |
| Sandager (1996) | Proj. | IE | GP | Bialaphos | 7.2% | 2.2-3.0% | Albino |
| Tingay et al. (1997) | A.t.[c] | IE | GP | Bialaphos | | 4.2% | |
| Cho et al. (1998) | Proj. | IE | Galena | Bialaphos | 1.0% | 0.5% | See section IV |
| | | Callus | Harrington | Hygromycin B | | 1.2% | |
| Cho et al. (1997) | Proj. | Green, regenerative tissue | GP | Bialaphos | | 0.9% | See section IV |
| | | | Galena | G418 | | 1.9% | |
| | | | Harrington | Hygromycin B | 1.7% | | Regenerable |
| Zhang et al. | Proj. | Shoot meristem cultures | Harrington | Bialaphos | | 0.8–1.6% | See section V. |

[a]Proj. = microprojectile bombardment
[b]Prot. = protoplast direct DNA uptake
[c]A.t. = *Agrobacterium tumefaciens*
[d]IE = immature embryo
[e]MDE = microspore-derived embryo
[f]GP = Golden Promise
[g]TF = transformation frequency = # independent events/# targets
[h]ETF = effective transformation frequency = # independent events of fertile plants/# targets
*Compared Particle Inflow Gun (PIG) with Bio-Rad PDS-1000 (PDS)

TABLE 2
Compositions of media used for barley tissue culture and transformation

| Medium | Basal Salts | Hormone/Other | Reference |
|---|---|---|---|
| Callus induction | MS | 2.5 mg/L dicamba or 2,4-D | Wan and Lemaux (1994)<br>Cho et al. (1998b) |
| DC | MS | 2.5 mg/L 2,4-D and 5.0 $\mu$M $CuSO_4$ | Cho et al. (1998b) |
| DBC1 | MS | 2.5 mg/L 2,4-D, 0.01 mg/L BAP and 5.0 $\mu$M $CuSO_4$ | Cho et al. (1998b) |
| DBC2 | MS | 2.5 mg/L 2,4-D, 0.1 mg/L BAP and 5.0 $\mu$M $CuSO_4$ | Cho et al. (1998b) |
| DBC3 | MS | 1.0 mg/L 2,4-D, 0.5 mg/L BAP and 5.0 $\mu$M $CuSO_4$ | Cho et al. (1998b) |
| FHG | Modified MS | 1.0 mg/L BAP | Hunter (1988)<br>Wan and Lemaux (1994) |
| Rooting | MS | hormone-free | Wan and Lemaux (1994) |

to 14 days. During the next two to three selection passages in the dark (10 to 20 days per passage), callus pieces showing vigorous growth were transferred to new selection plates and cultured until uniform growth was obtained. Bialaphos-resistant calli were transferred to FHG regeneration medium (see Table 2) containing 1 mg/L of the cytokinin, 6-benzylaminopurine (BAP) and 1 mg/L bialaphos under low light (45 to 55 $\mu$E, 16 hr light). Green plantlets were transferred to rooting medium (hormone-free callus-induction medium; see Table 2) supplemented with 1 mg/L bialaphos and then transferred to the greenhouse and grown to maturity.

Using this standard embryogenic callus induction method, Wan and Lemaux (1994) generated 56 independent PAT-positive callus lines from 711 half-IEs, giving a transformation frequency (TF), defined as the number of transformants/number of explants, of about 8%. Of the 56 transformed callus lines, approximately 80% were regenerable; 69% (31/45) of the regenerable lines yielded fertile, green plants, giving an 'effective transformation frequency' (ETF) of approximately 4%. This approach is reproducible with this genotype, but requires the availability of IEs from plants grown under controlled conditions. Using IEs of Golden Promise and following essentially the same protocol as Wan and Lemaux (1994) with slight modifications, several other reports were subsequently published describing the production of fertile, transgenic barley plants (Jensen et al., 1996; Jørgensen, 1996; Lemaux et al., 1996; Sandager, 1996). The results of these later experiments gave TFs/ETFs of approximately the same magnitude, 6.5 to 13% (see Table 1).

Because frequencies of callus induction are often low with certain cultivars; targets can be negatively impacted by bombardment conditions; and long-term cultures of barley lose regenerability, Wan and Lemaux, (1994) explored the use of two-week-old callus. Twenty-seven PAT-positive lines were identified; however, only 11% (3/27) gave rise to green plants, 11% were nonregenerable and 78% (21/27) yielded albino plants; the reason for the increased albinism was not identified. Because of the difficulty in tracking individual callus pieces, TF frequencies were not calculated. Microspore-derived embryos from cultured anthers of the cultivar Igri, and microspores alone were also used as bombardment targets by Wan and Lemaux (1994). The TF for microspore-derived embryos was 0.3%; 8 of 2409 microspore-derived embryos yielded independently transformed PAT-positive callus lines. The ETF was 0.1%; only two of 2409 explants yielded green plants and the remainder yielded albino plants. Microspores were not stably transformed, although microspores expressing GUS were observed. Transmission of the introduced genes (*bar, uidA,* and barley yellow dwarf virus coat protein gene) to $T_1$ progeny was confirmed by DNA blot hybridization for events from all three explants. Physical segregation of the transgenes, in most events studied which had a single insert, was 3:1, consistent with Mendelian inheritance although additional studies have indicated that the expected segregation did not occur in some events during generation advance (Bregitzer, unpublished data). This segregation ratio was observed in the plants generated from microspore-derived embryos, indicating that the cultured tissue became diploid prior to transformation. Cotransformation of *uidA* was approximately 85%. In callus, coexpression frequency of PAT [the herbicide resistance enzyme encoded by *bar* (Thompson et al., 1987], assessed by resistance to the herbicide Basta™, and GUS, determined histochemically (Jefferson et al., 1987), was about 50%.

Successful generation of fertile transgenic plants using microparticle bombardment of Igri microspores directly was described by Jähne et al., (1994). Isolated microspores were bombarded (PDS-1000 He, Bio-Rad) with CaMV 35S-driven *bar* and actin-driven *uidA* and cultured with no selection. Two to three weeks post-bombardment, cultures with visible microcalli were transferred to selection medium containing 3–5 mg/L phophinothricin (also Basta™ or glufosinate the active component of bialaphos). One week after culture on selection medium, plates were transferred to the light. Regenerating plants were cultured on hormone-free regeneration medium containing 3–10 mg/L Basta. The ETF was reported to be one fertile, transformed plant per $10^7$ microspores. All transformed plants were reported to have normal morphology and to be self-fertile. Four of twelve (33%) independent events contained *bar* and *uidA* and, of the two plants examined, all $T_1$ progeny were resistant to Basta and contained the transgene(s) as determined by DNA blot hybridization. This finding indicates that the transformed plants resulted from haploid cells. This approach has the disadvantages of being genotype-limited, of resulting in very low efficiencies of generating stably transformed plants relative to other published methods and of being dependent on the rather difficult, and often irreproducible, process of microspore isolation. It has the potential advantage of creating homozygous transgenic plants in the first generation.

The remaining 1994 report utilized bombardment of a different explant source. Ritala et al., (1994) reported successful transformation of the two-rowed, Finnish cultivar, Kymppi, utilizing bombardment of the axis side of IEs without selection or embryogenic callus induction. A total of 282 IEs (1 mm) were isolated and the embryonic axis bombarded (PDS-1000 He) with 35S-driven *nptII*. One day post-bombardment, embryos were transferred to modified MS medium with 0.4 mg/L BAP and 35 g/L maltose and no selection agent. After three weeks, green shoots were transferred to hormone-free rooting medium. During *in vitro* rooting, NPTII activity was analyzed in leaf material from 227 plantlets by a dot blot method. One *nptII*-positive, chimeric $T_0$ plant was recovered; plants derived from seeds of four tilters contained the introduced gene by polymerase chain reaction (PCR) and Southern analysis and expressed NPTII, giving an ETF of 0.4%. The transgene segregated in the $T_2$ progeny in a Mendelian fashion (Ritala et al., 1994; Ritala et al., 1995); *nptII* was stably inherited and expressed in $T_3$ progeny from the single original transformant (Ritala et al., 1995). Without a method to identify transformed sectors easily, the tedium of this approach makes it unlikely to become a routine procedure. However, it proves that it is possible to introduce DNA through bombardment into cells of the IE embryonic axis of barley and these transformed cells can give rise to lineages that result in germline transfer (see also Gordon-Kamm et al., this volume).

Another later report on transgenic barley plants produced using bombardment of IEs described the use of an antibiotic, hygromycin B, as a selection agent, rather than bialaphos or phosphinothricin (Hagio et al., 1995). The availability of other selection schemes can be important when introducing a second trait into a plant which has already been transformed and expresses the first selection gene or because certain selective agents work more effectively in a particular cultivar. Whole IEs of four barley cultivars, the two-rowed varieties Golden Promise, Haruna Nijo and New Golden and the six-rowed variety Dissa, were bombarded (PDS-1000 He, Bio-Rad) with *hpt* driven by 35S-*adh*1 intron and *adh/adh*1 intron-driven *uidA* and cultured on callus-induction medium containing 2,4-D and BAP. One week after bombardment, embryos were placed on callus-induction medium plus hygromycin B (10–30 mg/L). The ETF, as judged by number of DNA blot-positive events in $T_0$ plants per number of IEs, varied amongst the cultivars: 0.6%, Golden Promise; 0.2%, Haruna Nijo; 0.7%, Dissa; 0%, New Golden. Individual IEs were not tracked to assure independence of events. Stable transmission of *hpt* to $T_1$ progeny was confirmed by DNA blot hybridization in two lines, one line from Haruna Nijo and one from Dissa. Notably no albino plants were observed from any transformed lines.

Insights into a potential mitigator of genotype dependence came from a report in which DNA delivery parameters were examined. Six two-rowed spring cultivars, Golden Promise, Dera, Dorina, Corniche, Salome and Femina, (Koprek et al., 1996) were bombarded using both a Bio-Rad PDS-1000 and a particle inflow gun (Finer et al., 1992). Whole IEs, with the embryonic axis removed, from the six cultivars were placed on callus-induction medium containing 2,4-D and, after one day, scutella were bombarded with ubiquitin-driven *bar* and *uidA* constructs. Helium pressures for the PDS-1000 and particle inflow gun were varied between 800 and 1200 psi and 60 and 90 psi, respectively, and the numbers of scutella capable of callus induction following bombardment were quantitated. The callus response of the cultivars other than Golden Promise was found to be severely reduced by the bombardment conditions used by other groups; Golden Promise was unaffected. Stable transformation experiments were performed in which two days post-bombardment, scutella were transferred to selection medium (3 mg/L bialaphos). Embryogenic structures were observed on Golden Promise and Dera three weeks post-bombardment; development of somatic embryoids occurred later and at a lower frequency for the other genotypes. Using the PDS-1000 only Golden Promise, Dera and Corniche were transformed, the ETF of Dera and Corniche being 0.3%. Using the particle inflow gun and less destructive conditions, stably transformed plants were recovered from Corniche, Salome and Femina at higher ETFs, 1.1%, 1.6% and 0.6%, respectively, despite the fact that numbers of GUS-expressing foci using the particle inflow gun were significantly lower in transient assays. Scanning electron microscopic analysis of scutella showed that bombardment with the particle inflow gun had less physical impact on the recalcitrant embryos than with the PDS-1000, consistent with callus-induction data. Callus of Dorina never formed somatic embryoids after bombardment with either the PDS-1000 or particle inflow gun devices, despite that fact it remained friable, pale yellow and compact for more than six months. Molecular analysis by DNA blot hybridization confirmed the integration of the introduced genes in the genomes of $T_0$ plants and the transmission of the transgenes to $T_1$ progeny. These results indicate that IEs from different cultivars have differing capacities to withstand microparticle bombardment. An assessment of callus-induction frequency following bombardment provides useful information on the likelihood that a particular bombardment condition might adversely impact culture response and transformation efficiency.

In the first report of successful use of microspores in transformation, plants homozygous for the transgene were generated in the $T_0$ generation of the two lines studied (Jähne et al., 1994), suggesting that the initial transformed cells were haploid. In a second report of transformation of barley microspores of cv. Igri, six independent lines were obtained following microparticle bombardment of isolated microspores with *bar* and *uidA* and selection on bialaphos (Yao et al., 1997). Plants from two of the lines were haploid and sterile; one was trisomic and partially sterile; and three were diploid (one was sterile). DNA hybridization analysis of $T_1$ progeny from the fertile diploid lines indicated that the genes were stably transmitted but the plants were hemizygous for the transgene. The differences in the nature of the transgenic plants from the two studies were likely due to differences in the developmental stage of the microspores at the time of bombardment and DNA integration. The 28-day cold pretreatment of Jähne, et al., (1994) maintained the microspores at the uninucleate stage in contrast to the mannitol pretreatment of Yao et al., (1997), during which the microspores continued DNA synthesis and/or nuclear division prior to bombardment.

In most successful reports of barley transformation, either *bar*, *nptII* or *hpt* were used as the selection genes and *uidA* as the screenable marker. An alternative selection method has been explored in order to avoid the problems associated with the use of herbicide or antibiotic resistance genes. The method makes use of an *Escherichia coli* gene involved in amino acid synthesis, which is feedback-insensitive, e.g. aspartate kinase or *lysC,* (Perl et al., 1993). In a successful scheme used to generate transformed potato, wild-type cells cultured on millimolar concentrations of lysine and threonine die of methionine starvation due to the fact that the endogenous aspartate kinase is inhibited by the lysine and threonine in the medium. In barley the introduction of ubiquitin promoter-driven *lysC* into IEs by particle bombardment, followed by selection on lysine- and threonine-containing medium, led to the isolation of regenerable cell lines. The plants deriving from these lines were shown by PCR and enzyme assays to contain the introduced construct (Holm et al., 1996). While this approach has the advantage of not introducing herbicide or antibiotic resistance genes, it is likely to be limited in the genotypes which are amenable and the media in which the selection system can be used.

Alternative screenable markers have also been explored. The product of the luciferase gene (DeWet et al., 1987) can be detected in transgenic barley using a low-light imaging system. It has the advantages of being nondestructive and of perhaps not being subject to the high rate of gene silencing often seen with *uidA* (W. Harwood, personal communication). Attempts to use this system as a direct screen for transformants using bombardment of IEs proved very labor-intensive because of the large amounts of tissue which had to be maintained and screened (W. Harwood, personal communication). In addition the equipment needed for visualization is expensive. Another nondestructive screenable marker gene is that for the green fluorescence protein (*gfp*) and versions of the gene, optimized for expression in plants, have been constructed (Chiu et al., 1996; Pang et al., 1996; Haseloff et al., 1997; Leffel et al., 1997). Although this system has many positive characteristics, it is possible that expression of this version of the protein might interfere with the ability of microspores to divide (K. Kasha, personal communication) and may lead to the observed fertility problems in regenerated plants expressing the green fluorescence protein (M.-J Cho, S. Zhang and T. Koprek, unpublished results).

Although successful introduction and transient expression of DNA in protoplasts had been reported earlier in barley, (see Section IIIA), reports of successful stable transformation of barley utilizing protoplasts came after the use of bombardment of other explants. Polyethylene glycol-mediated stable transformation of protoplasts and the generation of fertile, transgenic plants was reported by Funatsuki et al., (1995). In these studies, protoplasts were isolated from embryogenic cell suspensions of Igri, a plasmid containing *nptII* under control of the rice *Act1* promoter was introduced with polyethylene-glycol, and the protoplasts collected by centrifugation, embedded in molten nonselective medium and cultured with feeder cells (Funatsuki et al., 1992). Developing colonies were excised and incubated in liquid medium containing 25 mg/L of the antibiotic, G418 (geneticin). Surviving calli were transferred to solid selection medium containing G418 followed by regeneration in the absence of G418; both green and albino plants were produced. Eleven plants from four transgenic lines set seed. DNA blot hybridization and NPTII ELISA analyses showed that the *nptII* gene was transmitted in the one line tested and that it was expressed in the $T_1$ progeny from the two lines tested. Another report of transgenic barley plants involved the electroporation of protoplasts isolated from cultured calli derived from microspores (Salmenkallio-Marttila et al., 1995). Protoplasts from the two-rowed spring cultivar, Kymppi, were transformed with CaMV 35S-driven *nptII.* From $16.5 \times 10^6$ protoplasts, 42 green plants were regenerated without selection and screened for NPTII activity by gel assay; three plants were NPTII-positive. DNA blot hybridization analysis of the transformed plants showed that several copies of the *nptII* gene were integrated into high molecular weight, genomic DNA and that the introduced gene was stably inherited and expressed in progeny. This analysis also showed that all plants likely originated from a single original transformed cell, since all hybridization patterns were identical. Although in these two reports the use of protoplasts was shown to lead to fertile transformed plants, in general these methods involve the difficult-to-reproduce isolation of protoplasts and the sometimes difficult regeneration of plants from such sources.

All of the methods described above, which utilized direct DNA introduction methods (into IEs, IE-derived callus, microspores, microspore-derived embryos and protoplasts), provide a means to obtain fertile, stably transformed barley plants of certain cultivars that contain and express the transgene. Although reproducible, these methods still require improvements in frequencies of regenerable transformed lines, reduction in albinism, extension to commercially important cultivars and more stable transgene expression (see Sections IV A,B,C).

### C. *Stable Transformation via* Agrobacterium-*mediated Transfer*

Direct DNA introduction methods have the advantage of not requiring the construction of relatively complicated vector systems, but the disadvantage that they do not provide easy control over the numbers of transgene copies integrated into the host genome or the location into which the transgene integrates. The number of copies of the transgene has been implicated in co-suppression mechanisms (Matzke and Matzke, 1995), which lead to loss of transgene expression. In general, it is stated that *Agrobacterium*-mediated transformation leads in some cases to fewer copies of the transgene being integrated into the genome and therefore might provide certain advantages over direct DNA introduction methods. *Agrobacterium*-mediated gene delivery was not utilized in cereals until recently because it was believed that this method was not efficient or reproducible. However, success was reported in rice (Chan et al., 1992, 1993; Hiei et al., 1994; Park et al., 1996), maize (Gould et al., 1991; Ishida et al., 1996), wheat (Cheng et al., 1997) and recently a report was published of successful transformation of barley utilizing *Agrobacterium*-mediated transfer of DNA (Tingay et al., 1997) using a modification of the original published method in rice.

In the approach with barley, the scutellar surface of IEs of Golden Promise was injured by bombardment (PDS-1000, Bio-Rad) with 1 $\mu$m gold particles at 1100 psi. The bombarded IEs were placed in a liquid culture of *Agrobacterium tumefaciens* strain AGL1, carrying a binary vector with *bar* and *uidA* driven by *Ubi1* and *Act1* promoters, respectively, and incubated for 2–3 days. After the co-cultivation, embryos were transferred to callus-induction medium supplemented with 3 mg/L bialaphos and 150 mg/L Timentin™. After approximately eight weeks of selection, bialaphos-resistant calli were transferred to FHG supplemented with 1 mg/L BAP and 3 mg/L bialaphos. Fertile plants were recovered from 54 independently transformed lines resulting from 1282 IEs, giving an ETF of about 4%, in the same range as that with microparticle bombardment of IEs. Expression of both *bar* and *uidA* was confirmed in the $T_0$ and $T_1$ generations of selected plants. In general, integration patterns in this study were simple with 18 of 27 (66%) independent events having three copies or less of the transgene. This is compared to 11 out of 22 (50%) events with three copies or less with microparticle bombardment (Wan and Lemaux, 1994). Although relevant comparisons were not made by Tingay et al., (1997), stated advantages of the *Agrobacterium* approach in other species were lower copy numbers of transgenes (Klee et al., 1987; Songstad et al., 1995) in stably transformed tissues and preferential integration into transcriptionally active chromosomal regions (Czernilofsky et al., 1986; Koncz et al., 1989). However, despite these anticipated advantages, there was still evidence in this study of loss and non-Mendelian segregation of transgene expression in the $T_1$ generation in some events. No additional data were presented in order to assess the issue of improvement of transgene or transgene expression stability. In addition to *Agrobacterium*-mediated gene delivery, other methods are being explored to improve transgene and transgene expression stability (see section IV.D).

## IV. Improvements in Barley Transformation

### A. *Rationale for Improved Transformation Systems*

Despite the successful development of transformation methods for certain barley cultivars, their utility for improvement of commercial barley cultivars is still hampered by significant problems.

*Lack of reproducible transformation systems for commercial cultivars*
Initial reports of successful transformation of barley were highly dependent on the use of particular cultivars, which are amenable to *in vitro* growth and the subsequent regeneration of fertile, green plants. Golden Promise, for example, can be successfully manipulated *in vitro* as embryogenic callus far more efficiently than most other genotypes examined (Lührs and Lörz, 1987; Bregitzer, 1992; Koprek et al., 1996). Despite the importance of these amenable genotypes for the development of transformation strategies, the ability to deploy germplasm with useful novel genes in response to urgent needs is one of the main goals of transformation technologies. For instance the barley germplasm which forms the foundation for North American breeding programs is different from Golden Promise in virtually every aspect important to commercial production and utilization. Therefore, utilization of Golden Promise as a source of transgene-derived traits probably will require multiple cycles of backcrossing and selection prior to the identification of commercially valuable genotypes. The obligatory transfer of transgenes from Golden Promise to unrelated 6-rowed germplasm that is the basis for malting cultivars in the midwestern U.S. would require multiple cycles of hybridization and selection. Such a lengthy and costly process would be necessary to preserve allelic combinations that are critical to commercial success in these important barley cultivars. The lack of culturing techniques which insure long-term regenerability of transformed tissue as well as other aspects of the transformation process, hinders progress in achieving the objective of transformation of commercial germplasm.

*Potential for somatic mutation and heritable changes introduced during* in vitro *culture*

Plants derived from *in vitro* culture frequently accumulate heritable genetic changes, which in barley are generally manifested as moderate to severe negative alterations in critical agronomic and biochemical characteristics. The elements of the *in vitro* environment which induce these changes are poorly understood, but appear to be correlated with critical facets of the transformation process, such as the establishment of *in vitro*-cultured tissue, introduction of DNA and subsequent selection of transformed tissue. Somaclonal variation (SCV), the accumulation of genetic and epigenetic changes, likely will hinder various stages of the transformation process, from the regeneration of fertile, green plants, to their use as parents in a breeding program. In barley, analysis of the agronomic performance of non-transgenic tissue culture-generated plants and transgenic barley plants showed significant SCV for important determinants of agronomic performance (see Section IV.D.1). Although backcrossing can often eliminate such problems, this requires considerable additional time and effort and would be problematic if the induced mutations were closely linked to the transgene.

*Transgene and transgene expression instability*

One limitation of current methods of transformation is that insertions occur randomly and the location of the insertion might not be optimal for gene expression. This has been noted in plants derived from direct DNA introduction methods, but has also occurred with *Agrobacterium*-mediated methods despite the stated preference for insertion into transcriptionally active regions (Czernilofsky et al., 1986; Koncz, et al., 1989). Another problem, more common with direct DNA methods, is that the introduced DNA is often present in multiple, tandemly arrayed copies. Having multiple copies of the introduced genes, closely linked in the genome, can lead both to gene inactivation and genetic instability (Flavell, 1994; Matzke and Matzke, 1995). While position effects and multiple-copy-induced gene silencing occur, they are not likely to be the only mechanisms by which transgenes are silenced. For example, neither the presence of tandemly arrayed copies of the transgene nor position effects is likely to explain variability in expression among progeny derived from the same parent (*e.g.* Wan and Lemaux, 1994; Zhang et al., 1996a). Differences in the basal state and stability of the genome in different target tissues and from different *in vitro* culturing regimes are also likely to be a significant contributing factor and might explain sibling differences.

### *B. Development of Transformation Systems for Recalcitrant Cultivars using IEs or IE-derived Tissue as a Target*

To maximize the utility of genetic engineering technologies in classical breeding programs, it is preferable to transform commercial germplasm directly, rather than introgressing the transgene into these varieties from an amenable genotype. The amenable spring barley cultivar, Golden Promise, was used by several groups (Wan and Lemaux, 1994; Hagio et al., 1995; Jensen et al., 1996; Jørgensen et al., 1996; Koprek et al., 1996; Sandager, 1996; Tingay et al., 1997; Cho et al., 1998a), while others used the amenable winter variety, Igri, (Jähne et al., 1994; Wan and Lemaux, 1994; Funatsuki et al., 1995). In addition there have been a few reports of success with cultivars from Europe and Japan. A single transformant of the Finnish variety, Kymppi (Ritala et al., 1994; Salmenkallio-Marttila et al., 1995) and success with certain European (Koprek et al., 1996) and Japanese varieties (Hagio et al., 1995) have been reported. However, no success has been reported with North American varieties. Previous attempts to transform commercially important North American varieties, e.g. Galena, Harrington, Moravian III and Morex, using published procedures (Wan and Lemaux, 1994), resulted in the identification of independently transformed lines which were either nonregenerable or yielded only albino plants (Wan, Y., Cho, M.-J. and Jiang, W. unpublished results). Changing the level of selection agent or shortening the time of selection led to the regeneration of green plants, but none were transformed (Wan, Y. and Lemaux, P.G., unpublished results).

This lack of success was likely due to factors inherent to the biological properties of the embryo or scutellum and the technical limitations of bombardment. First, in many genotypes the number of embryos or scutella which give rise to callus is low, and this number can be further reduced by microparticle bombardment (Koprek et al., 1996; see section III.B). However, callus response, *per se*, is not sufficient since high frequency is not indicative of high frequency regeneration ability (Bregitzer, 1992; Ohkoshi et al., 1991; Hanzel et al., 1985; Lührs and Lörz, 1987; Koprek et al., 1996; Cho, M.-J, unpublished results). Callus response embodies several aspects, such as perception of hormone stimuli, cell division, and the triggering of particular developmental fates that are not necessarily related to the ability to regenerate plants. This is not surprising, since callus response is dependent on the ability of scutellar cells to dedifferentiate, whereas regeneration requires the opposite response. Therefore, the number of stable transformants will also be partially dependent on a second factor, the number of scutella that gives rise to totipotent cells under *in vitro* culture conditions. This is under genetic control and will vary among cultivars. A third factor reducing the numbers of transformed, regenerable cells is the limited numbers of totipotent cells into which DNA can be introduced using microparticle bombardment or *Agrobacterium*-mediated transfer. Although for bombardment, this number can often be increased by altering physical or biochemical parameters, these changes are often detrimental to the cells or tissue (Koprek et al., 1996). A fourth limitation for transformation efficiency results from the limited ability of *in vitro*-derived cereal tissues to give rise to plants after

intermediate step between callus maintenance and regeneration or for callus initiation. Since initial callus-induction frequency was low and callus growth rate was slow on the medium containing 2,4-D, BAP and higher levels of copper, the intermediate medium (DBC2 or DBC3) was used to improve regenerability of callus grown on medium containing auxin only. It was noted that direct regeneration on FHG of calli maintained on medium containing dicamba as the only hormone generally produced only a few soft, bright green callus sectors, which gave rise to a few, or sometimes no, plantlets. The use of the intermediate step appeared to improve regeneration by allowing the small number of green, totipotent cells to continue proliferating and ultimately to convert the tissue from an embryogenic state (Figure 1A) into a state that more closely resembled the morphology of a shoot meristem culture (Figure 1B; see Section IV.C.3). The latter tissue could be triggered to produce multiple shoots from the green, regenerative tissues. From some calli green plants could not be produced in the absence of the intermediate step; this was likely because of insufficient numbers of green cells to form plantlets.

To test the hypothesis that use of the intermediate step would lead to improved regenerability, both nontransgenic and transgenic callus were cultured with and without an intermediate step and the effects on shoot regeneration were determined (Cho et al., 1998b). Two month-old, dark-grown nontransgenic calli of Golden Promise and Morex were initiated and maintained on callus-induction medium with either 2,4-D or dicamba alone. Tissue was then transferred to FHG either directly or with an intermediate culturing period of 3 to 4 weeks or longer on DBC2 or DBC3 (depending on culture morphology) under dim-light conditions (10 to 20 $\mu$E; 16 h light/8 h dark). Following transfer to FHG, green sectors were grown under higher light intensity light (30–50 $\mu$E). Similarly, regenerative capacity was tested for transgenic callus lines of Golden Promise obtained 2 to 3 months after selection in the dark on medium containing bialaphos and dicamba as the only hormone (Wan and Lemaux, 1994; Lemaux et al., 1996). Culturing the tissue directly on FHG produced only a single or a few shoots per green sector with transgenic and nontransgenic Golden Promise or no green shoots in the case of nontransgenic Morex. From both the nontransgenic and transgenic callus lines, multiple shoots were produced on FHG, after an intermediate growth period on DBC2 or DBC3. For example, the increased shoot regeneration frequency of tissue deriving from the intermediate step was from 2.4- to11.4-fold higher than that with tissues initiated and maintained on 2,4-D and dicamba alone without the intermediate step.

During the use of the intermediate step, it was noted that some tissues yielded higher numbers of shoots when cultured on DBC2, some on DBC3. This variation in the response of tissue to the two different intermediate-step media is probably due to differences in the developmental state of the particular tissue. In general, DBC2 medium, containing a higher level of 2,4-D and a lower level of BAP, appears to generate a better response from callus with smaller-sized, less organized, green sectors. Culturing tissues on DBC2 medium inhibits shoot formation, allowing green, regenerative structures with less organization to be proliferated until they achieve a size appropriate for regeneration. Callus with larger-sized, more organized, green structures can be maintained and proliferated in that state on DBC3 and regenerated on FHG. Despite the improvements in regenerability seen with the use of the intermediate step, non-regenerable transgenic callus cultures were still produced. This was likely either because the original transformed cell was not regenerable or because the proliferating cells lost regenerability early in the selection process.

2. *Transformation of recalcitrant cultivars using the intermediate step*

In addition to improvements in the culturability of tissue from recalcitrant varieties, a number of other changes in the transformation protocol were investigated as a prelude to transformation attempts of recalcitrant cultivars. First, the effects of different bombardment conditions on callus-induction response were studied. The recalcitrant genotypes, Galena, Morex and Harrington, produce more watery and less embryogenic callus than Golden Promise and have softer scutellar tissue. This was observed using comparative scanning electron microscopic analyses following bombardment of Golden Promise and the recalcitrant genotypes Corniche (Koprek et al., 1996) and Galena (Koprek and Lemaux, unpublished data); Golden Promise incurs much less damage. Decreasing the rupture pressure slows microparticles and lessens the impact on target tissue. In published transformation procedures (Wan and Lemaux, 1994; Hagio et al., 1995; Jørgenson, 1996; Lemaux et al., 1996; Sandager, 1996, Cho et al., 1998a), bombardments were carried out at a rupture pressure of 1100 psi. Although Golden Promise appears to tolerate this condition without a reduction in callus-induction frequency, the callus-induction frequency of other genotypes is dramatically affected (Koprek et al., 1996), including recalcitrant, commercially important cultivars such as Galena, Harrington and Morex (Koprek and Lemaux, unpublished results). Therefore, rupture pressure was decreased to 600 to 900 psi in the transformation protocols for the recalcitrant cultivars. Secondly, IEs and 10- to 14-day-old callus were treated with osmoticum before bombardment as described earlier for maize (Vain et al., 1993). The use of osmoticum containing equimolar (0.2M) mannitol and sorbitol might ameliorate cell death following bombardment, as evidenced by increased numbers of GUS-expressing cells in transient assays of osmoticum-treated versus non-treated controls (Jiang, Cho and Lemaux, unpublished results). Osmotic treatment might also allow deeper penetration of particles with less tissue damage, which could be associated with the reduction in cell turgor pressure caused by the treatment (Kemper, et al., 1996). Although increased numbers of transiently expressing cells do not necessarily correlate with stable transformation efficiencies (Gordon-Kamm et al., 1990; Koprek et al., 1996), data from other laboratories on maize indicated that osmotic protection does increase stable transformation frequencies (Vain et al., 1993). Thirdly, changes in the selection protocols were made because tissue from recalcitrant varieties might be more sensitive to damage and slower to recover from bombardment in terms of cell division and callus induction.

The modified protocol which enabled transformation of Harrington and Galena was as follows (Cho et al., 1998b). IEs of Galena or 10- to 14-old callus tissues of Harrington were treated with osmoticum, bombarded (900 psi for Galena; 600 psi for Harrington, Bio-Rad PDS-1000 He) with a *bar*-containing plasmid for Galena and an *hpt*-containing plasmid for Harrington and transferred to either 2.5 mg/L 2,4-D and 5.0 $\mu$M $CuSO_4$ (DC, see Table 2) for Galena or DBC1 for Harrington, both in the absence of selection. These two media were chosen because they gave optimal initial callus induction for these two genotypes; different selection schemes were chosen in order to increase the flexibility in identifying transformants. At the second transfer (10 to 14 d post-bombardment), selection was initiated at 4 mg/L bialaphos for Galena and at 20 mg/L hygromycin for Harrington. Seven to fourteen days after the second transfer, calli were moved to dim light; at the third transfer DBC2 medium was used for both genotypes. From the fourth transfer, selection pressure for Harrington was increased to 30 mg/L hygromycin until small green sectors were observed, after which selection pressure was eliminated. Putatively transformed tissues of Galena were transferred to bialaphos-containing DBC2 or DBC3 medium, depending on callus morphology (see Section IV.B.2). Bialaphos-resistant calli with green sectors, generally observed after the fifth transfer (Figure 2A), were proliferated on the same medium with bialaphos. Fully developed regenerative structures (Figure 2B) were regenerated on FHG with bialaphos for Galena, without hygromycin for Harrington, transferred to rooting medium and plantlets transferred to soil approximately 3 weeks after transfer to FHG.

Using this modified transformation protocol, Galena was transformed with *bar* and *uidA*, both under the control of the maize *ubiqutin* promoter (Christensen and Quail, 1996). From 799 bombarded IEs, eight independent events were obtained which contained and expressed *bar*; four were regenerable giving an ETF of 0.5%. All four regenerable lines contained *bar*; three lines contained *uidA* by PCR analysis. Functional expression of *uidA* and *bar* (Figure. 1C) was observed in $T_0$ leaf tissue of all transgenic lines containing the genes. To test for functional expression of *bar* in the $T_1$ generation, a regeneration test following the protocol of Wan and Lemaux (1994) was performed by plating $T_1$ embryos on bialaphos-containing (3 mg/L) rooting medium. In the two transgenic lines tested, $T_1$ IEs germinated in the presence of bialaphos; immature $T_1$ seeds had GUS activity. For both *bar* and *uidA*, a segregation ratio of 3:1 for transgene expression was observed, consistent with the presence of a single dominant gene in a selfed population. Two regenerable Harrington lines transformed with *hpt* and *uidA* were obtained; the one line that has matured had a 3:1 segregation ratio for GUS expression.

Another group has explored means for transforming commercial varieties grown in Saskatchewan. A collaborative research project between the Crop Development Centre, University of Saskatchewan and the Plant Biotechnology Research Institute, National Research Council of Canada, Saskatoon has focused on the development of transformation technologies for barley varieties of commercial significance to Saskatchewan (R. Chibbar, personal communication). The approach is based on an enhanced regeneration system in which cultured scutella, isolated from IEs, show an enhanced ability to form embryogenic callus and somatic embryos. Although the system appears to work for a wide variety of genotypes, it must be optimized for each genotype matching the ideal physiological and developmental state of immature scutella with appropriate types and levels of growth regulators, nutrients and environmental conditions. In addition to the culturing and explant changes, this group has also tested the effects on regeneration of pre-isolation storage of harvested spikes (0, 2, and 4 days at 4°C). Hulless genotypes had the lowest regeneration with no pre-isolation storage and higher frequencies after 2 or 4 days of pre-isolation storage. Malting genotypes had the highest regeneration with little or no pre-isolation storage. Bombardment and selection protocols were also modified. The DNA coating procedure and timing of bombardment were altered to achieve a greater number and more even distribution of GUS-expressing foci. Following bombardment, Basta™ was used instead of L-PPT for *bar*-containing transformants; 5 mg/L G418 or 25 mg/L paromomycin was used two days post-bombardment for *npt*II transformants. Regeneration medium containing 10 mg/L G418 or 50 mg/L paromomycin was used in the latter case. Using these modified conditions, two lines have been confirmed by DNA hybridization analysis: integration of *uidA* into a hulless feed barley line (H4) and of the *npt*II gene into a malting barley line (H14). PCR and DNA hybridization analysis have confirmed stable maintenance of the transgenes in these lines during field tests and data relating to yield and other agronomic characters are currently being collected on these lines.

*3. Development of long-term, green regenerative tissues*

For success with recalcitrant varieties, it is necessary to increase the numbers of totipotent cells in cultures that are capable of sustained division. In addition such tissue must be capable of giving rise to fertile, green plants over prolonged periods because of the requirement for selection of transformed tissue following DNA introduction (typically a two or four month-period).

Using standard *in vitro* protocols, cultured barley tissues typically lose much of their regenerative capacity in that timeframe (Ziauddin and Kasha, 1990; Hang and Bregitzer, 1993; Bregitzer et al., 1995a; Jiang et al., 1998).

Several recent findings provide new insight into the development of improved methods for culturing barley and other cereals. First, the appropriate use of certain phytohormones, i.e. 2,4-D and BAP, improves callus quality and significantly enhances the period of regenerability for Golden Promise and three commercially important North American cultivars, Galena (Jiang et al., 1998), Harrington, and Morex (Cho and Jiang, unpublished results). The use of BAP was previously shown to prevent senescence in cytokinin-treated leaf samples of intact bean plants (Fletcher et al., 1969) and to prevent the production and action of ethylene in cut flowers of carnation (Elsinger, 1977). The effect of BAP on the quality of *in vitro* response and prolonging regenerability might be mediated through its effects on ethylene documented in these earlier studies. Secondly, dramatic improvements in the regenerability of several wheat (Purnhauser, 1991; Ghaemi et al., 1994) and barley (Dahleen, 1996) cultivars have been observed in response to levels of cupric sulfate which were higher (*e.g.* range of 0.5–100 $\mu$M) than the basal 0.1 $\mu$M level in the culture media. The basis for this response is unknown, but it has been suggested that 10–50 $\mu$M cupric sulfate in hydroponic nutrient solutions inhibited the formation of the ethylene precursor, 1-aminocyclopropane-1-carboxylic acid in rice (Lidon et al., 1995). As previously stated, ethylene has been identified as a potent modulator of *in vitro* regeneration response in a number of plant species, including barley (for review, see Cho and Kasha, 1989; Biddington, 1992; Evans and Batty, 1994). Lastly, in barley, cultured meristematic tissues (see section IV.C.2) have a low incidence of albinism and good long-term regenerability properties (Zhang, unpublished results). Because it is possible to achieve similar tissue types from IE-derived tissue with certain culture conditions, it seemed possible that such cultured tissue might lead to improved transformation success.

Potential improvements to culturing methods of IEs were investigated by isolating intact IEs (about 1.0 to 2.5 mm) from spikes of Galena, Golden Promise, Harrington and Salome of growth chamber- or greenhouse-grown plants and placing them scutellum-side down in the dark (Jiang et al., 1998). The medium used was callus-induction medium with varying levels of 2,4-D, BAP and copper (see Table 2). Germinating shoots and roots were removed 5 to 7 d post-initiation and after 2–3 weeks embryogenic-looking callus was selected and transferred to dim light (approximately 10 to 20 $\mu$E, 16 h-light). Green sectors were observed 5 to 20 d after exposure to light (Figure 1A); a higher percentage of tissue from Golden Promise produced green sectors than from the recalcitrant genotypes. Highly regenerative, green structures were identified and transferred to either DBC2 or DBC3 (Figure. 1B).

Four-month-old green, regenerative tissue from several recalcitrant genotypes, *e.g.* Galena, Harrington and Salome, produced multiple green shoots (11 to 17 green shoots per piece of green tissue) when transferred to solid FHG (Hunter 1988) or rooting medium (Wan and Lemaux, 1994) and exposed to low light (approximately 30–50 $\mu$E); no albino plants were observed (Cho et al., 1998b). Increasing the culture time on intermediate medium for an additional 3 to 4 weeks resulted in the production of greater numbers of shoots from the small meristem-like structures. In contrast, callus of Golden Promise, as young as two months, maintained in the dark on callus-induction medium with 2.5 mg/L dicamba or 2.5 mg/L 2,4-D produced no green shoots and 0.35 to 1.15 green shoots per callus piece, respectively. No green shoots were produced from 2-month-old callus of Morex maintained on 2,4-D. Green, regenerative tissues of Galena, Golden Promise, Harrington and Salome, maintained for more than a year on DBC2 or DBC3, gave rise to fertile, green plants.

Morphologically, the green tissues generated by this protocol looked strikingly similar to those derived from excised shoot apices of barley cultured on 2,4-D and BAP (see Section IV.C.3a). Molecular analysis using immunolocalization confirmed visual observations. The expression pattern of a gene associated with maintenance of the meristematic state in barley shoots (Zhang et al., 1997), a *knotted*1 (Vollbrecht et al., 1991) homologue, was studied in tissue derived from the excised shoot apex (Figure 3B, Section IV.C.3b) and from green tissue (Figure 2C) The pattern was similar in these two tissues but different from that seen in callus initiated and maintained on auxin alone (2,4-D or dicamba) (Zhang, unpublished results).

*4. Use of green, regenerative tissues for transformation*

Green regenerative tissues generated and maintained on DBC2 or DBC3 medium (see Section IV.B.1, Table 2) were used directly as transformation targets (Cho, Jiang and Lemaux, unpublished data). For bombardment of Galena, Golden Promise and Harrington, 3 to 4 mm pieces of green tissue, removed 3 to 4 months after initiation of the original culture, were osmotically treated on DBC2 or DBC3 medium (Section IV.B.1) and bombarded the same day, uncut side-up, with three different constructs, *ubi-bar/ubi-uidA, ubi-nptII/ubi-uidA* and *act-hpt/ubi-uidA*. The Bio-Rad PDS-1000 He was used at 900 or 1100 psi; rupture pressures for green regenerative tissues was less critical than for IEs since their compact, nodular structure resulted in less damage as evidenced by the ability of these tissues to continue proliferation following bombardment (Cho, unpublished results). Sixteen to 18 hr post-bombardment, green tissues were transferred without osmoticum and selective agent to dim light on DBC2 or DBC3 medium for Golden Promise and Harrington or DBC3 for Galena. Selection with all agents started 3–4 weeks post-bombardment to allow for proliferation of transformed cells in the absence of cell death resulting from wounding or selection. At that time, tissue was broken into pieces (3–5 mm) and transferred to either DBC2 or DBC3 medium (depending on morphology) with 3 mg/L bialaphos (*bar*), 25 mg/L hygromycin B (*hpt*) or 30 mg/L G418 (*npt*II). Starting with the third round of selection, levels of selective agent were increased to 4–5 mg/L bialaphos, 30 mg/L hygromycin B or 40 mg/L G418. Putative transformants, identified by their relatively fast growth on selective medium, were transferred to rooting medium with 2 to 3 mg/L bialaphos (*bar*) or without selective agent for *hpt* or *nptII* transformants. Selection with hygromycin B or G418 resulted in a tighter selection for green tissues than bialaphos, based on the numbers of escaped (nontransformed but growing) tissues. However, frequently albino plants were observed during regeneration of selected tissues, especially from hygromycin B and to a lesser extent those from G418 selection, whereas no albino plants were observed from tissues selected with bialaphos. This implies that hygromycin B and G418 selection might cause more stress on the *in vitro*-cultured cells. Using this transformation protocol, Galena (*npt*II) and Golden Promise lines (*bar*) were produced and confirmed by PCR and yielded green plants; three transformed Harrington lines from hygromycin B selection have also been confirmed.

The transformation of recalcitrant varieties using the green tissue system has certain advantages over IEs. First, the use of this tissue does not require a constant source of IEs from plants grown under controlled growth conditions. Second, green regenerative tissues do not require special care during bombardment. Third, these tissues incur negligible losses of regenerability compared to the IE-derived callus during the 2 to 3 months of culturing required for selection of transformed tissue. For a period of up to three months, no regenerability losses were seen in any genotype, after three months some loss was seen in Morex, with lesser losses in regenerability seen in the recalcitrant genotypes, Galena, Harrington and Salome. Minimal losses were observed with Golden Promise tissue for periods up to two years. Fourth, albinism problems rarely occur except under certain selection conditions. To date all cultivars tested, i.e. Galena, Golden Promise, Harrington, Morex and Salome, gave rise to green regenerative tissues that give rise to regenerative cultures, which appear amenable to this transformation protocol. Fifth, it is possible that plants deriving from green tissues incur less somaclonal variation than those from IE-derived embryogenic callus (Zhang et al., 1998b) (see Section IV.D); this hypothesis must be further tested.

### *C. Development of Transformation Systems for Recalcitrant Cultivars using Shoot Meristems as Target Tissue*

Certain methodologies have prevailed in the successful generation of fertile transformed barley plants, as well as other cereals and grasses, *e.g.* rice, wheat, maize, oat, sorghum, rye, triticale and turfgrass (Vasil, 1994). First, there have been preferred DNA delivery modes, namely the direct methods of microparticle bombardment, electroporation and PEG-mediated introduction. Historically, these methods were pursued because the *Agrobacterium*-based delivery systems used for dicotyledonous plants could not be efficiently and reproducibly applied to cereals and grasses (Potrykus, 1990a,b); however, recently these methods have proven to be successful (see Section III.C.). Second, previous published effort in cereals and grasses has largely focused on a limited number of target tissues, such as the scutellum of the IE, microspores, or tissues derived from these starting materials that undergo somatic embryogenesis (Haccius, 1978). The advantages of using these embryogenic target tissues are several-fold. They are easily obtained and manipulated, except that often plants must be grown under controlled growth conditions in order to assure reproducibility and efficiency of the procedures (*e.g.* Wan and Lemaux, 1994). In addition, transgenic plant regeneration through embryogenesis obviates the problem of chimerism in which potential loss of transgenes occurs in the $T_1$ generation due to the lack of germline transmission. Lastly, the use of these tissues has yielded 'workable' ETFs in certain genotypes which are amenable to these *in vitro* conditions (see Sections III.B; IV.B).

*1. Rationale for use of shoot meristems*

The use of *in vitro*-derived target tissues for transformation has potential shortcomings which limit its utility. The first is that cells in embryogenic callus accumulate heritable genetic variability, termed somaclonal variation (SCV) (Larkin and Scowcroft, 1981), which results from tissue culture-induced genomic instability (Phillips et al., 1994) and can cause adverse effects on phenotypic and field-performance characteristics in tissue culture-derived plants (see section IV.D.1). The second hurdle to the efficient transformation of commercial germplasm is the genotype-dependence of the *in vitro* response of scutella from IEs and microspores. For example, the cultivar Golden Promise can be reproducibly transformed using IEs (Wan and Lemaux, 1994; Hagio, et al., 1995; Jensen, et al., 1996; Jørgensen, 1996; Lemaux et al., 1996; Sandager, 1996; Koprek et al., 1996; Cho et al., 1998a); the variety Igri can be reproducibly transformed using microspores (Jähne et al., 1994) or microspore-derived embryos (Wan and Lemaux, 1994). However, extension of published methodologies to commercially important varieties (see Section IV.A) has not been trivial and they are not likely to have wide applicability for commercial germplasm. The third potential hurdle to the use of embryogenic target tissue relates to transgene and transgene expression instability in succeeding generations (see Section IV.D.3). Occasionally, in transgenic cereals, as with other crop species, the introduced gene itself is physically eliminated in subsequent generations

(Spencer et al., 1992; Srivastava et al., 1996). In addition to physical loss of the transgene(s), there are numerous examples of loss of transgene expression (Finnegan and McElroy, 1994), which occurs usually at a higher frequency than the physical loss of the gene(s) itself in later generations (e.g. Wan and Lemaux, 1994; Bregitzer and Lemaux, unpublished data). Rogers and Rogers (1992) speculated on the existence of a system in barley which permits the identification and elimination of integrated foreign DNA from the genome based on methylation patterns. Molecular mechanisms explaining transgene expression variability or silencing are proposed to be due primarily to homology-dependent or repeat-induced gene silencing mediated at the transcriptional or post-transcriptional levels (reviewed by Matzke and Matzke, 1995). However, such mechanisms cannot explain all cases of transgene silencing, for example, variation observed among plants derived from the same transgene event (e.g. Wan and Lemaux, 1994; Zhang et al., 1996a). In addition, there is a lack of direct correlation between the severity of silencing and the extent of sequence similarity (Conner et al., 1997), a lack of linear correlation between the number of copies integrated and the severity of transgene silencing (e.g. ten Hoopen et al., 1996), and in some cases evidence of endogenous gene silencing in null-segregant progeny of transgenic events (Conner et al., 1997).

*2. Use of shoot apical meristems as a transformation target*

Because current published methods for generating *in vitro*-derived embryogenic target tissue are likely to be directly or indirectly responsible for at least part of the technological and biological impediments mentioned in the preceding paragraph, changes in this aspect of the transformation process are needed. It is possible that the use of a target tissue with a different developmental fate or, optimally, the complete elimination of *in vitro*-cultured tissue in transformation could be beneficial.

The shoot apical meristem (SAM) is a potential target tissue with a different developmental fate. The SAM is a small group of cells at the tip of the plant, which provides all the cells responsible for subsequent aboveground plant development. The SAM is composed of cells which ultimately give rise to germ cells (Steeves and Sussex, 1989). The fundamental concept in using the SAM as a target tissue for transformation is that a cell within the SAM can be transformed and, during plant development, yield transgenic sectors which lead to passage of the integrated transgene to progeny (Potrykus, 1992). If successful, this approach avoids the adverse impacts of the *in vitro* culture process on transgenic plants and is also likely to be genotype-independent.

An optimal approach to avoid *in vitro* culturing would be to transform the SAM in cereals directly, as described for the inflorescence meristem of *Arabidopsis* using vacuum infiltration (Bechtold et al., 1993). However, the process by which this occurs in the inflorescence meristems of *Arabidopsis* is not fully understood. In addition, the technology has not been extended to the SAM in *Arabidopsis* and the extension of the methodology to the inflorescence meristem or the SAM of cereals has not yet been reported.

Previous attempts to demonstrate stable transformation in cereals using SAMs have involved electrophoresis (Ahokas, 1989), microinjection (Simmonds et al., 1992), *Agrobacterium* infection (Chen and Dale, 1992; Park et al., 1996) and particle bombardment (Oard et al., 1990). To obtain the SAM for such uses, *in vitro*-germinated seedlings or early-stage IEs were used; the latter being technically easier to use since the SAM is exposed without having to physically manipulate it. In barley, a microtargeting gun (Sautter et al., 1991) was used to demonstrate transient gene expression in exposed SAMs (Zhang, S., Sautter C., Potrykus I., unpublished data), as was reported for the vegetative SAM of wheat (Bilang et al., 1993). The advantage of using microtargeting for such experiments is that the meristem is small and this device permits the delivery of a larger number of particles to a small area (100–200 $\mu$m) (Sautter, et al., 1991). Despite this advantage, at present this approach is limited by several factors. First, the procedure has a low through-put. For example, in a day, a single individual can bombard no more than ca. 100 SAMs, since only one SAM can be bombarded at a time. Secondly, this approach is limited by the low frequency with which DNA can be introduced into a particular cell type and the depth to which particles can penetrate without damaging the meristem or changing the fate of the cells. Lastly, this approach is limited by the developmental characteristics of the cereal SAM (Steeves and Sussex, 1989).

There have been a few reports of success using microparticle bombardment of SAMs to accomplish germline transformation of cereals. In barley, Ritala et al. (1994) reported successful transformation of the Finnish cultivar, Kymppi, by bombarding the embryonic-axis of IEs and regenerating plantlets in the absence of selection (see Section III.B). One chimeric $T_0$ plant was identified and non-chimeric $T_1$ progeny were generated (Ritala et al., 1994; Ritala et al., 1995). That this approach resulted in stable, germline-transmitted transgene integration, albeit at a low frequency, proves that it is possible to introduce DNA through bombardment into germline cells of the immature embryonic axis of barley.

Recently two approaches in maize, which utilize the apical meristems of coleoptilar-stage embryos, were used (Lowe et al., 1995; Lowe et al., 1997, see Gordon-Kamm et al., this volume). In the first approach, the bombarded embryos underwent germination, followed by hormonally-induced shoot proliferation during selection. This approach resulted in stable periclinal or homogeneously transformed plants from an inbred and a sweetcorn hybrid. The second approach involved bombardment of directly germinated coleoptilar- or transitional-stage embryos, followed by various treatments to reorganize the meristem and alter transgenic sector fate. The latter approach yielded frequent transgenic sectors in plants that typically did not persist or transmit to progeny. However, a small number of transformants was identified in which stable integration of the transgene was verified in $T_1$ progeny but, as was reported by Ritala (1995) for barley, the frequency of events which transmitted to progeny was low.

3. *Use of* in vitro*-cultured shoot meristems as transformation targets*
Another possible target tissue, which is similar in some respects to the SAM, is the *in vitro*-derived shoot meristem. These cultures are derived from shoot apices of maize (Zhong et al., 1992; Lowe et al., 1995; Zhang et al., 1998), oat (Zhang et al., 1996b) and barley (Zhang et al., 1998) and are cultured on auxin- and cytokinin-containing media. This medium arrests the development of the SAM in barley, oat and maize but allows the proliferation of axillary meristems (AXMs) from which adventitious meristems (ADMs) develop (Zhang et al., 1998). DNA can be introduced into cells of the proliferating ADMs that were cultured under selection through multiple rounds of proliferation to insure that meristematic domes are composed entirely of transformed cells. Non-chimeric, transformed plants can be generated from the transgenic *in vitro*-proliferated shoot meristems in maize (Zhong et al., 1996) and in barley and oat (Zhang et al., unpublished results).

The use of continuously proliferating ADMs attempts to capitalize on the advantages of using shoot meristems, while potentially avoiding problems with chimerism. One advantage of this target tissue over IEs and microspores (or tissues derived from these sources) relates to the fundamental nature of the cells that respond during *in vitro* culture. Cells in the immature scutellum, for example, are believed to undergo de-differentiation *in vitro*, followed by sustained divisions to yield embryogenic callus, which ultimately forms somatic embryos. *In vitro* shoot meristematic cells might be fundamentally different in that they likely proliferate from pre-existing meristematic cells in the AXM (Zhang et al., 1998a), which undergo little or no de-differentiation. Proliferation might require only a simple redirection of cells in the shoot meristem, not a de-differentiation as likely occurs with cells in the immature scutellum or microspore.

The fundamental difference between cells comprising ADMs and those in embryogenic callus might have a significant impact on various aspects of the transformation process. First, the use of *in vitro*-proliferated meristematic domes might reduce the adverse effects of putting tissue through a callus phase, as reflected by comparative analyses of the genomic DNA methylation state and the agronomic field performance of plants derived from the *in vitro* culture of meristems and IEs (see Section IV.D.2). The reduced impact seen in the methylation patterns of plants arising from cultured meristems is possibly due to the fact that ADMs arise from a simple redirection of cells in the meristem and not a de-differentiation (Zhang et al., 1998a). Second, *in vitro* culturing of meristems might be less genotype-dependent than the culturing of IEs or microspores since it does not appear to require cellular de-differentiation (see Section IV.C.3a). Third, since these cultures are derived from the germination of dry seeds, this target tissue does not require the maintenance of donor plants grown under controlled growth conditions. Lastly, since cultures can be maintained for more than a year without loss of regenerability (Zhang, S., unpublished results), the work- and resource-intensive re-initiation of new cultures suitable for transformation is not required.

*a. Development of* in vitro *shoot meristem culturing methods for recalcitrant commercial cultivars*

Methods were described previously for culturing shoot apices of maize and oat (Zhong et al., 1992; Lowe et al., 1995; Zhang et al., 1996b). Culturing shoot apices of barley was attempted in a similar manner using the shoot meristem proliferation medium described for oat and maize. Continuous proliferation of AXMs using meristem proliferation medium was successful for Golden Promise, the cultivar used in previous successful transformation efforts with IEs (Wan and Lemaux, 1994; Hagio et al., 1995; Jensen et al., 1996; Jørgensen, 1996; Koprek et al., 1996; Sandager, 1996; Cho et al., 1998a); Golden Promise yielded long-term cultures capable of regenerating large numbers of plants over extended periods. Attempts to culture other commercially important genotypes, i.e. Harrington, Morex and Crystal, using the same medium, resulted in only very limited proliferation of the AXMs. Previously published experimentation on the *in vitro* culture of callus derived from microspores and IEs showed that maltose (Finnie et al., 1989; Scott and Lyne, 1994a, b; Bregitzer, unpublished data) and $CuSO_4$ (Dahleen, 1996; Cho et al., 1998b; see Section IV.B) improved the appearance and regenerability of cultures. Therefore, these two components were tested individually and together in meristem proliferation medium to improve the culturability of AXMs. Maltose alone reduced the production of brown tissue, the presence of which correlated with poor *in vitro* growth; $CuSO_4$ alone simply promoted plant development. In combination, copper and maltose dramatically improved the shoot meristem proliferation efficiency and increased its long-term culturability. For example, in the presence of both copper and maltose, 80–90% of Harrington shoot apices cultured gave rise to proliferating AXMs, 50–70% of shoot apices of Crystal responded and 30–40% of Morex responded. The addition of these two media components expanded the genotypes from which ADM cultures could be produced and maintained. However, the fact that the response rates were not 100% and were different for the various genotypes [as was observed in maize (Lowe et al., 1995)] indicates that all factors controlling *in vitro* response are not understood and the precise conditions for maximal proliferation of all genotypes are not known. Further study of the *in vitro* proliferation process is likely to reveal additional factors which will improve response.

*b. Molecular characterization of axillary shoot meristem proliferation and adventitious shoot meristem formation*

The *in vitro* growth and development of plant tissue have been characterized as having two distinct developmental routes, organogenesis and embryogenesis. Organogenesis is the process by which totipotent cells or tissues produce a unipolar structure, namely a shoot or root primordium, the vascular system of which is often connected to the parent tissue (Thorpe, 1994). In contrast, somatic embryogenesis occurs when a bipolar structure, containing a root and shoot axis with a closed, independent vascular system, is produced (Haccius, 1978). Distinction between the two routes has thus far been descriptive based on microscopic observation and thin-section analyses (*e.g.* Thorpe, 1994). Practical differentiation between these two routes depends on a culturist's trained eye. However, these physical observations often are not sufficient to distinguish between the two routes of development and inaccurate conclusions have sometimes been made.

The terms organogenic and embryogenic have been used for decades; however, these two processes have not been dissected at the molecular level. In order to define more precisely the mechanisms by which plants regenerate from ADMs, molecular markers, i.e. maize *cdc2ZM* and *knotted1*, were used to characterize these tissues (Figure 3A) and determine the precise developmental mode by which plants arose (Zhang et al., 1998a). The *cdc2* gene encodes a cyclin-dependent kinase which plays a central role in the eukaryotic cell cycle; its expression is correlated with actively proliferating cells. In cultured meristematic domes, immunolocalization analysis of the expression patterns of *cdc2* suggests that cell division is triggered in the enlarged AXMs, which give rise to new ADMs. The maize homeobox-containing protein, Knotted1 (KN1) is expressed primarily in the shoot meristems. Down-regulation of its expression occurs during the differentiation of lateral organs (Smith et al., 1992, 1995; Jackson, et al., 1994). Expression of KN1-homologues in uncultured barley shoot apices (Zhang et al., 1998a) was very similar to that of KN1 in uncultured maize shoot apices (Smith et al., 1992, 1995), suggesting that expression of the homologue(s) could be used as a molecular marker to study shoot meristem formation during *in vitro* growth of barley tissues. ADM formation during *in vitro* AXM proliferation was analyzed by immunolocalization. Based on these analyses, ADMs appear to form directly from KN1-homologue(s)-expressing meristematic cells in the enlarged meristematic domes, providing molecular evidence that the *in vitro* morphogenic route in this system is shoot meristematic cell proliferation, with no intermediate embryogenic callus phase (Zhang et al., 1998a) (Figure 3B).

*c) Transformation of a recalcitrant commercial cultivar using* in vitro-*derived shoot meristems*

Based on the improvements in the culturing of AXMs from the commercially important varieties, Harrington, Crystal and Morex (see Section IV.C.3a), a transformation protocol was developed for one of these varieties, Harrington. Previous attempts to transform this variety using published protocols with IEs (Wan and Lemaux, 1994) were not successful (see Section IV.B). Nine-month old cultured ADM tissue of Harrington was used in microparticle bombardment experiments using both the Bio-Rad PDS-1000 He and the particle inflow gun devices (Finer et al., 1992). Gene constructs containing different selectable genes (*bar, nptII, hpt*) coupled to either the maize *ubiquitin* or rice *actin* promoters were used; plating on the appropriate selection agents, bialaphos (3–5 mg/L), G418 (30–50 mg/L), and hygromycin B (25–30 mg/L), respectively, began 3–4 weeks post-bombardment. Three to five months after selection commenced, resistant shoot meristem tissue was obtained from bialaphos and G418. Multiple shoots developed from the resistant tissues. No albino shoots were observed with bialaphos and G418; however, some albino shoots were observed during selection with hygromycin B, indicating that this selective agent causes more stress on the cultured tissue than G418. Shoots were transferred to Magenta boxes in the absence of selection for G418 or in the presence of bialaphos at the same concentration as for initial selections. $T_0$ plants were transferred to the greenhouse (Figure 3C) and molecular analyses of two independent transformants confirmed integration of the transgenes in the $T_0$ plants; PCR and transgene expression analyses confirmed transmission of the transgenes to progeny (Zhang et al., unpublished results).

## D. Impact of Transformation Method on Fidelity and Quality of Transgenic Plants

Nearly all transformation procedures utilize *in vitro* cultured tissue. Cultured somatic tissues have been shown to accumulate genetic and epigenetic changes, termed somaclonal variation (SCV). First noticed as a hindrance to clonal propagation of certain horticultural crops, SCV was subsequently hailed as a potential source of valuable genetic diversity (Larkin and Scowcroft, 1981); however, the majority of recovered plants with SCV were found to have undesirable traits. Except in cases where specific mutations to selective agents were sought (e.g. Gengenbach et al., 1977; Rines and Luke, 1985; Vidhyasekaran et al., 1990), the isolation and identification of useful variants were largely a matter of serendipity and hard work. In contrast, genetic engineering technologies offer the opportunity to make very precise genomic changes utilizing genes from a wide variety of sources. This realization has justifiably decreased interest in SCV as a breeding tool, but has also created on unjustified complacency about its potential negative effects.

SCV is an inherent characteristic of *in vitro* culture systems and will likely influence the application of genetic engineering to plant breeding. The impact might be observed in at least four ways: 1) genotypic and temporal restrictions on regeneration of fertile, green plants (see sections IV.B, C); 2) alterations of the expression of carefully selected, commercially valuable characteristics (see Section IV.D.1); 3), gross genomic alterations, including ploidy changes (see Section IV.D.2); and 4) instability in transgene expression and inheritance (see Section IV.D.3).

*1. Somaclonal variation and field performance of transgenic plants derived from embryogenic callus*

The most successful current systems for the *in vitro* manipulation of cereals, including barley, require plant regeneration from *in vitro*-derived, dedifferentiated tissues. Many barley cultivars possess the ability to produce plants from *in vitro*-derived dedifferentiated tissue (Lührs and Lörz, 1987; Ullrich et al., 1991; Bregitzer, 1992), but the efficiency of green plant regeneration is poor for most varieties. Furthermore, the ability of cultured barley tissue to regenerate fertile, green plants typically declines rapidly with age. The recovery of green plants from embryogenic callus of barley more than a few weeks old can be difficult even from amenable genotypes such as Golden Promise. The relationship of SCV to this gradual loss of totipotency is likely caused by the accumulation of genomic alterations in cultured cells (for review, Kaeppler and Phillips, 1993b), which interfere with the physiological processes necessary for redifferentation and viability. Enhanced regenerability should be made possible by discovering and ameliorating the elements in the *in vitro* environment that induce genomic changes. Furthermore, this hypothesis predicts that plants recovered from cultured tissues that are highly totipotent, either as a result of using amenable genotypes or favorable *in vitro* conditions, will possess less SCV.

During the first few weeks after the establishment of *in vitro* cultured barley tissue, regenerability typically declines drastically and widespread genomic alterations can be observed, either directly in culture cells or in regenerated plants. Reports include observations of heritable alterations in the methylation patterns of microspore-derived plants (Devaux et al., 1993) and alterations in the methylation patterns of plants derived from different *in vitro* culturing methods (Zhang et al., 1998b; see Section IV.D.2). The alterations in methylation patterns observed by S. Zhang et al., (unpublished results) are positively correlated with time in culture and negatively correlated with the regeneration potential of the cultures. Other studies have documented a time-dependent accumulation of cytogenetic aberrations in, and a loss of regenerability of, immature embryo-derived barley callus (Ziauddin and Kasha, 1990; Hang and Bregitzer, 1993; Bregitzer et al., 1995a). The auxin 2,4-D has been both positively and negatively correlated with good regenerability and normal karyotype and ploidy (Ziauddin and Kasha, 1990; Bregitzer et al., 1995b, H.W. Choi and M.-J. Cho, unpublished results). Several other reports have suggested a relationship between totipotency and normal karyotype based on comparisons of regenerable and non-regenerable cultures (Singh, 1986; Wang et al., 1992). In addition to changes in nuclear DNA, extensive plastid DNA alterations in albino, microspore-derived barley plants have been documented (Day and Ellis, 1985; Dunford and Walden, 1991). Subsequent work has provided evidence that such alterations occur during the redifferentiation process (Mouritzen and Holm, 1994).

Modern barley cultivars owe their superior characteristics to complex allelic interactions which are poorly understood, but easily disturbed by recombination between unrelated genomes (see Section II). Crosses which yield progeny that are superior to both parents almost always are the result of knowledgeable, careful choices by the breeder that preserve existing alleles at the majority of critical loci. Accordingly, the uncontrolled generation of SCV *in vitro* would be expected to produce primarily negative phenotypic changes. Phenotypically inferior, tissue culture-derived plants may be as difficult to use in breeding programs as exotic germplasm, in that recovery of the transgenic trait in an elite background will require multiple cycles of backcrossing and selection. Therefore, the extent to which SCV interferes with the deployment of transgenic lines for breeding purposes will partially depend on the extent and nature of phenotypic alterations and on their heritability.

Numerous studies have documented phenotypic manifestations of SCV in tissue culture-derived barley lines. Ullrich et al. (1991) studied variation in 18 cultivars or breeding lines of barley and observed considerable morphological variation, the frequency of which was dependent on the particular genotype or line. In another field study, protoplast-derived barley plants from the cultivars Igri and Dissa showed significant, negative changes in height, heading date, fertility, spike length, and spikelet density (Kihara et al., 1998). Bregitzer and Poulson (1995) analyzed 30 tissue culture-derived lines from six North American barley cultivars, including Golden Promise and Morex, in replicated field trials. Agronomic performance, as judged by heading date, height, lodging, grain yield, test weight and percentage plump kernels, was significantly reduced in the majority of lines. Further analyses revealed that malting qualities in many of these lines were also negatively impacted (Bregitzer et al., 1995c).

In addition to the apparent mutagenicity caused by the basic *in vitro* culturing process, other aspects of the transformation process, relating to bombardment and selection, likely imposed additional stresses, causing increased SCV in transgenic barley. The original transgenic lines of Wan and Lemaux (1994), carrying *uidA, bar* and a gene encoding the coat protein of barley yellow dwarf virus, were characterized by abnormally slow growth and development during the initial generation and during seed increases in the greenhouse. Poor growth of this kind had not been previously observed in plants of any genotype that had derived from *in vitro* culturing alone (P. Bregitzer, unpublished data) and was early evidence of a significant problem with the quality of the transgenic barley plants.

Further examination of these first transgenic barley lines in field plots provided some disturbing data. Morphological abnormalities were more common in plants derived from transgenic Golden Promise callus lines relative to previous observations of tissue culture-derived Golden Promise plants (Bregitzer and Poulson, 1995). Agronomic data told a similar story. Heading date was delayed, plants were shorter and lower yielding, and the seed weight was reduced in the transgenic lines, relative to those derived from the tissue culture process alone. The agronomic performance of the transgenic plants was more significantly reduced than that for plants derived from nontransgenic tissue culture material. For example, grain yield of Golden Promise assessed in transgenic plants was approximately 50% of the uncultured Golden Promise control in field trials. Study of null-segregants showed the reductions were not from the effects of transgene insertion or expression (Bregitzer et al., 1998). In contrast, Golden Promise plants derived from similar tissue culture methodologies, but in the absence of transformation, had grain yields that were 91% of the uncultured Golden Promise control. Although these data came from different studies, the comparison to a common control and the magnitude of the differences leave little doubt that the transformation process contributed to elevated SCV. These observations are consistent with a similar study of rice, in which control plants and protoplast-derived nontransgenic plants had similar agronomic characteristics. Protoplast-derived transgenic plants, however, were markedly inferior, *e.g.* yield was only 10% of the uncultured control (Schuh et al., 1993).

The aspects of the transformation process that are responsible for inducing additional SCV are as yet uncharacterized. There are a number of potentially mitigating factors specific to the transformation process that might be responsible, such as selection with metabolic inhibitors and the resultant death and release of potentially toxic substances from nontransgenic cells. Observations of an increased frequency of albino plants regenerated from transgenic tissues subjected to selection with bialaphos relative to those not subjected to selection are consistent with the idea that selection increases SCV (Lemaux et al., unpublished data).

The field performance of the transgenic barley lines was grossly inferior to the elite germplasm favored by plant breeders. However, the difficulty or ease with which these plants could be used in a breeding program will be determined not only by their agronomic characteristics, but by the nature and heritability of these characteristics. Two qualitative tissue culture-derived variants in barley, a semi-dwarf phenotype isolated in the Morex background and a chlorophyll-deficient phenotype isolated in the Golden Promise background, show monogenic inheritance with the variant allele being recessive (Bregitzer, unpublished data). Additionally examination of the agronomic characteristics of transgenic-derived lines showed that the $T_2$ and $T_4$ generations were indistinguishable in their heading date, height, yield and seed weight (Bregitzer, et al., 1998), indicating that the observed SCV was heritable in self-pollinated progeny and no amelioration of the SCV with generation advance was observed. Experiments are in progress to study the heritability of SCV in transgenic progeny from crosses of transgenic and nontransgenic parents; however, the impacts of heritable variation, such as that observed by Bregitzer et al. (1998), are obvious. Studies to identify and ameliorate the impacting factors are necessary.

*2. Comparative analysis of genomic stability in plants derived from different target tissues and* in vitro *proliferation processes*

Genome stability of plants derived from *in vitro* culture is of critical importance for the direct applicability of transformation technologies to plant breeding. Analysis of previously published studies of barley plants derived from embryogenic callus showed considerable variation in terms of karyotype, DNA stability and field performance (see Sections IV.C.1, IV.D.1). Collectively, the effects of SCV in transgenic plants are likely to impact the application of genetic engineering technologies to the improvement of commercial germplasm; however, the precise nature of the factors leading to these changes is not well understood.

While there appears to be a relationship between increasing SCV and loss of regenerability, the use of regenerability as a means of analyzing the *in vitro* culture process and attempting to ameliorate its effects is tedious. Therefore, a study was initiated to develop molecular methods which could be used to predict the severity of the effects of *in vitro* culture on plant quality and performance. This study involves a comparison of the methylation state of the genome in uncultured control plants compared to those derived from different *in vitro* culture methods (S. Zhang, S. Zhang, M.-J. Cho, P. Bregitzer and P.G. Lemaux, unpublished results). The culture methods chosen included those which have been used previously to transform barley, i.e. standard embryogenic callus induction from IEs (i.e. Wan and Lemaux, 1994), embryogenic callus from the modified culturing method (Cho et al., 1998; see Section IV.B.3) and *in vitro* shoot meristem cultures (see Section IV.C.3).

Initial attempts to develop a quantitative method, based on molecular hallmarks that might be useful for monitoring the detrimental changes that occur during *in vitro* culture, were made using RAPD analysis. However, as previously noted (Devaux et al., 1993), few changes were observed, indicating that sequence variations did not occur in the genomic regions targeted by the primers used. Subsequently, genomic DNA was analyzed using methylation-sensitive enzymes. Single-plant-derived seed of two cultivars, Golden Promise and Morex, was used as the initial source material for initiating *in vitro* cultures. Seed from a single plant was used to avoid complications during later analyses caused by potential heterogeneity within the cultivar. Two time points were chosen for regeneration of plants, one and three months for the two embryogenic culture methods and three and six months for the shoot meristem cultures.

Specific DNA probes, identified with the help of individual researchers involved in the North American Barley Genome Mapping Project, were used in DNA hybridization analyses on the digested genomic DNA from regenerated plants. In Golden Promise plants, the methylation patterns in the majority of the plants derived from the standard embryogenic callus route were different from the control plants with most of the five probes used to date. The frequency of methylation changes was more marked in plants regenerated at three months than at one month for the embryogenic cultures. In contrast analysis of the results from plants derived from shoot meristem cultures, even at six months, showed that hybridization patterns were relatively stable compared to those deriving from the standard embryogenic culturing procedure. In most Golden Promise plants, the hybridization patterns of the DNA from plants obtained from the shoot meristem cultures were identical to that of single plant-derived seed (Zhang et al., 1998b). Plants derived from the modified culture method (see Section IV.B.2), showed intermediate levels of methylation polymorphism between DNA from plants deriving from shoot meristem culture and the standard embryogenic approach. The lower frequency of methylation polymorphism in the plants derived from the modified method compared to the standard embryogenic method might be due to the fact that in the modified method early conversion of tissue from an embryogenic mode of growth to a more meristematic mode of growth occurs following the intermediate step (see Section IV.B.1., Figure 1B and C). To date, the data analyzed derives from one genotype and a limited number of probes. Additional probes are being used for DNA hybridization analysis of the Golden Promise plants; data on the Morex plants is being gathered. The existence of potential genotype times environment interactions cannot be ruled out until all of this data is collected and fully analyzed.

The frequency of methylation polymorphism increased with time in culture and, as methylation variation increased, the regenerability of the cultures decreased. For example, it was not possible to obtain the desired number of Morex plants using the standard embryogenic culture procedure at the three-month time point; plants were, however, obtained using the other two culture methods. The implication is that the Morex cultures did not initially contain large numbers of totipotent cells and these rapidly lost regeneration potential using the standard embryogenic culture route. It could be that the genome of plants deriving from cultures which were propagated in the meristematic state is more stable than that from plants derived from embryogenic cultures. Therefore totipotent cells do not lose regeneration capacity as rapidly. This is consistent with earlier observations in maize of a greater degree of variability in plants derived from embryogenic cultures compared to those from organogenic cultures (Armstrong and Phillips, 1988).

The reduction in methylation polymorphism might be related to the mode of *in vitro* growth, i.e. tissues undergoing less dedifferentiation might experience less stress, which manifests itself as a lower rate of methylation change. It has been proposed (Phillips et al., 1994) that methylation changes are a major factor in SCV. The lower rate of methylation change in plants deriving from the meristematic cultures is likely to lead to reduced SCV, and therefore to lower impacts on field performance. The correlation between frequency of methylation change and field performance is currently being assessed in field trials of progeny from the regenerated plants derived from the three methods. Preliminary results of agronomic evaluations on $T_1$ progeny were consistent with the methylation data derived from Golden Promise, in that plant height and yield show at least some correlation with the degree of methylation variation. However, there was evidence of genotypic interactions; the trends in agronomic performance of Morex-derived plants were not as clear. Additional studies of agronomic performance and methylation status of these Golden Promise and Morex lines, will be necessary before firm conclusions can be drawn.

The conclusion from the preliminary analyses of methylation polymorphism and agronomic performance data is that the use of shoot meristem cultures or early conversion of embryogenic cultures to a meristematic growth mode might reduce SCV. This represents a departure from many published procedures for *in vitro* culturing and transformation since much effort has been expended on developing and characterizing embryogenic cultures of cereals as useful targets for transformation efforts (e.g. Vasil, 1987). These data on methylation stability and SCV suggest that the focus justifiably might shift to cultures which are propagated in a more meristematic state. In addition the availability of molecular tools, which can both define the meristematic state, compared to the embryogenic state (see Section IV.C), and which can be used to monitor the effects of *in vitro* culture on genomic stability and possibly field performance, makes this a feasible approach, which can be pursued for other crop species.

A substantial increase in SCV occurs in response to transformation, documented to date in field tests of transformed rice and barley plants (see

Section IV.D.1). Whether or not the meristematic cultures will produce transgenic plants that show a marked reduction in methylation variation, and presumably in agronomic variation, remains speculative, pending further experimentation. The observed increase in SCV indicates that aspects of the transformation process are also involved in increasing the severity of SCV. Although the nature of these impacts is unknown, they could include bombardment and selection stress. If increased frequency of methylation changes is found to correlate with decreased field performance in the current study, it would then be possible to use methylation polymorphism as a tool for analyzing the effects of particular aspects of the transformation process on SCV. Once negative factors are identified, changes can be made in the transformation protocol in an attempt to ameliorate these effects.

*3. Effects of the state of in vitro-cultured tissue on stability of transgenes and transgene expression*

Stable physical transmission and expression of transgenes are critical for efficient application of transformation technologies to plant breeding. However, the use of current transformation methods, mediated by either T-DNA or direct gene transfer, often leads to plants in which the transgene itself or its expression is unstable (for review, Finnegan and McElroy, 1994). Evidence of both physical loss of the transgene or transgene expression instability has been well-documented in barley. In the original report of stable transformation, 69% (24/35) of the $T_0$ plants, representing 21 independent lines, did not show Mendelian inheritance of expression in $T_1$ progeny (Wan and Lemaux, 1994). In addition, evidence of loss of the actual transgenes in $T_1$ or later generation progeny has also been documented in a subset of the events (T. Koprek, R. Williams-Carrier, R. Fessenden and P.G. Lemaux and P. Bregitzer, unpublished results). The mechanisms involved in the physical loss (partial or complete) or amplification of transgenes in later generation plants are not well characterized. However, mechanisms to explain transgene expression variability have been put forward and these involve, to a large extent, homology-dependent or repeat-induced gene silencing (for review, Matzke and Matzke, 1995). These mechanisms can be post-transcriptional (co-suppression) or transcriptional, often associated with DNA methylation at repeat sequences.

These explanations appear not be adequate to explain all cases of transgene expression variability, such as that seen among clonal progeny of single $T_0$ plants or the differing rates at which variation occurs among clonal progeny. These explanations are also not consistent with the observed lack of direct correlation between the severity of silencing and the extent of sequence similarity (Conner et al., 1997), with the lack of linear correlation between the number of copies integrated and the severity of transgene silencing (*e.g.* Wan and Lemaux, 1994) or the evidence of endogenous gene silencing in null-segregant progeny of transgenic events (Conner et al., 1997).

Most current plant transformation methods use *in vitro*-proliferated target cells that incur methylation pattern changes (see Section IV.D.2), chromosome rearrangements/deletions (see Section IV.D.1), ploidy changes (Hang and Bregitzer, 1993; H.W. Choi and M.-J. Cho, unpublished results) and genetic mutations (Phillips et al., 1994). These changes could lead to a breakdown in normal cellular control mechanisms, which can further exacerbate transgene instability. That genomic instability of *in vitro*-cultured plant cells might affect the stability of transgenes, and their expression, is exemplified by a study of the integration site structure in transformants created by direct gene transfer to rice protoplasts (Takano et al., 1997). Transgenes were found in regions with inverted structures and large genomic duplications; very limited sequence homology between plasmid and target DNA implicated illegitimate recombination in the process. Since only three junctions were studied and each contained extensive rearrangements, it was suggested that the process of *in vitro* culturing of the target protoplasts was responsible for predisposing the cells to illegitimate recombination and genomic rearrangements. In addition, in a study on gene targeting, the target loci were deleted in up to 20% of the calli deriving from tobacco leaf protoplasts (Risseeuw et al., 1997). A possible explanation is that T-DNAs became unstable during *in vitro* propagation and underwent deletions. In stressed cells, such as callus, genomic instability occurs at particular loci and these sites become favored targets for T-DNA integration by illegitimate recombination methods. Analysis of the results from these two studies indicates that *in vitro* culturing leads to genomic instability and that using cells with relatively stable genomes might improve transgene stability and expression.

In barley stability of the transgene itself and its expression have been studied in transgenic plants deriving from independently transformed lines generated with the standard embryogenic culture approach (Wan and Lemaux, 1994), using Southern, PCR and *in vivo* biochemical assays to determine the physical presence or expression of the transgenes (P. Bregitzer, R. Williams-Carrier, P.G. Lemaux, unpublished results). Plants from some lines were characterized as showing Mendelian inheritance of the transgene and its expression through four generations, but some lines showed significantly fewer transgenic segregants than expected. In the unstable transgenic lines, as well as in a number of earlier generation plants from other transgenic lines, it is likely that the variation in stability among the different transgenic events is related at least in part to the site of transgene integration. The influence of the site could be mediated by the fact that different genomic methylation patterns are present in embryogenic versus mature plant tissue, leading to a situation where a transgene, integrated into a site not methylated in the embryo, is subsequently methylated in a mature tissue. Another mediator could be the fact that demethylation can occur during *in vitro* culture (Phillips et al., 1994). This could lead to a situation in which a transgene integrates into a site that has been demethylated in the $T_0$ plant, for example; however, the site becomes remethylated in subsequent generations. The propensity for and speed with which any one site will undergo demethylation and remethylation could be different in different transgenic plants, leading to variation in the propensity of an integrated transgene to undergo silencing due to methylation. In addition, variable rates of remethylation of sequences (demethylated in response to *in vitro* culture) among progeny plants could lead to variability in transgene silencing among progeny plants. A final explanation may be rooted in a more traditional interpretation of variable gene expression, that of epistatic allelic interactions. For instance, heterozygous and heritable alterations in methylation patterns have been observed in regenerated maize plants at multiple genomic sites (Kaeppler and Phillips, 1993a). Progeny of such plants will necessarily segregate, leading to variability for heritable methylation patterns among progeny plants. Epistatic interactions between the alternately methylated forms of certain alleles, and transgenes, conceivably could lead to variable expression of transgenes among progeny derived from the same transgenic plant. One or more of these mechanisms could result in the observed variability of transgene expression among $T_0$-derived progeny (Wan and Lemaux, 1994; Zhang et al., 1996a).

Analyses must be carried out to look at the characteristics of transgene integration sites in lines that are stable over multiple generations versus those showing physical or expression instability. A correlation between the degree of site methylation and the stability or instability of the transgene and its expression could be useful in predicting transgene stability. In barley, these analyses could take advantage of the fact that cereal genomes are known to contain large percentages of repeat DNA (80% in barley), probably generated through non-random amplification of DNA segments that are hypermethylated (Moore, 1995). This highly repeated, hypermethylated DNA is interspersed with undermethylated sequences which contain genes. Certain of these repeated elements in the barley genome are amplifiable, prone to methylation and localized to distinct restriction fragment size classes (Moore et al., 1991). It has also been noted that recombination in large cereal gemones is predominantly confined to regions distal to the centromere (Moore et al., 1991). Stability of the transgene and its expression could be compared to the proximity of the transgene to repeat DNA and to the centromere. A correlation between the stability of the transgene and the characteristics of its integration site could enable the development of diagnostic tools that are useful in predicting the stability of the transgene and its expression in early-stage transformants.

The use of different *in vitro* manipulation methods can lead to different levels of genomic instability in *in vitro*-cultured plant cells based on our data looking at methylation polymorphism (see Section IV.D.2). It is possible that the use of transformation target tissues with more stable genomes will lead to greater stability of the transgene and its expression. Although this will not ameliorate all instances of transgene instability, it is likely to provide incremental improvement. Preliminary data from both transgenic barley (S. Zhang, M.-J. Cho, T. Koprek, unpublished results) and oat (S. Zhang and M.-J. Cho, unpublished results, see Somers et al., this volume) supports this hypothesis.

## V. Utilization of Transformation Systems for Barley

Despite the hurdles, which remain to optimize the utility of barley transformation (see Section IV.A), a number of projects have been initiated which utilize these technologies either to study basic physiological, biochemical or genetic processes or to create improved barley cultivars. The extent to which existing genetic engineering technologies will be useful depends upon the nature of the application.

### A. *Transient Expression Systems*

The hurdles to the application of the technology will not affect their utility for transient assay systems, which do not depend on stable integration or stability of transgenes or their expression. A transient assay was developed that can be used to assess resistance strategies to, for example, the fungal pathogen *Erysiphe graminis* f.sp. *hordei* (Nelson and Bushnell, 1997). Epidermal cells of barley coleoptiles were bombarded with host genes, believed to be involved in the response to the fungal pathogen. Cells receiving DNA were identified using an anthocyanin reporter gene and these cells could be monitored for effects of transient expression of the potential resistance genes on fungal development. A similar strategy was used in developing a transient assay to assess strategies for gene tagging in barley (McElroy et al., 1997). Scutellar cells of immature barley embryos were bombarded with several genes, the *uidA* gene interrupted by the defective maize transposable element, *Ds*, the gene for the transposase responsible for the movement of the Ds element, and the maize anthocyanin gene. To determine efficiency of excision of *Ds* from *uidA*, the number of cells synthesizing anthocyanin were counted, followed by histochemical GUS staining and counting of cells expressing GUS. Using this assay, it was determined that the maize *Ds* element is capable of excision in barley cells, thereby setting the stage for the creation of barley lines in which gene tagging can be done. Similar studies are being conducted at the Technical University of Munich with another maize transposable element, *En/Spm* (T. Golds, personal communication).

### *B. Expression of Barley Genes in Heterologous Species*

In addition to transient assays, the functionality and utility of barley genes can be studied in heterologous systems prior to stable introduction into barley. For example, Kristensen et al., (1997) used transgenic tobacco plants to demonstrate that a leaf peroxidase from barley, induced by fungal pathogens, was expressed in transgenic plants; however, growth retardation and no enhanced resistance were observed in the transgenic plants. In another study Xu et al. (1996) generated stably transformed rice plants expressing the barley LEA protein, HVA1 and demonstrated that they had a higher tolerance to water deficiency and salinity and maintained higher growth rates than control plants under stress conditions. In another study, genes involved in fructan synthesis in barley were isolated. Fructans are an important storage carbohydrate in many plant species and also appear to play a role in protection of the plant to a variety of stresses, e.g. drought, cold and nutrient. The isolated fructan biosynthetic genes from barley were used to transform tobacco and chicory, and their expression led to the formation of branched fructans of the graminan type in tobacco, a species otherwise unable to make these sugars, and in chicory, a plant naturally producing only unbranched fructans, rather than the branched type (Sprenger et al., 1997).

Transgenic plants from heterologous species can also be used to dissect the regulatory regions from barley genes. Raho et al., (1996) used transgenic tobacco to dissect the properties of the 5′ upstream region of the barley heat shock gene, *Hvhsp*17 gene by fusing it to the screenable marker gene *uidA*. The promoter was found to respond to heat, to be tissue-specific and to be induced by abscissic acid and certain metal ions. Grosset et al., (1997) showed that the promoter for a barley $\alpha$-amylase inhibitor subunit gene was unable to direct the synthesis of the marker gene, *uidA*, in transgenic tobacco seed but was able to direct expression in bombarded maize endosperm in a tissue-specific and developmentally regulated manner. Fernandez et al., (1993) used transgenic tobacco expressing GUS under the control of the promoter of the barley $\alpha$-thionin gene to demonstrate temporal and tissue-specific control of the promoter in endosperm tissue. In addition, Marris et al., (1988) showed that an endosperm-specific $B_1$-hordein promoter of barley directs tissue- and development-specific chloramphenicol acetyl transferase expression in transgenic tobacco seeds. Endosperm-specific GUS expression of this promoter was also demonstrated in transient assays performed in IEs of barley (Cho and Lemaux, 1997; Cho et al., 1998a,c).

### *C. Stable Transformation of Barley*

While transient assays or stable transformation of heterologous species can be used to answer some questions, the use of stably transformed barley plants is necessary in many cases to either bring a concept to commercial application or to answer fundamental questions about the biochemistry, physiology or development of barley. In order to answer biochemical or developmental questions, engineered plants often need to be of a particular genotype, *e.g.* in order to perform complementation in a specific mutant background. To derive optimal value for improvements in agronomic performance or to change food or feed characteristics, it would be optimal to modify commercial germplasm, rather than to undertake the time-consuming and sometimes impossible task of backcrossing into commercial cultivars and maintaining transgene expression (see Section IV.A). In other cases the particular cultivar or its phenotype is not so critical, for example, the use of transgenic barley to produce value-added products.

### *1. Pest resistance*

In the area of pest resistance, there are a multitude of potential targets for genetic engineering strategies. To date, only limited work has been published in this area. Barley yellow dwarf virus, transmitted by aphids, is one of the most prevalent viruses of small grains, causing significant losses in the United States and Australia (Lister RM, 1995). The coat protein gene of this luteovirus, pathovar P-PAV, was stably introduced into Golden Promise, a susceptible cultivar (Wan and Lemaux, 1994). Eight plants from six lines showed moderate to high levels of resistance to the P-PAV isolate and this resistance correlated with the presence of the coat protein transgene (McGrath et al., 1997). Some $T_2$ plants, derived from resistant parents containing the coat protein gene, were also found to be highly resistant; field tests of this germplasm have not been conducted. Yellow mosaic disease of barley is caused by two viruses, barley yellow mosaic virus and barley mild mosaic virus, which are transmitted by the soil-borne fungus, *Polymyxa graminis*. The virus causes significant yield losses in winter barley in northern Europe and Asia. Although natural sources of resistance are available for both barley yellow dwarf and yellow mosaic viruses, the resistance is not complete and/or durable in some cases and new strains are emerging, which overcome the natural resistance. Genetic engineering approaches are aimed at enhancing the natural resistances. Attempts are being made at the University of Hamburg (H. Lörz, personal communication) and Plant Breeding International (R. Stratford, personal communication) to engineer resistance to yellow mosaic disease; however, successful demonstration of resistance has not been documented. Transformation of barley has been undertaken with a gene from grape, *Vitis vinifera* L., which is responsible for the synthesis of the stilbene-type phytoalexin, resveratrol, which is made in a variety of plants upon attack by pathogens and appears to have antifungal activity. Enhanced expression of this gene, under the control of the stilbene synthase promoter, was observed in $T_1$ progeny of barley and analysis of the pathology in these plants following infection with the fungal pathogen, *Botrytis cinerea*, showed decreased damage due to the pathogen (Leckband and Lörz, 1997).

To date a number of genes have been identified and mapped through efforts such as those of the North American Barley Genome Mapping Project (Hayes et al., 1996). Efforts are underway to clone the mapped genes involved in agronomic, pest-resistance and quality traits. As these genes become available, additional opportunities will arise for utilizing transgenic barley to assess resistance strategies. These approaches will be aided by the observed synteny in cereal genomes (Devos et al., 1995), which allows quick movement from a gene identified in one cereal species to its counterpart in another. In addition genes identified in one cereal (wild or cultivated) can be used directly for approaches in barley. In fact, given the fundamental similarities in the biochemical nature of many characteristics in plants and their pests, successful approaches identified in dicotyledonous species should prove useful in engineering improvements in barley.

### 2. *Nutritional quality*

Genetic engineering can be used to modify the nutritional quality of feed barley. Conventional breeding and utilization of mutations introduced through classical breeding have improved feed quality to a certain extent, for example increasing its lysine content (Munck, 1993) and decreasing the phytic acid content of the barley grain (V. Raboy, personal communication). Through detailed studies of the developing barley grain, a detailed understanding of the biochemical and molecular basis of the nutritional quality was developed; it has also provided genes that encode key proteins that determine quality. Shewry et al., (1994) showed that certain of these genes could be used in transgenic tobacco and in transient expression assays in cereal protoplasts to change the outcome of the biochemical pathways in the seed. More recently, Brinch-Pedersen et al., (1996) introduced into barley the *E. coli* genes encoding lysine feed-back-insensitive forms of two enzymes involved in the synthesis of lysine and methionine. Elevated levels of the two enzymes were measured in $T_1$ plants and mature $T_1$ seeds of the transgenics showed a two-fold increase in free lysine, arginine and asparagine; no differences were observed in the composition of total amino acids. While no attempts were made to optimize the expression of the bacterial genes by modifying codon preferences, this study shows that using bacterial genes might be of utility in modifying the levels of amino acids in barley seeds.

In order to modify specifically the characteristics of the grain, it is optimal to target expression of transgenes specifically to the barley grain. Studies have been conducted in transgenic barley to express GUS and wheat thioredoxin (Gautier et al., 1998), driven by the B- and D-hordein promoters (Brandt *et al.,* 1985; Sorensen et al., 1986); to date expression of GUS has been localized specifically in the endosperm tissue of developing barley grains (Cho et al., 1998a,c; M.-J. Cho, J.H. Wong, E.C. Marx, W. Jiang, B.B. Buchanan and P.G. Lemaux, unpublished results).

### 3. *Malting and brewing characteristics*

Most classical breeding efforts in barley have been aimed toward the improvement of malting cultivars, partly due to the higher value of this commodity as a crop. Malting is one of the most ancient of processes and its biochemical and physiological nature is relatively well understood (McElroy and Jacobsen, 1995). Malting consists of steeping dry barley seeds, followed by their controlled germination. During this time there is an activation of the metabolic apparatus of the embryo, particularly the aleurone layer, resulting in activation or synthesis of cell wall-, starch- and protein-degrading enzymes. The cell wall-degrading enzymes, termed $\beta$-glucan endohydrolases, make the nutritional reserves of the endosperm accessible during later fermentation, increasing extractability and fermentability of the malt. They also remove the viscous (1-3, 1-4)-$\beta$-glucans, which can cause problems during the filtrations steps. Hydrolytic degradation is terminated by heating and drying the malt, which is then hydrated and subjected to a variety of temperatures. This is the step during which most of the hydrolytic breakdown of starch occurs. Because of the heating phases, the hydrolytic enzymes are often inactivated before they complete their degradative role, causing problems during the subsequent filtration step after which the 'mash' is boiled, mixed with hops and used for fermentation.

Because this process is so well understood at the biochemical level and many of the genes involved in the process have been isolated, modifying these biochemical processes is an ideal goal for genetic engineering. These efforts are just beginning. One attempt at the biochemical modification of malting properties involved the introduction of a hybrid bacterial gene from *Bacillus* sp., encoding a thermostable (1,3–1,4)-$\beta$–glucanase which had been modified to match the codon usage of the barley (1,3–1,4)-$\beta$-glucanase isoenzyme EII gene (Jensen et al., 1996). Both the modified and unmodified bacterial genes were fused to a barley high-pI $\alpha$-amylase signal peptide which was placed downstream of the barley (1,3–1,4)-$\beta$-glucanase isoenzyme EII promoter; activity from the modified genes was demonstrated in barley aleurone protoplasts. In addition, the modified gene was cloned downstream of the barley high pI $\alpha$-amylase promoter and signal peptide and fertile plants were regenerated. Approximately 75% of the grains harvested from the $T_0$ plants synthesized the thermostable enzyme during germination.

Efforts of another group were also aimed at using a heterologous source of the gene to engineer a thermostable $\beta$-glucanase (Mannonen et al., 1997). A thermotolerant $\beta$-glucanase gene from the fungus, *Trichoderma reesei*, was engineered into the Finnish cultivar Kymppi. Analysis of preliminary results showed that the enzyme was active and thermotolerant when extracted from transgenic seeds and assayed *in vitro*; however, the molecular mass and isoelectric point of the bacterial protein appeared to be modified either by proteolytic cleavage in barley cells during processing of the protein or during extraction of the proteins from the seeds. *In vitro* assays indicated that the enzyme was active under conditions used during the mashing process and that the levels of β-glucanase in the transgenic seed was about 0.025% of the total soluble protein. Although the levels of enzyme were low, the amount appeared to be sufficient to reduce the soluble glucans during mashing to a level that improve filterability of the final wort.

Studies at the University of Adelaide also focus on enhancing the synthesis of β-glucan endohydrolases but in this approach they utilized endogenous barley genes (Stewart et al., 1997). Protein engineering strategies were developed to increase the thermotolerance of the barley (1-3, 1-4)-β-glucanase. A variety of mutations were introduced into the (1-3,1-4)-β-glucanase protein, based on the three-dimensional structure of the protein and three mutant proteins showed increased thermostability at elevated temperatures. The genes are currently being stably introduced into barley and the thermostability of the β-glucanase activity will be measured.

Kihara et al., (1997) have used an engineered gene for barley β-amylase that showed increased thermostability when isolated from *E. coli* (Kihara, et al., 1997). This enzyme plays a critical role in starch breakdown in the barley grain and is particularly heat labile. Stably transformed barley plants containing the gene were generated and the seeds of some of these plants had more thermostable β-amylase activity.

In separate studies, efforts are underway at the John Innes Institute to introduce a bacterial glucoamylase gene, aimed at modifying malting qualities (W. Harwood, personal communication), and at the Crop Development Centre and Plant Biotechnology Research Institute in Saskatoon Saskatchewan to insert two different antisense constructs for the wheat starch branching enzyme 1 (R. Chibbar, personal communication).

### 4. *Production of value-added materials*

In addition to its use as a food and feed source, barley starch can also be used as an important raw material for industrial application, both for food and non-food purposes. Of particular interest is the use of starch as a non-petroleum chemical feed stock for the manufacture of biodegradable polymers, such as plastics and as a non-cellulose stock for the paper industry. Since starch is not only totally biodegradable but also inexpensive, renewable and readily available, the replacement of certain petroleum products with starch-based products has spurred interest in usage of plant starches as an industrial raw material. Since barley storage and handling characteristics are well understood and superior to other potential starch sources, like potato, efforts to bioengineer barley to improve it as a starch source are underway at Stockholm University (C. Jansson, personal communication). This approach to modifying the starch composition of barley through the starch-branching enzymes is focused on the central enzymes and genes involved in starch synthesis in barley endosperm, *e.g.* $\alpha$-1,4-glucan, 6-glucosyl transferase, which catalyzes the formation of $\alpha$-1,6 glycosidic bonds in amylopectin. The genes and promoters for this enzyme have been cloned and the promoters are currently being analyzed.

### 5. *Fundamental studies*

Perhaps due to the sizable effort needed to generate transgenic barley, little work has been done using transgenic barley as a system to study basic plant biology; however, as transformation systems become more 'user-friendly' and amenable to the academic laboratory, interest in fundamental studies wll increase. The use of transgenic plants to study fundamental aspects of plant growth and development can provide new insights which are more difficult, and perhaps intractable, using classical analytical methods.

An example of the power of the technology is seen in a recent paper by Williams-Carrier et al., (1997). This report described the use of transgenic barley to study the effects of ectopic expression of the maize homeobox gene, *knotted.* The ectopic expression of this cDNA gene was found to phenocopy a naturally occurring mutant of barley, *Hooded*, in producing ectopic inflorescence on the barley awn. In addition new insights were gained on the regulation of these pivotal developmental genes. Although expression of the knotted protein was driven by a constitutive promoter, maize *ubiquitin*, which supports expression of GUS throughout the barley plant including the awn, the product of *knotted* and its mRNA were found only in very defined areas of the inflorescence, which coincided with the formation of ectopic meristems. In this case the effects of the *knotted* promoter and any of its introns as participating in this regulation could be dismissed. The implication from these studies is that in transgenic barley the expression of the Knotted protein is being regulated by a very specific, post-translational mechanism which likely is involved in its regulation in maize and in the regulation of the barley homologue(s).

In another study Kjærulff (1995) used an antisense construct for the PSA-E polypeptide of the photosystem I complex of barley under the control of the maize *ubiquitin* promoter. The level of mRNA for this protein, which facilitates ferredoxin reduction and possibly the cyclic electron transport around photosystem I, was reduced by 62% in transgenic barley indicating that antisense approaches can be used in barley and that such approaches might be useful for studying and manipulating the photosynthetic process.

The NADP/thioredoxin system plays a role in the mobilization of storage proteins and in activating and inactivating enzymes and proteins during the germination of cereals. The thioredoxin *h* gene has been introduced into barley under the control of endosperm-specific promoters. The effects of the expression of the thioredoxin gene in the endosperm are being studied, as it relates to the redox state of proteins in the endosperm and the properties of the seed (M.-J. Cho, J.H. Wong, E.C. Marx, W. Jiang, B.B. Buchanan and P.G. Lemaux, unpublished results).

Two types of studies have been undertaken which specifically address issues of concern with transgenic plants. The first involves the removal of unwanted transgenes, such as selectable markers. There are several systems that are currently being developed to accomplish this, including the bacterial recombinational systems *cre-lox* (e.g. Bayley et al., 1992) and *flp-frt* (*e.g.* Lyznik et al., 1993). These systems permit the removal of transgenes through recombinational processes mediated by the presence of specific small recognition sequences (*lox, frt*) and the action of the cognate recombinational enzyme (*lox, flp*). Another possible mechanism for separating the transgene(s) of choice from selectable marker genes utilizes transposable elements from other species, *e.g.* the maize transposable elements, *Ac/Ds* and *Spm*. In this strategy the gene of interest is placed inside the recognition sites for the transposition enzyme; the selection gene remains outside the recognition sites. In the presence of the transposase, the gene of interest can be moved away from the remainder of the introduced DNA allowing the desired transgene to be physically separated from the remainder of the introduced DNA during subsequent generations. This system is currently being developed in barley using maize *Ac/Ds* (McElroy et al., 1997; T. Koprek and P.G. Lemaux, unpublished data).

The second area of study relating to transgenic plants involves transgenic pollen dispersal under field conditions. Mannonen et al., (1998) assessed the dispersal distance using a homozygous transgenic line as the pollen source and three open-flowering, cytoplasmically male-sterile barley lines as recipients. Preliminary analysis of the results from this experiment showed that in the dominant wind direction cross-pollination occurred out to as far as 50 meters and that the number of seed from cross-pollinated plants decreased with the distance according to a logarithmic model. Proof that the plants were pollinated by the transgenic lines has not yet been completed. It should be noted, however, that male-sterile lines of barley are not normally present in the field and that the frequency of outcrossing in male-fertile barley in the field is less that 0.2% (Fehr and Hadley, 1980) and no wild relatives of barley exist in the continental US with which it can outcross (von Bothmer et al., 1991).

## VI. Future Prospects for Production and Use of Transgenic Barley

Genetic transformation has enormous potential for producing novel and useful cultivars of all crop species, including barley. Transformation technologies also have the potential to reduce dependence on expensive and environmentally destructive synthetic chemicals, to enhance the nutritional and commercial utility of barley, and to enable basic biochemical and physiological studies. To maximize this potential, it must be possible to introduce DNA efficiently without regard to the genetic background of the target tissue, without causing unintended alterations in endogenous gene expression, and in such a manner that endogenous DNA is stably expressed and inherited.

Current technologies for barley transformation provide the basic framework for genetic manipulation via transformation and are useful for certain objectives. Successful production of transformed barley plants has been accomplished in several labs by a variety of methods (see Sections III and V). Transformation systems based on IE-derived embryogenic callus have been the most successful, as measured by the number of transformed plants produced and the number of genotypes from which transgenic plants have been obtained. However, these published methods fall far short of optimal utility in many respects, and serious problems related to regenerability, somaclonal variation and stable transgene expression remain.

Successful production of transgenic barley has depended on the retention and expression of totipotency by transgenic tissues regardless of the particular culture system used. Rapid loss of regenerability from cultured barley tissue has made the development of transformation protocols particularly challenging and, in fact, totipotency appears to be a critical weak link in systems for barley transformation. Even in the most efficient *in vitro* culture systems, the vast majority of cells do not give rise to plants. Improvements have been made by amending particular nutritional and hormonal components of the growth media, by excising scutella or by developing alternative tissue culture systems, such as shoot meristem cultures. However, continued progress will probably depend on an increased understanding of the genetic and developmental phenomena that occur *in vitro*.

Loss of totipotency and the generation of somaclonal variation are both likely related to genomic disturbances that occur *in vitro*. The most severe alterations, such as the loss of a chromosome, are surely incompatible with viability in a diploid species such as barley. Many other aberrations, alone or in combination, could be expected to disrupt the physiological functions necessary for redifferentiation. Less severe alterations may not prevent plant regeneration but may result in plants with a variety of phenotypic alterations, presumably a result of changes in endogenous gene expression. Such changes complicate subsequent breeding efforts and basic studies, and may be partly responsible for problems with transgene and transgene expression stability and inheritance.

Solutions to these problems are not immediately apparent, but there are a number of clues which suggest reasonable research directions. Developing an understanding and respect for the limitations of current transformations technologies, and for the implications that these limitations have on the utility of the transformation systems and the resulting transgenic plants, is the first step toward solutions to the problems. It has been a relatively short time since barley transformation passed from a theoretically possible to a realized goal. Further progress will be dependent on a willingness to recognize and eliminate the remaining obstacles, which currently limit the applications of transformation technologies.

## Acknowledgements

The authors thank their many 'barley colleagues' who shared ideas and data prior to publication; Drs. Dennis Dolan and Dave Thomas, Coors Brewing Company and Mr. Scott Heisel, American Malting Barley Association, for providing helpful information; Mr. Frank Hagie, Applied Phytologics, Inc., for careful reading of aspects of the manuscript; Ms. Rosalind Williams-Carrier for help with the graphics and Ms. Barbara Alonso for excellent word-processing help.

## References

Ahokas, H. (1989) Transfection of germinating barley seeds electrophoretically with exogenous DNA. Theor. Appl. Genet. 77: 469–472.

Armstrong, C.L, and Phillips, R.L. (1988) Genetic and cytogenetic variation in plants regenerated from organogenic and friable embryogenic tissue cultures of maize. Crop Sci. 28: 363–369.

Bayley, C.C., Morgan, M., Dale, E.C., and Ow, D.W. (1992) Exchange of gene activity in transgenic plants catalyzed by the *Cre-lox* site-specific recombination system. Plant Mol. Biol. 18: 353–362.

Bechtold, N., Ellis, J., and Pelletier, G. (1993) In planta *Agrobacterium*-mediated gene transfer by infiltration of adult *Arabidopsis thaliana* plants. C. R. Acad. Sci. Ser. III 316: 1194–1199.

Biddington, N.L. (1992) The influence of ethylene in plant tissue culture. Plant Growth Reg. 11: 172–187.

Bilang, R., Zhang, S., Leduc, N., Iglesias, V.A., Gisel, A., Simmonds, J., Potrykus, I., and Sautter, C. (1993) Transient gene expression in vegetative shoot apical meristems of wheat after ballistic microtargeting. Plant J. 4: 735–744.

Brandt, A., Montembault, A., Cameron-Mills, V., and Rasmussen, S.K. (1985) Primary structure of a B1 hordein gene from barley. Carlsberg Res. Comm. 50: 333–345.

Bregitzer, P. (1992) Plant regeneration and callus type in barley: effects of genotype and culture medium. Crop Sci. 32: 1108–1112.

Bregitzer, P., Campbell, R.D., and Wu, Y. (1995a) *In vitro* response of barley (*Hordeum vulgare* L) callus: effect of auxins on plant regeneration and karyotype. Plant Cell Tiss. Org. Cult. 43: 229–235.

Bregitzer, P., Campbell, R.D., and Wu, Y. (1995b) Plant regeneration from barley callus: effects of 2,4-dichlorophenoxyacetic acid and phenylacetic acid. Plant Cell Tiss. Org. Cult. 43: 229–235.

Bregitzer, P., and Poulson, M. (1995) Agronomic performance of barley lines derived from tissue culture. Crop. Sci. 35: 1144–1148.

Bregitzer, P., Poulson, M., and Jones, B.L. (1995c) Malting quality of barley lines derived from tissue culture. Cereal Chem. 72: 433–435.

Bregitzer, P., Halbert, S.E., and Lemaux, P.G. (1998) Somaclonal variation in the progeny of transgenic barley. Theor. Appl. Genet. 96: 421–425.

Brinch-Pedersen, H., Galili, G., Knudsen, S., and Holm, P.B. (1996) Engineering of the aspartate family biosynthetic pathway in barley (*Hordeum vulgare* L.) by transformation with heterologous genes encoding feedback-insensitive aspartate kinase and dihydrodipicolinate synthase. Plant Mol. Biol. 32: 611–620.

Chan, M., Lee, M.T., and Chang, H. (1992) Transformation of indica rice (*Oryza sativa* L.) mediated by *Agrobacterium*. Plant Cell Physiol. 33: 577–583.

Chan, M., Chang, H., Ho, S., Tong, W., and Yu, S. (1993) *Agrobacterium*-mediated production of transgenic rice plants expressing a chimeric $\alpha$-amylase promoter/$\beta$-glucuronidase gene. Plant Mol. Biol. 22: 491–506.

Chen, D.F., and Dale, P.J. (1992) A comparison of methods delivering DNA to wheat: the application of wheat dwarf virus DNA to seeds with exposed apical meristems. Transgen. Res. 1: 93–100.

Cheng, M., Fry, J.E., Pang, S., Zhou, H., Hironaka, C.M., Duncan, D.R., Conner, T.W., and Wan, Y. (1997). Genetic transformation of wheat mediated by *Agrobacterium tumefaciens*. Plant Physiol. 115: 971–980.

Chiu, W.L., Niwa, Y., Zeng, W., Hirano, T., Kobayashi, H., and Sheen, J. (1996) Engineered GFP as a vital reporter in plants. Curr. Biol. 6: 325–330.

Cho, M.-J., Choi, H.W., Buchanan, B.B., and Lemaux, P.G. (1998a) Inheritance of tissue-specific expression of *uidA* in transgenic barley plants. Theor. Appl. Genet. (accepted).

Cho, M.-J., Jiang, W., and Lemaux, P.G. (1998b) Transformation of recalcitrant cultivars through improvement in regenerability and decreased albinism. Plant Science, in press.

Cho, M.-J., Ha, C.D., Buchanan, B.B., and Lemaux, P.G. (1998c) Subcellular targeting of barley hordein promoter-*uidA* fusions in transgenic barley seed. 1998 Cong. In Vitro Biol., p. 1023.

Cho, M.-J., and Lemaux, P.G. (1997) Rapid PCR amplification of chimeric products and its direct application to the *in vivo* testing of recombinant DNA construction strategies. Mol. Biotech. 8: 13–16.

Cho, U., and Kasha, K.J. (1989) Ethylene production and embryogenesis from anther cultures of barley (*Hordeum vulgare* L.). Plant Cell Rep. 8: 415–417.

Christensen, A.H., and Quail, P.H. (1996) Ubiquitin promoter-based vectors for high-level expression of selectable and/or screenable marker genes in monocotyledonous plants. Transgen. Res. 5: 1–6.

Conner, J.A., Stein, T., Tantikanjana, C., Kandasamy, M.K., Nasrallah, J.B., and Nasrallah, M.E. (1997) Transgene-induced silencing of *S*-locus genes and related genes in *Brassica*. Plant J. 121: 809–823.

Czernilofsky, A.P., Hain, R., Herrera-Estrella, L., Goyvaerts, E., Baker, B.J., and Schell, J. (1986) Fate of selectable marker DNA integrated into the genome of *Nicotiana tabacum*. DNA 5: 101–113.

Dahleen, L.S. (1996) Plant regeneration from barley immature embryos improved by increasing copper levels. Plant Cell Tiss. Org. Cult. 43: 267–269.

Day, A., and Ellis, T.H.N. (1985) Deleted forms of plastid DNA in albino plants from cereal anther culture. Curr. Genet. 9: 671–678.

Devaux, P., Kilian, A., and Kleinhofs, A. (1993) Anther culture and *Hordeum bulbosum*-derived barley doubled haploids: mutations and methylation. Mol. Gen. Genet. 241: 674–79.

Devos, K.M., Moore, G., and Gale, M.D. (1995) Conservation of marker synteny during evolution. Euphytica. 85:367–372.

DeWet, U.R., Wood, K.V., DeLuca, M., and Helinski, D.R. (1987) Firefly luciferase gene:

structure and expression in mammalian cells. Mol. Cell Biol. 7: 725–737.
Dunford, R., and Walden, R.M. (1991) Plastid genome structure and plastid-related transcript levels in albino barley plants derived from anther culture. Curr. Genet. 20: 339–347.
Eisinger, W. (1977) Role of cytokinins in carnation flower senescence. Plant Physiol. 59: 707–709.
Evans, J.M., and Batty, N.P. (1994) Ethylene precursors and antagonists increase embryogenesis of *Hordeum vulgare* L. anther culture. Plant Cell Rep. 13: 676–678.
FAO (1992) FAO Yearbook 1992. 46: 71–93.
Fehr, W.R., and Hadley, H. (1980) Hybridization of crop plants. Amer. Soc. Agron. Crop. Sci., Madison, WI.
Fernandez, J.A., Moreno, M., Carmona, M.J., Castagnaro, A., and Garcia-Olmedo, F. (1993) The barley alpha-thionin promoter is rich in negative regulatory motifs and directs tissue-specific expression of a reporter gene in tobacco. Biochim. Biophy. Acta. 1171: 346–348.
Finer, J.J., Vain, P., Jones, M.W., and McMullen, M.D. (1992) Development of the particle inflow gun for DNA delivery to plant cells. Plant Cell Rep. 11: 323–328.
Finnie, S.J., Powell, W., and Dyer, A.F. (1989) The effect of carbohydrate composition and concentration on anther culture response in barley (*Hordeum vulgare* L.) Plant Breed. 103: 110–118.
Finnegan, J., and McElroy, D. (1994) Transgene inactivation: plants fight back! Bio/Technology 12: 883–888.
Flavell, R.B. (1994) Inactivation of gene expression in plants as a consequence of specific sequence duplication. Proc. Nat. Acad. Sci. USA. 91: 3490–3496.
Fletcher, R.A. (1969) Retardation of leaf senescence by benzyladenine in intact bean plants. Planta. 89: 1–8.
Funatsuki, H., Lazzeri, P.A., and Lörz, H. (1992) Use of feeder cells to improve barley protoplast culture and plant regeneration. Plant Sci. 85: 251–254.
Funatuski, H., Kuroda, M., Lazzeri, P.A., Müller, E., Lörz, H., and Kishinami, I. (1995) Fertile transgenic barley generated by direct DNA transfer to protoplasts. Theor. Appl. Genet. 91: 707–712.
Gasser, C.A., and Fraley, R.T. (1989) Genetically engineering plants for crop improvement. Science. 244: 1293–1299.
Gautier, M., Lullien-Pellerim, V., deLamotte-Guéry, F., Guimo, A., and Joudrier, P. (1998) Characterization of wheat thioredoxin *h*: cDNA and production of an active *Triticum aestivum* protein in *Escherichia coli*. Eur. J. Biochem. 252: 314–324.
Gengenbach, B.G., Green, C.E., and Donovan, C.M. (1977) Inheritance of selected pathotoxin resistance in maize plants regenerated from cell cultures. Proc. Nat. Acad. Sci. USA. 74: 5113–5117.
Ghaemi, M., Sarrafi, A., and Alibert, G. (1994) The effect of silver nitrate, colchicine, cupric sulfate and genotype on the production of embryoids from anthers of tetraploid wheat (*Triticum turgidum*). Plant Cell Tiss Org. Cult. 36: 355–359.
Gordon-Kamm, W.J., Spencer, T.M., Mangano, M.L., Adams, T.R., Daines, R.J., Start, W.G., O'Brien, J.V., Chambers, S.A., Adams, W.R. Willetts, N.G., Rice, T.B., Mackey, C.J., Krueger, R.W., Kausch, A.P., and Lemaux, P.G. (1990) Transformation of maize cells and regeneration of fertile transgenic plants. Plant Cell. 2: 603–618.
Grosset, J., Alary, R., Gautier, M., Menossi, M., Martinez-Izquierdo, J.A., and Joudrier, P. (1997) Characterization of a barley gene coding for an alpha-amylase inhibitor subunit (CMd protein) and analysis of its promoter in transgenic tobacco plants and in maize kernels by microprojectile bombardment. Plant Mol. Biol. 34: 331–338.
Haccius, B. (1978) Question of unicellular origin of non-zygotic embryos in callus cultures. Phytomorphology. 28: 74–81.
Hagio, T., Hirabayashi, T., Machii, H., and Tomotsune, H. (1995) Production of fertile transgenic barley (*Hordeum vulgare* L.) plant using the hygromycin-resistance marker. Plant Cell Rep. 14: 329–334.
Hang, A., and Bregitzer, P. (1993) Chromosomal variations in immature embryo-derived

calli from six barley cultivars. J. Hered. 84: 105–108.

Hänsch, R., Koprek, T., Heydemann, H., Mendel, R.R., and Schulze, J. (1996) Electroporation-mediated transient gene expression in isolated scutella of *Hordeum vulgare*. Physiol. Plant. 98: 20–27.

Hanzel, J.J., Miller, J.P., Brinkmann, M.A., and Fendos, E. (1985) Genotype and media effects on callus formation and regeneration in barley. Crop. Sci. 25: 27–31.

Harlan, J.R. (1979) Barley: Origin, Botany, Culture, Winter Hardiness, Genetics, Utilization, Pests, Agricultural Handbook No. 338, United States Department of Agriculture, Washington D.C., pp. 10–36.

Haseloff, J., Siemering, K.R., Prahser, D.C., and Hodge, S. (1997) Removal of a cryptic intron and subcellular localization of green fluorescent protein are required to mark transgenic *Arabidopsis* plants brightly. Proc. Nat. Acad. Sci. USA. 94: 2122–2127.

Hayes, P.M., Chen, F.Q., Kleinhofs, A., Kilian, A., and Mather, D.E. (1996) Barley genome mapping and its applications in methods of genome analysis in plants. In: Jauhar, P.P. (ed.), Methods of Genome Analysis in Plants, pp. 229–249. CRC Press, Inc.

Hiei, Y., Ohta, S., Komari, T., and Kumashiro, T. (1994) Efficient transformation of rice (*Oryza sativa* L.) mediated by *Agrobacterium* and sequence analysis of the boundaries of the T-DNA. Plant J. 6: 271–282.

Holm, P.B., Knudsen, S., Mouritzen, P., Negri, D., Olsen, F.L., and Roue, C. (1994) Regeneration of fertile barley plants from mechanically isolated protoplasts of the fertilized egg cell. Plant Cell 6: 531–543.

Holm, P.B., Brinch-Pedersen, H., Olsen, O., Jørgensen, A.B., Sandager, L.D., Sørensen, L.D., and Knudsen, S. (1996) Transformation of barley for improved malting quality. Internat. Conf. Agric. Biotec., Saskatoon, Saskatchewan, Canada, pp. 11–14.

Hunter, C.P. (1988) Plant regeneration from microspores of barley, *Hordeum vulgare*. Ph.D. thesis, Wye College, University of London, Ashford, Kent.

Ishida, Y., Saito, H., Ohta, S., Hiei, Y., Komari, T., and Kumashiro, T. (1996). High efficiency transformation of maize (*Zea mays* L.) mediated by *Agrobacterium tumefaciens*. Nature Biotechnology 14: 745–750.

Jackson, D., Veit, B., and Hake, S. (1994) Expression of maize *knotted1* related homeobox genes in the shoot apical meristem predicts patterns of morphogenesis in the vegetative shoot. Development 120: 405–413.

Jähne, A., Becker, D., Brettschneider, R., and Lörz, H. (1994) Regeneration of transgenic, microspore-derived, fertile barley. Theor. Appl. Genet. 89: 525–533.

Jefferson, R.A., Kavanagh, T.A., and Bevan, M.W. (1987) Gus fusions beta glucuronidase as a sensitive and versatile gene fusion marker in higher plants. EMBO J. 6: 3901–3907.

Jensen, L.G., Olsen, O., Kops, O., Wolf, N., and Thomsen, K.K. (1996) Transgenic barley expressing a protein-engineered, thermostable (1,3-1,4)-$\beta$-glucanase during germination. Proc. Nat. Acad. Sci. USA. 93: 3487–3491.

Jiang, W., Cho, M.-J., and Lemaux, P.G. (1998) Improved callus quality and prolonged regenerability in model and recalcitrant barley (*Hordeum vulgare* L.) genotypes. Plant Biotechnology, 15: 63–69.

Jørgensen, A.B. (1996) Barley transformation for sense and antisense expression of two defense response genes induced by powdery mildew. Thesis #L8349, Section for Plant Pathology, Department of Plant Biology, Carlsberg Research Laboratory, The Carlsberg Research Center, Copenhagen, April 1996.

Junker, B., Zimmy, J., Lührs, R., and Lörz., H. (1987) Transient expression of chimeric genes in dividing and nondividing cereal protoplasts after PEG induced DNA uptake. Plant Cell Rep. 6: 329–332.

Kaeppler, S.M., and Phillips, R.L. (1993a) Tissue culture-induced DNA methylation variation in maize. Proc. Nat. Acad. Sci. USA. 90: 8773–8776.

Kaeppler, S.M. and Phillips, R.L. (1993b) DNA methylation and tissue culture-induced variation in plants. In Vitro Cell. Dev. Biol. 29: 125–130.

Kartha, K.K., Chibbar, R.N., Georges, F., Leung, N., Caswell, K., Kendall, E., and Qureshi,

J. (1989) Transient expression of chloramphenicol acetyltransferase (CAT) gene in barley cell cultures and immature embryos through microprojectile bombardment. Plant Cell Rep. 8: 429–432.

King, S.P., and Kasha, K.J. (1994) Optimizing somatic embryogenesis and particle bombardment of barley (*Hordeum vulgare* L.) immature embryos. In Vitro Cell Dev. Biol. 30P: 117–123.

Kemper, E.L., da Silva, M.J., and Arruda, P. (1996) Effect of microprojectile bombardment parameters and osmotic treatment on particle penetration and tissue damage in transiently transformed cultured immature maize (*Zea mays* L.) embryos. Plant Sci. 121: 85–93.

Kihara, M., Okada, Y, Kuroda, H., Saeki, K., Ito, K., and Yoshigi, N. (1997) Generation of fertile transgenic barley synthesizing thermostable $\beta$-amylase. 26th European Brewing Convention Congress, Maastricht, Netherlands, October 20–21, 1997. J. Inst. Brewing. 103: 153.

Kihara, M., Takahashi, S., Funatsuki, H., Ito, K. (1998) Field performance of the progeny of protoplast-derived barley, *Hordeum vulgare* 1. Breeding Sci. 481–4.

Kjærulff, S. (1995) Antisense repression of *PsaE* mRNA in transgenic barley (*Hordeum vulgare* L.). In: Pathis, P. (ed.), Photosynthesis: from Light to Biosphere, pp. 151–154. Kluwer Academic Publishers.

Kle, H., Horsch, R., and Rogers, S. (1987) *Agrobacterium*-mediated transformation and its further application to plant biology. Ann. Rev. Plant Physiol. 38: 467–486.

Knudsen, S., and Müller, M. (1991) Transformation of the developing barley endosperm by particle bombardment. Planta. 185: 330–336.

Koncz, C., Martini, N., Mayerhofer, R., Koncz-Kalman, Z., Körber, H., Redei, G.P., and Schell, J. (1989). High-frequency T-DNA-mediated gene tagging in plants. Proc. Nat. Acad. Sci. USA. 86: 8467–8471.

Koprek, T., Haensch, R., Nerlich, A., Mendel, R.R., and Schulze, J. (1996) Fertile transgenic barley of different cultivars obtained by adjustment of bombardment conditions to tissue response. Plant Sci. 119: 79–91.

Kristensen, B.K., Brandt, J., Bojsen, K., Thordal-Christensen, H., Kerby, K.B., Collinge, D.B., Mikkelsen, J.D., and Rasmussen, S.K. (1997) Expression of a defense-related intercellular barley peroxidase in transgenic tobacco. Plant Sci. 122: 173–182.

Larkin, P.J., and Scowcroft, W.R. (1981) Somaclonal variation – a novel source of variabilty from cell cultures for plant improvement. Theor. Appl. Genet. 60: 197–214.

Lazzeri, P.A., Brettschneider, R., Lührs, R., and Lörz, H. (1991). Stable transformation of barley via PEG-induced direct DNA uptake into protoplasts. Theor. Appl. Genet. 81: 437–444.

Leckband, G., and Lörz, H. (1998) Transformation and expression of a stilbene synthase gene of *Vitis vinifera* L. in barley and wheat for increased fungal resistance. Theor. Appl. Genet. 1004–1012.

Leduc, N., Matthys-Rochon, E., Rougier, M., Mogensen, L., Holm, P., Magnard, J., and Dumas, C. (1996). Isolated maize zygotes mimic *in vivo* embryonic development and express microinjected genes when cultured *in vitro*. Dev. Biol. 177: 190–203.

Lee, B.T., Murdock, K., Topping, J., Kreis, M., and Jones, M.G.K. (1989) Transient gene expression in aleurone protoplasts isolated from developing caryopses of barley and wheat. Plant Mol. Biol. 13: 21–29.

Lee, B.T., Murdoch, K., Topping, J., Jones, M.G.K., and Kreis, M. (1991) Transient expression of foreign genes introduced into barley endosperm protoplasts by PEG-mediated transfer or into intact endosperm tissue by microprojectile bombardment. Plant Sci. 78: 237–246.

Leffel, S.M., Mabon, S.A., and Stewart, Jr. C.N. (1997) Applications of green fluorescent protein in plants. Biotechniques. 23: 912–918.

Lemaux, P.G., Cho, M.-J., Louwerse, J., Williams, R., and Wan, Y. (1996) Bombardment-mediated transformation methods for barley. Bio-Rad. US/EG Bull. 2007: 1–6.

Lidon, F.C., Da Graca Barreiro, M., and Santos Henriquez, F. (1995) Interactions between biomass production and ethylene biosynthesis in copper-treated rice. J. Plant Nutr. 18: 1301–1314.

Lister, R.M. (1995) Distribution and economic importance of barley yellow dwarf. In: D'Arcy, C.J., and Burnett, P.A. (eds), Barley yellow dwarf: 40 years of progress, pp. 29–53. American Phytopathological Society Press, St. Paul MN.

Lowe, K., Bowen, B., Hoerster, G., Ross, M., Bond, D., Pierre, D., and Gordon-Kamm, W. (1995). Germline transformation of maize following manipulation of chimeric shoot meristems. Bio/Technology. 13: 677–682.

Lowe, K., Sandahl, M.R.G., Miller, M., Howerster, G., Church, L., Tagliani, L., Bond, D., and Gordon-Kamm, W. (1997). Transformation of the maize apical meristem: transgenic sector reorganization and germline transmission. In: Tsaftaris, A.S. (ed.), Genetics, Biotechnology and Breeding of Maize and Sorghum, pp. 94–97. Thomas Graham House, Cambridge, UK.

Lührs, R., and Lörz, H. (1987) Plant regeneration *in vitro* from embryogenic cultures of spring- and winter-type barley (*Hordeum vulgare* L.) varieties. Theor. Appl. Genet. 75: 16–25.

Lupotto, E. (1984) Callus induction and plant regeneration from barley mature embryos. Ann. Bot. 54: 523–529.

Lyznik, L.A., Mitchell, J.C., Hirayama, L., and Hodges, T.K. (1993) Activity of yeast Flp recombinase in maize and rice protoplasts. Nuc. Acids Res. 21: 969–975.

Mann, C. (1997) Reseeding the green revolution. Science. 277: 1038–1043.

Mannonen, L., Ritala, A., Nuutila, A.M., Kurtén, U., Kauppinen, V., Aspegren, K., Teeri, T.H., Aikasalo, R., and Tammisola, J. (1997) Thermotolerant fungal glucanase in malting barley. 26th European Brewing Convention Congress, Maastricht, Netherlands, October 20–21, J. of the Inst. of Brewing 103: 148–149.

Marris, C., Gallois, P., Copley, J., and Kreis, M. (1988) The 5′ flanking region of a barley B hordein gene controls tissue and developmental specific CAT expression in tobacco plants. Plant Mol. Biol. 10: 359–366.

Matzke, M.A., and Matzke, A.J.M. (1995) How and why do plants inactivate homologous (trans) genes? Plant Physiol. 107: 679–685.

McElroy, D., and Jacobsen, J. (1995) What's brewing in barley biotechnology? Bio/ Technology 13: 245–249.

McElroy, D., Louwerse, J.D., McElroy, S.M., and Lemaux, P.G. (1997) Development of a simple transient assay for *Ac-Ds* activity in cells of intact barley tissue. Plant J. 11: 157–165.

McGrath, P.F., Vincent, J.R., Lei, C., Pawlowski, W.P., Torbert, K.A., Gu, W., Kaeppler, H.F., Wan, Y., Lemaux, P.G., Rines, H.R., Somers, D.A., Larkins, B.A., and Lister, R.A. (1997) Coat protein-mediated resistance to isolates of barley yellow dwarf in oats and barley. Eur. J. Plant Path., 103: 695–710.

Medford, J. (1992) Vegetative apical meristems. Plant Cell 4: 1029–1039.

Mendel, R.R., Clauss, E., Hellmund, R., Schulze, J., Steinbiß, H.H., and Tewes, A. (1990) Gene transfer to barley. In: Nijkamp, H.J.J., Van der Plas, L.W.W., and Van Aartrijk, J. (eds), Progress in Plant Cellular and Molecular Biology, pp 73–78. Kluwer Academic Publishers, Dordrecht.

Mendel, R.R., Müller, B., Schulze, J., Kolesnikov, V., and Zelenin, A. (1989) Delivery of foreign genes to intact barley cells by high-velocity microprojectiles. Theor. Appl. Genet. 78: 31–34.

Michel, M., (1995) Ubertragung von *H. bulbosum* - Resistenzgenen gegenüber mehltau und gerstengelbmosaikvirus in die kulturgerste. Vort Pflanzenzuchtg. 31: 78–79.

Moore, G. (1995) Cereal genome evolution: Pastoral pursuits with 'Lego' genomes. Curr. Opin. Genet. Dev. 5: 717–724.

Moore, G., Cheung, W., Schwarzacher, T., and Flavell, R. (1991) BIS 1, a major component of the cereal genome and a tool for studying genomic organization. Genomics. 10: 469–476.

Mornhinweg, D.W., Porter, D.R., and Webster, J.A. (1995) Registration of STARS-9301B barley germplasm resistant to the Russian wheat aphid. Crop. Sci. 35: 603.

Mouritzen, P., and Holm, P.B. (1994) Chloroplast genome breakdown in microspore cultures of barley (*Hordeum vulgare* L.) occurs primarily during regeneration. J. Plant Physiol. 144: 586–593.

Munck, C.M. (1993) Whole-crop utilization of barley, including potential new uses. In: MacGregor, A.W., and Bhatty, R.S. (eds), Barley: Chemistry and Technology, pp. 437–474. Amer. Assoc. Cereal Chem., St. Paul, MN.

Murashige, T., and Skoog, F. (1962) A revised medium for rapid growth and bioassays with tobacco tissue cultures. Physiol. Plant. 15: 473–497.

Nelson, A.J., and Bushnell, W.R. (1997) Transient expression of anthocyanin genes in barley epidermal cells: Potential for use in evaluation of disease response genes. Transgen. Res. 6: 233–244.

Oard, J.H., Paige, D.F., Simmonds, J.A., and Gradziel, T.M. (1990) Transient gene expression in maize, rice and wheat cells using an airgun apparatus. Plant Physiol. 92: 334–339.

Ohkoshi, S., Komatsuda, T., Enomoto, S., Taniguchi, M., and Ohyama, K. (1991) Variations between varieties in callus formation and plant regeneration from immature embryos of barley. Bull Nat. Inst. Agrobiol. Resour. 6: 189–207.

Pang, S., DeBoer, D.L., Wan, Y., Ye, G., Layton, J.G., Neher, M.K., Armstrong, C.L., Fry, J.E., Hinchee, M.A., and Fromm, M.E. (1996) An improved green fluorescent protein gene as a vital marker in plants. Plant Physiol. 112: 893–900.

Park, S.H., Pinson, S.R.M., and Smith, R.H. (1996) T-DNA integration into genomic DNA of rice following *Agrobacterium* inoculation of isolated shoot apices. Plant Mol. Biol. 32: 1135–1148.

Perl, A., Galili, E., Shaul, I., Ben-Tzvi, I., and Galili, G. (1993) Bacterial dihydrodipicolinate synthase and desensitized aspartate kinase: two novel selectable markers for plant transformation. Bio/Technology. 11: 715–718.

Phillips, R.L., Kaeppler, S.M., and Olhoft, P. (1994) Genetic instability of plant tissue culture: breakdown of normal controls. Proc. Nat. Acad. Sci. USA. 91: 5222–5226.

Pickering, R.A., Hill, A.M., and Kynast, R.G. (1997) Characterization by RFLP analysis and genomic *in situ* hybridization of a recombinant and a monosomic substition plant derived from *Hordeum vulgare* L. X. *H. bulbosum* L. crosses. Genome 40: 195–200.

Pickering, R.A., Hill, A.M., Michel, M., and Timmerman-Vaughan, G.M. (1995) The transfer of a powdery mildew resistance gene from *Hordeum bulbosum* to barley (*H. vulgare* L.) chromosome 2 (2I). Theor. Appl. Genet. 91: 1288–1292.

Potrykus, I. (1990a) Gene transfer to cereals: an assessment. Bio/Technology 8: 535–542.

Potrykus, I. (1990b) Gene transfer to plants: assessment and perspectives. Physiol. Plant. 79: 125–134.

Potrykus, I. (1992) Micro-targeting of microprojectiles to target areas in the micrometre range. Nature 355: 568–569.

Purnhauser, L. (1991) Stimulation of shoot and root regeneration in wheat *Triticum aestivum* callus cultures by copper. Cereal Res. Comm. 19: 419–423.

Raho, G., Lupotto, E., Hartings, H., Torre, P.A.D., Perrotta, C., and Armiroli, N. (1996) Tissue-specific expression and environmental regulation of the barley *Hvhsp* 17 gene promoter in transgenic tobacco plants. J. Exp. Bot. 47: 1587–1594.

Rasmusson, D.C. (1985) Barley. American Society of Agronomy, Number 26.

Rines, H.W., and Luke, H.H. (1985) Selection and regeneration of toxin-insensitive plants from tissue cultures of oats (*Avena sativa*) susceptible to *Helminthosporium victorea*. Theor. Appl. Genet. 71: 16–21.

Risseeuw, E., Franke-Van Dijk, M.E.I., and Hooykaas, P.J.J. (1997) Gene targeting and instability of *Agrobacterium* T-DNA loci in the plant genome. Plant J. 11: 717–728.

Ritala, A., Aikasalo, R., Aspegren, K., Salmenkallio-Marttila, M., Akerman, S., Mannonen, L., Kurtén, U., Puupponen-Pimia, R., Teeri, T.H., and Kauppinen, V. (1995) Transgenic barley by particle bombardment. Inheritance of the transferred gene and characteristics of transgenic barley plants. Euphytica. 85: 81–88.

Ritala, A., Apegren, K., Kurtén, U., Salmenkallio-Marttila, M., Mannonen, L., Hannus, R., Kauppinen, V., Teeri, T.H., and Enari, T. (1994) Fertile transgenic barley by particle bombardment of immature embryos. Plant Mol. Biol. 24: 317–25.

Ritala, A., Mannoen, L., Aspegren, K., Salmenkaillo-Marttila, M., Kurtén, U., Hannus, R., Mendez-Lozano, J., Teeri, T.H., and Kauppinen, V. (1993) Stable transformation of barley tissue culture by particle bombardment. Plant Cell Rep. 12: 434–440.

Roemer, T., Scheibe, A., Schmidt, J., and Woermann, E. (1953) Handbuch der Landwirtschaft II, Pflanzenbaulehre, pp. 67–77. Paul Parey Verlag, Berlin and Hamburg.

Rogers, S.W., and Rogers, J.C. (1992) The importance of DNA methylation for stability of foreign DNA in barley. Plant Mol. Biol. 18: 945–961.

Salmenkallio-Marttila, M., Aspegren, K., Åkerman, S., Kurtén, U., Mannonen, L., Ritala, A., Teeri, T.H., and Kauppinen, V. (1995) Transgenic barley (*Hordeum vulgare L.*) by electroporation of protoplasts. Plant Cell Rep. 15: 301–304.

Sandager, L. (1996) Engineering of barley for enhanced levels of lipid transfer protein 1 in the seed. Thesis #890564, Institut for Molekylær Biologi, Laboratoriet for Genekspression, Aarhus Universitet, Aarhus.

Sautter, C., Waldner, H., Neuhaus-Uri, G., Galli, A., Neuhaus, G., and Potrykus, I. (1991) Micro-targeting: high efficiency gene transfer using a novel approach for the acceleration of microprojectiles. Bio/Technology 9: 1080–1085.

Schaller, C.W., Rasmusson, D.C., and Qualset, C.O. (1963) Sources of resistance to yellow dwarf virus in barley. Crop. Sci. 3: 342–344.

Schildbach, R. (1994) Malting barley worldwide. Brauwelt Internat. 4: 292–310.

Schuh, W., Nelson, M.R., Bigelow, D.M., Orum, T.V., Orth, C.E., Lynch, P.T., Eyles, P.S., Blackhall, N.W., Jones, J., Cocking, E.C., and Davey, M.R. (1993) The phenotypic characterisation of R-2 generation transgenic rice plants under field conditions. Plant Sci. 89: 69–79.

Scott, P., and Lyne, R.L. (1994a) The effect of different carbohydrate sources upon the initiation of embryogenesis from barley microspores. Plant Cell Tiss. Org. Cult. 36: 129–133.

Scott, P., and Lyne, R.L. (1994b) Initiation of embryogenesis from cultured barley microspores: a further investigation into the toxic effects of sucrose and glucose. Plant Cell Tiss. Org. Cult. 37: 61–65.

Sears, R.G., and Deckard, E.L. (1982) Tissue culture variability in wheat: callus induction and plant regeneration. Crop Sci. 22: 546–550.

Shewry, P.R., Tatham, A.S., Halford, N.G., Barker, J.H.A., Hannappel, U., Gallois, P., Thomas, M., and Kreis, M. (1994) Opportunities for manipulating the seed protein composition of wheat and barley in order to improve quality. Transgen. Res. 3: 3–12.

Simmonds, J.A., Stewart, P., and Simmonds, D. (1992). Regeneration of *Triticum aestivum* apical explants after microinjection of germ line progenitor cells with DNA. Physiol Plant. 85: 197–206.

Singh, R.J. (1986) Chromosomal variation in immature embryo derived calluses of barley (*Hordeum vulgare*. L). Theor. Appl. Genet. 72: 710–716.

Smith, L.G., Greene, B., Veit, B., and Hake, S. (1992) A dominant mutation in the maize homeobox gene, *Knotted-1*, causes its ectopic expression in leaf cells with altered fates. Development 116: 21–30.

Smith, L.G., Jackson, D., and Hake, S. (1995) Expression of *knotted1* marks shoot meristem formation during maize embryogenesis. Dev. Genet. 16: 344–348.

Songstad, D.D., Somers, D.A., and Griesbach, R.J. (1995) Advances in alternative DNA delivery techniques. Plant Cell Tiss. Org. Cult. 40: 1–15.

Sørensen, M.B., Müller, M., Skerritt, J., and Simpson, D. (1996) Hordein promoter methylation and transcriptional activity in wild-type and mutant barley endosperm. Mol. Gen. Genet. 250: 750–760.

Spencer, T.M., O'Brien, J.V., Start, W.G., Adams, T.R., Gordon-Kamm, W.J., and Lemaux, P.G. (1992) Segregation of transgenes in maize. Plant Mol. Biol. 18: 201–210.

Sprenger, N., Schellenbaum, L., Van Dun, K., Boller, T., and Wiemken, A. (1997) Fructan synthesis in transgenic tobacco and chicory plants expressing barley sucrose:fructan 6-fructosyltransferase. FEBS Lett. 400: 355–358.

Srivastava, V., Vasil, V., and Vasil, I.K. (1996) Molecular characterization of the fate of transgenes in transformed wheat (*Triticum aestivum* L.). Theor. Appl. Genet. 92: 1031–1037.

Steeves, T.A., and Sussex, I.M. (1989) Patterns in Plant Development, 2nd edition. Cambridge University Press, Cambridge.

Stewart, R.J., Hodges, S., Hrmova, M., Garrett, T.P.J., Varghese, J.N., Hoj, P.B., and Fincher, G.B. (1997) Protein engineering of thermostable barley beta-glucanases and their commercial evaluation. (1997 American Association of Cereal Chemists Annual Meeting, San Diego, California, U.S.A., October 12–16) Cereal Foods World, v.42, n.8, (1997): 662–663.

Stiff, C.M., Kilian, A., Zhou, H., Kudrna, D.A., and Kleinhofs, A. (1995) Stable transformation of barley callus using biolistic particle bombardment and the phosphinothricin acetyltransferase (*bar*) gene. Plant Cell Tiss. Org. Cult. 40: 243–248.

Takano, M., Egawa, H., Ikeda, J., and Wakasa, K. (1997) The structures of integration sites in transgenic rice. Plant J. 11: 353–361.

ten Hoopen, R., Robbins, T.P., Fransz, P.F., Montijn, B.M., Oud, O., Gerats, A.G.M., and Nanninga, N. (1996) Localization of T-DNA insertions in petunia by fluorescence *in situ* hybridization: physical evidence for suppression of recombination. Plant Cell. 8: 823–830.

Thompson, C.J., Novva, N.R., Tizard, T., Crameri, R., Davies, J.E., Lauwereys, M., and Botterman, J. (1987) Characterization of the herbicide-resistance gene *bar* from *Streptomyces hygrtoscopicus*. EMBO J. 6: 2519–2523.

Thorpe, T.A. (1994) Morphogenesis and regeneration. In: Vasil I.K., and Thorpe, T.A. (eds), Plant Cell and Tissue Culture, pp. 17–36. Kluwer Academic Publishers, Dordrecht.

Tingay, S., McElroy, D., Kalla, R., Fieg, S., Wang, M., Thornton, S., and Brettell, R. (1997) *Agrobacterium tumefaciens*-mediated barley transformation. Plant J. 11: 1369–1376.

Toyoda, H., Yamaga, T., Matsuda, Y., and Ouchi, S. (1990) Transient expression of the *β*-glucuronidase gene introduced into barley coleoptile cells by microinjection. Plant Cell Rep. 9: 299–302.

Ullrich, S.E., Edmiston, J.M., Kleinhofs, A., Kudrna, D.A., and Maatougui, M.E.H. (1991) Evaluation of somaclonal variation in barley. Cereal Res. Comm. 19: 245–260.

Vain, P., McMullen, M.D., and Finer, J.J. (1993) Osmotic treatment enhances particle bombardment-mediated transient and stable transformation of maize. Plant Cell Rep. 12: 84–88.

Vasil, I.K. (1987) Developing cell and tissue culture systems for the improvement of cereal and grass crops. J. Plant Physiol. 128: 193–218.

Vasil, I.K. (1994) Molecular improvement of cereals. Plant Mol. Biol. 25: 925–937.

Vidhyasekaran, P., Ling, D.H., Borromeo, E.S., Zapata, F.J., and Mew, T.W. (1990) Selection of brown spot-resistant rice plants from *Helminthosporium oryzae* toxin-resistant calluses. Ann. Appl. Biol. 117: 515–523.

Vollbrecht, E., Veit, B., Sinha, N., and Hake, S. (1991) The developmental gene *Knotted-1* is a member of a maize homeobox gene family. Nature. 350: 241–243.

von Bothmer, R., Jacobsen, N., Baden, C., Jorgensen, R.B., and Linde-Laursen, I. (1995). An ecogeographical study of the genus *Hordeum*. In: von Bothmer, R. (ed.), Systematic and Ecogeographic Studies on Crop Genepools, 7 (2nd edition), pp. 1–19. International Plant Genetic Resources Institute, Rome.

Wan, Y., and Lemaux, P.G. (1994) Generation of large numbers of independently transformed fertile barley plants. Plant Physiol. 104: 37–48.

Wang, X., Lazzeri, P.A., and Lörz, H. (1992) Chromosomal variation in dividing protoplasts derived from cell suspensions of barley (*Hordeum vulgare* L.) Theor. Appl. Genet. 85: 181–185.

Williams-Carrier, R., Lie, Y.S., Hake, S., and Lemaux, P.G. (1997) Ectopic expression of the maize *kn1* gene phenocopies the *Hooded* mutant of barley. Development 124: 3737–3745.

Xu, J., and Kasha, K.J. (1992) Transfer of a dominant gene for powdery mildew resistance and DNA from *Hordeum bulbosum* into cultivated barley (*H. vulgare*). Theor. Appl. Genet. 84: 771–777.

Xu, D., Duan, X., Wang, B., Hong, B., Ho, T., and Wu, R. (1996) Expression of a late embryogenesis abundant protein gene, *HVA1*, from barley confers tolerance to water deficit and salt stress in transgenic rice. Plant Physiol. 110: 249–257.

Yao, Q.A., Simion, E., William, M., Krochko, J., and Kasha, K.J. (1997) Biolistic transformation of haploid isolated microspores of barley (*Hordeum vulgare* L.). Genome 40: 570–581.

Zhang, J., Tiwari, V.K., Golds, T.J., Blackhall, N.W., Cocking, E.C., Mulligan, B.J., Power, J.B., and Davey, M.R. (1995) Parameters influencing transient and stable transformation of barley. (*Hordeum vulgare* L.) protoplasts. Plant Cell Tiss. Org. Cult. 41: 125–138.

Zhang, S., Warkentin, D., Sun, B., Zhong, H., and Sticklen, M. (1996a) Variation in the inheritance of expression among subclones for unselected (*uidA*) and selected (*bar*) transgenes in maize (*Zea mays* L.). Theor. Appl. Genet. 92: 752–761.

Zhang, S., Williams-Carrier, R., Jackson, D., and Lemaux, P.G. (1998a) Expression of CDC2Zm and KNOTTED1 during axillary shoot meristem *in vitro* proliferation and adventitious shoot meristem formation in maize and barley. Planta 204: 542–549.

Zhang, S., Zhang, S., Cho, M.-J., Bregitzer, P., and Lemaux, P.G. (1998b) Comparative analysis of genomic DNA methylation status and field performance of plants derived from embryogenic calli and shoot meristematic cultures. Prox. IX Internatl. Cong. Plant Tiss. Cell Cult. Jerusalem, Israel (in press).

Zhang, S., Zhong, H., and Sticklen, M.B. (1996b) Production of multiple shoots from shoot apical meristems of oat (*Avena sativa* L.). J. Plant Physiol. 148: 667–671.

Zhong, H., Srinivasan, C., and Sticklen, M.B. (1992) *In-vitro* morphogenesis of corn (*Zea mays* L.). I. Differentiation of multiple shoot clumps and somatic embryos from shoot tips. Planta 187: 483–489.

Zhong, H., Sun, B., Warkentin, D., Zhang, S., Wu, R., Wu, T., and Sticklen, M.B. (1996) The competence of maize shoot meristems for integrative transformation and inherited expression of transgenes. Plant Physiol. 110: 1097–1107.

Ziauddin, A., and Kasha, K.J. (1990) Long-term callus cultures of diploid barley (*Hordeum vulgare*) II. Effect of auxins on chromosomal status of cultures and regeneration of plants. Euphytica. 48: 279–286.

# 10. Transgenic Cereals: *Avena sativa* (oat)

DAVID A. SOMERS
*Department of Agronomy and Plant Genetics, University of Minnesota,*
*411 Borlaug Hall, 1991 Upper Buford Circle, St. Paul, Minnesota 55108 USA.*
*E-mail: somers@biosci.cbs.umn.edu*

ABSTRACT. Genetic engineering of allohexaploid oat (*Avena sativa* L.) has been substantially improved over the past five years. This chapter documents recent progress made in the molecular improvement of oat. New tissue culture systems have been developed that reduce the labor and time required to produce transgenic plants. These allow a broad range of genotypes, including varieties currently in production, to be genetically engineered. Selectable markers have been identified that reduce the potential for ecological risk upon outcrossing of transgenic plants with wild oat. Some problems with transgene expression are observed which must be resolved before the extant genetic engineering systems become fully useful for oat improvement. Examples of applications of genetic engineering to oat improvement are presented.

## Introduction

The oat (*Avena sativa* L.) grain is utilized for both human food and animal feed. Annual worldwide production is estimated at between 35–45 million metric tons, which ranks oat sixth in world cereal production behind wheat, maize, rice, barley and sorghum (Schrickel, 1986, Murphy and Hoffman, 1992, Hoffman, 1995). Commensurate with the relatively smaller production of oat, only a small number of research efforts are devoted to molecular improvement of oat. Most of these programs are in public institutions. Nevertheless, progress in developing genetic engineering systems useful for improvement of oat have kept pace with species that are the focus of larger and more diversified research efforts because, at least initially, oat was somewhat more amenable to tissue culture and genetic engineering than other cereals such as wheat and barley.

The goal of this chapter is to document the progress in genetic engineering of oat achieved since 1992 when the crop was first genetically engineered (Somers et al., 1992). Tissue culture of oat has been extensively reviewed (Bregitzer et al., 1995, Rines et al., 1992). Oat genetic engineering also has been reviewed (Somers et al., 1994, 1996a,b). However, new oat transformation systems, progress in characterizing transgene integration and inheritance, and applications of oat genetic engineering have since been

*I.K. Vasil (ed.), Molecular Improvement of Cereal Crops, 317–339*

reported. In this review, the current status of basic research and oat improvement using genetic engineering are described.

## Economic Importance of Oat and Oat Improvement

Oat is a cool season annual allohexaploid ($2n = 6\times = 42$) that may be of either a winter (vernalization requiring) or spring type (Murphy and Hoffman, 1992). Of the two types, spring oats are planted on the largest area worldwide. Major growing regions of spring oat include the cooler temperature regions of the northern hemisphere; in particular, northern European countries, Russia and some other former Soviet Republics, and Canada and north central United States. Other major production occurs in areas of the southern hemisphere and where the milder winters allow spring type oats to be grown as a winter season crop. These areas include southern Australia and a region in South America that includes southern Brazil, Uruguay, and Argentina. Besides the grain, oat as a crop may be valued as a source of forage and bedding for livestock, as a companion crop for forage legume establishment, and as a component of a crop rotation system to halt pest, weed and pathogen build-up. While the total annual production of oat is low compared to wheat, maize and rice, the crop is an important component of a number of agricultural systems.

Oat grain serves dual use as both a human food, chiefly as oatmeal or a component of breakfast cereal products, and as a high quality feed for horses and other livestock. Many aspects of the nutritional properties of oat for both uses have been reviewed (Peterson, 1992, Cuddeford, 1995, Ranhotra and Gelroth, 1995, Frolich, 1996, Gerger and Ink, 1996, Paton and Fedec, 1996). Probably the most recent trend of economic importance to oat production is the realization that consumption of oat has positive human health effects (Welch, 1995, Gerger and Ink, 1996). Documentation that dietary fiber of oat has cholesterol-lowering properties and promotion of oat as a nutraceutical likely will maintain an expanding market for oat as a food ingredient. Beyond the cholesterol-lowering effects of oat dietary fiber, other fractions of the groat (caryopses removed from the hull) also are considered to have positive effects on human health (Peterson, 1992). The fatty acid composition of oat oil is considered healthful compared to other vegetable oils although there is little commercial utilization of oat oil (Youngs, 1986). Oat groats contain antioxidant substances, such as tocols, known to have cancer-preventing and cholesterol-lowering properties (Peterson, 1995). The granular structure of oat starch has unique properties compared to other sources of starch providing the opportunity for commercial exploitation (Paton, 1986). Oat groat protein is lower in the prolamin fractions than other cereals such as wheat (Burrows and Altosaar, 1995). The low levels of these proteins (avenins) found in oat enable some coeliac sufferers to consume oat as a source of dietary starch and protein without negative effects (Welch, 1995). Other important traits in oat and strategies for their manipulation both by conventional and molecular methods have been reviewed by Burrows and Altosaar (1995). The unique beneficial properties of oat as a food either are or may be the focus of present and future oat improvement programs. These nutraceutical traits likely will be enhanced through a combination of conventional plant breeding and genetic engineering approaches.

Feed use of oat was estimated to consume 75% of the crop world wide during 1980 through 1985 (Murphy and Hoffman, 1992). A significant portion of oat production is used as high-quality feed for race horses; a utilization in which the energy value of the feed is of premier importance (Cuddeford, 1995, Campbell, 1996). Oat is a high energy feed because it is relatively high in oil compared to many other cereals. Oil is deposited in the germ and endosperm of the groat (Youngs, 1986) rather than being largely confined to the embryo as in corn, wheat or barley. Breeding oat for animal feed use has often focused on increasing the energy content of oat through increasing groat oil levels. This aspect of oat improvement has resulted in development of divergent goals in oat improvement programs because the characteristics of oat for human consumption and animal feed differ. In some cultivars, oil content exceeds levels at which food produced from these oats would be considered low fat. Moreover, high oil content may lead to rancidity problems in oat or oat-derived food products. Thus, two breeding goals have emerged for this crop (Burrows, 1986). One aim is to enhance the nutraceutical properties of oat for human consumption and lower oil content while the other is to produce high energy oat feed for race horses and other livestock by increasing oil content.

The greatest problems encountered in oat cultivar development, and the broader acceptance of oat as a crop by producers, are agronomic. Foremost among these is that oat yields often are lower than those for other cereals; thus oat acreage is declining in favor of higher value crops. Low yields are further compounded by the problem that oat is susceptible to a broad array of diseases. Fungal pathogens, notably rusts, have caused reduced production in disease-prone areas and subsequent expansion of oat production in rust-free regions (Clifford, 1995). Likewise, barley yellow dwarf virus (BYDV) also limits oat production in some regions. Resistance genes to these and other diseases are used in breeding programs to manage these disease problems. However, there is some concern that sufficient numbers of new disease resistance genes are not available in the oat gene pool for maintenance of disease resistance. Thus the future of oat production in areas prone to disease is uncertain. Producing disease-resistant cultivars of oat is a vital goal of conventional breeding programs, and considering the severity of the problem and its threat to stable oat production, disease resistance should be a major focus of genetic engineering programs as well.

## Special Considerations regarding Genetically Engineered Oat

A challenge to oat improvers and specifically to the use of genetic engineering is that domesticated oat is sexually compatible with its weed relatives *A. sterilis* and *A. fatua*. These weeds infest the crop in many oat growing regions of the world. Although oat is largely a self-pollinating species, gene flow from the crop to these weeds via cross pollination has been documented. Thus, the potential for introgression of transgenic traits from the crop into the weeds is high. Genetic engineering traits for oat improvement must be considered with the realization that the transgenic traits will with high probability be transferred to the weedy relatives. The initial use of herbicide resistance to develop and investigate oat transformation (Somers et al., 1992, Gless et al., 1996, 1998a) was justifiably criticized for this reason (Gressel, 1992). The usefulness of herbicide resistance engineered into a crop, e.g. glufosinate-resistant wheat (Vasil et al., 1992, 1993), would be nullified by the presence of wild oats resistant to the same herbicide due to introgression into the wild oats from a genetically engineered oat carrying the same herbicide resistance. Therefore, transgenic traits and selectable markers must be considered in terms of their potential to pose ecological risks to managed or unmanaged ecosystems. Unfortunately and probably commensurate with the fact that oat is a minor crop, there is a paucity of risk assessment data on the possible outcomes and the impacts of transgene flow between oat and wild oat and the risk potential of specific transgenic traits.

## Oat Tissue Cultures used for Transformation

Four genetic engineering systems have been reported for oat that differ only in the source and type of explants and tissue cultures used as totipotent cells (Table 1). These include embryogenic tissue cultures initiated from immature embryos, mature embryos, and seedling leaf bases, and organogenic tissue cultures initiated from a shoot apical meristem culture system. All systems utilize high velocity particle bombardment (Klein et al., 1987) for DNA delivery into oat tissue culture cells. Because the extant oat genetic engineering systems are dependent on tissue cultures, it is important to briefly review the oat tissue culture literature.

Nonregenerable oat tissue cultures and their contribution to development of regenerable tissue cultures have been previously reviewed (Rines et al., 1992, Bregitzer et al., 1995). Oat plants were first regenerated from tissue cultures in 1976 (Cummings et al., 1976, Lörz et al., 1976). Plant-regenerating tissue cultures were initially established from immature embryos. These regenerable oat tissue cultures appear to arise from the embryo axis (Bregitzer, 1995) and not from cells in the scutellum of the immature embryo as shown for maize (Springer et al., 1979). This difference in tissue

TABLE 1
Oat transformation systems that have produced fertile, transgenic plants. All systems used microprojectile bombardment

| Date of first report | Source of tissue culture | Selectable marker(s) | Authors |
|---|---|---|---|
| 1992 | immature embryos | *bar* | Somers et al. |
| 1995 | immature embryos | *nptII* | Torbert et al. |
| 1996 | mature embryos | *nptII* | Torbert et al. |
| 1998a | leaf base | *bar* | Gless et al. |
| 1997 | shoot meristems | *bar, nptII, hpt* | Zhang, Cho and Lemaux, personal communication |

culture initiation behavior continues to influence development of oat tissue cultures for genetic engineering. Following the benchmark discovery of plant regeneration from oat tissue cultures initiated from immature embryos, a number of alternative totipotent explants have been reported. These include mature embryos (Cummings et al., 1976, Heyser and Nabors, 1982, Nabors et al., 1982, 1983, Torbert et al., 1996, 1998b), tissues from mature embryos like the mesocotyl (Bregitzer et al., 1989, Heyser and Nabors, 1982, Nabors et al., 1982), roots, leaf base (Chen et al., 1995a, b, Gless et al., 1996, 1998b) and shoot apical meristems (A. Nassuth and I. Altosaar 1989, Personal Communication, Zhang et al., 1996).

Regenerable tissue cultures of oat may be distinguished as embryogenic or organogenic. The first regenerable tissue cultures derived from immature embryos appeared to be highly organized and organogenic (Rines et al., 1992). Meristems are clearly visible in the tissue cultures (Cummings and Green, 1976, Heyser and Nabors, 1982, Rines and McCoy, 1981). Nabors et al. (1982) demonstrated initiation of both organogenic and embryogenic callus types from mature embryos of cv Park. Bregitzer et al. (1989) observed embryo-like structures and friable, embryogenic callus at low frequency within immature embryo-derived organogenic tissue cultures initiated from a specific genotype referred to as GAF/Park. Selective subculture of these callus sectors resulted in cultures that stably exhibited a friable, embryogenic appearance and retained plant regeneration capacity for longer than one year in culture. Bregitzer et al. (1989, 1991) demonstrated that embryo-like structures in this callus type were indeed somatic embryos by culturing individual isolated embryos through conversion into plantlets. They also observed that inclusion of benzyl aminopurine (BAP) in the regeneration medium caused isolated embryos to change from bipolar structures that simultaneously produced shoot and root apices to structures that proliferated shoot meristems which once elongated had to be placed on a special medium for rooting. This observation, combined with the fact that both callus types initiated from immature embryos, suggests that the differentiation of embryogenic and organogenic cultures in oat may be less defined than in other systems and is not tightly regulated by hormonal levels.

Tissue cultures produced directly from the mature embryo isolated from the caryopsis 16 h after imbibition (Torbert et al., 1996, 1997b) (Figure 1a) or from leaf bases of young seedlings (Chen et al., 1995a,b, Gless et al., 1996, 1998b) also are regenerable. While it is likely that the tissue cultures are similar to the friable, embryogenic callus initiated from immature embryos, these newer systems overcome a major disadvantage of using immature embryos as an explant for tissue culture initiation. Growth of oat plants for immature embryo isolation year round is costly, requires sophisticated growth facilities and may subject donor plants to physiological variation that can cause fluctuations in tissue culture initiation frequency. Additionally, isolation of immature embryos from oat is labor intensive because fertilization along the panicle is not as synchronized as in wheat or barley spikes or maize ears. This characteristic of oats dictates that greater number of donor plants must be produced for immature embryo isolation compared to these other species. Alternatively, production of a large number of mature seed is inexpensive and provides a uniform source of explant for tissue culture initiation, possibly avoiding variation experienced in culture initiation from immature embryos.

Meristematic tissue cultures established from shoot apical meristems also appear to be a promising source of totipotent cells for oat transformation. Zhang et al. (1995) reported multiple shoot production and regeneration of fertile plants using apical meristem explants of four oat genotypes. The tissue cultures are initiated on MS medium (Murashige and Skoog, 1962) containing 2,4-D and BAP and appear to be masses of proliferating adventitious meristems (Zhang et al., 1995). The development and investigation of the different types of oat tissue culture will allow comparisons of genotype specificity, effects of tissue culture-induced genetic variation, and ease and efficacy of transformation among a diversity of oat tissue culture systems.

In some cereals, the genotype of the explant influences tissue culture response and thus constrains the diversity of genotypes that may be genetically engineered. The ability to genetically engineer only one or a few genotypes within a species profoundly influences how the genetic engineering system will be integrated into the crop improvement program. These problems have plagued oat genetic engineering to some extent because the genetic engineering system utilizes tissue cultures as totipotent cells. Variation in tissue culture initiation from immature embryos of different oat genotypes has been reviewed (King et al., 1976, Bregitzer et al., 1995). One genotype, which has specifically developed for tissue culture responsiveness, continues to be the basis of our oat transformation research. This line, referred to as GAF/Park and its derivatives, resulted from an $F_4$ selection of a cross between Garland by *Avena fatua* and Park (Rines and Luke, 1985). GAF/Park has been further purified and homogeneously responding lines are available. Research on tissue culture initiation from GAF/Park has been focused on immature and mature embryos as explants (Bregitzer, 1989, Gana et al., 1995, Torbert et al., 1996, 1998b). In some cases modification of phytohormone composition of the initiation medium has increased callus initiation frequency from immature embryos of different genotypes (Bregitzer et al., 1995, Gana et al., 1995, Milach et al., 1992). Thus, it appears possible that with manipulation of medium composition other more agronomic genotypes may be genetically engineered. However, the cost and labor required for initiation of regenerable tissue cultures from immature embryos will likely discourage further extensive efforts for identifying conditions for tissue culture initiation from immature embryos of other genotypes. Alternatively, mature embryos, and leaf base and shoot meristem explants have been reported to initiate transformable callus from a greater number of genotypes compared with the immature embryo explant (Cho et al., 1998, Gless et al., 1996, 1998a,b, Torbert et al., 1996, 1998b,c, S. Zhang, M-J. Cho and P.G. Lemaux, personal communication). This is encouraging because the convenience of these explants compared to immature embryos will definitely stimulate further improvements in producing tissue cultures for genetic engineering of different oat genotypes to the point where leading agronomic cultivars will be genetically engineered at frequencies comparable to the model systems.

## Genetic Engineering Systems

A summary of the oat transformation systems is shown in Table 1. The initial oat genetic engineering system, as all others described to date, used microprojectile bombardment to introduce DNA into friable, embryogenic callus and suspension cultures initiated from immature embryos of GAF/Park (Somers et al., 1992). The plasmid pBARGUS was used. It contains the *bar* gene from *Streptomyces hygroscopicus* encoding the enzyme phosphinothricin (PPT) acetyl transferase (PAT) which confers plant resistance to PPT-containing herbicides. This marker was used to develop the initial transformation system because of its demonstrated usefulness for tissue culture selection of transgenic cereal cells (Fromm et al., 1990). The plasmid also carried the *uidA* gene fused to the maize alcohol dehydrogenase I promoter which conferred $\beta$-glucuronidase activity (GUS) in transgenic tissue cultures and in seed (Kyozuka et al., 1991). The oat tissue cultures used for genetic engineering were initiated from just a few immature embryos and in some cases were over one-year old. This approach proved unworkable because the majority of genetically engineered plants were sterile, presumably due to tissue culture-induced genetic variation.

Subsequent improvements of this system, such as increasing the number of different tissue cultures that were bombarded and decreasing the culture age to less than one year, improved both the production of transgenic tissue cultures and the fertility of regenerated plants (Table 2). Kuai et al. (1993) also reported development of an oat transformation system using the *bar* gene as a selectable marker in tissue cultures initiated from immature embryos.

A requisite refinement of the original transformation system was to replace the *bar* gene as the selectable marker so that introgression with wild oat would not transfer herbicide resistance to the weed species once transgenic oat were introduced into the environment. The *E. coli* neomycin phosphotransferase (*nptII)* gene in combination with the antibiotic paromomycin was developed as an efficient selectable marker (Tables 1 and 2, Figure 1b). Plasmids used are described by Klein et al. (1989) and Torbert et al. (1995). While the adoption of *nptII* as a selection system appeared to somewhat reduce the numbers of transgenic tissue cultures selected following microprojectile bombardment compared to PPT selection, no selection escapes have been detected over the past two years of using this system (Torbert et al. 1998b). Tight selection of transgenic oat is a significant advantage of the *nptII* selection because it reduces the effort required to identify transgenic regenerated plants.

A promising new visual selection system has been reported by Kaeppler et al. (1997). The green fluorescence protein (*gfp*) gene was modified for expression in monocot cells and used in microprojectile bombardment experiments with oat. Transgenic tissue culture were readily identified by

TABLE 2

Comparison of regeneration and fertility of transgenic plants produced by microprojectile bombardment of oat tissue cultures initiated from different explants

| Explant | Number of transgenic tissue cultures tested | Number of transgenic tissue cultures exhibiting Plant regeneration | Number of transgenic tissue cultures exhibiting Fertility |
|---|---|---|---|
| Immature embryo<br>Torbert et al., 1995 | 88 | 32 | 17 (19) |
| Mature embryo<br>Torbert et al., 1998b | 85 | 49 | 35 (41) |
| Leaf base<br>Glass et al., 1998a | 11 | 10 | 7 (64) |
| Shoot meristems<br>Zhang et al., pers. comm. | 7 | 7 | 6 (86) |

Numbers in parenthesis indicate the percent of transgenic tissue cultures that produced fertile plants

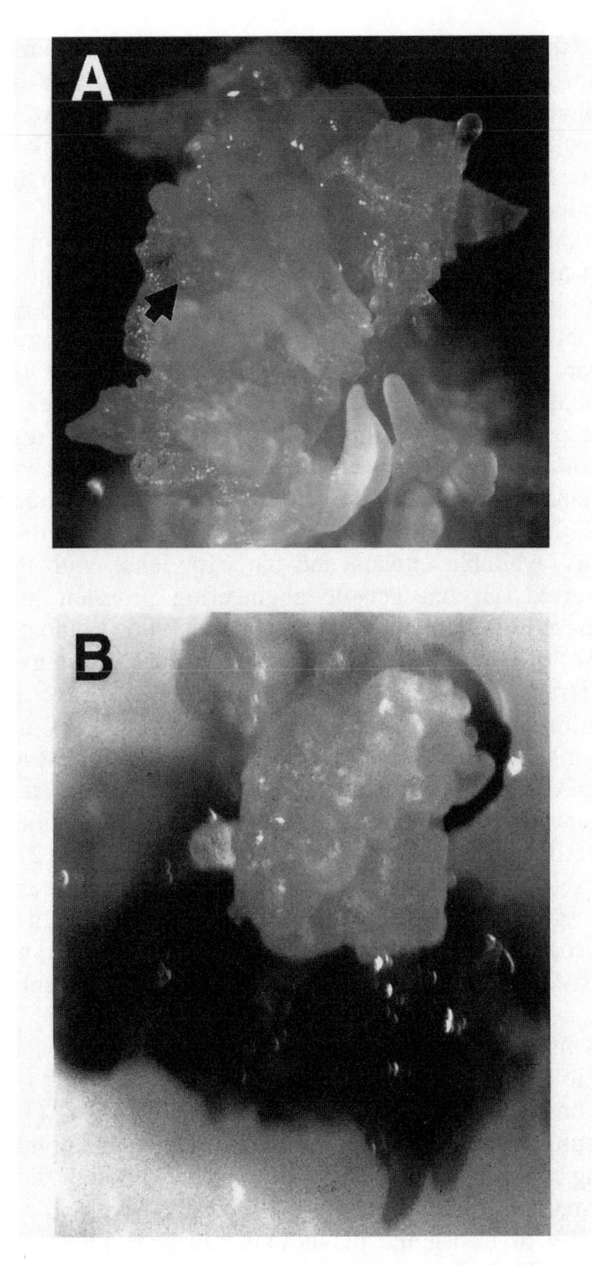


*Fig. 1* (A) Embryogenic callus initiated from mature embryos of GAF/Park seed following eight weeks of incubation. (B) A putatively transgenic tissue culture growing on paromomycin-containing selection medium six weeks after microprojectile bombardment

their expression of the *gfp* gene as fluorescent tissue cultures. Regenerated fertile plants and progeny have since been produced that also express *gfp* (H. Kaeppler, personal communication). These results are very important because visual selection of transgenic tissue cultures likely will reduce metabolic stress on transgenic tissue cultures compared to that imparted by biochemical selection using herbicides or antibiotics.

Mature embryo-derived tissue cultures have been shown to be useful for oat transformation. Torbert et al. (1996, 1998b,c) demonstrated that 8–9 week-old tissue cultures initiated from mature embryos isolated from seed imbibed for 16 h produced about the same number of transgenic tissue cultures per microprojectile bombardment treatment as recovered from bombardment of immature embryo-derived tissue cultures. Plant regeneration frequency and the fertility of the transgenic plants regenerated from mature embryo-derived tissue cultures were greater than observed from the immature embryo-derived tissue cultures (Table 2). As already mentioned, because of the ease of the cost-effectiveness of using this explant as a continuously available explant and our experience with this system, we have converted our oat genetic engineering program to using mature embryo-derived tissue cultures (Torbert et al. 1996, 1998b,c).

Leaf bases of young oat seedlings also are a promising explant for tissue culture and transformation, and have the potential to further reduce the time in tissue culture required to produce transgenic plants. As mentioned, plant regeneration has been achieved from leaf base-derived tissue cultures from a number of oat genotypes (Chen et al. 1995a,b, Gless et al. 1996, 1998b). Gless et al. (1998a) reported successful transformation using direct microprojectile bombardment of leaf base explants (Table 2). Using the *bar* gene as a selectable marker, about 200 bombardment leaf base explants yielded 10 transgenic lines of which 7 produced fertile plants (Table 2). The ability to produce transgenic plants via bombardment of a primary explant (and not a tissue culture) will further stimulate improvements in oat genetic engineering.

Progress in transforming oat via microprojectile bombardment of shoot meristematic cultures also has been reported (Cho et al., 1998, S. Zhang, M-J. Cho and P.G. Lemaux, personal communication) (Table 2). Shoot meristem cultures were initiated from the cv Garry and bombarded after 26 weeks using either the *bar*, *hpt* (hygromycin phosphotransferase) or *nptII* as selectable markers. Transformants were obtained with all selection agents used. Fertility of transgenic plants and transgene inheritance have been demonstrated (Table 2).

One strategy to describe efficiency of a transformation system is to report the number of fertile transgenic plants produced per microprojectile bombardment. This method of determining transformation efficiency describes the labor required to produce fertile, transgenic plants. However, other transformation workers prefer to use the number of fertile transgenic plants produced per bombarded explant as a measure of transformation efficiency. Thus, it is difficult to compare transformation systems that use established tissue cultures versus leaf bases as targets for microprojectile bombardment. For example, in our early work with immature embryo-derived tissue cultures, 0.5–0.8 g callus was bombarded per treatment (Somers et al., 1992, Torbert et al., 1995) and all callus tissue that was used for a complete experiment (up to 30 bombardments) might be produced from a single embryo. Since the callus is friable it is impossible to determine the units of the tissue culture that are bombarded. Obviously the ideal method for determining transformation efficiency would involve determining the number of transformants produced per cells bombarded. However, this method would be very difficult to apply to organized explants. Estimates of the various transformation systems were derived with the assistance of the authors of each report, and I want to express my gratitude to them in text for allowing me to present their data. While the output of each system varies considerably when expressed on a per explant basis, the systems are quite similar when the data are expressed as per bombardment treatment and range from about 0.5 to more than 2 independent fertile transgenic plants produced per bombardment. These results indicate that the systems, at least at this level of comparison, are all about equally effective in terms of the production of fertile transgenic plants. Among the various systems, there is significant variation in the numbers of transgenic tissue cultures that retain the capacity to regenerate fertile plants (Table 2), which likely is influenced by the oat genotypes used, types of tissue culture, the selection system, and duration in tissue culture required to produce transgenic plants.

## Transgene Integration, Inheritance and Stability of Expression

The mechanism of genomic integration of DNA delivered into plant cells using microprojectile bombardment is not well understood. When analyzed in transgenic tissue cultures and regenerated plants, multiple transgene copies and rearranged fragments are usually observed to be integrated into one or two random loci in the recipient genome (Pawlowski and Somers, 1996). This trend also is observed in transgenic oats. There is little published information on transgene integration in oat beyond initial reports by Pawlowski et al. (1996). In an analysis of 23 transgenic plants (Figure 2), the introduced plasmids were integrated as multiple rearranged copies. Examples of these Southern analyses are shown in Figure 2. The plasmid pBARGUS was used in this transformation experiment. All transgenic lines except one exhibited unit length *bar* sequences, whereas two lines did not have unit-length *uidA* sequences. On average, there were 4 copies of the *uidA* gene and about 3 copies of *bar* per genome equivalent as well as numerous larger and smaller transgene-hybridizing fragments. Analyses of

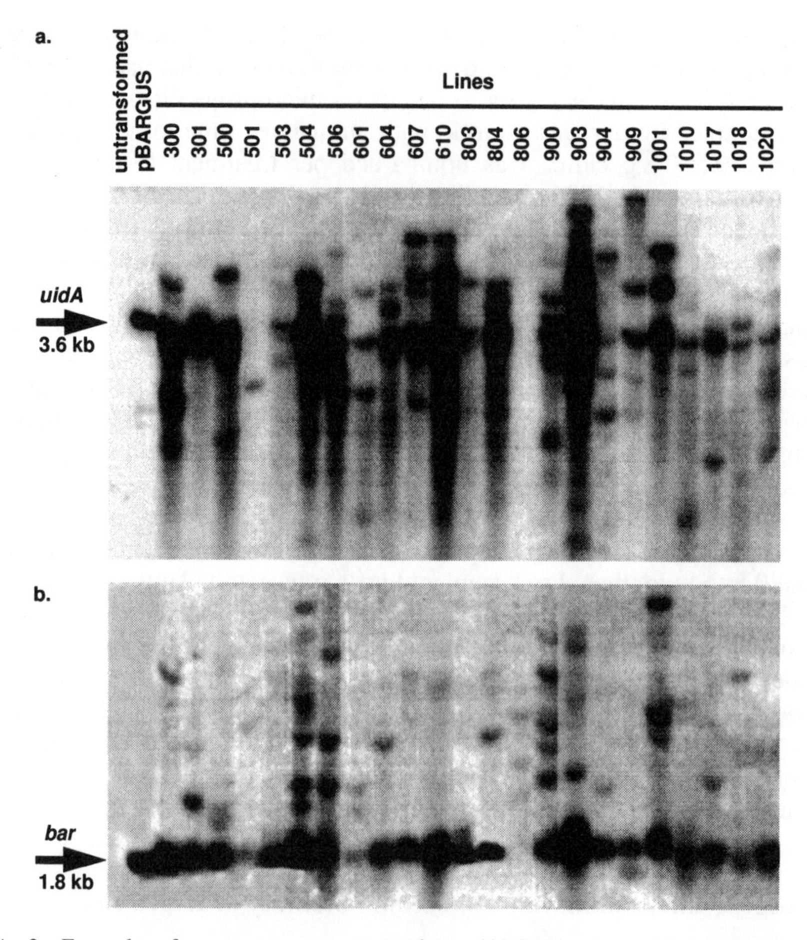


*Fig. 2* Examples of transgene rearrangement in oat (A) DNA extracted from the transgenic plants was digested with restriction enzymes that liberate the unit length *uidA* gene and (B) the *bar* gene

segregation of the transgenes in progeny of the 23 lines indicated that all transgene-hybridizing fragments cosegregated as a single Mendelian locus in 19 lines (Pawlowski, 1996). This observation indicates that all transgene fragments were integrated into a single genomic site in these lines. Segregation analysis indicated that one other line also was single locus but did not produce homozygotes suggesting that the transgenic locus was linked to a lethal mutation causing lack of transgene transmission. Another line segregated for two transgene loci; one active and the other inactive. The remaining two lines appeared to be homozygous, i.e. their hybridization pattern was invariant among all progeny tested whether or not the progeny expressed the transgenic phenotypes. As mentioned, the mechanisms of the integration process and the generation of the rearranged transgene fragments are not understood in oat and warrant further study. However, creation of a hemizygous transgenic line that segregates as a single disomic locus is typical of most transformation systems including those of other hexaploids such as wheat (Weeks et al., 1993, Vasil et al., 1992, 1993, Nehra et al., 1994, Srivastava et al., 1996). The origin of homozygous regenerated transgenic plants also will be further investigated.

An essential prerequisite for the successful use of genetic engineering in the improvement of any crop is to understand the genetic behavior of transgenes in the host plant. The transformation literature reports numerous examples of stable, heritable transgene expression but also is replete with cases of transgene silencing; i.e. instability or turning off of transgene expression. These phenomena are widely documented and several mechanisms have been proposed (Finnegan and McElroy, 1994, Flavell, 1994, Matzke and Matzke, 1995). Moreover, there are reports that indicate that transgene silencing is a frequent observation in plants genetically engineered using microprojectile bombardment (Register et al., 1994, Armstrong et al., 1995). Transgene silencing is frequently observed in transgenic oat plants and their progeny. In the population of 27 transgenic lines characterized by Pawlowski et al. (1996), expression and inheritance of the *bar* and *uidA* transgenes were investigated using: (a) Southern blot analyses to determine cosegregation of the transgenes and their transgene phenotypes (Figure 3), (b) cosegregation of the PAT (Pawlowski et al., 1996) and GUS (Jefferson, 1987, Kosugi et al., 1990) phenotypes since the transgenes were always linked in the different lines, and (c) detection of phenotypic instability indicated by segregation distortion or reversible inactivation of the transgenes. Overall, only one line out of the 27 exhibited perfect expression and Mendelian inheritance of both transgenes and their phenotypes. The remainder of the lines exhibited some level of transgene silencing even though in the majority of lines phenotypically normal and stable homozygous transgenic progeny were identified.

This frequency of transgene silencing appears to be somewhat higher than reports from other plants transformed with microprojectile bombardment (Pawlowski and Somers, 1996) and provides some guidelines to keep in mind for future applications of genetic engineering to oat improvement. One indication is that, because less than 5% of the transgenic events exhibited perfect transgene expression (Pawlowski et al., 1996), production of numerous (probably more than 20) independent transgenic lines is required to insure that a stable line is produced. This estimate is based on the expression of both a selectable marker and a reporter gene that are expected to exhibit little homology with sequences in the oat genome. It should be anticipated that expression of other genes, genes with homology to oat genes or other genetic manipulations, such as antisense, may require even greater numbers of independent transgenic oat lines to identify stable events. Estimating that about one fertile transgenic plant can be produced

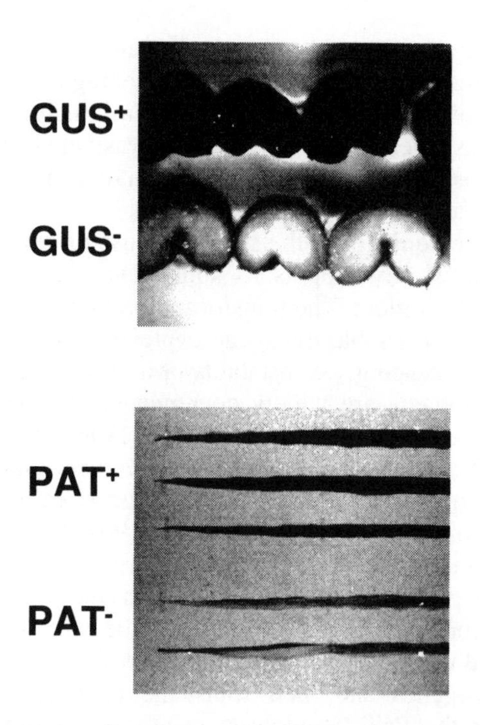


*Fig. 3* Transgene expression phenotypes used to follow transgene expression

per microprojectile bombardment treatment, and taking into account the low frequency of stable transgene-expressing events, it appears that a stable line may be produced for about every 20 or more microprojectile bombardment treatments. These results indicate that more research is required to reduce the high frequency of transgene silencing seen in the oat transformation system. Another observation from this study was that the expression of one transgene is a poor predictor of expression of the other transgene even though they likely are contiguously linked in the integration locus. The final and surprising result was the apparent instability of transgene silencing observed in the majority of transgenic oat lines. This instability, manifested by generational alteration of expression of either transgene, e.g. detection of phenotypically positive progeny produced by phenotypically negative plants and alteration of segregation distortion to normal segregation, raises significant concern about the agronomic usefulness of transgenic oat. While the mechanisms of these phenomena are not understood, their high frequency suggests that transgene expression in the progeny of most lines cannot be expected to be stable. These plants have all been grown in controlled growth chambers presumably under similar environmental conditions over several generations (Pawlowski, 1996). Whether the frequency of transgene silencing in transgenic oat may be influenced by environmental stresses as reported for other transgenic plants (Brandle et al., 1995) remains to be determined. The role the allohexaploid nature of oat in relationship to the high frequency of transgene silencing observed has yet to be determined.

A major confounding problem with the investigation of transgene silencing in transgenic oat and most other plant systems is that the transformation of totipotent cells requires a tissue culture phase (Christou, 1992). Selection of transgenic tissue cultures generally takes from 12–24 weeks and thus for the oat system the total duration in tissue culture can be 32–48 weeks before plants are produced. In oat, the types of tissue culture-induced genetic variation (somaclonal variation) are well characterized (McCoy et al., 1982, Rines et al., 1986). Culture age and the genotype influence the frequency of somaclonal variants isolated from oat tissue cultures. Thus, it is difficult to separate the effects of the various types of tissue culture-induced genetic changes described by Phillips et al. (1994) from transgene silencing. It is possible that the effects are related to the oat genotype but until a higher number of other oat genotypes are transformed and characterized, it will be difficult to determine how these are related.

## Promoter Sequences and Genes for Agronomic Improvement of Oat

Promoters and other gene sequences that specify or enhance transgene expression in the host organism must be identified and characterized to enable their use to regulate transgene expression in the appropriate tissue or stage of development for a particular genetically engineered trait. Investigations of transgene expression patterns controlled by specific promoters are among the first reports of transgenic oat plants beyond the study of expression and inheritance of marker genes. These studies involve fusion of promoter sequences to the *uidA* reporter gene. Transgenic plants and their progeny are then stained for histochemical GUS activity. Torbert et al. (1998a) described the expression pattern conferred by the *Commelina* yellow mottle virus (CoYMV) promoter in transgenic oat. CoYMV is a member of the badnaviruses which infect both monocots and dicots (Lockhart, 1990, Lockhart and Olszewski, 1994). This family of viruses appears to be an important source of useful promoters. The rice tungro virus promoter has been characterized in transgenic rice (Bhattacharyya-Pakrasi et al., 1993, Yin and Beachy, 1995). Both the CoYMV (Torbert et al., 1997) and the rice tungro virus confer transgene expression in the vascular tissue of regenerated plants. In a more recent study, the sugarcane bacilliform virus (SCBV) promoter was shown to confer constitutive transgene expression in oat (Tzafrir, 1998). Further characterization of promoters will likely be required to develop an array of DNA sequences that confer desired expression levels and patterns required for transgenic traits. Tissue specificity of promoter analyzed in transgenic oat is summarized in Table 3.

TABLE 3
GUS expression in transgenic plants

| Tissue | 35S | Promoter | | |
|---|---|---|---|---|
| | | Adh1 | SCBV | CoYMV |
| leaf | + | – | + | + vascular |
| root | + | root tips | + | + vascular |
| stem | + | – | + | + vascular |
| floret | + | – | + | + vascular |
| anther | – | – | +[a] | – |
| ovary | + | – | + | + |
| endosperm | ? | + | + | – |
| embryo | ? | + | + | – |

[a]Expression observed in the central portion of the anther

To date, there are no published reports of successful applications of genetic engineering for trait manipulation in oat improvement; however, several efforts are underway. Investigations of transgenic disease resistance are focused on mechanisms of resistance to BYDV and fungal diseases. Expression of pathogen-derived transgenes is the main strategy being investigated for BYDV resistance (McGrath et al., 1997, Koev et al., 1998). Preliminary reports indicate that expression of BYDV open reading frames, including coat protein, in transgenic oat may confer a minor decrease in viral titre in fertile transgenic plants (McGrath et al., 1997, Koev et al., 1996). However, considerably more research is required to determine the stability of resistance and to develop agronomically significant levels of BYDV resistance. Genes encoding antifungal proteins such as glucanase and chitinase are being introduced into oat to investigate fungal pathogen resistance. Both the BYDV and fungal disease resistance efforts will be long-term considering the relatively unknown mechanisms for resistance to these diseases and the slow process of oat genetic engineering.

The other obvious applications of oat genetic engineering involve modification of groat composition. Again there are no published reports of examples of genetically engineered manipulation of groat composition or processing properties, although numerous possibilities have been reviewed (Burrows and Altosaar, 1995). The allohexaploid nature of the oat genome raises the possibility of an interesting manipulation of groat composition using antisense. For any particular enzyme involved in synthesis of a groat component other than protein, it is anticipated that there will be three homoeologous genes in the allohexaploid genome. The similarity of DNA sequence among these homoeologous copies is expected to be high. The question remains whether the level of sequence identity is sufficient for using a single homoeolog to antisense-inhibit translation from all three homoeologous transcripts. The answer to this question is currently not known for most genes, but if such an antisense approach were successful it would lead to the creation of a dominant null phenotype for the metabolic enzyme targeted. A model pathway to test these types of manipulations would be in attempting to manipulate oat starch composition. Genes encoding numerous enzymes in starch biosynthesis have been isolated and characterized from a number of different cereals and will be useful for isolation and manipulation of the homoeologous genes in oat. Undoubtedly, the greater long-term utility of genetic engineering to oat improvement will be the expansion of the gene pool. Genes from other organisms will be used to alter metabolic pathways and confer disease and pest resistance.

## Conclusions

In formulating conclusions regarding the status of oat genetic engineering, it is important to consider technical limitations and their potential solutions. All extant oat genetic engineering systems utilize tissue cultures as sources of totipotent target cells. Initiation of totipotent tissue cultures of oat is subject to variation due to the explant used, its physiological state and most importantly, its genotype. Recent progress in identification of alternative totipotent explants such as the mature embryo, leaf bases and shoot meristems appears to have reduced the genotype constraints to oat genetic engineering previously imposed by the use of immature embryo-derived tissue cultures. Recent reports of oat genetic engineering using these different explant sources indicate that a number of oat cultivars are being genetically engineered. The ability to genetically engineer oat cultivars currently in production is extremely important for the rapid incorporation of transgenic traits into oat improvement programs.

Tissue-culture-induced genetic variation will continue to be a problem with the current transformation technologies. Because the extent and frequency of genetic changes in oat has been shown to vary by age and genotype of the tissue culture (McCoy et al., 1982, Dahleen et al., 1991), it is possible that the ability to transform a broader array of genotypes may decrease the negative impact of tissue culture-induced genetic variation on the 'quality' of the transgenic plants, thereby minimizing the breeding efforts required to 'clean up' this mutational load. The identification of the new oat tissue culture systems that reduce age of the tissue culture before microprojectile bombardment also likely will decrease the frequency of this tissue culture-induced genetic variation. Nevertheless, all current systems use tissue culture selection and until a non-tissue culture transformation system is developed for oat, tissue culture-induced genetic variation will remain to be a problem. An exciting possibility is that alternative types of oat tissue cultures, such as those described by Zhang et al. (1996), may be subject to reduced frequencies of tissue culture-induced genetic variation. However, the frequency of tissue culture-induced genetic variation detected in plants regenerated from organogenic and embryogenic maize tissue cultures was quite similar (Armstrong et al., 1988) indicating that the usefulness of new oat tissue cultures for reducing genetic variation will need to be further evaluated.

The production efficiency of fertile, transgenic oat plants is adequate for the production of new germplasm using genetic engineering. However, when the low proportion of plants that exhibit stable transgene expression is taken into account, it becomes evident that while the current oat genetic engineering systems reviewed have the potential to produce some useful transgenic plants, the numbers of such plants will be low. One solution to this problem is to continue to improve the production capacity of the existing transformation systems. One prospect for this is to improve DNA delivery frequencies either through improvement in microprojectile delivery technology or by using *Agrobacterium*-mediated DNA delivery. Recent success in the genetic engineering of cereals using *Agrobacterium* (see chapter by Komari and Kubo, this volume) indicate that oat should also be transformable using this system. Whether using *Agrobacterium* will increase the production of transgenic plants or the frequency of stable transgene expression in oat remains to be seen. Other areas for improving the frequency of transgenic plant production would be to identify alternative selection systems, media for improved plant regeneration and culture systems that give rise to a higher frequency of fertile regenerated plants. While all of these improvements are likely to come about as a result of continued improvements of the transformation systems, most practical solutions will come from continued research investigating transgene expression and development of strategies that insulate transgene expression from transgene silencing. This is a major area of basic research in plant molecular genetics. Considering the efficiency of the oat transformation system, it should be quite straight forward to evaluate these new molecular strategies for improving transgene expression in oat.

In summary, the prospects for continued improvement of oat genetic engineering are bright. It seems likely that with the current transformation technologies, genetic engineering will be possible with most oat cultivars. Thus, integration of genetic engineering with plant improvement programs will be considerably more rapid than in other genotype-constrained species. With this in mind, and the efforts currently focused on transgenic approaches to improve disease resistance and modify groat composition and other agronomic traits, it is likely that transgenic oat varieties will become commercially available in the next 5–10 years.

## Acknowledgments

I wish to express my sincere gratitude to Christine Gless, Horst Lörz and Alwine Jahne-Gartner (University of Hamburg) and Shibo Zhang, Myeong-je Cho and Peggy Lemaux (University of California, Berkeley) for providing their prepublication results on oat transformation. I thank Howard Rines and Deon Stuthman for their critical review of the manuscript. The work in my laboratory was supported in part by The Quaker Oats Company. Minnesota Agricultural Experiment Station Publication No. 971130034.

## References

Armstrong, C.L., Parker, B.G., Pershing, J.C., Brown, S.M., Sanders, P.R., Duncan, D.R., Stone, T., Dean, D.A., DeBoer, D.L., Hart, J., Howe, A.R., Morrish, F.M., Pajeau, M.E., Petersen, W.L., Reich, B.J., Rodriguez, R., Santino, C.G., Sato, S.J., Schuler, W., Sims, S.R., Stehling, S., Tarochione, L.J., and Fromm, M.E. (1995) Field evaluation of European corn borer control in progeny of 173 transgenic corn events expressing an insecticidal protein from *Bacillus thuringiensis*. Crop Sci. 35: 550–557.

Armstrong, C.L., and Phillips, R.L. (1988) Genetic and cytogenetic variation in plants regenerated from organogenic and friable, embryogenic tissue cultures of maize. Crop Sci. 28: 363–369.

Bhattacharyya-Pakrasi, M., Peng, J., Elmer, J.S., Laco, G., Shen, P., Kaniewska, M.B., Kononowicz, H., Wen, F., Hodges, T.K., and Beachy, R.N. (1993) Specificity of a promoter from the rice tungro bacciliform virus for expression in phloem tissues. Plant J. 4: 71–79.

Brandle, J.E., McHugh, S.G., James, L., Labbe, H., and Miki, B.L. (1995) Instability of transgene expression in field grown tobacco carrying the *csr1-1* gene for sulfonylurea herbicide resistance. Bio/Technology 13: 944–998.

Bregitzer, P.P., Bushnell, W.R., Rines, H.W., and Somers, D.A. (1991) Callus formation and plant regeneration from somatic embryos of oat (*Avena sativa* L.). Plant Cell Rep. 10: 243–246.

Bregitzer, P.P., Milach, S.K.C., Rines, H.W., and Somers, D.A. (1995) Somatic embryogenesis in oat (*Avena sativa* L.). In: Bajaj, Y.P.S. (ed.) Biotechnology in Agriculture and Forestry, 31: 53–62, Somatic Embryogenesis and Synthetic Seed. Springer-Verlag, Berlin.

Bregitzer, P.P., Somers, D.A., and Rines, H.W. (1989) Development and characterization of friable, embryogenic oat callus. Crop Sci. 29: 798–803.

Burrows, V.D. (1986) Breeding oats for food and feed: conventional and new techniques and materials In: Webster, F.H. (ed.) Oat Chemistry and Technology, pp. 13–43. American Society of Cereal Chemists, St. Paul.

Burrows, V.D., and Altosaar, I. (1995) Biotechnology and oat improvement – progress and prospects. In: Welch, R.W. (ed.) The Oat Crop: Production and Utilization, pp. 533–560. Chapman & Hall, New York.

Campbell, G.L. (1996) Oat and barley as livestock feed – the future. In: Scoles, G., and Rossnagel, B. (eds) Proc. Internat. Oat Conf. & Internat. Barley Genet. Symp., pp. 77–81. University Extension Press, University of Saskatchewan.

Chen, H., Xu, G., Loschke, D.C., Tomaska, L., and Rolfe, B.G. (1995a) Efficient callus formation and plant regeneration from leaves of oats (*Avena sativa* L.). Plant Cell Rep. 14: 393–397.

Chen, Z., Zhuge, Q., and Sundqvist, C. (1995b) Oat leaf base: tissue with an efficient regeneration capacity. Plant Cell Rep. 14: 354–358.

Cho, M-J., Zhang, S., and Lemaux, P.G. (1998) Transformation of shoot meristem tissues of oat using three different selectable markers. Abstract, Society for In Vitro Biology (in press).

Christou, P. (1992) Genetic transformation of crop plants using microprojectile bombardment. Plant J. 2: 275–281.

Clifford, B.C. (1995) Diseases, pests and disorders of oats. In: Welch, R.W. (ed.) The Oat Crop: Production and Utilization, pp. 252–278. Chapman & Hill, New York.

Cuddeford, D. (1995) Oats for animal feed. In: Welch, R.W. (ed.) The Oat Crop: Production and Utilization, pp. 321–358. Chapman & Hill, New York.

Cummings, D.P., Green, C.E., and Stuthman, D.D. (1976) Callus induction and plant regeneration in oats. Crop Sci. 16: 465–470.

Dahleen, L.S., Stuthman, D.D., and Rines, H.W. (1991) Agronomic trait variation in oat lines derived from tissue culture. Crop Sci. 31: 90–94.

Finnegan, J., and McElroy, D. (1994) Transgene inactivation: plants fight back! Bio/Technology 12: 883–888.

Flavell, R.B. (1994) Inactivation of gene expression in plants as a consequence of specific sequence duplication. Proc. Nat. Acad. Sci. USA 91: 3490–3496.

Frolich, W. (1996) Novel food – is there a market for barley and oat. In: Scoles, G., and Rossnagel, B. (eds), Proc. Internat. Oat Conf. & Internat. Barley Genet. Symp., pp. 65–71. University Extension Press, University of Saskatchewan.

Fromm, M.E., Morrish, F., Armstrong, C., Williams, R., Thomas, J., and Klein, T.M. (1990) Inheritance and expression of chimeric genes in the progeny of transgenic maize plants. Bio/Technology 8: 833–839.

Gana, J.A., Sharma, G.C., Zipf, A., Saha, S., Roberts, J., and Wesenberg, D.M. (1995) Genotype effects on plant regeneration in callus and suspension cultures of *Avena.* Plant Cell Tiss. Org. Cult. 40: 217–224.

Gerger, C.J., and Ink, S.L. (1996) The future of oat and barley as functional foods. In: Scoles, G., and Rossnagel, B. (eds), Proc. Internat. Oat Conf. Internat. Barley Genet. Symp., pp. 45–49. University Extension Press, University Saskatchewan.

Gless, C., Lörz, H., and Jahne-Gartner, A. (1996) Development of a highly efficient in vitro regeneration system from leaf bases of oat and barley as a prerequisite for transformation experiments. In: Slinkard, A., Scoles, G., and Rossnagel, B. (eds). Proc. Internat. Oat Conf. & Internat. Barley Genet. Symp., Poster Session Vol. 2: 410–411. University Extension Press, University of Saskatchewan.

Gless, C., Lörz, H., and Jahne-Gartner, A. (1998a) Transgenic oat plants obtained at high efficiency by microprojectile bombardment of leaf base segments. J. Plant Physiol. 152: 151–157.

Gless, C., Lörz, H., and Jahne-Gartner, A. (1998b) Establishment of a highly efficient regeneration system from leaf base segments of oat (*Avena sativa* L.). Plant Cell Rep. 17: 441–445.

Gressel, J. (1992) Indiscriminate use of selectable markers – sowing wild oats? Trends Biotech. 10: 382.

Hiei, Y., Ohta, S., Komari, T., and Kumasho, T. (1994) Efficient transformation of rice (*Oryza sativa* L.) mediated by *Agrobacterium* and sequence analysis of the boundaries of the T-DNA. Plant J. 6: 271–282.

Heyser, J.W., and Nabors, M.W. (1982) Long term plant regeneration, somatic embryogenesis, and green spot formation in secondary oat (*Avena sativa* L.) callus. Z. Pflanzenphysiol. 107: 153–166.

Hoffman, L.A. (1995) World production and use of oats In: Welch, R.W. (ed.) The Oat Crop: Production and Utilization, pp. 34–61. Chapman & Hill, New York.

Ishida, Y., Saito, H., Ohta, S., Hiei, Y., Komari, T., and Kumashiro, T. (1996) High efficiency transformation of maize (*Zea mays* L.) mediated by *Agrobacterium tumefaciens.* Nature Biotechnology 14: 745–751.

Jefferson, R.A. (1987) Assaying chimeric genes in plants: the GUS gene fusion system. Plant Mol. Biol. Rep. 5: 397–405.

Kaeppler, H.F., Menon, G.K., Nuutila, A.M., and Skadsen, R.G. (1997) Delivery and expression of the green fluorescent protein (GFP) gene in oat, barley and maize and utilization as a selectable marker. Agronomy Abstracts, p. 149.

King, I.P., Thomas, H., and Dale, P.J. (1986) Callus induction and plant regeneration from oat cultivars. In: Lawes, D.A., and Thomas, T. (ed.) Proc. 2nd Internat. Oats Conf. pp. 46–47. Aberystwyth, Wales. Martinus Nijhoff Publishers, Dordrecht, Netherlands.

Klein, T.M., Wolfe, E.D., Wu, R., and Sanford, J.C. (1987) High velocity microprojectile for delivery of nucleic acids into living cells. Nature 327: 70–73.

Klein, T.M., Kornstein, L., Sanford, J.C., and Fromm, M.E. (1989) Genetic transformation of maize cells by particle bombardment. Plant Physiol. 91: 440–444.

Koev, G., Mohan, B.R., Dinesh-Kumar, S.P., Torbert, K.A., Somers, D.A., and Miller, W.A. (1998) Extreme reduction of disease in oats transformed with the 5′ half of the barley yellow dwarf virus-PAV genome. Phytopathology 88: 1013–1019.

Kosugi, S., Ohashi, Y., Nakajima, K., and Arai, Y. (1990) An improved assay for $\beta$-glucuronidase in transformed cells: methanol almost completely suppressed endogenous $\beta$-glucuronidase activity. Plant Sci. 70: 133–140.

Kuai, B.K., Dalton, S.J., Bettany, A.J.E., and Morris, P. (1993) Transformation of oats via microprojectile bombardment. J. Exp. Bot. Abstracts p. 44.

Kyozuka, J., Fujimoto, H., Izawa, T., and Shimamoto, K. (1991) Anaerobic induction and tissue specific expression of maize *Adh1* promoter in transgenic rice plants and their progeny. Mol. Gen. Genet. 228: 40–48.

Lockhart, B.E.L. (1990) Evidence for a double-stranded circular genome in a second group of plant viruses. Phytopathol. 80: 127–131.

Lockhart, B.E.L., and Olszewski, N.E. (1994) Badnavirus group. In: Webster, R.G., and Gronoff, A. (eds), The Encyclopedia of Virology, pp. 139–143. Academic Press Inc., San Diego.

Lörz, H., Harms, C.T., and Potrykus, I. (1976) Regeneration of plants from callus in *Avena sativa* L. Z. Pflanzenzuchtg. 77: 257–259.

Matzke, M.A., and Matzke, A.J.M. (1995) How and why do plants inactivate homologous (trans)genes? Plant Physiol. 107: 679–685.

McCoy, T.J., Phillips, R.L., and Rines, H.W. (1982) Cytogenetic analysis of plants regenerated from oat (*Avena sativa*) tissue culture: High frequency of partial chromosome loss. Canad. J. Genet. Cytol. 24: 37–50.

McGrath, P.F., Vincent, J.R., Lei, C-H., Pawlowski, W.P., Torbert, K.A., Gu, W., Kaeppler, H.F., Wan, Y., Lemaux, P.G., Rines, H.R., Somers, D.A., Larkins, B.A., and Lister, R.M. (1997) Coat protein-mediated resistance to isolates of barley yellow dwarf in oats and barley. Eur. J. Plant Pathol. 103: 695–710.

Milach, S.C.K., Rines, H.W., Somers, D.A., Gu, W., and Grando, M. (1992) Improvements in embryogenic callus culture in oat (*Avena sativa* L). In: Internat. Crop Sci. Cong. Abstracts, p. 61. Iowa State University, Ames, Iowa.

Murashige, T., and Skoog, F. (1962) A revised medium for rapid growth and bioassays with tobacco tissue cultures. Physiol Plant 15: 473–497.

Murphy, J.P., and Hoffman, L.A. (1992) The origin, history and production of oat. In: Marshall, H.G., and Sorrells, M.E. (eds), Oat Science and Technology, pp. 1–28. American Society of Agronomy, Madison, Wisconsin.

Nabors, M.W., Heyser, J.W., Dykes, T.A., and DeMott, K.J. (1983) Long duration, high-frequency plant regeneration from cereal tissue culture cells. Planta. 157: 385–391.

Nabors, M.W., Kroskey, C.S., and McHugh, D.W. (1982) Green spots are predictors of high callus growth rates and shoot formation in normal and salt stressed tissue cultures of oat (*Avena sativa* L). Z. Pflanzenphysiol. 105: 341–349.

Nehra, N.S., Chibbar, R.N., Leung, N., Caswell, K., Mallard, C,. Steinhauer, L., Baga, M., and Kartha, K.K. (1994) Self-fertile transgenic wheat plants regenerated from isolated scutellar tissues following microprojectile bombardment with two distinct gene constructs. Plant J. 5: 285–297.

Paton, D. (1986) Oat starch: physical, chemical and structural properties. In: Webster, F.H. (ed.) Oat Chemistry and Technology, pp. 93–118. American Society of Cereal Chemists, St. Paul.

Paton, D., and Fedec, P. (1996) Non-food uses of oat and barley. In: Scoles, G., and Rossnagel, B. (eds), Proc. Internat. Oat Conf. & Internat. Barley Genet. Symp., pp. 58–64. University Extension Press, University of Saskatchewan.

Pawlowski, W.P. (1996) Transgene integration, expression and inheritance in genetically engineered oat. PhD thesis, University of Minnesota.

Pawlowski, W.P., and Somers, D.A. (1996) Transgene inheritance in plants genetically engineered by microprojectile bombardment. Molec. Biotech. 6: 17–30.

Pawlowski, W.P., Torbert K.A., Rines, H.W., and Somers, D.A. (1996) Expression and inheritance of transgenes in genetically engineered oat. In: Scoles, G., and Rossnagel, B. (eds), Proc. Internat. Oat Conf. & Internat. Barley Genet. Symp., Posters Vol. 2, pp. 432–434. University Extension Press, University of Saskatchewan.

Peterson, D.M. (1992) Compositional and nutritional characteristics of oat grain and products. In: Marshall, G.H., and Sorrells, M.E. (eds) Oat Science and Technology, pp. 265–292. American Society of Agronomy, Madison.

Peterson, D.M. (1995) Oat tocols: concentration and stability in oat products and distribution within the kernel. Cereal Chem. 72: 21–24.

Phillips, R.L., Kaeppler, S.M., and Olhoft, P. (1994) Genetic instability of plant tissue cultures: breakdown of normal controls. Proc. Nat. Acad. Sci. USA 91: 5222–5226.

Ranhotra, G.S., and Gelroth, J.A. (1995) Food uses of oats. In: Welch, R.W. (ed.) The Oat Crop: Production and Utilization, pp. 409–432. Chapman & Hill, New York.

Register, III, J.C., Peterson, D.J., Bell, P.J., Bullock, W.P., Evans, E.J., Frame, B., Greeland, A.J., Higgs, N.S., Jepson, I., Kiao, S., Lewnau, C.J., Sillick, J.M., and Wilson, H.M. (1994) Structure and function of selectable and non-selectable transgenes in maize after introduction by particle bombardment. Plant Mol. Biol. 25: 951–961.

Rines, H.W., Johnson, S.S., and Phillips, R.L. (1986) Tissue culture induced variation in oats. In: Lawes, D.A., and Thomas, H. (eds), Proc. 2nd Internat. Oats Conf., pp. 34–38. Martinus Nijhoff, Dordrecht, Netherlands.

Rines, H.W., and Luke, H.H. (1985) Selection and regeneration of toxin insensitive plants from tissue cultures of oats (*Avena sativa*) susceptible of *Helminthosporium victoriae*. Theor. Appl. Genet. 71: 16–21.

Rines, H.W., and McCoy, T.J. (1981) Tissue culture initiation and plant regeneration in hexaploid species of oats. Crop Sci. 21: 837–842.

Rines, H.W., Phillips, R.L., and Somers, D.A. (1992) Applications of tissue culture to oat improvement. In: Marshall, G.H., and Sorrells, M.E. (eds). Oat Science and Technology, Agronomy No. 33, pp. 777–791. American Society of Agronomy, Madison.

Schrickel, D.J. (1986) Oat production value and use: In: Webster, F.H. (ed.) Oat Chemistry and Technology, pp. 1–11. American Society of Cereal Chemists, St. Paul.

Somers, D.A., Rines, H.W., Gu, W., Kaeppler, H.F., and Bushnell, W.R. (1992) Fertile, transgenic oat plants. Bio/Technology 10: 1589–1594.

Somers, D.A., Torbert, K.A., Pawlowski, W.P., and Rines, H.W. (1994) Genetic engineering of oat. In: Henry, R.J. (ed.) Improvement of Cereal Quality by Genetic Engineering, pp. 37–46. Plenum Press, New York.

Somers, D.A., Rines, H.W., Torbert, K.A., Pawlowski, W.P., and Milach, S.K.C. (1996a) Genetic transformation in *Avena sativa* L. (oat). In: Bajaj, Y.P.S. (ed.) Biotechnology in Agriculture and Forestry, 38: 178–190. Plant Protoplasts and Genetic Engineering VII, Springer-Verlag, Berlin, Heidelberg.

Somers, D.A., Torbert, K.A., Pawlowski, W.P., and Rines, H.W. (1996b) Progress in genetic engineering of oat. In: Scoles, G., and Rossnagel, B. (eds). Proc. Internat. Oat Conf. & Internat. Barley Genet. Symp., pp. 230–234. University Extension Press, University of Saskatchewan.

Springer, W.D., Green, C.E., and Kohn, K.A. (1979) A histological examination of tissue culture initiation from immature embryos of maize. Photoplasma 101: 269–281.

Srivastava, V., Vasil, V., and Vasil, I.K. (1996) Molecular characterization of the fate of transgenes in transformed wheat (*Triticum aestivum* L.). Theor. Appl. Genet. 92: 1031–1037.

Torbert, K.A., Rines, H.W., and Somers, D.A. (1995) Use of paromomycin as a selective agent for oat transformation. Plant Cell Rep. 14: 635–640.

Torbert, K.A., Gopalraj, M., Medberry, S.L., Olszewski, N.E., and Somers, D.A. (1998a) Expression of the *Commelina* yellow mottle virus promoter in transgenic oat. Plant Cell Rep. 17: 284–287.

Torbert, K.A., Rines, H.W., and Somers, D.A. (1996) Mature embryos: an alternative tissue culture explant for efficient transformation of oat. In: Scoles, G., and Rossnagel, B. (eds) Proc. Internat. Oat Conf. & Internat. Barley Genet. Symp., Posters 2: 438–439. University Extension Press, University of Saskatchewan.

Torbert, K.A., Rines, H.W., and Somers, D.A. (1998b) Transformation of oat using mature embryo-derived tissue cultures. Crop Sci. 38: 226–231.

Torbert, K.A., Rines, H.W., Kaeppler, H.F., Menon, G.R., and Somers, D.A. (1998c) Genetically engineering elite oat cultivars. Crop Sci. 38: 1685–1687.

Tzazfir, I., Torbert, K.A., Lockhart, B.E.L., Somers, D.A., and Olszewski, N.E. (1998) The sugarcane bacilliform badnavirus promoter is active in both monocots and dicots. Plant Molec. Biol. 38: 347–356.

Vasil, V., Castillo, A.M., Fromm, M.E., and Vasil, I.K. (1992) Herbicide resistant fertile transgenic wheat plants obtained by microprojectile bombardment of regenerable, embryogenic callus. Bio/Technology 10: 667–674.

Vasil, V., Srivastava, V., Castillo, A.M., Fromm, M.E., and Vasil, I.K. (1993) Rapid production of transgenic wheat plants by direct bombardment of cultured immature embryos. Bio/Technology 11: 1553–1558.

Welch, R.W. (1995) Oats in human mutation and health. In: Welch, R.W. (ed.) The oat crop: production and utilization. pp. 433–479. Chapman & Hill, New York.

Weeks, J.T., Anderson, O.D., and Blechl, A.E. (1993) Rapid production of multiple independent lines of fertile transgenic wheat (*Triticum aestivum*). Plant Physiol. 102: 1077–1084.

Yin, Y., and Beachy, R.N. (1995) The regulatory regions of the rice tungro bacilliform virus promoter and interacting nuclear factors in rice (*Oryza sativa* L.) Plant J. 7: 969–980.

Youngs, V.L. (1986) Oat lipids and lipid-related enzymes. In: Webster, F.H. (ed.) Oat Chemistry and Technology, pp. 205–223. American Society of Cereal Chemists, St. Paul.

Zhang, S., Warkentin, D., Sun, B., Zhong, H., and Sticklen, M. (1996) Variation in the inheritance of expression among subclones for subselected (*uidA*) and selected (*bar*) transgenes in maize (*Zea mays* L.). Theor. Appl. Genet. 92: 752–761.

Zhang, S., Zhong, H., and Sticklen, M.B. (1996) Production of multiple shoots from shoot apical meristems of oat (*Avena sativa* L). J. Plant Physiol. 148: 667–671.

# 11. Transgenic Cereals: *Secale cereale* and *Sorghum bicolor* (rye and sorghum)

ANA M. CASTILLO and ANA M. CASAS
*Departamento de Genética y Producción Vegetal, Estación Experimental d e Aula Dei (CSIC), Apdo. 202 – 50080 Zaragoza, Spain.*
*E-mails: amcast@eead.csic.es;acasas@eead.csic.es*

## 1. Introduction

### *1.1 Rye*

Rye, *Secale cereale* L., was first domesticated in Central Asia, Afghanistan, Tibet and Iran. Although world rye grain production decreased from 33.4 million tons in 1990 to 22.6 million tons in 1995, European rye production slightly increased during this period (FAO 1996). Sixty to 70% of the world production of rye is in Central and Eastern Europe, mainly in Russia, Poland, Germany and Byelorussia. Rye is the most winter-hardy of all the winter cereals and is tolerant of low fertility. It is grown in light sandy soils of low pH, under adverse environmental conditions including drought, cold temperatures and at high altitude. Among the cereals, rye has the lowest requirements for fertilizers. It is much prized for making the characteristically dense and slightly sour bread. A rye diet has to be avoided or carefully monitored in young monogastric animals owing to its lower protein content and lower digestion coefficient in comparison to triticale or wheat. However, a mixture of rye grain is fed to adult monogastric animals (Rakowska 1996). It is also used as green forage and for alcohol production.

Rye production is based on population varieties and hybrids. In recent years, grain yield from hybrids has been about 1 t/ha more than from population varieties (Karpenstein-Machan and Maschka 1996). Hybrid and seed production is facilitated by cytoplasmic-genic male sterility (CMS). Although about 20 sources of CMS have been described, only the 'Pampa' type (P) (Geiger and Schnell 1970) has gained commercial importance. There is a need to introduce new CMS sources to increase cytoplasmic genetic variability and reduce vulnerability to diseases and pests. Pampa male sterility is easy to maintain, since non-restorer genotypes are frequently found in all European breeding material. Although effective restorer genes are extremely rare in European germplasm, they have been found in

*I.K. Vasil (ed.), Molecular Improvement of Cereal Crops, 341–360*

Argentinean and Iranian germplasm. Nevertheless, transfer of these genes into current European breeding populations by classical backcrossing requires much time and effort (Geiger and Miedaner 1996).

Breeding objectives vary depending on the end-use of the product. Breeding objectives for the improvement of the feeding value of rye include high protein content, low pentosan content, and low viscosity of soluble pentosans. Pentosans are the main antinutritive compounds of rye (Rakowska 1996). Nevertheless, breeding for good baking performance should consider total and soluble pentosan-rich types and low content of secalins. Pentosans increase dough yield and loaf volume, preserve freshness and increase shelf-life (Weipert 1996).

Susceptibility to lodging and preharvest sprouting produced by frequent rain during grain maturation cause reduction in yield. Molecular marker assisted selection may be a very useful tool for the improvement of these characters in rye (Anderson et al., 1993, Plaschke et al., 1993).

Fungal diseases, particularly leaf rust, brown rust, head blight and powdery mildew, can reduce grain yield by up to 40%, although some genetic improvement in disease resistance can be expected from selection (Kobylanski and Solodukhing, 1996, Miedaner et al., 1996, Sperling et al., 1996).

### *1.2 Sorghum*

Sorghum, *Sorghum bicolor* L. Moench, was domesticated in East Africa several thousand years ago and now ranks fifth in importance among the world's cereals. In 1995 it was grown on some 44 million ha, with a production of 54 million tons (FAO 1996). Of the world sorghum cultivated area, over 80% is in developing countries (22 million ha in Africa, 14.7 million ha in Asia). The crop is often grown under stressful conditions and in acid soils. Sorghum is grown as feed for livestock and poultry in Europe and North America, but it is an important source of human nutrition in Africa and Asia (Doggett, 1988). It is considered to be the second most important cereal after maize, south of the Sahara in Africa. It is used also for brewing beer and for making brooms. When sorghum is used as feed for poultry and pigs, it needs to be supplemented with lysine, threonine and methionine.

The parasitic witchweed (*Striga* spp.) is one of the most serious causes of sorghum crop loss in rainfed agriculture of the semi-arid tropics. The crop is also a favorite target of birds. Sorghum is unique among the cereals for its capacity to produce the polymeric polyphenols known as 'tannins'. Sorghums relatively rich in tannin are generally more resistant to grain-eating birds, to grain molds and weathering and possibly to other production constraints, than are tannin-free sorghums. However, high tannin sorghums generally have lower growth rates and efficiency of feed utilization (Butler 1990). Resistance to aphids, stem borers and sorghum midge are important in the developing world, where the farmers must rely largely upon genetic resistance to these pests.

### *1.3 Target Traits for Transformation*

Genetic improvement of rye and sorghum for agronomic and quality traits, up until recently, has been carried out by traditional plant breeding. The new biotechnological approaches are expected to play a major role in the genetic improvement of these important crops.

Crops resistant to fungal diseases have been obtained by the introduction of pathogenesis-related (PR) proteins including $\beta$-glucanases, chitinases (Lin et al., 1995), and ribosome inactivating proteins or phytoalexin expression. Most of these results have been obtained in tobacco or potato (Logemann et al., 1992, Hain et al., 1993, Zhu et al., 1994, Jach et al., 1995). Resistance obtained with these proteins is effective against a limited spectrum of pathogens. Another strategy for fungal disease resistance is based on inhibition of fungal growth and reproduction by controlled generation of necrotic lesions at infection sites, analogous to the hypersensitive response. More durable resistance to a broader spectrum of fungi is expected with this strategy (Strittmatter et al., 1995).

Concerning resistance to insect pests, recent results in maize and rice have been achieved by introduction of proteinase inhibitors (Duan et al., 1996, Xu et al., 1996) or synthetic genes from *Bacillus thuringensis* coding for toxins (Kozier et al., 1993, Armstrong et al., 1995, Wünn et al., 1996).

Modification of seed protein composition, and increasing the content of essential amino acids could also be suitable targets for transformation in sorghum (Larkins et al., 1990).

Mariani et al. (1990, 1992) described a system to selectively produce male sterile as well as restorer plants. It is based on the specific expression of the *RNase* gene from *Bacillus amyloliquefaciens (barnase)* in the anther tapetum, under control of the TA29 promoter. The ribonuclease destroys the anther tapetal cells and results in male sterile plants. The product of the *barstar* gene is a protein that specifically binds to the ribonuclease, resulting in fertility restoration. This approach provided a new strategy for the production of hybrids as described by Lasa and Bosemark (1993) and has been successfully applied to important crops such as oilseed rape, cabbage, chicory, cotton and maize (see Salamini and Motto, 1993). This strategy could be applied in rye for hybrid production as an alternative to cytoplasmic-genic male sterility, reducing the cost of breeding.

World sorghum production has decreased during the last ten years (it was 57 million tons in 1988) probably because it is often grown under stressful conditions such as in arid areas or in acid soils. Biotechnology, through increasing the tolerance of the crop to acid soils, could allow the expansion of sorghum production in the future. In this regard, it has been shown recently that overproduction of citrate synthase results in aluminium tolerance in transgenic plants of tobacco and papaya (De la Fuente et al., 1997). The demand for sorghum for food is expected to remain strong in those countries where there are limited possibilities to produce alternative staples.

### *1.4 Early Transformation Attempts*

Transformation of rye was reported by De la Peña et al. (1987) by the injection of DNA into developing floral tillers using a hypodermic syringe. Plasmids containing the aminoglycoside phosphotransferase II gene (*APH3′II*) were used. The seeds set on the injected tillers were screened for kanamycin resistance. Out of 4023 seeds, 10 were resistant, and three showed APH3′II activity. The authors postulated that DNA was transported by the plant vascular system into the germ cells. However, no clear evidence of the integration of the gene into the rye genome, or of the stable transmission of the gene to the progeny, was provided. These results have not been reproduced in rye or in any other crop.

The first reports of genetic transformation of sorghum described the introduction of DNA into protoplasts by electroporation and selection of transformed cells, without achieving plant regeneration. Ou-Lee et al. (1986) showed transient expression of the *cat* (chloramphenicol acetyl transferase) gene under control of the 35S cauliflower mosaic virus promoter (CaMV 35S), or a *copia* long terminal repeat promoter of *Drosophila*. Later, Battraw and Hall (1991) obtained stably transformed sorghum calli. These authors reported the expression of the neomycin phosphotransferase II (*nptII*) and *β*-glucuronidase (*uidA*) genes, under control of the CaMV 35S promoter. All transformation attempts produced kanamycin resistant callus clones which contained fragments that hybridized to the *nptII* and/or *uidA* gene probes.

A key to successful transformation of all major cereals (Vasil, 1994, Casas et al., 1995), including rye and sorghum (Casas et al., 1993, 1997, Castillo et al., 1994), was the development of microprojectile bombardment devices for DNA delivery into cells with high morphogenic potential (Sanford, 1988, Klein et al., 1992). This technology was first applied to sorghum by Hagio et al. (1991). A non-regenerable cell suspension was bombarded with tungsten particles coated with DNA encoding for *β*-glucuronidase (GUS), under control of the maize alcohol dehydrogenase 1 promoter and first intron, and the selectable marker genes hygromycin phosphotransferase (*hph*) or *nptII*, allowing selection in the presence of antibiotics. Although sorghum calli showed significant growth inhibition at a concentration of 50 mg/L hygromycin, a high level of tolerance to kanamycin was observed, with strong inhibition of growth only at concentrations as high as 500 mg/L. Transformed colonies resistant to both antibiotics were obtained. However, transcripts from the introduced foreign genes accumulated to detectable levels and were expressed in only two of ten transgenic cell lines analyzed.

## 2. Genetic Transformation

### *2.1 Transformation of Rye*

#### *2.1.1 Plant material and tissue culture*

To establish a transformation system, first it is necessary to develop a reliable and efficient method for regeneration of normal and fertile rye plants. Regenerable embryonic cultures of rye had been established from different explants including immature embryos (Lu et al., 1984, Zimny and Lörz, 1989), immature inflorescences (Krumbiegel-Schroeren et al., 1984, Rakoczy-Trojanowska and Malepszy, 1993), and immature leaves (Linacero and Vázquez, 1986). In most of these studies, regeneration capacity was evaluated after one month in culture. In most of the early work on transformation of cereals, it was necessary to maintain the embryogenic and regeneration capacity for three to four months, during which selection of the transgenic cells was carried out, as reported in maize (Fromm et al., 1990, Walters et al., 1993), or wheat (Vasil et al., 1992).

Since it has been documented that morphogenic competence depends largely on the genotype (Krumbiegel-Schroeren et al., 1984, Linacero and Vàzquez, 1990), we evaluated the capacities of induction of embryogenic calli, maintenance and regeneration using immature leaves, immature inflorescences and immature embryos from the cultivars Ailes, Merced, Wrens Abruzzi and Winter Grazing Bolt. Immature leaves and immature inflorescences were cultured following protocols developed by Linacero and Vázquez (1986, 1990). From immature leaves, friable, non-embryogenic calli were obtained in most of the cases. Frequencies of production of embryogenic calli around 90%, and of regeneration from 1 to 50% were produced from immature inflorescences. However, these calli could not be maintained for more than 3 months in culture. Immature embryos (0.8–1.3 mm) were cultured on RMS2 or RMS3 media (Castillo et al., 1994). Calli were transferred to the same medium every three to four weeks. Three types of calli were obtained from immature embryos: (1) compact, organized and yellow *embryogenic callus*, (2) soft and organized embryogenic callus, and (3) friable and mucilaginous non-embryogenic callus. By careful selection of embryogenic callus at the time of subculture, the soft and organized embryogenic callus could be maintained for over a year.

Between 50 to 70% of the calli remained embryogenic at each subculture. Several media differing in growth regulator composition were assayed for regeneration. The highest regeneration rates were produced on the Murashige and Skoog (1962) medium with vitamins but without any growth regulators (MS0), in which 22–30% of the calli from cultivar Wrens Abruzzi regenerated plants after four months in culture. The rates of regeneration declined to 10–15% after ten months. Among cereals, rye seemed to be one of the most recalcitrant for regeneration. It has been suggested that *in vitro* response of immature embryos and immature inflorescences is a complex trait controlled by many genes acting in different ways, and that the genes controlling regeneration ability are recessive (Rakoczy-Trojanowska and Malepszy, 1993, 1995). Considering all these results, we decided to use immature embryos or calli derived from them as explants for transformation.

#### *2.1.2 DNA delivery – microprojectile bombardment*

All our transformation experiments were performed with the cultivar Wrens Abruzzi, which showed the greatest capacity for maintenance and regeneration of embryogenic callus. Immature embryos (6–12 d after culture initiation) and embryogenic callus of varying ages (18–25 d, 45 d, or 5 months after culture initiation) were cultured on RMS2 or RMS3. Tissue was bombarded with gold particles (1.0 $\mu$m), coated with plasmid DNA using the Biolistic PDS-1000/He device (Castillo et al., 1994). The plasmid pAHC25 (Christensen and Quail, 1996) was used for transformation. This plasmid contains the bacterial *uidA* reporter gene (Jefferson et al., 1987), and the selectable *bar* gene (Thompson et al., 1987) from *Streptomyces hygroscopicus*, encoding for the enzyme phosphinothricin acetyltransferase (PAT), which confers resistance to phosphinothricin (PPT), the active ingredient of the herbicide Basta; both genes are driven by the maize ubiquitin promoter (*Ubi1*), first exon and first intron (Christensen and Quail, 1996).

DNA samples were prepared as described by Sanford et al. (1993), with some modifications (Castillo et al., 1994). Bombardment conditions were: pressure 1100 psi; 1 or 2 shots per Petri dish; 250 $\mu$g gold per Petri dish, independently of the number of shots; amount of DNA 0.8 $\mu$g per shot. Transient expression of *uidA* gene was assayed histochemically, 24 hours after bombardment, following protocols described by Jefferson (1987). An average of 40 to 60 blue spots were obtained per explant; no differences in the number of blue spots were found between one and two shots per dish.

#### *2.1.3 Selection of transformed callus and plant regeneration*

A reliable selection procedure is a prerequisite for development of an efficient transformation system. Considering that only a small percentage of the transiently expressing cells are likely to be stably transformed, it is important to use a selection protocol that confers a growth advantage to the transformed cells, not interfering with developmental processes and maintaining the competence for regeneration.

Two to 15 days after bombardment, the embryos and calli were transferred to MS medium containing 2 mg/L of 2,4-D (MS2) and 50 to 30 mg/L of PPT (P5 to P30) (Castillo et al., 1994). Calli were transferred to fresh selection medium every 3 weeks. Two different strategies were used for selection: (a) increasing the concentration of PPT step-wise from P5 or P10 to P12–P30, and (b) keeping the same concentration of PPT (P5 or P10) in the medium. After two to three subcultures at high concentrations, two more subcultures on P5 were made, before transferring the resistant calli to regeneration medium. Fourteen independent, putative transgenic callus lines were obtained out of 1383 explants. The resistant calli displayed slower growth rates on the selection medium than control calli on non-selective medium. The concentrations of PPT used in this study are higher than the ones used in calli derived from embryos in wheat (Vasil et al., 1993), or calli derived from cell suspensions of maize (Fromm et al., 1990), or oat (Somers et al., 1992). Rye callus seems to be more resistant to PPT than other cereals.

The resistant callus lines were transferred to the regeneration medium, containing Basta, the ammonium salt of PPT (Castillo et al., 1994). To define the optimal concentration of the herbicide for regeneration, control calli were plated in MS0 without Basta and medium containing 1, 5 or 10 mg/L of the herbicide. Regeneration was completely inhibited in medium containing 10 mg/L of Basta, so 5 and 10 mg/L were used for proliferation and regeneration of putative transformed lines. Out of the 14 resistant lines, two (VI-10 and VI-16) – both derived from embryos bombarded 18–25 days after culture – regenerated healthy plants. Plants were vernalized at 4°C for 3–4 weeks in Magenta boxes containing MS0 medium, transferred to soil and hardened off in a growth chamber. Flowering transgenic plants were produced 8–9 months after culture initiation.

### *2.1.4 Analysis of transgenic tissues and plants*

Out of the 14 resistant callus lines, three tested positive for histochemical GUS activity and six showed PAT activity (Castillo et al., 1994). All of the PAT positive callus lines were selected with the step-wise increase of PPT concentration, and five were taken up to P20–P30. All primary regenerants ($R_0$) of lines VI-10 and VI-16 tested positive for PAT activity and showed expression of the *uidA* gene in leaves, roots, pollen grains and ovaries. Topical application of Basta solution (0.001%) to the apical one-third portion of young leaves, of both control and PAT positive plants, was carried out in vernalized plants (3–4 leaves). In the control plants, necrotic areas were observed within 3–4 days, followed by wilting by the seventh day, whereas leaves from PAT positive plants remained healthy.

All plants of both the transformed lines were morphologically similar to seed-derived rye plants (Castillo et al., 1994). All the transgenic plants were male and female fertile. Each spike from $R_0$ transgenic plants was bagged together with a wild-type spike. To shorten the time for evaluation, embryos were rescued 15 to 20 days after pollination and cultured on MS0 for germination. The analysis of the progeny is shown in Table 1. There was female and male transmission of the *bar* and *uidA* genes to the progeny and both genes segregated as dominant Mendelian traits. PAT activity was seen in leaves of 14 out of 30 plants from line VI-10 and 17 out of 33 plants from line VI-16. A very strong GUS activity was observed in leaves of almost all the PAT positive plants. Topical application of Basta to PAT positive plants showed full resistance of 9 plants from both the lines, and partial resistance in 5 and 8 plants from lines VI-10 and VI-16, respectively. These plants with partial resistance showed low levels of PAT activity.

Genomic DNA was isolated from leaves of $R_0$ and $R_1$ plants of each transformed line (VI-10 and VI-16), as well as from control plants. EcoRI digested, and undigested DNA, was probed with a $^{32}P$-labeled 0.5 kb *bar* fragment. EcoRI digestion releases the *bar* coding sequence out of the plasmid, generating a 1.4 kb fragment that was found in digested DNA from plants of lines VI-10 and VI-16, but was absent in control DNA. The patterns of hybridization in the two transgenic lines were different, indicating two independent transformation events. Hybridization patterns in $R_1$ plants were similar to those of $R_0$ plants, indicating that no rearrangements occurred during the first sexual cycle. Hybridization also occurred in the high molecular weight region of undigested DNA, indicating the integration of the transgene into rye nuclear genome. Stable integration of the *uidA* gene was also confirmed by Southern hybridization.

TABLE 1
Analysis of $R_1$ plants from transgenic lines of rye

| Transgenic lines | No. plants analyzed | Expression | | |
|---|---|---|---|---|
| | | PAT (+/–)[1] | GUS (+/–) | Basta (R/PR)[2] |
| VI-10 | 30 | 14/16 | 13/17 | 9/5 |
| VI-16 | 33 | 17/16 | 14/19 | 9/8 |
| Total | 63 | 31/32 | 27/36 | 18/31 |

[1](+/–) Positive and negative activity for PAT and GUS; [2](R/PR) Resistance and partial resistance to Basta application

## 2.2 *Transformation of Sorghum*

### 2.2.1 *Plant material and tissue culture*

Plant regeneration from cultured immature embryos (Dunstan et al., 1978, 1979), mature embryos (Cai et al., 1987), shoot segments of seedlings (Bhaskaran and Smith, 1988), immature inflorescences (Brettell et al., 1980, Cai and Butler, 1990) and leaf segments (Wernicke and Brettell, 1980) of sorghum has been described. Also, Gendy et al. (1996) described a method for direct somatic embryogenesis and rapid plant regeneration from epicotyl transverse thin cell layers. We focused on immature embryos and immature inflorescences assuming that morphogenesis from these explants occurs predominantly via somatic embryogenesis rather than organogenesis (Dunstan et al., 1978, Brettell et al., 1980, Boyes and Vasil, 1984). We also considered the accessibility of bombarded microprojectiles to responsive cells, which are located within epidermal or subepidermal cell layers, and not inside the cultured tissue as it is the case for shoot tips (Bhaskaran et al., 1988). Protocols for regeneration via the embryogenesis pathway for both types of explants developed in the laboratory of the late Dr. Larry G. Butler (Department of Biochemistry, Purdue University, USA), were used with some modifications. These are described in Casas et al. (1993, 1997).

We evaluated immature embryos and inflorescences of several sorghum genotypes (CS3541, M91051, SRN39, IS4225, Tx430, P898012, P954035, and PP290), representing a range of genetic backgrounds and a variety of agronomic types, for morphogenic competence. Substantial differences were found among genotypes in their response to culture conditions that assure embryogenic callus induction, callus proliferation and plant regeneration. The most responsive genotypes were selected for transformation. Thus, we focused on immature embryos of P898012 and P954035, two drought tolerant cultivars, and immature inflorescences of M91051, a food grain, and SRN39, a Striga-resistant cultivar.

### 2.2.2 *DNA delivery – microprojectile bombardment*

Immature sorghum embryos (1–2 mm), or segments of 4 mm from immature inflorescences (1.5–3.0 cm), were used as explants for bombardment.

Microprojectile-mediated DNA delivery was accomplished with the Biolistic PDS 1000/He system. The plasmids used (pPHP620, pPHP687, pPHP1528 and pPHP3528) were provided by Pioneer Hi-Bred International (Johnston, IA). The plasmid pPHP620 contains the *uidA* and *bar* genes. pPHP687 contains the maize anthocyanin regulatory elements *R* (Ludwig et al., 1989) and *C1* (Paz-Ares et al., 1987). pPHP1528 contains the *luc* gene (Ow et al., 1986) encoding for firefly luciferase, and pPHP3528 contains the *bar* gene. In each plasmid, the genes are driven by a double CaMV 35S promoter (Kay et al., 1987), with the ΩRNA leader sequence (Gallie et al.,

1987) and the first intron of the maize alcohol dehydrogenase gene (Callis et al., 1987). Coating of DNA onto tungsten or gold microprojectiles followed published protocols (Sanford et al., 1993). Prior to bombardment, partial desiccation of the explants was carried out as described in Casas et al. (1993, 1997).

Optimal parameters for transformation were established by evaluating the transient expression of GUS in immature embryos and immature inflorescences, 1 to 5 days following bombardment. In addition, in immature embryos, the accumulation of anthocyanin after expression of the maize transcriptional activators appeared to be a better way to evaluate the efficiency of transformation, non-destructively. The optimal conditions for DNA delivery were: pressure 1100 psi; for immature embryos, 1.5–3 $\mu$m gold or 1.7 $\mu$m tungsten, and for immature inflorescences, 1 $\mu$m gold or tungsten. Particles were used at a concentration of 500 $\mu$g per shot with gold or 125 $\mu$g per shot with tungsten; 1 or 2 $\mu$g DNA per shot; 1 shot per Petri dish with embryos and 2 shots with inflorescences. With the optimized bombardment parameters, the average GUS expression was 6 or 17 blue spots per embryo, in P898012 or P954035 embryos, respectively, or 12 blue spots per segment in SRN39 immature inflorescences. A much higher value was obtained when anthocyanin accumulation was evaluated, with an average of 84 or 113 red spots per embryo, in P898012 or P954035 embryos, respectively (Casas et al., 1993).

### 2.2.3 *Selection of transformed callus and plant regeneration*

Initially, the antibiotic kanamycin was evaluated as a selective agent for sorghum. Nevertheless, as has been described for other cereal species, control calli grew well on high concentrations of this antibiotic (up to 400 mg/L), but finally lost their embryogenic competence. It has been reported that sorghum tissues release phenolic compounds into the medium (Oberthur et al., 1983, Cai and Butler, 1990). This phenomenon was enhanced in the presence of kanamycin. For this reason, we decided to use bialaphos as an alternative selective agent, since it had been shown to be extremely useful for other cereals (Gordon-Kamm et al., 1990, Christou et al., 1991).

The selection strategy is a compromise that involves the level of explant organization, the type and concentration of the selective agent, and the timing and duration of the selection pressure. To establish an acceptably efficient transformation system, it was critical to develop a separate selection strategy for each explant type. For immature embryos (Casas et al., 1993), selection was applied immediately after bombardment, with 1 mg/L bialaphos, in the induction medium. Two weeks later, the calli that had just initiated were transferred to maintenance medium containing 3 mg/L bialaphos. A minimum of three more transfers, two weeks each, to media with the same concentration of herbicide appeared to be necessary to identify putative transgenic callus. In the case of immature inflorescences (Casas et al., 1997), even 1 mg/L bialaphos inhibited callus induction. In control, non-bombarded immature inflorescences, embryogenic callus formation is not visible until 3–4 weeks after culture initation, and somatic embryos are formed after 5–6 weeks (Brettell et al., 1980, Boyes and Vasil, 1984, Cai and Butler, 1990). The selection protocol developed for immature inflorescences was two weeks culture of primary explants in the absence of herbicide to allow embryogenic callus initiation and proliferation, and subsequent subculture on induction and maintenance media containing 3 mg/L bialaphos. Calli were subcultured on maintenance medium for at least three months while selecting for surviving and embryogenic cells. Calli that survived recurrent selection on maintenance medium were subsequently transferred to regeneration medium supplemented with 3 mg/L bialaphos. Five to seven cm tall shoots were transferred to rooting medium containing 1 mg/L bialaphos.

From 1400 bombarded immature embryos, 56 transgenic plants were regenerated, representing 4 different transformation events (1119, 1409, 2251 and 2441; Table 2).

Many calli derived from immature inflorescences survived selection on 3 mg/L bialaphos during the proliferation and maintenance stage of culture, but it was necessary to reduce the selection pressure in order to regenerate plants. Out of 60 bombarded plates with segments of immature inflorescence of SRN39, 191 plants were obtained on media with 1 mg/L bialaphos or without herbicide, and only five of those plants were found to be transgenic, representing two independent transformation events (1702 and 1752). Histological analysis showed that, in the presence of the herbicide, plant regeneration from these explants was mainly via organogenesis through the differentiation of shoot meristems (Casas et al. 1997).

The entire regeneration process from plating in tissue culture to acclimation in the greenhouse took 4–8 months, depending on the explant and selection strategy applied.

TABLE 2

Phenotypic and molecular analysis of transgenic sorghum plants

| | | | | Expression | | | |
|---|---|---|---|---|---|---|---|
| Plant | Origin[1] | Genotype | Fertility[2] | Phenotype | PAT | GUS | LUC |
| 1119 | i.e. | P898012 | fertile | normal | yes | no | – |
| 1409 | i.e. | P898012 | sterile | abnormal | yes | no | – |
| 2251 | i.e. | P898012 | fertile | normal | yes | – | no |
| 2441 | i.e. | P898012 | fertile | normal | yes | – | yes |
| 1702 | i.i. | SRN39 | fertile | normal | yes | no | – |
| 1752 | i.i. | SRN39 | m.s. | normal | yes | no | – |

[1]i.e. immature embryo; i.i. immature inflorescence; [2]m.s. – male sterile

*2.2.4 Analysis of transgenic tissues and plants*

Expression of the introduced genes was evaluated in young $R_0$ plants (20 cm tall, 3 to 5 leaves), and their progeny. GUS activity could not be detected later than three weeks after bombardment, either in calli or in any of the regenerated plants. For this reason, herbicide resistance was chosen for initial screening of putative transformants. It was performed by local application of a 0.6% aqueous solution of the herbicide Ignite/Basta (Hoechst), onto the surface of young leaves. The effect of the herbicide was evaluated for at least four days after treatment. The leaves of control or untransformed plants suffered substantial necrosis within 48 hr and eventually died. Often, the symptoms spread to the surrounding leaves, leading to the death of the entire tiller. All the plants derived from embryos were resistant to the herbicide, and were shown to contain the *bar* gene by Southern blot analysis. However, most of the plants derived from inflorescences were escapes and died after application of the herbicide. Plant 1702 came from a tiller that was produced by a herbicide susceptible plant, indicating that it was derived from a chimeric sector in the original regenerant. Southern analysis of more than 29 herbicide susceptible plants, derived from different bombardments, indicated that susceptibility was not due to silencing of the transgene. Thus, selection during regeneration is essential.

Five of the six transformation events produced normal plants, similar to seed-derived sorghum plants (Table 2). Southern analysis (Casas et al., 1993, 1997) indicated the presence of the introduced genes, with different patterns of integration. The number of insertions of the transgene *bar* ranged from one (1119 and 1702) to three (2441). In vitro *PAT assay* for acetylation of PPT (Spencer et al., 1990) provided further evidence for expression of the transgene in all the transgenic plants. As mentioned above, we did not observe GUS expression in any of the transformed plants that were studied, even though the presence of the *uidA* gene was detected by Southern analysis (Casas et al., 1997). Digestion of genomic DNA with an enzyme that linearizes the plasmid, and cuts outside of the *uidA* coding sequence, detected one (1702) to five (1119) insertions of this gene. Although we did not look for the intact full length of the *uidA* cassette, the lack of expression implicates the likelihood of transgene inactivation (Finnegan and McElroy, 1994, Matzke and Matzke, 1995). The other two transformations (2251 and 2441) were performed using the reporter gene *luc*, and one of the transgenic lines gave rise to plants expressing luciferase.

For five normal transformation events, the plants were selfed or back-crossed to seed derived control plants. The seeds were collected and planted on soil. When the seedlings were two weeks old, they were sprayed with a diluted solution of herbicide and, resistant and susceptible seedings were scored. In all the cases the herbicide resistance data agreed with Mendelian segregation for one or two independent loci, with the exception of the progeny from 1702 (Table 3). Transmission of the *bar* gene to $R_2$ and $R_3$

generations, and expression in progenies of those five transgenic events, was also confirmed.

## 3. Conclusions and Future Prospects

Transgenic rye and sorghum plants have been obtained by microprojectile bombardment of both immature embryos and/or immature inflorescences. However, the results reported in this communication suggest that it is still too early for this technology, in its present state, to be suitable for routine use in a crop improvement program, because of the low efficiency and genotype-dependence of the system.

There are two points that we consider very critical for transformation. The first is the establishment of regenerable tissues. It is necessary to use donor plants grown under optimal environmental conditions, since the capacity for embryogenesis and plant regeneration depends greatly on the environment. We note that three out of the six PAT positive callus lines of rye were obtained from cultures initiated at the same time from field grown material, and regeneration of plants was achieved only from two of these three lines. Initial attempts in both crops were carried out using established embryogenic calli, two months old in sorghum, and up to five months old in rye. Due to the drastic loss of regeneration capacity in rye with the time in culture, and also the low transient expression of the *uidA* gene in sorghum, bombardment should be done as soon as possible. For this reason we used primary explants. There are several advantages that justify their use: i) cells of immature explants show a high degree of competence for embryogenesis; ii) one can target individual cells before the process of differentiation begins, reducing the probability of obtaining chimeras, and iii) the use of established embryogenic callus as target increases the length of time in tissue culture and might result in an increased loss of totipotency.

TABLE 3

Analysis of $R_1$ plants from transgenic lines of sorghum

| $R_0$ plants | | $R_1$ plants | |
|---|---|---|---|
| Plant | bar Insertions | R/S | Segregation |
| 1119 | 1 | 432:117 | 3:1 |
| 1702 | 1 | 277:1 | 256:1 |
| 1752 | 2 | 78:29 | 3:1 |
| 2251 | 2 | 81:27 | 3:1 |
| 2441 | 3 | 70:26 | 3:1 |

R/S = Resistant or susceptible to Basta application

Second, is the establishment of a selection process. Step-wise selection seemed to be more efficient for rye than keeping the concentration constant throughout selection. High concentration of PPT (20 to 30 mg/L) is required for efficient selection of the callus lines. Nevertheless, plants are more sensitive to herbicide than callus tissues. No plants were regenerated in medium containing 10 mg/L of Basta from any of the PAT negative lines, which showed the importance of keeping the selection during regeneration phase. Similar results were obtained for sorghum, since gradual increases in the concentration of bialaphos up to 3 mg/L were efficient for selecting transgenic embryo-derived cell lines. Regarding immature inflorescences, it is critical to maintain selection pressure during regeneration in order to produce true transgenic plants.

A comparison of selective agents for use with the selectable marker gene *bar* in maize and wheat transformation, revealed that bialaphos is a more potent selective agent than glufosinate in maize (Dennehey et al., 1994), or than PPT in wheat (Altpeter et al., 1996). Bialaphos could be used as an alternative selective agent for rye.

There are several constraints that should be considered when endeavoring to improve the transformation system. Regenerable tissue cultures are an important component of the transformation system. Manipulations in the conditions of tissue culture should aim to increase regeneration frequency. Increased levels of micronutrients, such as $Cu^{2+}$ and $Ag^{2+}$ in the regeneration medium have been shown to enhance regeneration rates in wheat, maize, barley and triticale (Purnhauser et al., 1987, Vain et al., 1989, Purnhauser and Gyulai 1993, Dahleen 1995, Castillo et al., 1998). *Partial desiccation* of the callus before transfer to regeneration medium has been reported to be useful in rice and barley (Tsukuhara and Kirosawa 1992; Castillo et al., 1998). These changes may be helpful to other gramineous species, including rye and sorghum.

Sorghum transformation was achieved using genes under control of the CaMV 35S promoter. The use of stronger monocot promoters such as actin1 (McElroy et al., 1990) from rice or *ubi1* from maize, may give higher transient expression in sorghum tissues, as demonstrated in rye, and thus facilitate optimization of DNA delivery conditions. Higher transient expression may also result from a better distribution of the microprojectiles, using baffles or meshes, and also the use of osmotic pretreatments (Vain et al., 1993).

There are other reporter genes encoding for proteins such as firefly luciferase or the green fluorescent protein from the jellyfish *Aequorea victoria* (Sheen et al., 1995), which allow a non-destructive screening and selection of the putative transgenic tissues and plants. Recently, visible selection of transformed cells based on luciferase activity in cassava (Raemakers et al., 1996) and sugarcane (Bower et al., 1996) has been reported. This system does not require the use of a selectable marker and it would be interesting to test it in rye and sorghum. In one transformation experiment, transgenic sorghum plants expressing the luciferase protein were obtained, although more data are required.

Although transgenic plants with herbicide resistance genes were successfully produced in these studies, the current system is not yet optimized for routine transformation of rye or sorghum. The environmental risk of the possible transfer of the transgenes (especially herbicide resistance genes) due to gene flow from the transgenic plants to wild relatives, such as *Secale* spp. or *Sorghum halepense* (Johnsongrass) (Arriola and Ellstrand, 1996), will clearly limit the future use of these plants. Alternative selection systems based on antibiotics might be more environmentally acceptable.

For a practical application of this technology, another possibility that should be considered is the selective elimination of the selectable marker away from a trait of interest (Yoder and Goldsbrough, 1994). Promising alternatives are also co-transformation and subsequent segregation, as demonstrated recently in tobacco and rice using *Agrobacterium* transformation (Komari et al., 1996), site-specific recombination (Odell et al., 1994, Bar et al., 1996), or gene replacement (Morton and Hooykaas, 1995).

## Acknowledgements

The work on rye transformation was conducted under the guidance of Professor Indra K. Vasil (Laboratory of Plant Cell and Molecular Biology, University of Florida, Gainesville, FL, USA). Sorghum transformation was conducted under the guidance of Professors Paul M. Hasegawa and Ray A. Bressan (Center for Plant Environmental Stress Physiology, Purdue University, West Lafayette, IN, USA). The authors acknowledge the valuable help they received from Drs Vimla Vasil, Andrzej Kononowicz, Usha Zehr and Dwight Tomes. Rye research was supported by funds provided to Professor Vasil by Monsanto Co. (St. Louis, MO). Sorghum research was supported by Pioneer Hi-Bred International (Johnston, IA), and grants from the Consortium for Plant Biotechnology Research Inc., the McKnight Foundation and USAID Grant DAN 254-G-00-002-00 through the International Sorghum and Millet Collaborative Research Support Program.

## References

Altpeter, F., Vasil, V., Srivastava, V., Stöger, E., and Vasil, I.K. (1996) Accelerated production of transgenic wheat (*Triticum aestivum* L.). Plant Cell Rep. 16: 12–17.

Armstrong, C.L., Parker, G.B., Pershing, J.C., Brown, S.M., Sanders, P.R., Duncan, D.R., Stone, T., Dean, D.A., DeBoer, D.L., Hart, J., Howe, A.R., Morrish, F., Pajeau, M.E., Petersen, W.L., Reich, B.J., Rodriguez, R., Santino,C.G., Sato, S.J., Shuler, W., Sims, S.R., Stehling, S., Tarochione, L.J., and Fromm, M.E. (1995) Field evaluation of European corn borer control in progeny of 173 transgenic corn events expressing an insecticidal protein from *Bacillus thuringiensis*. Crop Sci. 35: 550–557.

Anderson, J.A., Sorrells, M.E., and Tansksley, S.D. (1993) RFLP analysis of genomic regions associated with resistance to preharvest sprouting in wheat. Crop Sci. 33: 453–459.

Arriola, P.E., and Ellstrand, N.C. (1996) Crop-to-weed gene flow in the genus *Sorghum* (Poaceae): Spontaneous interspecific hybridization between johnsongrass, *Sorghum halepense*, and crop sorghum, *S. bicolor*. Amer. J. Bot. 83: 1153–1160.

Bar, M., Leshem, B., Gilboa, N., and Gidoni, D. (1996) Visual characterization of recombination at *FRT-gusA* loci in transgenic tobacco mediated by constitutive expression of the native FLP recombinase. Theor. Appl. Genet. 93: 407–413.

Battraw, M., and Hall, T.C. (1991) Stable transformation of *Sorghum bicolor* protoplasts with chimeric neomycin phosphotransferase II and $\beta$-glucuronidase genes. Theor. Appl. Genet. 82: 161–168.

Bhaskaran, S., and Smith, R.H. (1988) Enhanced somatic embryogenesis in *Sorghum bicolor* from shoot tip culture. In Vitro Cell Dev. Biol. 24: 65–70.

Bhaskaran, S., Neumann, A.J., and Smith, R.H. (1988) Origin of somatic embryos from cultured shoot tips of *Sorghum bicolor* (L.) Moench. In Vitro Cell Dev. Biol. 24: 947–950.

Bower, R., Elliott, A.R., Potier, B.A.M., and Birch, R.G. (1996) High-efficiency, microprojectile-mediated co-transformation of sugarcane, using visible or selectable markers. Mol. Breed. 2: 239–249.

Boyes, C.J., and Vasil, I.K. (1984) Plant regeneration by somatic embryogenesis from cultured young inflorescences of *Sorghum arundinaceum* (Desv.) Stapf. var. Sudanense (Sudan grass). Plant Sci. Let. 35: 153–157.

Brettell, R.I.S., Wernicke, W., and Thomas, E. (1980) Embryogenesis from cultured immature inflorescences of *Sorghum bicolor*. Protoplasma 104: 141–148.

Butler, L.G. (1990) The nature and amelioration of the antinutritional effects of tannins in sorghum grain. In: Ejeta, G., Mertz, E.T., Rooney, L., Schaffert, R., and Yoe, J. (eds), Proc. Internat. Conf. Sorghum Nutritional Quality, pp. 191–205. Purdue University, West Lafayette, Indiana.

Cai, T., Daly, B., and Butler, L. (1987) Callus induction and plant regeneration from shoot portions of mature embryos of high tannin sorghums. Plant Cell Tiss. Org. Cult. 9: 245–252.

Cai, T., and Butler, L. (1990) Plant regeneration from embryogenic callus initiated from immature inflorescences of several high-tannin sorghums. Plant Cell Tiss. Org. Cult. 20: 101–110.

Callis, J., Fromm, M., and Walbot, V. (1987) Introns increase gene expression in cultured maize cells. Genes Dev. 1: 1183–1200.

Casas, A.M., Kononowicz, A.K., Zehr, U.B., Tomes, D.T., Axtell, J.D., Butler, L.G., Bressan, R.A., and Hasegawa, P.M. (1993) Transgenic sorghum plants via microprojectile bombardment. Proc. Nat. Acad. Sci. USA 90: 11212–11216.

Casas, A.M., Kononowicz, A.K., Bressan, R.A., and Hasegawa, P.M. (1995) Cereal transformation through particle bombardment. Plant Breed Rev. 13: 235–264.

Casas, A.M., Kononowicz, A.K., Haan, T.G., Zhang, L., Tomes, D.T., Bressan, R.A., and Hasegawa, P.M. (1997) Transgenic sorghum plants obtained after microprojectile bombardment of immature inflorescences. In Vitro Cell Dev. Biol. 33P: 92–100.

Castillo, A.M., Vasil, V., and Vasil, I.K. (1994) Rapid production of fertile transgenic plants of rye (*Secale cereale* L.). Bio/Technology 12: 1366–1371.

Castillo, A.M., Egaña, B., Sanz, J.M., and Cistué, L. (1998) Somatic embryogenesis and plant regeneration from barley cultivars grown in Spain. Plant Cell Rep. (in press).

Christensen, A.J., and Quail, P.H. (1996) Ubiquitin promoter-based vectors for high-level expression of selectable and/or screenable marker genes in monocotyledonous plants. Transgen. Res., 5: 213–218.

Christou, P., Ford, T.L., and Kofron, M. (1991) Production of transgenic rice (*Orzya sativa* L.) from agronomically important Indica and Japonica varieties via electric discharge particle acceleration of exogenous DNA into immature zygotic embryos. Bio/Technology 9: 957–96.

Dahleen, L.S. (1995) Improved plant regeneration from barley callus cultures by increased copper levels. Plant Cell Tiss. Org. Cult. 43: 267–269.

De la Fuente, J.M., Ramírez-Rodríguez, V., Cabrera-Ponce, J.L., and Herrera-Estrella, L. (1997) Aluminium tolerance in transgenic plants by alteration of citrate synthesis. Science 276: 1566–1568.

De la Peña, A., Lörz, H., and Schell, J. (1987) Transgenic rye plants obtained by injecting DNA into young floral tillers. Nature 325: 274–276.

Dennehy, B.K., Petersen, W.L., Ford-Santino, C., Pajeau, M., and Armstrong, C.L. (1994) Comparison of selective agents for use with the selectable marker gene *bar* in maize transformation. Plant Cell Tiss. Org. Cult. 36: 1–7.

Doggett, H. (1988) Sorghum (2nd Edition). Longman Scientific & Technical, Essex.

Duan, X., Li, X., Xue, Q., Abo-El-Saad, M., Xu, D., and Wu, R. (1996) Transgenic rice plants harboring an introduced potato proteinase inhibitor II gene are insect resistant. Nature Biotechnology. 14: 494–498.

Dunstan, D.I., Short, K.C., and Thomas, E. (1978) The anatomy of secondary morphogenesis in cultured scutellum tissues of *Sorghum bicolor*. Protoplasma 97: 251–260.

Dunstan, D.I., Short, K.C., Dhaliwal, H., and Thomas, E. (1979) Further studies on plantlet production from cultured tissues of *Sorghum bicolor*. Protoplasma 101: 355–361.

FAO (1996) FAO Production Yearbook 1995, Vol. 49. Food and Agriculture Organization of the United Nations, Rome.

Finnegan, J., and McElroy, D. (1994) Transgene inactivation: Plants fight back. Bio/ Technology 12: 883–888.

Fromm, M.E., Morrish, F., Armstrong, C., Williams, C., Thomas, J., and Klein, T.M. (1990) Inheritance and expression of chimeric genes in the progeny of transgenic maize plants. Bio/Technology 8: 833–839.

Gallie, D.R., Sleat, D.E., Watts, J.W., Turner, P.C., and Wilson, T.M.A. (1987) The 5′-leader sequence of tobacco mosaic virus RNA enhances the expression of foreign gene transcripts in vitro and in vivo. Nuc. Acids Res. 15: 2871–2888.

Geiger, H.H., and Schnell, F.W. (1970) Cytoplasmic male sterility in rye (*Secale cereale* L.). Crop Sci. 10: 590–593.

Geiger, H.H., and Miedaner, T. (1996) Genetic basis and phenotype stability of male fertility restoration in rye. In: Geiger, H.H., Deimling, S., Miedaner, T., and Schilling, A.G. (eds). Proc. Internat. Symp. Rye Breed Genet. Vort Pflanzenzüchtg 35: 27–38.

Gendy, C., Séne, M., Van Le, B., Vidal, J., and Tran Thanh Van, K. (1996) Somatic embryogenesis and plant regeneration in *Sorghum bicolor* (L.) Moench. Plant Cell Rep. 15: 900–904.

Gordon-Kamm, W.J., Spencer, T.M., Mangano, M.L., Adams, T.R., Daines, R.J., Start, W.G., O'Brien, J.V., Chambers, S.A., Adams, W.R., Willets, N.G., Rice, T.B., Mackey, C.J., Krueger, R.W., Kausch, A.P., and Lemaux, P.G. (1990) Transformation of maize cells and regeneration of fertile transgenic plants. Plant Cell. 2: 603–618.

Hagio, T., Blowers, A.D., and Earle, E.D. (1991) Stable transformation of sorghum cell cultures after bombardment with DNA-coated microprojectiles. Plant Cell Rep. 10: 260–264.

Hair, R., Reif, H.J., Krause, E., Langebartels, R., Kindl, H., Vornam, B., Wiese, W., Schmelzer, E., Schreier, P.H., Stöcker, R.H., and Stenzel, K. (1993) Disease resistance results from foreign phytoalexins expression in a novel plant. Nature 361: 153–156.

Jach, G., Görnhardt, B., Mundy, J., Logemann, J., Pinsdorf, E., Leah, R., Schell, J., and Maas, C. (1995) Enhanced quantitative resistance against fungal disease by combinatorial expression of different barley antifungal proteins in transgenic tobacco. Plant J. 8: 97–109.

Jefferson, R.A. (1987) Assaying chimeric genes in plants: the GUS gene fusion system. Plant Mol. Biol. Rep. 5: 387–405.

Jefferson, R.A., Kavanagh, T.A., and Bevan, M.W. (1987) GUS fusions: $\beta$-glucuronidase as a sensitive and versatile gene fusion marker in higher plants. EMBO J. 6: 3901–3907.

Karpenstein-Machan, M., and Maschka, R. (1996) Progress in rye breeding. In: Geiger, H.H., Deimling, S., Miedaner, T., and Schilling, A.G. (eds). Proc. Internat. Symp. Rye Breed Genet. Vort Pflanzenzüchtg 35: 7–14.

Kay, R., Chan, A., Daly, M., and McPherson, J. (1987) Duplication of the CaMV 35S promoter creates a strong enhancer for plants. Science 236: 1299–1302.

Klein, T.M., Arentzen, R., Lewis, P.A., and Fitzpatrick-McElliott, S. (1992) Transformation of microbes, plants and animals by particle bombardment. Bio/Technology 10: 286–291.

Kobylansky, V.D., and Solodukhina, O.V. (1996) Genetic bases and practical breeding utilization of heterogenous resistance of rye to brown rust. In: Geiger, H.H., Deimling, S., Miedaner, T., and Schilling, A.G. (eds). Proc. Internat. Symp Rye Breed Genet. Vort Pflanzenzüchtg 35: 155–164.

Komari, T., Hiei, Y., Saito, Y., Murai, N., and Kumashiro, T. (1996) Vectors carrying two separate T-DNAs for co-transformation of higher plants mediated by *Agrobacterium tumefaciens* and segregation of transformants free from selection markers. Plant J. 10: 165–174.

Koziel, M.G., Beland, G.L., Bowman, C., Carozzi, B., Crenshaw, R., Crossland, L., Dawson, J., Desai, N., Hill, M., Kadwell, S., Launis, K., Lewis, K., Maddox, D., McPherson, K., Meghji, M.R., Merlin, E., Rhodes, R., Warren, G.W., Wright, M., and Evola, S.V. (1993) Field performance of elite transgenic plants expressing an insecticidal protein derived from *Bacillus thuringiensis*. Bio/Technology 11: 194–200.

Krumbiegel-Schroeren, G., Fingemaize, J., Schroeren, V., and Binding, H. (1984) Embryoid formation and plant regeneration from callus of *Secale cereale*. Z. Pflanzenzüchtg 92: 89–94.

Larkins, B.A., Lending, C.R., Wallace, J.C., Galili, G., and Lopes, M.A. (1990) Application of biotechnology for improving cereal protein quality. In: Ejeta, G., Mertz, E.T., Rooney, L., Schaffert, R., and Joe, J. (eds), pp. 155–163. Proc. Internat. Conf. Sorghum Nutritional Quality. Purdue University, West Lafayette, Indiana.

Lasa, J.M., and Bosemark, N.O. (1993) Male sterility. In: Hayward, M.D., Bosemark, N.O., and Romagosa, I. (eds), Plant Breeding: Principles and Prospects, pp. 213–228. Chapman & Hall.

Lin, W., Anuratha, C.S., Datta, K., Potrykus, I., Muthukrishman, S., and Datta, S.K. (1995) Genetic engineering of rice for resistance to sheath blight. Bio/Technology 13: 686–691.

Linacero, R., and Vázquez, A.M. (1986) Somatic embryogenesis and plant regeneration from leaf tissues of rye (*Secale cereale* L.). Plant Sci. 44: 219–222.

Linacero, R., and Vázquez, A.M. (1990) Somatic embryogenesis from immature inflorescences of rye. Plant Sci. 72: 253–258.

Logemann, J., Jach, G., Tommerup, H., Mundy, J., and Schell, J. (1992) Expression of a barley ribosome-inactivating protein leads to increased fungal protection in transgenic tobacco plants. Bio/Technology 10: 305–308.

Lu, C., Chandler, S.F., and Vasil, I.K. (1984) Somatic embryogenesis and plant regeneration from cultured immature embryos of rye (*Secale cereale* L.). J. Plant Physiol. 15: 237–244.

Ludwig, S.R., Habera, K.F., Dellaporta, S.L., and Wessler, S.R. (1989) *Lc*, a member of the maize *R* gene family responsible for tissue-specific anthocyanin production, encodes a protein similar to transcriptional activators and contains the myc-homology region. Proc. Nat. Acad. Sci. USA 86: 7092–7096.

Mariani, C., De Beuckeleer, M., Truettner, J., Leemans, J., and Goldberg, R.B. (1990) Induction of male sterility in plant by a chimaeric ribonuclease gene. Nature 347: 737–741.

Mariani, C., Gossele, V., De Beuckeleer, M., De Block, M., Goldberg, R.B., De Greef, W., and Leemans, J. (1992) A chimeric ribonuclease-inhibitor genes restores fertility to male sterile plants. Nature 357: 384–387.

Matzke, M.A., and Matzke, A.J.M. (1995) How and why do plants inactivate homologous (trans)genes? Plant Physiol. 107: 679–685.
McElroy, D., Zhang, W., Cao, J., and Wu, R. (1990) Isolation of an efficient actin promoter for use in rice transformation. Plant Cell 2: 163–171.
Miedaner, T., Gand, G., and Geiger, H.H. (1996) Genetics of resistance and aggressiveness in the winter rye/*Fusarium chulmorum* head blight pathosystem. In: Geiger, H.H., Deimling, S., Miedaner, T., and Schilling, A.G. (eds). Proc. Internat. Symp. Rye Breed Genet. Vort Pflanzenzüchtg 35: 165–174.
Morton, R., and Hooykaas, P.J.J. (1995) Gene replacement. Mol. Breed. 1: 123–132
Murashige, T., and Skoog, F. (1962) A revised medium for rapid growth and bioassays with tobacco tissue cultures. Physiol. Plant. 15: 473–497.
Oberthur, E.E., Nicholson, R.L., and Butler, L.G. (1983) Presence of polyphenolic materials, including condensed tannins, in sorghum callus. J. Agric. Food Chem. 31: 660–662.
Odell, J.T., Hoopes, J.L., and Vermerris, W. (1994) Seed-specific gene activation mediated by the Cre/lox site-specific recombination system. Plant Physiol. 106: 447–458.
Ou-Le, T.-M., Turgeon, R., and Wu, R. (1986) Expression of a foreign gene linked to either a plant-virus or a *Drosophila* promoter, after electroporation of protoplasts of rice, wheat, and sorghum. Proc. Nat. Acad. Sci. USA 83:6815–6819.
Ow, D., Wood, K.V., DeLuca, M., DeWet, J.R., Helinski, D.R., and Howell, S.H. (1986) Transient and stable expression of the firefly luciferase in plant cells and transgenic plants. Science 234: 856–859.
Paz-Ares, J., Ghosal, D., Wienand, U., Peterson, P.A., and Saedler, H. (1987) The regulatory *c1* locus of *Zea mays* encodes a protein with homology to *myb* proto-oncogene products and with structural similarities to transcriptional activators. EMBO J. 6: 3553–3558.
Plaschke, J., Börner, A., Xie, D.X., Koebner, R.M.D., Schlegel, R., and Gale, M.D. (1993) RFLP mapping of genes affecting plant height and growth habit in rye. Theor. Appl. Genet. 85: 1049–1054.
Purnhauser, L., Medgyesy, P., Czako, M., Dix, P.J., and Marton, L. (1987) Stimulation of shoot regeneration in *Triticum aestivum* and *Nicotiana plumbaginifolia* Viv. tissue cultures using the ethylene inhibitor $AgNO_3$. Plant Cell Rep. 6: 1–4.
Purnhauser, L., and Gyulai, G. (1993) Effect of copper on shoot and root regeneration in wheat, triticale, rape and tobacco tissue cultures. Plant Cell Tiss. Org. Cult. 35: 131–139.
Raemakers, C.J.J.M., Sofiari, E., Taylor, N., Henshaw, G., Jacobsen, E., and Visser, R.G.F. (1996) Production of transgenic cassava (*Manihot esculenta* Crantz) plants by particle bombardment using luciferase activity as selection marker. Mol. Breed. 2: 339–349.
Rakoczy-Trojanowska, M., and Malepszy, S. (1993) Genetic factors influence regeneration ability in rye (*Secale cereale* L.). I. Immature inflorescences. Theor. Appl. Genet. 86: 406–410.
Rakoczy-Trojanowska, M., and Malepszy, S. (1995) Genetic factors influencing the regeneration ability of rye (*Secale cereale* L.). II. Immature embryos. Euphytica 83: 233–239.
Rakowska, M. (1996) The nutritive quality of rye. In: Geiger, H.H., Deimling, S., Miedaner, T., and Schilling, A.G. (eds). Proc. Internat. Symp. Rye Breed. Genet. Vort Pflanzenzüchtg 35: 85–96.
Salamini, F., and Motto, M. (1993) The role of gene technology in plant breeding. In: Hayward, M.D., Bosemark, N.O., and Romagosa, I. (eds). Plant Breeding: Principles and Prospects, pp. 138–159. Chapman & Hall.
Sanford, J.C. (1988) The biolistic process. Trends Biotech. 6: 299–302.
Sanford, J.C., Smith, F.D., and Russell, J.A. (1993) Optimizing the biolistic process for different biological application. Meth. Enzym. 217: 483–509.
Sheen, J., Hwang, S., Niwa, Y., Kobayashi, H., and Galbraith, D.W. (1995) Green-fluorescent protein as a new vital marker in plant cells. Plant J. 8: 777–784.
Somers, D.A., Rines, H.W., Gu, W., Kaeppler, H.F., and Bushnell, W.R. (1992) Fertile, transgenic oat plant. Bio/Technology 10: 1589–1594.

Spencer, T.M., Gordon-Kamm, W.J., Daines, R.J., Start, W.G., and Lemaux, P.G. (1990) Bialaphos selection of stable transformants from maize cell culture. Theor. Appl. Genet. 79: 625–631.

Sperling, U., Leßner, B., Scholz, M., Wehling, P., Gey, A.-K.,Geiger, H.H., and Miedaner, T. (1996) Qualitative and quantitative variation for resistance of winter rye to leaf rust. In: Geiger, H.H., Deimling, S., Miedaner, T., and Schilling, A.G. (eds). Proc. Internat. Symp. Rye Breed. Genet. Vort Pflanzenzüchtg 35: 175.

Strittmatter, G., Janssens, J., Opsomer, C., and Botterman, J. (1995) Inhibition of fungal disease development in plants by engineering controlled cell death. Bio/Technology 13: 1085–1089.

Thompson, C.J., Moyva, N.R., Tizard, R., Crameri, R., Davies, J.E., Lauwereys, M., and Botterman, J. (1987) Characterization of the herbicide-resistance gene *bar* from *Streptomyces hygroscopicus*. EMBO J. 6: 2519–2523.

Tsukahara, M., and Kirosawa, T. (1992) Simple dehydration treatment promotes plantlet regeneration of rice (*Oryza sativa* L.) callus. Plant Cell Rep. 11: 550–553.

Vain, P., Yean, H., and Flament, P. (1989) Enhancement of production and regeneration of embryogenic type II callus in *Zea mays* L. by $AgNO_3$. Plant Cell Tiss. Org. Cult. 18: 143–151.

Vain, P., McMullen, M.D., and Finer, J.J. (1993) Osmotic treatment enhances particle bombardment mediated transient and stable transformation of maize. Plant Cell Rep. 12: 84–88.

Vasil, I.K. (1994) Molecular improvement of cereals. Plant Mol. Biol. 25: 925–937.

Vasil, V., Castillo, A.M., Fromm, M.E., and Vasil, I.K. (1992) Herbicide resistant fertile transgenic wheat plants obtained by microprojectile bombardment of regenerable embryogenic callus. Bio/Technology 10: 667–674.

Vasil, V., Srivastava, V., Castillo, A.M., Fromm, M.E., and Vasil, I.K. (1993) Rapid production of transgenic wheat plants by direct bombardment of cultured immature embryos. Bio/Technology 11: 1553–1558.

Walters, D.A., Vetsch, C.S., Potts, D.E. and Lundquist, R.C. (1993) Transformation and inheritance of a hygromycin phosphotransferase gene in maize plants. Plant Mol. Biol. 18: 189–200.

Weipert, D. (1996) Pentosans as selection in plant breeding. In: Geiger, H.H., Deimling, S., Miedaner, T., and Schilling, A.G. (eds). Proc. Internat. Symp. Rye Breed Genet. Vort Pflanzenzüchtg 35: 109–120.

Wernicke, W., and Brettell, R. (1980) Somatic embryogenesis from *Sorghum bicolor* leaves. Nature 287: 138–139.

Wünn, J., Klöti, A., Burkhardt, P.K., Biswas, G.C.G., Launis, K., Iglesias, V.A., and Potrykus, I. (1996) Transgenic Indica rice breeding line IR58 expressing a synthetic *cryIA(b)* gene from *Bacillus thuringiensis* provides effective insect pest control. Bio/Technology 14: 171–176.

Xu, D., Xue, Q., McElroy, D., Mawal, Y., Hilder, V.A., and Wu, R. (1996) Constitutive expression of a cowpea trypsin inhibitor gene, *CpTi*, in transgenic rice plants confers resistance to two major rice insect pests. Mol. Breed. 2: 167–173.

Yoder, J.I., and Goldsbrough, A.P. (1994) Transformation systems for generating marker-free transgenic plants. Bio/Technology 12: 263–267.

Zhu, Q., Maher, E.A., Masoud, S., Dixon, R.A., and Lamb, C.J. (1994) Enhanced protection against fungal attack by constitutive co-expression of chitinase and glucanase genes in transgenic tobacco. Bio/Technology 12: 807–812.

Zimny, J., and Lörz, H. (1989) High frequency of somatic embryogenesis and plant regeneration of rye (*Secale cereale* L.). Plant Breed. 102: 89–100.

# 12. Transgenic Cereals: Triticale and Tritordeum

PILAR BARCELO[1]*, SONRIZA RASCO-GAUNT[1], DIRK BECKER[2] and JANUSZ ZIMNY[3]
[1]*Rothamsted Experimental Station. IACR-Rothamsted, Harpenden AL5 2JQ, UK.*
[2]*Institut für Allgemeine Botanik, AMP II, Universität Hamburg, Ohnhorststraße 18, D-22609 Hamburg, Germany.* [3]*Plant Breeding and Acclimatisation Institute-Radzików, PL-00950 Warszawa P.O. Box 1019, Poland*
**E-mail: Pilar.Barcelo@bbsrc.ac.uk*

## Triticale

### 1. *Introduction*

#### 1.1 *Historical view and importance of Triticale*

The first hybrid between wheat and rye (Wilson, 1875), and the first fertile Triticale (Rimpau, 1891), were obtained more than a century ago. In spite of these early discoveries, Triticale remained an academic curiosity for many years. This was related to the lack of techniques for chromosome doubling (to allow duplication of primary hybrids) and embryo rescue. Chromosome doubling became possible in the late 1930's with the discovery that colchicine inhibits the formation of the mitotic spindle resulting in the doubling of chromosome number (Blakeslee and Avery, 1937). The second breakthrough occurred in the 1940's with the development of embryo rescue methods for the recovery of hybrid embryos from seed with malformed endosperm. Once these techniques were in place, wheat and rye species with various ploidy levels were used to produce tetraploid (2n = 28, AARR or BBRR or ABRR or BARR), hexaploid (2n = 42, AABBRR), octoploid (2n = 56, AABBDDRR or AABBRRRR), and decaploid (2n = 70, AABBDDRRRR) Triticales (Villareal et al., 1990). The actual breeding was started in 1948, and it was twenty years later when CIMMYT (Mexico) produced the first acceptable Triticale cultivar 'Armadillo' (Villareal et al., 1990). It showed desirable characteristics like high fertility, good yield, early maturity, improved test weight and a gene for dwarfism. Over the years, other traits like nutritional quality, disease resistance and pre-harvest sprouting were also improved, and today Triticale varieties are cultivated on over one million hectares in Russia, Poland, USA, Canada, West Germany, Argentina, Mexico, France, Portugal and

*I.K. Vasil (ed.), Molecular Improvement of Cereal Crops, 361–385*

Spain. The most widely grown Triticale is the polish cultivar 'Lasko'. It is adapted to and has excellent yield potential not only in all the areas where wheat is grown, but also in marginal production environments, such as acid soils, high elevations in the tropics, semiarid conditions, and sandy soils (Villareal et al., 1990).

### *1.2* In Vitro *plant regeneration*

Regeneration systems were developed in the 1980's as a means of generating genetic variability for use in practical breeding. Also, regeneration was seen as a prerequisite for producing transformed plants. Plants were regenerated *via* organogenesis or *via* somatic embryogenesis from cultured young inflorescences (Eapen and Rao, 1985), leaf bases (Fedak, 1987), microspores (Bernard, 1980, Sosinov et al., 1981, Charmet and Bernard, 1984), immature embryos (Eapen and Rao, 1982, Nakamura and Keller, 1982, Stolarz and Lörz, 1986a,b, Babeli et al., 1988, Immonen, 1993) and cell suspension cultures (Zimny, 1992). Among these explants, the scutellum of immature embryos has been found to be most suitable for *in vitro* regeneration (Zimny, 1989).

One problem encountered frequently among the regenerants is albinism (Ono and Larter, 1976). This phenomenon is particularly true for androgenetic haploids where it is strongly dependent on the genotype, and in long-term cultures (Nakamura and Keller, 1982). Somaclonal variants for plant height, time of flowering, morphology of whole plants and spike, leaf waxiness, hairy neck and fertility have also been reported (Jordan and Larter, 1984, Brettell et al., 1986, Stolarz and Lörz, 1986a). In addition, changes in chromosome number and structure are known to occur, ranging from $2n = 36 + 6t$ (t-telocentric) to $42 + 3t$ (Armstrong et al., 1983), hypo-aneuploidy and hyperaneuploidy as well as acrocentric, telocentric and dicentric chromosomes (Nakamura and Keller, 1982), aneuploidy, plants with higher ploidy levels (Sosinov et al., 1981) and modifications in the C-banding pattern (Armstrong et al., 1983). In contrast to all these reports, Stolarz and Lörz (1986a) obtained regenerants with the expected chromosome number, and Brettell et al. (1986) found loss of chromosomes in very few lines. According to Nakamura and Keller (1982), the abnormalities are related to the age of the callus, and do not pre-exist in the explant. In contrast, Wang and Hu (1993) suggested that chromosome variation is present in the explants prior to culture.

Electrophoretic studies have shown that although all of the regenerants had the same prolamin banding pattern as the donor Triticale, variation existed in the intensity of the bands, especially those encoded by the rye genome (Jordan and Larter, 1984). In other experiments, Brettell et al. (1986) analysed the rRNA genes located at the *Nor* loci on chromosomes 1B, 6B and 1R. The rRNA gene loci were stable, when the chromosomes carrying these loci were present. Only in one plant the authors found reduction of a number of rDNA spacer sequences. This reduction was heritable and correlated with reduced C-banding at the position of *Nor* – R1 on chromosome R1. This phenomenon was a consequence of tissue culture since neither the parents nor six other regenerants from the same culture carried the alteration. Kaltsikes and Babeli (1992) tested somaclonal variation in plant families coming from Triticale lines differing in telomeric heterochromatin on chromosome arms 7RL and 6RS. They showed that rye telomeric heterochromatin influences somaclonal variation and its presence leads to increased amounts of genetic variability in progeny of regenerants. In our opinion, although somaclonal variation does occur in Triticale, it can be easily mistaken with the inherent instability of the genome.

As in other cereal species (see chapter by Vasil, this volume), regenerating plants from cell suspensions and protoplasts has been difficult. In the 1980's, cell suspensions were seen as the only source of transformable and manipulatable protoplasts, and therefore the achievement of regeneration from protoplasts was a major aim. Embryogenic callus obtained from immature embryos of Triticale was used to establish cell suspension cultures (Stolarz and Lörz, 1986b, Zimny, 1992). Protoplasts isolated from the suspensions formed globular structures that developed into somatic embryos but did not form plants (Stolarz, 1990).

In summary, plants have been regenerated from a variety of explants from Triticale with efficiencies comparable to those obtained in other temperate cereals like wheat, barley and Tritordeum. This has allowed the development of genetic manipulation techniques for the modification of agronomic traits.

## 2. *Genetic Transformation*

Transgenic cereals have been produced during the past decade by various methods, including the widely used biolistics procedure (see chapters by Klein and Jones, and by Komari and Kubo, this volume). However, in the early 1980's, DNA could be introduced into cereals only through direct delivery into protoplasts (Lörz et al., 1985, Ushimiya et al., 1986, Fromm et al., 1986). Very similar experiments were carried out in Triticale in which suspension-derived protoplasts were used for introducing the *uidA* gene by the addition of Ca++ and polyethylene glycol (PEG). Stable transformation was achieved by this method (Zimny and Rafalski, 1993) but transformed plants were never recovered due to regeneration problems. This motivated the development of other direct gene transfer techniques that rely on regeneration of explants rather than protoplasts (see chapters by Vasil, and Klein and Jones, this volume).

*2.1 Transformation by particle bombardment*

The line MAH 1590 was selected for its high embryogenic and regenerative response *in vitro* from some 500 different lines tested. Immature embryos were removed from caryopses at the late spherical coleoptile stage (Zimny and Lörz, 1989) and precultured in the following induction medium (see below): MS medium (Murashige and Skoog, 1962) supplemented with 30 $\mu M$ Dicamba (HI medium), 9 $\mu M$ 2,4-D (GI medium), 9 $\mu M$ 2,4-D and 10% coconut water (GIC medium), and 0.4 $\mu M$ 2,4-D and 10% coconut water (HR medium). The regeneration media used were the GRI medium based on 190–2 medium (Pauk et al., 1991) supplemented with 2.5 $\mu M$ kinetin and 2.5 $\mu M$ NAA, and GR2 medium based on 190-4 medium (Pauk et al., 1991) supplemented with 6.5 $\mu M$ kinetin and 2.5 $\mu M$ NAA. All media contained 3% sucrose and 0.2% Gelrite. Cultures were kept in growth chambers at 27°C and 16 h light (for further details of culture media see Zimny et al., 1995).

Immature embryos were precultured for two (group I) or six days (group II) on HI induction medium prior to bombardment (Zimny et al., 1995). Thirty embryos were used for each replicate. They were placed on the medium with the scutellum facing up, and were bombardment with gold particles coated with plasmid pDB1 (Figure 1, Becker et al., 1994). Two days after bombardment, three out of 30 explants were randomly chosen for GUS expression. Only the bombarded explants showed GUS activity, and no significant differences in transient expression were observed between the two groups (I, II) and between the two helium pressure levels (1300 or 1550 psi) used for bombardment.

Bombarded explants were transferred to fresh HI induction medium and cultured for one week (Zimny et al., 1995). Thereafter, cultures developing embryogenic callus were transferred to GIC medium for another week. Well-developed embryogenic calli were subcultured onto HR medium and cultured for a further week to complete a total of three-week induction period. At the end of induction period, just before the transfer of calli to regeneration medium, the frequency of embryos forming embryogenic callus was measured. The embryos of group I (bombarded two days after isolation) showed dramatic inhibition of callus development after particle bombardment in comparison with control cultures (not bombarded) and with cultures from group II (bombarded six days after isolation). The explants from group II showed normal development of embryogenic callus, similar to the non-bombarded controls. Callus development within group I stopped in the central part of the scutellum and embryogenic callus was observed only on the periphery of the scutellum in 38% of the explants in comparison to 67% in control cultures. Explants of group II showed a markedly higher embryogenic response (67%) although lower than the corresponding control (76%). Not only the efficiency but also the intensity of somatic embryogenesis was significantly higher in group II, in which embryogenic callus was produced all over the scutellum. Finally, a total of six transgenic plants (four independent transformation events) from group I and 16 (10 independent events) from group II were regenerated. This shows the importance of the quality of the embryogenic callus formed during the culture process for the production of transgenic plants.

After this assessment, calluses were transferred to GRI regeneration medium for germination of somatic embryos (Zimny et al., 1995). Subsequently, germinating somatic embryos were transferred to Magenta boxes containing GR2 medium supplemented with 3 mg/l phosphinothricin (PPT). Within two weeks the majority of regenerants stopped growing. Surviving plantlets were selected further on GR2 medium with 6 mg/l PPT. This level of selection was found to be the optimum to inhibit germination of somatic embryos among different concentrations of PPT tested (0, 2, 4, 6, 8 and 10 mg/l). Plants that survived this treatment were adapted to greenhouse conditions and screened by spraying with a solution of Basta containing 120 mg/l PPT. Of the thirty plants that survived Basta spraying, 22 showed GUS activity in leaves. These 22 transgenic plants corresponded to 14 independent transformation events. All regenerated plants were phenotypically normal except one that was very bushy and produced numerous spikes. All plants were grown to maturity and set seeds.

Mature pollen grains of all 30 $T_0$ plants that survived herbicide treatment were stained histochemically; 25 showed GUS activity in pollen grains although in three of the 25 plants GUS activity had not been detectable in leaves (Zimny et al., 1995). Mature pollen grains of six plants were tested histochemically for segregation of the *uidA* gene (Table 1). In one transgenic line (16.1), segregation of the *uidA* gene in pollen grains was as expected for a single locus insertion (1:1), whereas the remaining five transgenic lines showed deviations from the expected segregation. These deviations were always towards a decrease in the number of expressing gametes (plants 18.1, 55.3, 13.1 and 1.1) for single locus insertions and (plant 100.1) for a two locus insertion. In the latter, the expected gamete segregation is 3:1. Table 1 also shows the results obtained from GUS and phosphinothricin acetyltransferase (PAT) expression assays in $T_1$ progeny. The inheritance of the expression of the *uidA* gene in different plants ranged from 2:1 to 15.6:1 and for the *bar* gene 2.4:1 to 4.7:1.

Southern analysis of $T_0$ plants confirmed the integration of *uidA* and *bar* in the Triticale genome, in contrast to DNA from non-transformed plants which did not exhibit hydridisation to DIG-labelled *uidA* or *bar* coding regions (Zimny et al., 1995). In undigested genomic DNA of each tested putative transformant, hybridisation signals were observed in the high molecular weight DNA, indicating the integration of the marker genes into the genomic DNA. In order to confirm that the transformed plants contained intact copies of the *uidA* and *bar* genes, DNA was digested with *Bam*H I/*Sac* I or *Sal* I which release a 1.8 kb or 550 bp fragment with the coding region

TABLE 1

Segregation of GUS and/or PAT activity in pollen grains of $T_0$ Triticale plants and in leaves of $T_1$ progeny. About 500 pollen grains and 100 progeny plants were analysed per assay (From Zimny et al., 1995)

| | $T_0$ pollen grains | $T_1$ progeny | |
|---|---|---|---|
| Plant | GUS+:GUS– | GUS+:GUS– | PAT+:PAT– |
| 100.1 | 1.8 : 1 | 15.6 : 1 | n.d. |
| 16.1 | 1 : 1 | 11.4 : 1 | 3.5 : 1 |
| 18.1 | 1 : 2 | 4 : 1 | 4.7 : 1 |
| 55.3 | 1 : 2 | 2 : 1 | 2.4 : 1 |
| 13.1 | 1 : 1.6 | 3.2 : 1 | 4.3 : 1 |
| 1.1 | 1 : 1.4 | 2.2 : 1 | 3.1 : 1 |

of *uidA* or *bar*. These analyses confirmed that all plants contained at least one intact copy of both marker genes. In addition to these hydridisation signals, other larger fragments were observed, suggesting that deletions, rearrangements and/or methylations at one or both restriction sites had occurred. The integration pattern of both marker genes, determined by digesting DNA with *Nco* I, was complex in all transformants tested. In all cases more than one copy of each marker gene was integrated in the genomic DNA.

For the analysis of the transmission of the pattern of integration of the *uidA* and *bar* genes to the next generation we chose plant 1.1 (Zimny et al., 1995). From this $T_0$ plant, DNA was isolated from four $GUS^+/PAT^+$ and one $GUS^-/PAT^-$ $T_1$ progeny plant and hybridised with *uidA* and *bar* probes. All four positive progeny plants showed hybridisation signals corresponding to *bar* and *uidA* genes. The integration pattern of the four positive progenies was similar to the parental line. Further Southern blot analysis of several $T_1$ plants from four other $T_0$ transgenic lines showed loss of the *bar* gene in some cases and silencing of *uidA* in other cases.

The segregation of the *uidA* and *bar* genes in the $T_2$ progeny (tested histochemically or by Basta spraying of PAT and GUS positive plants) was highly deviant, ranging from 45:1 to 0:50 and 3:1 to 0:42, respectively (Zimny et al., 1995). Plants 1.1 and 55.3 showed the highest stability for PPT resistance (26:14 and 34:11) in $T_2$ generation. The other two lines tested, 9.3 and 16.1, totally lost their PPT resistance. The progeny of $T_1$ lines 9.3 and 55.3 did not show any GUS activity, but in the case of line 1.1, 97.8% was GUS positive. For line 16.1, we found the segregation ratio nearly 3:1 (39:14). The only conclusion we can make is that the traits determined by the introduced genes can be inherited independently. For some lines expression can disappear (9.3) and for others it remains stable (1.1) (see Table 2).

TABLE 2
The analysis of $T_2$ progeny of four transgenic $T_1$ Triticale plants

| line / enzyme | 1.1 | 9.3 | 16.1 | 55.3 |
|---|---|---|---|---|
| $GUS^+$:$GUS^-$ | 45:1 | 0:50 | 39:14 | 0:50 |
| $PAT^+$:$PAT^-$ | 26:14 | 0:42 | 0:42 | 34:11 |

In summary, 465 embryos were bombarded (40 of them were used for transient expression analysis) and about 4000 plantlets were regenerated. Three hundred plants survived the stepwise selection (on media containing 3 or 6 mg/l PPT) and 30 of them survived spraying with the herbicide. Twenty-five plantlets were GUS positive, which were derived from at least 14 independent transformation events. On average, more than 3 independent transgenic plants were obtained per 100 bombarded immature embryos. Pollen and progeny analyses provided evidence for inheritance of the introduced genes to the next generation (Zimny et al., 1995).

### 2.2 *Localisation of transgenes by* in situ *hybridisation*

The practical use of transgenic plants is frequently complicated by the instability of transgenes and their expression. Therefore, it is important to understand the factors affecting the stability and expression of inserted genes. One significant factor seems to be the position of transgene in the genome, known as the 'position effect'. This position effect can be studied by *in situ* hybridisation which can identify the physical position of transgene within the genome. Routine isotopic *in situ* hybridisation protocols (ISH) were established in many laboratories during the 1970's (Gall and Pardue, 1969). The original ISH techniques using isotopic probes is sensitive and very useful for detecting single-copy DNA sequences. Also, non-isotopic fluorescent *in situ* hybridisation (FISH) can be used to identify chromosomes, detect chromosomal abnormalities or to determine specific sequences on chromosomes. DNA or RNA sequences are first labelled with reporter molecules (Trask, 1991). The probe and the target chromosomes are denatured and complementary sequences in the probe and target are allowed to re-anneal. After washing and incubation signal is visible at the site of probe hybridisation. For signal detection enzymatic reaction or fluorochromes can be used.

For *in situ* hybridisation, root tips were obtained from seedlings after synchronisation of cell divisions with hydroxyurea (Pan et al., 1993). They were pre-treated with 4 $\mu M$ APM (amiprophos mehyl) for 3 h and with ice water for 20 h before fixation in 3:1 ethanol:glacial acetic acid. The root tips were digested with 2% cellulase (Onozuka RS) and 2% pectinase at 37°C for 2–3 h. Chromosome preparations were made by the air-drying method

(Olin-Fatih and Heneen, 1992). To obtain the DNA probe, pDB1 plasmid was labelled with biotin-4-dATP by nick translation. Probe mixture was applied to the slides. Biotinylated probes were detected with FITC-avidin DCS and amplified with biotinylated anti-avidin D and another incubation with FITC-avidin DCS. Finally, the slides were mounted in anti-fade solution containing DAPI and propidium iodide (for details see Pedersen et al., 1996, 1997).

Four transgenic $T_0$ plants were analysed by *in situ* hybridisation. The integrated transgenes were detected in all four Triticale lines. The strongest signals were seen in the lines 1.1 and 55.3. However, the two other lines, 16 and 24, had several integration sites on the same chromosome arm. In line 24 there were two integration sites close together on the same arm of a rye chromosome. Rye chromosomes could easily be identified by the stronger staining with propidium iodide and DAPI. After examination and photography of metaphases hybridised with pDB1 probe, the preparations were hybridised with GAA-satelite sequences for identifying chromosomes with positive signals as described by Pedersen and Linde-Laursen (1994). In the other three lines the transformed chromosomes were 5A for line 55.3, 6A for line 16 and 4B for line 1.1. In the case of Triticale three of the four lines were homozygous for the transgene signals, while line 1.1 was hemizygous and therefore the signal was only found on one of the homologs of chromosome 4B. No correlation between *in situ* hybridisation and Southern blot analysis or segregation data has been found (see Table 3).

### *2.3 Anther culture and double haploid production*

The production of homozygous lines by traditional crosses requires up to seven generations of backcrosses. By using anther culture or pollination with maize pollen, combined with spontaneous or induced chromosome doubling, homozygosity can be achieved within no more than two generations.

TABLE 3

Triticale: Indication of homologous chromosomes with transgenes, copy number of transgenes based on Southern hybridisation analyses, and GUS and PAT activities established by x-gluc and spraying test (from Pedersen et al., 1997)

| Lines | 1.1 | 16 | 24 | 55.3 |
|---|---|---|---|---|
| Homo/hemi | 1 | 2 | 2 | 2 |
| Copy number | 4 | 4 | 5 | 5 |
| GUS-activity | + | + | + | – |
| PAT-activity | + | + | + | + |

Homo/hemi –1. Hemizygous, 2. Homozygous

The production of haploid Triticale plants by anther culture was first reported by Ya-Ying et al., (1973). Later investigations have shown the influence of many factors on the development of microspores into fertile plants. These include the genotype, the developmental stage of the microspore, cold pretreatment of the anthers, the influence of the growth conditions of the donor plants, culture medium components, and seasonal effects (Bernard, 1977, 1980, Sosinov et al., 1981, Charmet and Bernard, 1984, Kozdój and Zimny, 1993). In general, liquid (with Ficoll) or agar-solidified N6 (Chu, 1978), B5 (Gamborg et al., 1968) and MB (Bernard, 1977) media supplemented with 2,4-D for callus induction, and the same media without 2,4-D or supplemented with IAA and/or kinetin for regeneration, were used to culture cold pre-treated anthers containing uninucleate microspores. Among the regenerants different morphological variants were observed, mainly differing in the size of awns, straw thickness and leaf blade width. Aneuploids as well as plants of higher ploidy were also found (Sosinov et al., 1981). The phenomenon of albinism was noticed by many authors (Schumann, 1988, Bernard, 1977, 1980, Sosinov et al., 1981). It seems to be connected with the growth temperature of the donor plant and various culture conditions.

We have used androgenesis to obtain homozygous non-segregating transgenic lines. From 5000 anthers 295 green plants were regenerated. Ninety-nine of them were fertile transgenic lines and have been selfed and multiplied over three generations. A total of 1600 $TDH_3$ ($T_3$ generation of transgenic double haploid) plants showed transgene expression. It means that transgenic homozygous lines did not show any segregation of trait determined by the introduced *uidA* gene, and that the method of androgenesis can be used as expected to produce non-segregating stable transgenic lines.

### *2.4 Discussion*

The aim of the work reported here was to obtain transgenic plants of the allohexaploid cereal species Triticale by combining an efficient regeneration system (Stolarz and Lörz, 1986a,b, Zimny and Lörz, 1989) with a successful particle bombardment method (Zimny et al., 1995). A high transformation efficiency of at least 3.3%, comparable with data obtained for wheat (Nehra et al., 1994) and rice (Christou et al., 1991), was obtained. One of the most important steps in the recovery of transformants was the time when selection was applied. For species like rice, barley and wheat (Li et al., 1993, Wan and Lemaux, 1994, Altpeter et al., 1996a) selection applied at the beginning of the induction period appears to be efficient in selecting transgenic plants. It does not disrupt embryogenesis and regeneration. In our experiments, however, we found that selection applied at such early stages of development is very damaging to embryogenesis and, therefore, detrimental to transformation. In fact, our best transformation efficiencies were obtained when selection was applied after the period of induction of embryogenesis, as previously reported for Tritordeum (Barcelo et al., 1994b).

Another factor found to be crucial for transformation was the length of the preculture period prior to bombardment, where explants precultured for longer times were able to produce better callus and more transgenic plants than those precultured for just two days. This positive effect of the longer preculture appears to be due to a reduction in damage caused by bombardment in tissues already forming callus. The optimal length of preculture for least damage after bombardment is different for different species (see results in Tritordeum below).

In our experiments, the *uidA* gene in most cases showed distorted segregation in pollen grains. The expected Mendelian ratios in pollen grains for one and two loci are 1:1 and 3:1, respectively, and this was only observed in one of the plants (16.1). However, although segregation in pollen was atypical in five out of six plants analysed, segregation in $T_1$ progeny was the expected 3:1 and 15:1 in most cases. It is surprising that the plants in which the *uidA* gene was abnormally transmitted in pollen grains show transmission to the $T_1$ progeny as expected for one or two loci (100.1, 18.1, 55.3, 13.1, 1.1). The reason for deviation from 1:1 or 3:1 segregation in pollen grains is difficult to explain. One possible reason could be the error in distinguishing positive and negative pollen during counting, since the transformed pollen grains express a wide spectrum of blue colours and some could be ambiguously scored. Our results in Triticale are in agreement with what has been reported for other cereals, with most transgenic lines showing Mendelian ratios (Becker et al., 1994, D'Halluin et al., 1992, Nehra et al., 1994, Wan and Lemaux, 1994) and some lines distorted segregations (Rathore et al., 1993, Spencer et al., 1992, Walters et al., 1992, Barro et al., 1998).

The study on *in situ* hybridisation suggested that in Triticale transgenes were inserted randomly in all three (A-, B- and R-) genomes. It could, therefore, be used as a bridge species to transfer transgenes to other cereals. In summary, we have shown that *in vitro* cultures of Triticale can be genetically transformed to generate phenotypically normal and fertile transgenic plants (Zimny et al., 1995), and therefore this 'novel' cereal crop is now ready for further improvement through biotechnology.

## Tritordeum

### *3. Introduction*

#### *3.1 The beginning of Tritordeum as a new crop*

Interspecific and intergeneric hybridisation is a common tool for widening the genetic base of existing crops and for the creation of new ones. After the success with Triticale, breeders have been very interested in hybridising wheat and cultivated barley species. Although hybrids have been obtained many times between these two genera, the production of fertile amphiploids has always failed. This motivated the breeders to turn to wild barley species for hybridisation with wheat. The use of wild barleys gave rise to several amphiploids but a fertile amphiploid was obtained only when *Hordeum chilense* was crossed with *Triticum turgidum* (Martin and Sanchez-Monge, 1982). This amphiploid was called Tritordeum and showed from the beginning an unexpectedly high potential to become a new crop. Its most remarkable features were high fertility, good seed size and chromosomal stability (Martin and Cubero, 1981).

Another interesting characteristic found in the primary Tritordeums was the high protein content of the grains. Unfortunately, although the grain from high yielding Tritordeums still contains more protein than wheat, the difference is no longer significant. The amino acid profile of Tritordeum protein was very similar to that of wheat, rye and Triticale (Cubero et al., 1986), which are all deficient in certain essential amino acids such as lysine. However, Barro et al., (1991) found that Tritordeum generally exhibited higher levels of nitrate reductase activity than wheat, and a quicker and more effective $NO^{3-}$ extraction from the medium. This suggests that it may be possible to obtain Tritordeums with a higher capacity than wheat to extract N from the soil and that such excess of N could potentially be transferred to the grain and converted into protein.

These characteristics, together with the fact that Tritordeum showed a good response to breeding selection, were strong encouragement to examine the potential for breeding Tritordeum into a novel cereal crop. The response of Tritordeum to selection was considered to be surprisingly good because the genetic basis present in the first Tritordeums was very narrow, limited to just two *H. chilense* accessions and three wheat cultivars (Cubero et al., 1986, Martin, 1988). Therefore, the first priority of the breeding programme was to widen Tritordeum genetic background. For this, the collection of *H. Chilense* was increased by requesting material from gene banks and by collecting new accessions in expeditions to Chile organised by A. Martin's group. The main sources of wheat germplasm were the crossing block of durum wheats of CIMMYT and ICARDA and also Spanish cultivars. To date, hybrids have been obtained with 93 different *H. chilense* accessions and in total more than 160 primary amphiploids have been produced (Martin et al., 1996).

Once the actual breeding programme started, the priority traits to incorporate into Tritordeum were fertility, winter growth, free threshing, tough rachis, erect canopy and plant stature. Traits related to resistance to biotic and abiotic stresses, and yield and quality components were also considered (Martin et al., 1996). Breeding for these characteristics has been so far done in three cycles, starting with a classical genealogical scheme, followed by a single seed descent cycle to accelerate the programme, and completed by a third cycle of crossing between the more advanced Tritordeum lines.

To date, this programme has given rise to fully fertile lines showing early flowering, others with semi-dwarf and erect habit, and also lines with free threshing. However, it has proven to be much more difficult than initially expected to produce lines with all these interesting features combined. One particularly difficult combination is to produce fully fertile semi-dwarf lines, which was also difficult in the breeding of wheat and Triticale (Zillinsky, 1974). In terms of yield, breeding in Tritordeum has allowed an increase to up to 80% of that of commercial wheat cultivars, maintaining a protein content approximately 5% higher than in wheat.

Regarding other agronomic characteristics, Tritordeum shows remarkably higher resistance to drought and high temperatures than wheat and Triticale cultivars. This resistance appears to be due to a better osmotic adjustment so that Tritordeum maintains the normal level of photosynthesis and stomatal conductivity under stress circumstances when wheat and Triticale fail to do so (Gallardo and Fereres, 1989). This special feature of Tritordeum makes it a potential winter cereal crop to be grown in areas with dry and hot climates like the south of Europe and north Africa.

Two other well documented aspects of Tritordeum are its resistance to pathogens, particularly fungi and aphids (Martin et al., 1996) and its quality for breadmaking (Alvarez et al., 1995). Tritordeum gluten has viscoelastic properties similar to some bread wheat cultivars (Alvarez et al., 1992) but the overall quality is poorer than that of bread wheat. The fact that Tritordeum quality was in some instances as good as some wheats is very surprising. Tritordeum does not have the D genome in its genetic complement and therefore lacks some of the very important high molecular weight glutenin subunits (HMW-GS) associated with good viscoelastic properties (Vasil and Anderson, 1997). Two strategies were followed to introduce these quality genes into Tritordeum genome: a traditional approach where the complete or partial D genome was to be introduced by crossing, and a genetic engineering approach where individual genes coding for HMW-GS were to be introduced by transformation. Both methods have been successful and Tritordeum breadmaking quality has been modified by these approaches (A. Martin, personal communication, (Rooke et al., in press).

### *3.2 Cell culture and transformation*

Our objective in starting work on tissue cultures of Tritordeum in 1987 was to increase the existing genetic variability by the production of somaclonal variants. Primary explants such as immature embryos and inflorescences proved to be highly amenable to tissue culture and large populations of fertile plants were produced (Barcelo et al., 1989). As none of the regenerated plants showed any heritable morphological variation, the idea of producing somaclonal variants was abandoned.

## 4. *Genetic Transformation*

### *4.1 PEG-mediated DNA delivery into suspension-derived protoplasts*

The uptake and expression of DNA by protoplasts was the first direct gene transfer method clearly demonstrated to function in plants. The original and still most effective chemical uptake method is the use of polyethylene glycol (PEG, see Chapter by Shillito, this volume). Once DNA has been introduced into the naked cells or protoplasts the cell wall can be synthesised and the cells induced to organise into tissues and finally into plants. Protoplasts can be isolated from any plant tissue, but in cereals the sole source of dividing protoplasts are cell suspensions (Vasil and Vasil, 1992).

Callus derived from the scutellum of Tritordeum was used to establish cell suspensions (Barcelo et al., 1993). Protoplasts isolated from the suspensions, three days after culture, were used for PEG-mediated transformation. Some GUS expressing lines were selected on G418, but no transgenic plants were ever recovered (unpublished results).

### *4.2 DNA delivery by electroporation of tissues*

Electroporation was first used for introducing DNA into protoplasts (see Chapter by Shillito, this volume). Subsequently, Dekeyser et al., (1990) adapted the technique to deliver DNA to intact leaf base tissues of rice, maize, wheat and barley. The procedure has been used to obtain transgenic rice (Li et al., 1991, Xu and Li, 1994) and maize (D'Halluin et al., 1992, Laursen et al., 1994) plants.

Three Tritordeum lines (HT 28, HT 31 and HT 174) were grown in a growth chamber under 16/8 h light/dark photoperiod (350 $\mu$mol $m^{-2}$. $s^{-1}$) and 80% relative humidity. Immature inflorescence cultures were used as target explants, which have been shown to be highly regenerative and transformable by particle bombardment (Barcelo et al., 1989, 1994b). Inflorescences ranging in size from 0.5 to 1.5 cm were cut into ~1 mm pieces (Barcelo et al., 1989) and cultured for one day prior to electroporation in the culture medium described by He and Lazzeri (1998). Twenty pieces of inflorescence were placed into each electroporation cuvette (BioRad 4 mm electrode distance) containing 180 $\mu$l EB (electroporation buffer) and 20 $\mu$l plasmid pAHC25 (Figure 1, Christensen and Quail, 1996) at a concentration of 1 $\mu$g/$\mu$l. The transformation mixture was then incubated at 24°C or 37°C for 1 h prior to electroporation, which consisted of a single electric pulse of 550 V/cm from a 960-$\mu$F capacitor using a Gene Pulser Transfection Apparatus (Bio Rad). The explants were cultured as described by Rasco-Gaunt and Barcelo (1998) using PPT or Bialaphos as selection agents for three-week selection rounds starting at the point of regeneration, right after induction of embryogenesis. A total of 10 plants were recovered and two were shown to be transgenic expressing both *uidA* and *bar* genes. These plants appeared normal and set seed. Southern analysis of these plants showed that about 10 copies of the *uidA* gene were integrated. The factors found to be more relevant for this procedure of transformation were the voltage, pulse length, the volume and osmoticum of the electroporation buffer and the pre-electroporation incubation time and temperature.

In these experiments a total of 83 inflorescence explants were electroporated from which 2 transgenic plants were obtained. This gives a transformation efficiency of 2.4%, which is comparable to that obtained with particle bombardment (He and Lazzeri, in preparation).

Technically, tissue electroporation is simple in that the target material is suspended in a buffer containing DNA and then subjected to a high-voltage electrical pulse. However, the interaction between the numerous biological (e.g. plant tissue type, explant size and density, physiological status, pre-treatment), chemical (e.g. buffer composition, DNA concentration) and physical (e.g. pre- and post-electroporation temperature, cuvette size and type, buffer volume) parameters affecting transformation makes the system very complex to elucidate. The electroporation voltage is one of these crucial factors, since the minimum potential required for membrane poration that is not too high to cause excessive damage is difficult to identify. The physiological conditions of the target tissue prior to electroporation, such as explant age, pre-culture period and plasmolysis treatments are also very important factors and are hard to quantify.

This method has a major advantage over particle bombardment in that the equipment required is relatively non-specialised and needs little in the way of disposables. However, the complex factors mentioned above mean that this method for transformation is used only by very few laboratories and is not the method of choice for transforming Tritordeum.

### *4.3 DNA delivery by particle bombardment*

The lines of Tritordeum used as donors for particle bombardment were the same as those used for electroporation (lines HT 28, HT 31 and HT 174) but they were grown under greenhouse conditions (16-h photoperiod, 18°C/14°C day/night temperature). For each bombardment ~30 inflorescence pieces were placed in the centre of a 90 mm Petri dish containing induction medium. Explants were precultured in darkness at 24°C for 1 day prior to bombardment. The induction media used were L7D2, L7D4, L7P4 and L7P6 (Rasco-Gaunt and Barcelo, 1998).

The following plasmids were used for transformation (Figure 1): pAHC25 (Christensen and Quail, 1996) which contains the *uidA* and *bar* genes, both under the control of the maize ubiquitin promoter and the ubiquitin intron; pAct1-DGus (McElroy et al., 1990) containing the *uidA* gene under the control of the actin-1D promoter from rice; and pCaI-neo (constructed by S. Luetticke, University of Hamburg, modified from the plasmid pCaI-gus described in Callis et al., 1987) which contains the *neo* gene under the control of the CaMV 35S promoter and the maize *adh* I intron.

Immediately after bombardment, explants were spread over the surface of the medium in the original dishes and cultured at 24°C in darkness for three weeks. Explants from control dishes were then divided into two sets: one was regenerated without selection whereas the other set was subjected to selection. For selection and regeneration, calli were transferred to regeneration (Rz) medium (Rasco-Gaunt and Barcelo, 1998) supplemented with either 2 mg $l^{-1}$ L-phosphinothricin (L-PPT, the active ingredient of the herbicide BASTA) or 50 mg $l^{-1}$ G418 (geneticin disulphate) and cultured for a further three weeks. Surviving explants were transferred to a second regeneration medium (R) (Rasco-Gaunt and Barcelo, 1998) again containing 2 mg $l^{-1}$ L-PPT or 50 mg $l^{-1}$ G418. Successive three-week

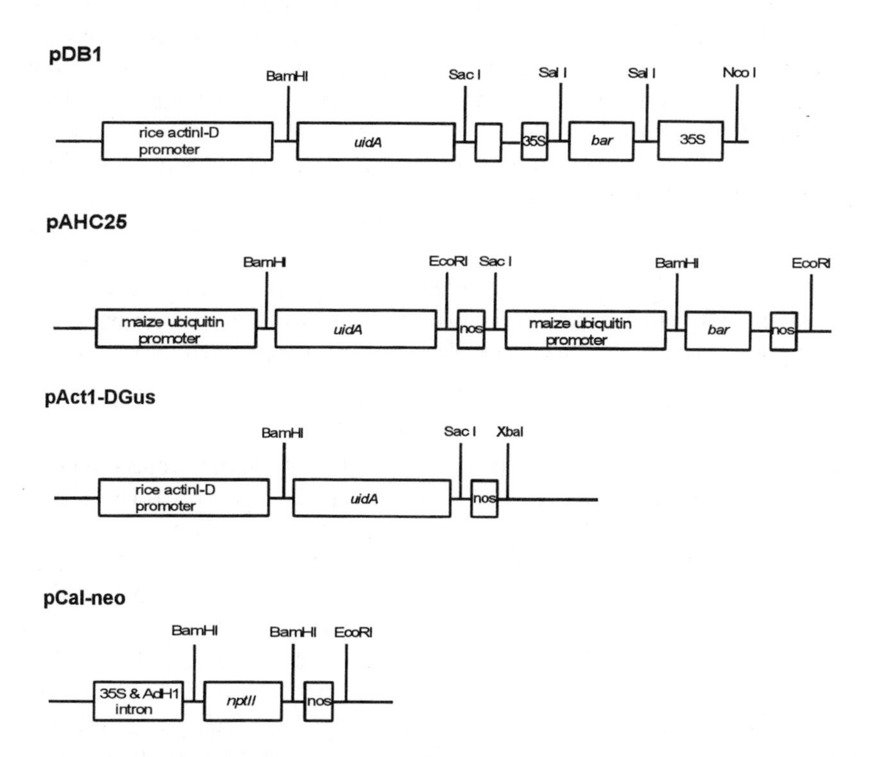


*Fig. 1* Constructs pDB1 (Becker et al., 1994), pAct-1-DGus (McElroy et al., 1990), pCaI-neo (Callis et al., 1987) and pAHC25 (Christensen and Quail, 1996)

selection passages were applied on this medium until all control cultures grown under selection were dead. Surviving plantlets from transformation treatments were transferred to soil and grown to maturity in the greenhouse.

Similar frequencies of transformation were obtained with either low or high bombardment pressures (Table 4). In contrast, the number of days the explants were cultured before bombardment had a significant effect on the recovery of transgenic plants and one and three days appeared to be optimal. The percentage of escapes was ~30% and 17 plants out of 24 regenerants were confirmed to be transformants. The patterns of integration found in the four plants analysed by Southern were relatively simple for the *neo* gene (Figure 2b), e.g. two plants with one and four copies, P3 and P10 respectively, and in both cases without rearrangements, and two other plants P2 and P5 with rearrangements and ~7 copies of the transgene. However, the same plants showed a much more complex integration pattern for the *uidA* gene as they all had rearranged copies (Figure 2a). This feature of simpler patterns of integration for the selected rather than for the non-selected gene has not been seen in other populations of transgenic Tritordeums produced subsequently (see below), where co-integration and co-expression frequencies for the *neo* and the *uidA* genes when delivered on two separate plasmids were very high and 88% of the plants contained and expressed both genes. From this population of plants the inheritance of one line was studied and it confirmed to a Mendelian segregation ratio expected for a one-locus integration.

In a second set of experiments, the objective was to produce Tritordeum plants carrying HMW-GS transgenes to investigate their effect on bread-

TABLE 4

Influence of preculture period *prior* to bombardment and of acceleration pressure (psi) on the production of transgenic plants from Tritordeum inflorescence cultures (from Barcelo et al., 1994b)

| | Acceleration pressure | | | | | | | |
|---|---|---|---|---|---|---|---|---|
| Preculture[a] | 650 a/b[b] | 900 a/b | 1100 a/b | 1350 a/b | 1550 a/b | Total a/b | T.plants/ bomb[c] | Freq.[d] |
| 0 | 0/4 | 2/20 | 0/4 | 0/4 | 0/20 | 2/52 | 2/9 | 0.2 |
| 1 | 1/6 | 2/18 | 0/6 | 1/6 | 5/18 | 9/54 | 9/9 | 1.0 |
| 3 | 1/5 | 2/21 | 1/5 | 0/5 | 0/21 | 4/57 | 4/9 | 0.4 |
| 6 | 0/3 | 1/3 | 0/3 | 0/3 | 1/3 | 2/15 | 2/5 | 0.4 |
| Totals | 2/18 | 7/62 | 1/18 | 1/18 | 6/59 | 17/178 | 17/32 | 0.5 |

[a]in vitro culture in days. [b]number of transgenic plants recovered per number of inflorescence bombarded (three to six inflorescence per bombardment, each divided into ~1 mm explants). [c]number of transgenic plants recovered per number of bombardment. [d]frequency of transgenic plants recovered per bombardment

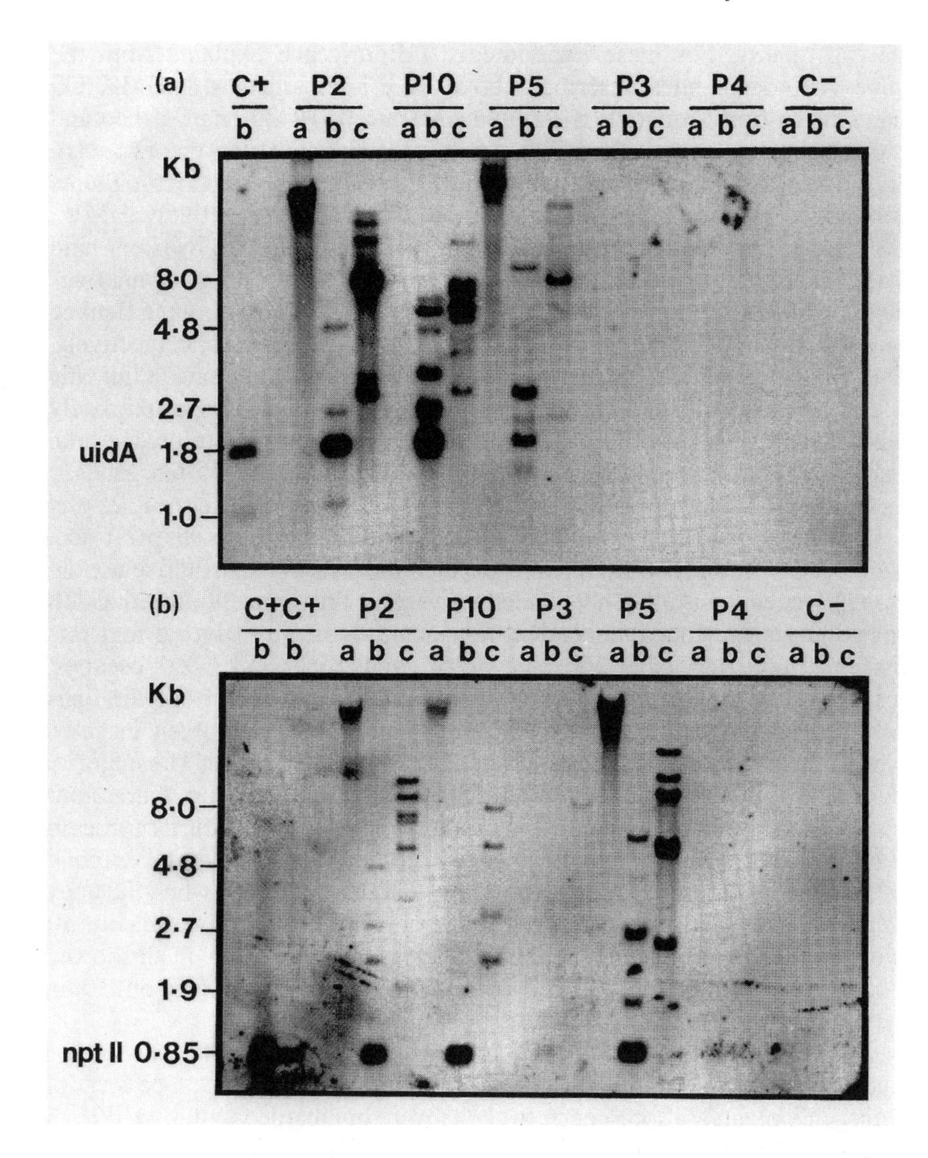


*Fig. 2* Southern blot analysis of Tritordeum transformants. a. and b. Hybridization with DIG-labelled *uidA* (a) and *npt II* (b) probe. In each set of three lanes, a denotes undigested DNA, b denotes DNA digested with *BamH* I / *Sac* I and *Bam*H I for the *uidA* and *npt II* probes, respectively, and c denotes DNA digested with *Xba* I and *Eco*R for the *uidA* and *nptt* II probes, respectively. P2, P3, P4, P5 and P10 are putative transformants, $C^+$ and $C^-$ positive and negative controls. From NPT II and GUS expression assays plant P2, P5 and P10 was scored NPT $II^+$, $GUS^+$, P3 NPT $II^+$, $GUS^-$, and P4 NPT $II^-$, $GUS^-$

making quality. For these experiments, inflorescence explants from the same Tritordeum lines described above were bombarded with HMW-GS genes in co-transformation with either plasmid pAHC25 or pCaI-neo and pAct1-DGus (see above, and Figure 1 for description of these constructs). Selection was applied using PPT (2 mg/1) or G418 (50 mg/1) for cultures bombarded with the pAHC25 or pCaI-neo constructs, respectively. HMW-GS 1Dx5 and 1Ax1 were delivered as plasmids pHMW1Dx5 and pHMW1Ax1 (Barro et al., 1997). The first contains a 8.7 Kb genomic fragment including the coding sequence of the *Glu-D1-1b* (1Dx5) gene flanked by approximately 3.8 Kb and 2.2 Kb and 5′ and 3′ sequences, respectively. The second plasmid contains a 7.0 Kb genomic fragment including the coding sequence of the *Glu-A1-1a* (1Ax1) gene flanked by approximately 2.2 Kb and 2.1 Kb of 5′ and 3′ sequences, respectively. Therefore, in both cases their own endosperm-specific promoter drove the HMW-GS genes.

A total of 777 inflorescence explants were bombarded with pAHC25 and pHMW1Ax1 or pHMW1Dx5 and selected on PPT, and 146 with pCaI-neo, pAct1-DGus and pHMW1Dx5 and selected on G418. Under PPT selection, from 97 regenerants nine independent transgenic lines were produced. G418 selection yielded four independent transgenic lines. This gives a transformation frequency of 1.2% and 2.7%, and 85% and 50% escapes, respectively. Considering co-transformation frequencies, 92% of the lines obtained had both markers and HMW-GS genes inserted, even in cases where transformation with three plasmids had been attempted. The majority of the lines showed simple patterns of integration, ranging from one to four copies of the HMW-GS genes. Regarding transgene expression, Tritordeum lines appeared to be less reliable than wheat (unpublished results). We found three out of nine lines with no expression of transgenes even when the copy number was low and intact copies of the transgene were present, in contrast with wheat lines produced with the same set of constructs, which all showed good levels of expression (Barro et al., 1997, Blechl and Anderson, 1996, Altpeter et al., 1996b).

A third set of experiments was aimed at studying the effect of auxin type and concentration applied during embryogenesis on the recovery of transgenic plants. Inflorescence explants were bombarded with pAHC25 or pCaI-neo (Figure 1) and were selected on PPT or G418, respectively. Inflorescence cultures gave rise to transgenic plants when cultured on either picloram or 2,4-D, although picloram was more efficient than 2,4-D in producing transgenic plants in both selection systems (see Table 5). This result may partly be explained by the fact that picloram is able to induce higher regeneration frequencies than 2,4-D. The highest frequency of production of transgenic plants was 3.2% on media containing 4 mg $1^{-1}$ picloram (Barro et al., 1998).

Overall PPT and G418 appeared to have very similar efficiency in selecting transgenic plants in Tritordeum: 1.2% transformation efficiency under PPT selection and 1.1% under G418 selection (averaged over the two auxins). However, the stringency of selection with G418 was better than with PPT: there were ~30% escapes under G418 and ~80% under PPT. In total, from the nine transgenic Tritordeum lines produced in this experiment, some ~30% appeared to be sterile, but this lack of fertility in some lines was related to poor pollen fertility rather than to the presence of the transgene because this was also observed in non-transformed control regenerants.

Nine transgenic lines of Tritordeum are at present being studied for the integration and expression of transgenes in $T_1$ and later generations by Martin Cannell (Rothamsted Experimental Station). The results of Southern analysis show that 79% (11) of the plants had low number of copies integrated (57% or 8 plants contained from 1 to 3, and 21% or 3 plants from 4 to 7 copies), and only 21% (3 plants) contained 10 or more transgene copies. Nine out of sixteen transgenic plants were obtained from cultures induced on 4 mg $l^{-1}$ picloram; of these seven (78%) contained low numbers of transgene copies.

Transgene segregation ratios were not significantly different (at the 5% probability level) to a Mendelian segregation ratio consistent for the presence of a single locus in 67% of the lines (8 out of 12). Three lines showed segregation ratios deviating towards the loss of transgene inheritance.

From the three Tritordeum lines containing plasmid pAHC25, only one expressed both *uidA* and *bar* genes, but this lack of expression was not due to lack of intact copies integrated or high complexity of the integration pattern. The remainder of the Tritordeum lines analysed did not appear to have irregularities of expression.

TABLE 5

Effect of auxin type and concentration, and selection system on the recovery of transgenic Tritordeum plants from bombarded inflorescence cultures (from Barro et al., 1998)

| Hormone | Reg. (%) | PPT selection TE (%) | G418 selection TE (%) | Totals TE (%) |
|---|---|---|---|---|
| 2,4-D | | | | |
| 2 mg/l | 44a | 0 | 1.2 | 0.5 |
| 4 mg/l | 45a | 2 | 0 | 1.0 |
| Picloram | | | | |
| 4 mg/l | 76b | 3.2 | 2.3 | 2.7 |
| 6 mg/l | 74b | 0 | 1.2 | 0.5 |
| Total | | 1.2 | 1.1 | 1.2 |

Means within species and columns with the same letter are not significantly different at the 0.05 level according to Fisher's Least Significant Difference (LSD) Test. TE, Transformation Efficiency

We have conducted a similar analysis of our wheat transformants (Barro et al., 1998). One of the more noticeable features emerging from this comparative study is that transgenes integrated in wheat appear to be more stable in primary transformants as well as in later generations (see also chapter by Vasil and Vasil, this volume), than when integrated in Tritordeum. This could be due to the fact that Tritordeum is a relatively newly-synthesised species where the genomes of wheat and barley have been placed together. This may create general instability in the genome and therefore in transgene expression.

### *4.4 Anther culture for the production of homozygous transgenic lines*

Tritordeum spikes containing anthers with microspores at the uninucleate stage were selected and either pre-treated for 10 days at 4°C (10 days pre-treatment) or used directly for culture (0 days pre-treatment). Isolated anthers were cultured in 3 cm Petri dishes containing 2.5 ml of induction medium containing 0.02% colchicine for three days in the dark at 28°C. The medium was then replaced with fresh induction medium without colchicine. Embryogenic tissues produced after 6–7 weeks were placed on regeneration medium without hormones or with 5 mg/I zeatin. The induction media contained various combinations of 2,4-D and kinetin and 10% Ficoll 400 (Barcelo et al., 1994a).

Two parameters, hormone combination in the induction period and length of cold pre-treatment, were tested for their effect on callus formation and plant regeneration. Cold pre-treatment appeared to have a stronger influence on the frequency of anther response and of regenerating cultures than the hormones used during induction. Particularly, 10 days cold pre-treatment resulted in a higher frequency of green plant recovery than five or less days of exposure. However, cold pre-treatment showed also a negative side, that is the induction of a higher proportion of albino plants. Therefore, a compromise between these two aspects of the technique has to be reached in order to maximise the production of green plants.

Duplication of the chromosome content of the haploid gamete is also a major step. It appears that 80–90% of the microspores in anthers exposed to 0.02% colchicine for three days duplicate their chromosome number. If duplication frequencies are lower, then other concentrations of colchicine and other durations of exposure must be tested (Barnabas et al., 1991).

### *4.5 Field performance of transgenic tritordeum*

The agronomic performance of transgenic Tritordeum lines under field conditions has been studied at the Institute of Sustainable Agriculture in Cordoba (Spain) by Pilar Hernandez in A. Martin's group. Six transgenic Tritordeum lines from the line HT 28 containing the *uidA* and *neo* genes inserted in the genome with different patterns of integration were chosen from a population of 17 transgenic lines produced by Barcelo et al., (1994b). Self-pollinated $T_1$ individuals from these six lines were analysed for expression of the *uidA* and *neo* genes and those expressing the transgenes were self-pollinated to produce $T_2$ progeny. Both $T_1$ and $T_2$ generations together with the controls were then investigated for their agronomic performance in the field during 1993–94 and 1994–95 seasons. As controls, seed derived plants and plants derived from bombarded non-transformed cultures were used.

The experiment was set up as a complete randomised design with four replicates with the plants sown, irrigated and maintained as the guidance for wheat farming in the area. The agronomic characters studied were grouped in those related to yield (tiller width and length, number of spikelets per spike and number of grains per spike, biomass and harvest index) and those related to life cycle (time to flowering, anthesis and maturity). The data obtained were analysed by the Analysis of Variance and by the Principal Component Analysis (PCA). The PCA shows the relationship among the lines on the basis of the characters studied and therefore groups the individual lines based on their similarities. With this analysis it was found that the transgenic lines always were grouped together and were separated from the two control lines which formed another group. This means that the transgenic lines studied were performing agronomically different from the controls and that it was not due to somaclonal variation, since the bombarded non-transformed control was similar in performance to the seed control. This result shows that transformation had clearly affected the agronomic performance of Tritordeum (at least for the population of transgenic Tritordeum chosen for this study). This finding was even more significant when the same results were observed in the second year of the study. In addition to these yield parameters, transformation appeared to delay both flowering and anthesis time (both were earlier in the controls). The analysis performed so far in transgenic Tritordeum is confined to the first two generations in which possible *in vitro* culture effects may still prevail and therefore to complete the analysis further generations will have to be studied (Hernandez, personal communication).

**Acknowledgements**

The authors acknowledge Dr Paul A. Lazzeri and Professor Antonio Martin for critical reading of the manuscript.

**References**

Altpeter, F., Vasil, V., Srivastava, V., Stöger, E., and Vasil, I.K. (1996a) Accelerated production of transgenic wheat (*Triticum aestivum* L.) plants. Plant Cell Rep. 16: 12–17.

Altpeter, F., Vasil, V., Srivastava, V., and Vasil, I.K. (1996b) Integration and expression of the high-molecular-weight glutenin subunit 1Ax1 gene into wheat. Nature Biotechnology. 14: 1155–1159.

Alvarez, J.B., Ballesteros, J., Sillero, J.A., and Martin, L.M. (1992) Tritordeum: a new crop of potential importance in the food industry. Hereditas. 116: 193–197.

Alvarez, J.B., Ballesteros, J., Arriaga, H.O., and Martin, L.M. (1995) The rheological properties and baking performance of flours from hexaploid Tritordeums. Cereal Sci. 21: 291–299.

Armstrong, K.C., Nakamura, C., and Keller, W.A. (1983) Karyotypic instability in tissue culture regenerants of *Triticale* (*x Triticosecale* Wittmack) cv. 'Welsh' from 6-month-old callus cultures. Z. Pflanzenzüchtg. 91: 233–245.

Babeli, P., Karp, A., and Kaltsikes, P.J. (1988) Plant regeneration and somaclonal variation from cultured immature embryos of sister lines of rye and triticale differing in their content of heterochromatin. 1. Morphogenetic response. Theor. Appl. Genet. 75: 929–936.

Barcelo, P., Vazquez, A., and Martin, A. (1989) Somatic embryogenesis and plant regeneration from Tritordeum. Plant Breed. 103: 235–240.

Barcelo, P., Cabrera, A., Hagel, C., and Lörz, H. (1994a) Production of doubled-haploid plants from tritordeum anther culture. Theor. Appl. Genet. 87: 741–745.

Barcelo, P., Hagel, C., Becker, D., Martin, A., and Lörz, H. (1994b) Transgenic cereal (Tritordeum) plants obtained at high efficiency by microprojectile bombardment of inflorescence tissue. Plant J. 5: 583–592.

Barcelo, P., Lazzeri, P.A., Hernandez, P., Martin, A., and Lörz, H. (1993) Morphogenic cell and protoplast cultures of tritordeum. Plant Sci. 88: 209–218.

Barnabas, B., Pfahler, P.L., and Kovacs, G. (1991) Direct effect of colchicine on the microspore embryogenesis to produce dihaploid plants in wheat (*Triticum aestivum* L.) Theor. Appl. Genet. 81: 675–678.

Barro, F., Fontes, A.G., and Maldonado, J.M. (1991) Organic nitrogen content and nitrite reductase activities in Tritordeum and wheat grown under nitrate or ammonium. Plant & Soil. 135: 251–256.

Barro, F., Rooke, L., Bekes, F., Grass, P., Tatham, A.S., Fido, R., Lazzeri, P.A., Shewry, P. R., and Barcelo, P. (1997) Transformation of wheat with HMW subunit genes results in improved functional properties. Nature Biotechnology 15: 1295–1299.

Barro, F., Cannell, M.E., Lazzeri, P.A., and Barcelo, P. (1998) The influence of auxins on transformation of wheat and tritordeum and analysis of transgene integration patterns in transformants. Theor. Appl. Genet. (in press).

Becker, D., Brettschneider, R., and Lörz, H. (1994) Fertile transgenic wheat from microprojectile bombardment of scutellar tissue. Plant J. 5: 299–307.

Bernard, S. (1977) Etude de quelques facteurs contribuant a la reussite de l'androgenese par culture d'antheres *in vitro* chez le *Triticale* hexaploide. Ann. Plantes. 27: 639–635.

Bernard, S. (1980) *In vitro* androgenesis in hexaploid *Triticale*: determination of physical conditions increasing embryoid formation and green plant production. Z. Pflanzenzüchtg. 85: 308–321.

Blakeslee, A.F., and Avery, A.G. (1937) Methods of inducing doubling of chromosomes in plants. J Heredity 28: 392–411.

Blechl, A.E., and Anderson, O.D. (1996) Expression of a novel high molecular weight glutenin subunit gene in transgenic wheat. Nature Biotechnology 14: 875–879.

Brettell, R.I.S., Denis, E.S., Scowcroft, W.R., and Peacock, W.J. (1986) Molecular analysis of somaclonal mutant of maize alcohol dehydrogenase. Mol. Gen. Genet. 202: 235–239.

Callis, J., Fromm, M.E., and Walbot, V. (1987) Introns increase gene expression in cultured maize cells. Genes Dev. 1: 1183–1200.

Charmet, G., and Bernard, S. (1984) Diallel analysis of androgenic plant production in hexaploid *Triticale* (x *Triticosecale* Wittmack). Theor. Appl. Genet. 69: 55–61.

Christensen, A.H., and Quail, P.H. (1996) Ubiquitin promoter-based vectors for high-level expression of selectable and/or screenable marker genes in monocotyledonous plants.

Transgen. Res. 5: 213–218.
Christou, P., Ford, T.L., and Kofron, M. (1991) Production of transgenic rice (*Oryza sativa* L.) plants from agronomically important indica and japonica varieties *via* electrical discharge particle acceleration of exogenous DNA into immature zygotic embryos. Bio/ Technology 9: 957–962.
Chu, C.C. (1978) The N6 medium and its applications to anther culture of cereal crops. Proceedings of the Symposium on Plant Tissue Culture. Peking, Science Press Peking.
Cubero, J.I., Martin, A., Millan, T., Gomez-Cabrera, A., and de Haro, A. (1986) Tritordeum: a new alloploid of potential importance as a protein source crop. Crop Sci. 26: 1186–1190.
Dekeyser, R.A., Claes, B., De Rycke, R.M.U., Habets, M.E., Van Montagu, M.C., and Caplan, A.B. (1990) Transient gene expression in intact and organised rice tissues. Plant Cell 2: 591–602.
D'Halluin, K., Bonne, E., Bossut, M., De Beuckeleer, M., and Leemans, T. (1992) Transgenic maize plants by tissue electroporation. Plant Cell 4: 1495–1505.
Eapen, S., and Rao, P.S. (1982) Callus induction and plant regeneration from immature embryos of rye and Triticale. Plant Cell Tis. Org. Cult. 1: 221–227.
Eapen, S., and Rao, P.S. (1985) Plant regeneration from immature inflorescence callus cultures of wheat, rye and *Triticale*. Euphytica. 34: 153–159.
Fedak, G. (1987) Chromosome irregularities in wheat and *Triticale* plants regenerated from leaf base callus. Plant Breed. 99: 151–154.
Fromm, M.E., Taylor, L.P., and Walbot, V. (1986) Stable transformation of maize after gene-transformation by electroporation. Nature 319: 178–182.
Gall, J.G., and Pardue, M.L. (1969) Formation and detection of RNA-DNA hybrid molecules in cytological preparations. Proc. Nat. Acad. Sci. USA 63: 378–383.
Gallardo, M., and Fereres, E. (1989) Resistencia a la sequia del Tritordeo (*Hordeum chilense* X *Triticum turgidum*) en relacion a la del trigo, cebada y Triticale. Invest. Agr. : Prod. Veg. 4: 361–375.
Gamborg, O.L., Miller, R.A., and Ojima, K. (1968) Nutrient requirements of suspension cultures of soybean root. Exp. Cell Res. 50: 151–158.
He, G.Y., and Lazzeri, P.A. (1998) Analysis and optimisation of DNA delivery into wheat scutellum and Tritordeum inflorescence explants by tissue electroporation. Plant Cell Rep. (in press).
Jordan, M.C., and Larter, E.N. (1984) Somaclonal variation in *Triticale* (x *Triticosecale* Wittmack) ev. Carman. Can. J. Genet. Cytol. 27: 151–157.
Kaltsikes, P.J., and Babeli, P.J. (1992) The effect of rye telomeric heterochromatin on the nature and size of variance in regenerated families of hexaploid triticale. J. Genet. Breed 46: 359–362.
Kozdój, J., and Zimny, J. (1993) Microspore development stages in chilled and unchilled anthers of *Triticale* (x *Triticosecale* Wittmack). Bull. Polish Acad. Biol. 41: 108–116.
Laursen, C.M., Krzyzek, R.A., Flick, C.E., Anderson, P.C., and Spencer, T.M. (1994) Production of fertile transgenic maize by electroporation of suspension culture cells. Plant Mol. Biol. 24: 51–61.
Li, L., Rongda, Q., de Kochko, A., Faquet, C., and Beachy, R.N. (1993) An improved rice transformation system using the biolostic method. Plant Cell Rep. 12: 250–255.
Li, B.J., Xu, X.P., Shi, H.P., and Ke, X.Y. (1991) Introduction of foreign genes into the seed embryo cells by electro-injection and the regeneration of transgenic rice plants. Science in China 34: 923–931.
Lörz, H., Baker, B., and Schell, J. (1985) Gene transfer to cereal cells mediated by protoplast transformation. Mol. Gen. Genet. 199: 178–182.
Martin, A. (1988) Tritordeum: the first ten years. Rachis. 7: 12–15.
Martin, A., and Cubero, J.I. (1981) The use of *Hordeum chilense* in cereal breeding. Cereal Res. Comm. 9: 317–323.
Martin, A., and Sanchez-Monge, E. (1982) Cytology and morphology of the amphiploid

*Hordeum chilense* X *Triticum turgidum* conv. *durum*. Euphytica. 31: 261–267.
Martin, A., Martinez-Araque, C., Rubiales, D., and Ballesteros, J. (1996) Tritordeum: Triticale's new brother cereal. In: Guedes-Pinto, H. et al., (eds) Triticale: Today and Tomorrow. Kluwer Academic Publishers.
McElroy, D., Zhang, W, Cao, J., and Wu, R. (1990) Isolation of an efficient actin promoter for use in rice transformation. Plant Cell 2: 163–171.
Murashige, T., and Skoog, F. (1962) A revised medium for rapid growth and bioassay with tobacco tissue cultures. Physiol. Plant. 15: 473–497.
Nakamura, C.H., and Keller, W.A. (1982) Callus proliferation and plant regeneration from immature embryos of hexaploid *Triticale*. Z. Pflanzenzüchtg. 91: 137–160.
Nehra, N.S., Chibbar, R.N., Leung, N., Caswell, K., Mallard, C., Steinhauer, L., Baga, M., and Kartha, K.K. (1994) Self-fertile transgenic wheat plants regenerated from isolated scutellar tissue following microprojectile bombardment with two distinct gene constructs. Plant J. 5: 285–297.
Olin-Fatih, M., and Heneen, W.K. (1992) C-banded karyopses of *Brassica campestris* var. *pekinensis, B. oleracea* and *B. napus*. Genome 35: 583–589.
Ono, H., and Larter, E.N. (1976) Anther culture of *Triticale.* Crop Sci. 16: 120–122.
Pan, W.H., Houben, A., and Schlegel, R. (1993) Highly effective cell synchronization in plant roots by hydroxyurea. Genome 36: 387–390.
Pauk, J., Manninen, O., Mattila, I., Salo, Y., and Pulli, S. (1991) Androgenesis in hexaploid spring wheat F2 populations and their parents using a multiple-step regeneration system. Plant Breed. 107: 18–27.
Pedersen, C., and Linde-Laursen, I. (1994) Chromosomal localizations of four minor rDNA loci and a marker microsatelite sequence in barley. Chromosome Res. 2: 65–71.
Pedersen, C., Rasmussen, S.K., and Linde-Laursen, I. (1996) Genome and chromosome identification in cultivated barley and related species of the *Triticeae* (*Poaceae*) by *in situ* hybridisation with the GAA – satelite sequence. Genome 39: 93–104.
Pedersen, C., Zimny, J., Becker, D., Gaertner, A., and Lörz, H. (1997) Localization of introduced genes on the chromosomes of transgenic barley, wheat and *Triticale* by fluorescence *in situ* by hybridization. Theor. Appl. Genet. 94: 749–757.
Rasco-Gaunt, S., and Barcelo, P. (1998) Immature inflorescence culture of cereals: a highly responsive system for regeneration and transformation. In: Hall, R.D. (ed.), Plant Cell Culture Protocols. Methods in Molecular Biology Series. Humana Press, Totowa, New Jersey (in press).
Rathore, K.S., Chowdhury, V.K., and Hodges, T.K. (1993) Use of bar as a selectable marker gene for the production of herbicide-resistant rice from protoplasts. Plant Mol. Biol. 21: 871–884.
Rimpau, W. (1891) Kreuzungsprodukte landwirtschaftlicher Kulturpflanzen. Landwirtschaftl. Jahrb. 20: 335–371.
Rooke, L., Barro, F., Tatham, A.S., Fido, R., Steele, S., Békés, F., Gras, P., Martin, A., Lazzen, P.A., Shewry, P.R., and Barcelo, P. Improved functional properties of Tritordeum by transformation with HMW Glutenin subunit genes. Theor. Appl. Genet. (in press).
Schumann, G. (1988) Untersuchungen zum Albinismus in Antherenkulturen von *Triticale*. Arch Züchtungsforsch 18, 2: 115–122.
Sosinov, A., Lukjanjuk, S., and Ignatova, S. (1981) Anther cultivation and induction of haploid plants in *Triticale*. Z. Pflanzenzüchtg. 86: 272–285.
Spencer, T.M., O'Brien, J.V., Start, W.G., and Adams, T.R. (1992) Segregation of transgenes in maize. Plant Mol. Biol. 18: 201–210.
Stolarz, A. (1990) Cell and protoplast culture, somatic embryogenesis and transformation studies in different forms of x *Triticosecale* Wittmack. In: Proc 2nd Internat. *Triticale* Symp., Passo Fundo, Brazil 286–289.
Stolarz, A., and Lörz, H. (1986a) Somatic embryogenesis, *in vitro* manipulation and plant regeneration from immature embryos of hexaploid *Triticale* (x *Triticosecale* Wittmack). Z

Pflanzenzüchtg. 96: 353–362.

Stolarz, A., and Lörz, H. (1986b) Somatic embryogenesis, cell and protoplast culture of *Triticale* (x *Triticosecale* Wittmack). In: Horn, W., Jensen, C.J., Odenbach, W., Schieder, O. (eds), Genetic Manipulation in Plant Breeding, pp. 499–501. W de Gruyter, Berlin.

Trask, B.J. (1991) Fluorescence *in situ* hybridization: application in cytogenetics and gene mapping. Technical Focus 7: 149–154.

Ushimiya, H., Fushimi, T., Hashimoto, H., Harada, H., Syano, K., and Sugawara, Y. (1986) Expression of a foreign gene in callus derived from DNA treated protoplasts of rice (*Oryza sativa* L.). Mol. Gen. Genet. 204: 204–207.

Vasil, I.K., and Anderson, O.D. (1997) Genetic engineering of wheat gluten. Trends. Plant Sci. 2: 292–297.

Vasil, I., and Vasil, V. (1992) Advances in cereal protoplast research. Physiol. Plant 85: 279–283.

Villareal, R.L., Varughese, G., and Abdalla, O.S. (1990) Advances in Spring Triticale Breeding. In: Wheat Program, pp. 43–87. CIMMYT, El Batan, Mexico.

Walters, D.W., Vetsch, C.S., Potts, D.E., and Lundquist, R.C. (1992) Transformation and inheritance of hygromycin phosphotransferase gene in maize plants. Plant Mol. Biol. 18: 189–200.

Wan, Y., and Lemaux, P. G. (1994) Generation of large numbers of independently transformed fertile barley plants. Plant Physiol. 104: 37–48.

Wang, Y., and Hu, H. (1993) Gamete composition and chromosome variation in pollen-derived plants from octoploid triticale x common wheat hybrids. Theor. Appl. Genet. 85: 681–687.

Wilson, A.S. (1875) On wheat and rye hybrids. Trans. Proc. Bot. Soc., Edinburgh. 12: 826–828.

Xu, X.P., and Li, B.J. (1994) Fertile transgenic indica rice plants obtained by electroporation of seed embryo cells. Plant Cell Rep. 13: 237–242.

Ya-Ying, W., Ching-San, S., Ching-Chu, W., and Nan-Fen, C. (1973) The induction of pollen plantlets of *Triticale* and *Capsicum annum* from anther culture. Sci. Sin. 16: 147–151.

Zillinsky, F.J. (1974) The development of Triticale. Adv. Agron. 315–349.

Zimny, J. (1989) Genotypic dependence of the somatic embryogenesis of *Triticale* (x *Triticosecale* Wittmack). In: Science for Plant Breeding pp. 7–13. XII Eucarpia Congress, Göttingen, Germany.

Zimny, J. (1992) Somatic embryogenesis of rye and *Triticale*. Keynote lecture at the Symposium: Regulation of plant somatic embryogenesis. Sumperk Proc. pp. 4–9.

Zimny, J., and Lörz, H. (1989) High frequency of somatic embryogenesis and plant regeneration of rye (*Secale cereale* L.). Z. Pflanzenzüchtng. 102: 89–100.

Zimny, J., and Rafalski, A. (1993) Transformowanie protoplastów pszen¿yta (x *Triticosecale* Wittmak). Biul Instyt Hod i Aklim Roœlin. 187: 127–132.

Zimny, J., Becker, D., Brettschneider, R., and Lörz, H. (1995) Fertile transgenic *Triticale* (*x Triticosecale* Wittmack). Mol. Breed. 1: 155–164.

# 13. The Grasses as a Single Genetic System

JEFFREY L. BENNETZEN
*Department of Biological Sciences, Purdue University, West Lafayette, IN 47906, USA.*
*E-mail: maize@bilbo.bio.purdue.edu*

ABSTRACT. Recent studies have indicated that many plant genomes have regions of common gene content and genetic map colinearity. This is particularly well documented in the grasses, where colinearity has been demonstrated both for recombinational maps using DNA markers and in the orthologous placement of comparable genes. Colinearity in regions of a centiMorgan or less (i.e., microcolinearity) has not been extensively investigated. Preliminary studies demonstrate that gene content, order and orientation are often conserved in small regions of related genomes, but there are frequent exceptions. Results indicate that grass genome colinearity will be a tremendously valuable tool that allows the synergistic pooling of knowledge and biological materials across a broad range of plant species. From this colinear perspective, genes from any grass species can be used for the improvement and understanding of any other grass. However, further investigations across a wide range of species and genomic locations will be needed to determine the limitations of the approach. In the interim, it is important to carefully investigate (rather than assume) the microcolinearity of the syntenic regions that are being analyzed.

## Introduction

The structural features of plant nuclear genomes are not well understood at any level (Bennetzen, 1998). We do know that genes are arranged in a linear order along chromosomes and that non-genic sequences, often tandem repeats (Peacock et al., 1981) or mobile DNAs (Bennetzen, 1996, Saedler and Gierl, 1996), can make up anywhere from 10–95% of a higher plant genome (Flavell et al., 1974, SanMiguel et al., 1996). Recent progress in recombinational mapping, using DNA markers, has made it possible to make genetic maps that are comparable between species (Bonierbale et al., 1988, Hulbert et al., 1990). Such studies have indicated that higher plants have very similar gene composition and regions of conserved gene order (Bennetzen and Freeling, 1993). This genomic colinearity is particularly extensive in the grasses (Moore et al., 1995).

Abbreviations: BAC, bacterial artificial chromosome; cM, centiMorgan; contig, contiguous physical map; RFLP, restriction fragment length polymorphism

*I.K. Vasil (ed.), Molecular Improvement of Cereal Crops, 387–394*

The similar gene content and frequent colinearity of grass genomes will allow novel and rapid advances in our understanding of genome evolution and, more importantly, will permit the cross-species utilization and understanding of plant genes (Bennetzen and Freeling, 1993, 1997). However, the ease and range of comparative genetic analysis of the grasses will depend upon the actual degree of colinearity. Chromosomal rearrangements do occur, and it is not yet clear how frequent large scale rearrangements (detected by recombinational maps or cytogenetics) are relative to small rearrangements. Recent studies provide reasons for optimism in colinear grass genetics, but also indicate good reasons for caution. This chapter will discuss progress in our understanding and exploitation of grass genome colinearity (and its exceptions).

## Comparative Maps with DNA Markers

The first genetic maps in plants lacked large numbers of markers due to a paucity of traits that could be easily mapped because of the reduced vigor of mutant stocks, frequent epistatic interactions, and variable penetrance of the scored traits. Even in some early studies, however, it was observed that comparable traits often appeared to be similarly linked between different plant species or on different chromosomes in a polypoid species (Rhoades, 1951). With the advent of restriction fragment length polymorphism (RFLP) DNA markers, genetic maps could be easily constructed with a wealth of loci only limited by the enthusiasm of the investigator. Moreover, the low-copy-number sequences that were useful as RFLP markers were usually found to cross-hybridize with sequences in other species. Hence, RFLP probes allowed the first comparative maps of genomes (Bonierbale et al., 1988, Hulbert et al., 1990).

With a large number of RFLP probes, and their facile use in the generation of genetic maps employing large progeny populations, investigators demonstrated that markers were not only syntenous (i.e., on comparable linkage groups in different genomes) but were often colinearly arranged within these syntenous linkage groups. Exceptions were noted, including duplications, inversions or translocations that often encompassed most of a chromosome arm (Bonierbale et al., 1988, Hulbert et al., 1990, Moore et al., 1995, Paterson et al., 1996, Prince et al., 1993). However, these large rearrangements were fairly rare and would not compromise comparison of similar regions within the rearrangement, but would only affect interpretations of homology at the boundary of the rearrangement.

Most comparative mapping with RFLP probes has been conducted in the grasses, particularly in species with agronomic importance. However, limited investigations in dicotyledenous species have indicated segments of colinearity among closely related species (Bonierbale et al., 1988, Fatokun et al., 1992, Kowalski et al., 1994, Prince et al., 1993, Teutonico and Osborn 1994, Weeden et al., 1992). In general, the comparative mapping data suggest that chromosomal rearrangements are more frequent, per unit evolutionary time, in dicots than in monocots, thereby making this approach less clearly valuable in broad-leaf plants. However, the vagaries of data collection and reporting have led to situations where, in two otherwise colinear regions, as many as 20–40% of the markers that do not fit any colinear pattern are simply ignored (Bennetzen and Freeling, 1997). Hence, the degree of colinearity reported between two species may be as much an indication of the philosophical leanings of the investigators as of the nature of the data.

## Morphological Traits on Comparative Maps: Orthology

In any comparison of DNA markers or other genetic traits, the appropriate comparisons are between orthologs; genes that share a direct vertical descent. Members of the same gene family are paralogs and may map at any of several locations in a plant genome. Some paralogs are members of a tandemly duplicated gene family, such as ribosomal DNAs (Dubcovsky and Dvorak, 1995), while others are found at unlinked sites. In some cases, these multiple gene copies may be found at colinear locations within a single genome, as in the A, B and D genomes of hexaploid wheat (Devos et al., 1993). These results indicate that the nuclear genome is derived from either a complete or segmental polyploidization (Rhodes, 1951, Helentjaris et al., 1988). With many unlinked gene family members, however, neither the mechanism nor the nature of the gene duplication is known. Particularly common, and problematic for comparative studies, are 'distantly tandem' duplications that yield two locations for the same marker separated by several centiMorgans on the same chromosome arm (Sanz-Alferez et al., 1995, Shoemaker et al., 1996).

Despite the paralog/ortholog complication raised by the numerous gene families of higher plants, many studies have shown that similar morphological or physiological traits often map to colinear locations in different plant species (Ahn et al., 1993, Fatokun et al., 1992, Paterson et al., 1995, Pereira and Lee, 1995, Teutonico and Osborn, 1994, Yu et al., 1996). In cases where colinearity is not observed, it is usually impossible to prove that the apparent lack of colinearity is not due to a paralogous comparison mistaken for an orthologous comparison. Even for the 'nomadic' ribosomal DNA repeats of the grasses, which do not usually map to colinear locations in different genomes (Dubcovsky and Dvorak, 1995), it is possible that small (only slightly amplified) rDNA repeat clusters might exist at the colinear locations. Hence, the use of standard genetic characteristics in comparative mapping generally supports genomic colinearity. Moreover, these results support the belief that colinearity can be used as a tool both to understand the evolution of important plant traits and for the isolation of the broadest range of allelic variation available for crop improvement (Bennetzen and Freeling, 1993, 1997).

## Microcolinearity

Most of the scientific value that can be derived from the colinearity of plant genomes requires that the conserved gene composition and marker order observed in low density recombinational maps also be true at the level of adjacent genes. In plants, as in other eukaryotes, recombination appears to be largely limited to genes, with a single gene usually accounting for 0.005 to 0.1 centiMorgans (cM) (Brown and Sundaresan, 1991, Civardi et al., 1994, Dooner, 1986). Hence, one cM will usually amount to about 30 genes, whether those genes are spread across 1200 kb in maize (Avramova et al., 1996), 140 kb in *Arabidopsis*, or many megabases near a wheat centromere. Because of the large populations that would be required, comparative recombinational maps sensitive to rearrangements at the level of less than a few tenths of a centiMorgan have not been constructed in plants. Hence, the first general investigations of microcolinearity, comparing recombinational maps in the Triticeae to physical maps of small segments of the rice genome, mainly indicated that most of the same markers were tightly linked in each species (Dunford et al., 1995, Kilian et al., 1995).

Direct physical comparisons between large cloned segments in maize, sorghum and rice (Avramova et al., 1996, Chen et al., 1997) indicated very similar gene content and extensive colinearity at the level of adjacent genes. As expected, genes were more tightly packed in the smaller genome species. However, most studies of microcolinearity also uncovered small rearrangements or differences in gene content in these same regions (Kilian et al., 1995, Foote et al., 1997, Leister et al., 1998). These exceptions indicate that small-scale rearrangements are much more abundant than are the large rearrangements that can be detected by standard recombinational maps or cytogenetic studies (Paterson et al., 1996).

In cases where gene content and marker order are conserved, colinearity can be used as a tool for finding the genes in a cloned chromosomal segment. This will often not be a trivial undertaking in complex plant genomes, like maize, where the majority of the DNA is made up of intergenic transposable elements (SanMiguel et al., 1996). Because many mobile elements are expressed (Avramova et al., 1995) and can also be found at very low copy numbers (Bennetzen 1996), neither complementarity to an RNA nor a low copy number can be used as definitive proof that a sequence within a chromosomal region is a gene. However, the conserved sequences and positions of genes allow these loci to be found within a sea of repetitive DNA by simple cross-hybridization of colinear clones (Avramova et al., 1996).

## Uses of Comparative Maps

Several publications have now described the potential value of a comparative approach to plant genetics, relying on common gene content and marker colinearity (Bennetzen and Freeling, 1993, 1997, Devos et al., 1995, Snape et al., 1996). These uses include the map-based isolation of genes from large-genome plants using small-genome plants as surrogates, the identification of significant (i.e., conserved) genetic characteristics in a chromosomal segment, and an ability to characterize the nature and timing of chromosomal evolution.

In the long run, comparative genetics will probably be most useful in providing a source of new alleles for crop improvement and in determining the nature of genetic changes that have given rise to significant developmental/physiological alterations in evolution. Moreover, the identification of orthologous loci in different species allows the investigator to use the information gained with those genes in any other system to enrich his own studies. Hence, the greatest beneficiaries of comparative genetics, particularly when extended into the comprehensive gene-identification provided through comparative genomics, will be the physiologists, biochemists and agronomists who want to study or utilize particular biological pathways. The commonalties between species will allow identification of the core properties of a pathway, while the variations will indicate how the pathway has been manipulated in evolution and could be manipulated by the genetic engineer. Taken to its culmination, comparative genetics will allow all biologists to work together in a single biological system, not isolated but rather illuminated by the vagaries of their study organism(s). In the grasses, where colinear regions are frequent, this synergistic approach to plant science has been described a the 'Unified Grass Genome' (Bennetzen and Freeling, 1993, 1997).

## Abuses of Comparative Maps

As with any new approach in science, the common gene content and map colinearity of grass genomes was first met with disbelief, rather rapidly followed by excessive credulity. A few examples of colinearity between genomes do not indicate that this is a universal phenomenon. Chromosomal rearrangements are not necessarily equally likely in all lineages of related organisms, as such changes are more likely to be punctuated than continuous. Hence, investigators should not assume colinearity of a region unless it has been tested. This skepticism is particularly important for questions of microcolinearity, where many exceptions to colinearity have already been noted (Kilian et al., 1995, Foote et al., 1997, Leister et al., 1997). Each case must be tested individually before colinearity or microcolinearity can be utilized.

When an investigator has recombinationally mapped many tightly linked markers in two orthologous regions, and shown their similar linkage, then the first requisite for the utilization of colinearity has been met. For map-based gene isolation, for instance, the investigator will need to check out the retention of colinearity at each step in the process. Colinearity can be a useful, even essential, tool for this and other gene identification/cloning processes, but one should not enter such a perilous process with blind confidence.

## The Final Comparative Map: Whole Genome 'Contigs' and Sequences

Eventually, genomics will provide the plant research community with the ultimate tool for comparative mapping, the full genome sequences. The complete sequence of the gene-containing regions of the *Arabidopsis* genome is only a few years away, and an ambitious program to sequence the rice genome has just been initiated (ftp:IIgenome1.bio.bnl.gov.pub/maize/rice.html). With this tool in hand, all of the conserved features (including genes) and their locations will be easily perceived when comparing two genomes. As these physical sequences are tied to traits on the genetic map and functional genomics tools become available, then investigators will have the full set of resources needed to understand the function and evolution of any given gene or biological pathway. Plant science will become less a matter of perspiration, and more of inspiration. Smaller research groups and individuals will be empowered, because it will not require a great array of personnel and facilities to undertake a major project, only a very good idea.

This great advance in the biological sciences will not be available immediately, but neither is it decades away (Bennetzen et al., 1998). Significant amounts of technology development and data generation are needed. In the interim, physical maps could be generated for many small genome plant species. Particularly feasible would be contiguous physical maps ('contigs') based on bacterial artificial chromosome (BAC) libraries of rice, sorghum and tomato. These contigs could be generated in one or two years, and could serve as the foundations for contig assembly in closely related plants with larger genomes (e.g., sorghum as a model for maize). With these tools in hand now, investigators could begin to utilize comparative genetics and genomics for the synergistic study and improvement of plants.

Comparative mapping has already proved informative in uncovering unexpected genetic relationships between different plant species. Several research groups are now trying to use this colinearity for the identification and isolation of important plant genes. The current status of this field requires investigators to take a cautious case-by-case approach to the utilization of colinearity and microcolinearity. With the development of more comprehensive tools, like comparable full-genome contigs, the routine and rather simple use of comparative plant genetics and the 'Unified Grass Genome' will soon become a reality.

## Acknowledgements

The writing of this manuscript was supported by grants from the United States Department of Agriculture (94-37300-0299 and 94-37310-0661).

## References

Ahn, S., Anderson, J.A., Sorrels, M.E., and Tanksley, S.D. (1993) Homoeologous relationships of rice, wheat and maize chromosomes. Mol. Gen. Genet. 241: 483–490.

Avramova, Z., SanMiguel, P., Georgieva, E., and Bennetzen, J.L. (1995) Matrix attachment regions and transcribed sequences within a long chromosomal continuum containing maize *adh1*. Plant Cell 7: 1667–1680.

Avramova, Z., Tikhonov, A., SanMiguel, P., Jin, Y.-K., Liu, C., Woo, S.-S., Wing, R.A., and Bennetzen, J.L (1996) Gene identification in a complex chromosomal continuum by local genomic cross-referencing. Plant J. 10: 1163–1168.

Bennetzen, J.L. (1996) The contributions of retroelements to plant genome organization, function and evolution. Trends Microbiol. 4: 347–353.

Bennetzen, J.L. (1998) The organization and evolution of angiosperm nuclear genomes. Curr. Opin. Plant Biol., in press.

Bennetzen, J.L., and Freeling, M. (1993) Grasses as a single genetic system: genome composition, collinearity and complementarity. Trends Genet. 9: 259–261.

Bennetzen, J.L., and Freeling, M. (1997) The unified grass genome: synergy in synteny. Genome Res. 7: 301–307.

Bennetzen, J.L., Kellogg, E.A., Lee, M., and Messing, J. (1998) A plant genome initiative. Plant Cell. 10: 488–493.

Bonierbale, M.W., Plaisted, R.L., and Tanksley, S.D. (1988) RFLP maps based on a set of common clones reveals modes of chromosomal evolution in potato and tomato. Genetics. 120:1095–1103.

Brown, J., and Sundaresan, V. (1991) A recombination hotspot in the maize *A1* intragenic region. Theor. Appl. Genet.. 81: 185–188.

Chen, M., SanMiguel, P., Oliviera, A.C., Woo, S.-S., Zhang, H., Wing, R.A., and Bennetzen, J.L. (1997) Microcolinearity in the *sh2*-homologous regions of the maize, rice and sorghum genomes. Proc. Nat. Acad. Sci. USA 94: 3431–3435.

Civardi, L., Xia, Y., Edwards, K.J., Schnable, P.S., and Nikolau, B.J. (1994) The relationship between genetic and physical distances in the cloned *a1-sh2* interval of the *Zea mays L.* genome. Proc. Nat. Acad. Sci. USA. 91: 8268–8272.

Devos, K.M., Millan, T., and Gale, M. (1993) Comparative RFLP maps of homeologous group 2 chromosomes of wheat, rye and barley. Theor. Appl. Genet. 85: 784–792.

Devos, K.M., Moore, G., and Gale, M.D. (1995) Conservation of marker synteny during evolution. Euphytica 85:367–372.

Donner, H.K. (1986) Genetic fine structure of the *bronze* locus in maize. Genetics. 113: 1021–1036.

Dubcovsky, J., and Dvorak, J. (1995) Ribosomal RNA multigene loci: nomads of the Triticeae genomes. Genetics. 140: 1367–1377.

Dunford, R.P., Kurata, N., Laurie, D.A., Money, T.A., Minobe, Y., and Moore, G. (1995) Conservation of fine-scale DNA marker order in the genomes of rice and the Triticeae. Nuc. Acids Res. 23: 2724–2728.

Flavell, R.B., Bennett, M.D., Smith, J.B., and Smith, D.B. (1974) Genome size and proportion of repeated nucleotide sequence DNA in plants. Biochem. Genet. 12: 257–269.

Foote, T., Roberts, M., Kurata, N, Sasaki, T., and Moore, G. (1997) Detailed comparative mapping of cereal chromosome regions corresponding to the *Ph1* locus in wheat. Genetics 147: 801–807.

Helentjaris, T., Weber, D.L., and Wright, S. (1988) Identification of the genomic locations of duplicate nucleotide sequences in maize by analysis of restriction fragment length polymorphisms. Genetics 118: 353–363.

Hulbert, S.H., Richter, T.E., Axtell, J.D., and Bennetzen, J.L. (1990) Genetic mapping and characterization of sorghum and related crops by means of maize DNA probes. Proc. Nat. Acad. Sci. USA 87: 4251–4255.

Kilian, A., Kudrna, D.A., Kleinhofs, A., Yano, M., Kurata, N., Steffenson, B., and Sasaki, T. (1995) Rice-barley synteny and its application to saturation mapping of the barley *Rpg1* region. Nuc. Acids Res. 23: 2729–2733.

Kowalski, S.P., Lan T.H., Feldmann, K.A., and Paterson, A.H. (1994) Comparative mapping of *Arabidopsis thaliana and Brassica oleracea* chromosomes reveals islands of conserved organization. Genetics 138: 499–510.

Leister, D., Kurth, J., Laurie, D.A., Yano, M., Sasaki, T., Devos, K., Graner, A., and Schulze-Lefert, P. (1998) Rapid reorganization of resistance gene homologues in cereal genomes. Proc. Nat. Acad. Sci. USA. 95: 370–375.

Moore, G., Devos, K.M., Wang, Z., and Gale, M. (1995) Grasses, line up and form a circle. Curr. Biol. 5: 737–739.

Paterson, A.H., Lin, Y.R., Li, Z., Schertz, K.F., Doebley, J.F., Pinson, S.R.M., Liu, S.-C., Stansel, J.W., and Irvine, J.E. (1995) Convergent evolution of cereal crops by independent mutations of corresponding genetic loci. Science 269: 1714–1717.

Paterson, A.H., Lan, T.-H., Reischmann, K.P., Chang, C., Lin, Y.-R., Liu, S.C., Burow, M.D., Kowalski, S.P., Katsar, C.S., DelMonte, T.A., Feldmann, K.A., Schertz, K.F., and Wendel, J.F. (1996) Toward a unified genetic map of higher plants transcending the monocot-dicot divergence. Nature Genetics 14: 380–382.

Peacock, W.J., Dennis, E.S., Rhoades, M.M., and Pryor, A.J. (1981) Highly repeated DNA sequence limited to knob heterochromatin in maize. Proc. Nat. Acad. Sci. USA 78: 4490–4494.

Pereira, M.G., and Lee, M. (1995) Identification of genomic regions affecting plant height in sorghum and maize. Theor. Appl. Genet.. 90: 380–388.

Prince, J.E., Pochard, E., and Tanksley, S.D. (1993) Construction of a molecular linkage map of pepper and a comparison of synteny with tomato. Genome 36: 404–417.

Rhoades, M.M. (1951) Duplicate genes in maize. Amer. Nat. 85: 105–110.

Saedler, H., and Gierl, A. (1996) Transposable Elements. Springer-Verlag, Berlin.

SanMiguel, P., Tikhonov, A., Jin, Y.-K., Motchoulskaia, N., Zakharov, D., Melake-Berhan, A., Springer, P.S., Edwards, K.J., Lee, M., Avramova, Z., and Bennetzen, J.L. (1996) Nested retrotransposons in the intergenic regions of the maize genome. Science 274: 765–768.

Sanz-Alferez, S., Richter, T.E., Hulbert, S.H., and Bennetzen, J.L. (1995) The *Rp3* disease resistance gene of maize: mapping and characterization of introgessed alleles. Theor. Appl. Genet. 91: 25–32.

Shoemaker, R.C., Polzin, K., Labate, J., Specht, J., Brummer, E.C., Olson, T., Young, N., Concibido, V., Wilcox, J., Tamulonis, J.P., Kochert, G., and Boerma, H.R. (1996) Genome duplication in soybean (*Glycine* subgenus *soja*). Genetics 144: 329–338.

Snape, J.W., Quarrie, S.A., and Laurie, D.A. (1996) Comparative mapping and its use for the genetic analysis of agronomic characters in wheat. Euphytica 89: 27–31.

Teutonico, R.A., and Osborn, T.C. (1994) Mapping of RFLP and qualitative trait loci in *Brassica rapa* and comparison to the linkage maps of *B. napus, B. oleracea, and Arabidopsis thaliana*. Theor. Appl. Genet. 89: 885–894.

Yu, G.X., Bush, A.L., and Wise, R.P. (1996) Comparative mapping of homoeologous group 1 regions and genes for resistance to obligate biotrophs in *Avena, Hordeum, and Zea mays*. Genome 39: 155–164.

# Index